THERMOCHEMISTRY AND ADVANCES IN CHEMISTRY RESEARCH

Thermochemistry and Advances in Chemistry Research

Anton G. Mirskiy
Editor

Nova Science Publishers, Inc.
New York

For permission to use material from this book please contact us:
Telephone 631-231-7269; Fax 631-231-8175
Web Site: http://www.novapublishers.com

LIBRARY OF CONGRESS CATALOGING-IN-PUBLICATION DATA

Library of Congress Cataloging-in-Publication Data
Thermochemistry and advances in chemistry research / editors, Anton G. Mirskiy.
p. cm.
Includes index.
ISBN 978-1-60692-376-4 (hardcover)
1. Thermochemistry. 2. Chemistry. I. Mirskiy, Anton G.
QD511.T456 2009
541'.36--dc22

2009002556

Published by Nova Science Publishers, Inc. ✝ New York

Contents

PREFACE

In thermodynamics and physical chemistry, thermochemistry is the study of the heat evolved or absorbed in chemical reactions. Thermochemistry, generally, is concerned with the heat exchange accompanying transformations, such as mixing, phase transitions and chemical reactions. This new book presents leading research from around the world in this important field. Recent developments in the thermochemistry of pre-nucleation clusters are discussed, as well as the atmospheric implications of the new thermochemical data. Furthermore, plasma surface treatments are widely used in industrial applications. Thus, thermochemical processes assisted by plasma in the manufacturing sector are presented. Also discussed in this book are other advances in the field of chemistry, including the effect of nanostructure on the optical properties of earth ions.

Chapter 1 - In both processes of positive (electron impact - EI) and negative ions formation (dissociative resonance capture – DRC) the molecular ion $M\cdot^{+}$ or $M\cdot^{-}$, respectively, appears in electronically excited state. This happens because at EI the energy of bombarding electron 70 eV is essentially higher than the ionization energy of *any* molecule (7-15 eV) while at DRC process, since this event is exothermic and resonance one, the *whole* energy of attaching electron is transferred to a molecule and then to a molecular ion.

What happens then with a molecular ion being in electronically excited state irrespective whether it is registered or not? The answer could be found in quasi-equilibrium theory (QET) of Eyring, the architect of statistical thermodynamics [1]. The essence of QET in a brief form, in addition to Ref.1, could be also found in Ref. 2 (negative ions) and Refs 3 and 4 (positive ions) (the literature is given in Chapters 1 and 3). The essence of QET is following. 1) The time necessary for dissociation of initially formed molecular ion is *large* compared with the time interaction of electron with molecule the former being necessary for formation and excitation of molecular ion. This time is $\sim10^{-16}$ s. The shortest time being observable for dissociation (this is a period of a single vibration) is about 10^{-14} s for C-H bond, for other bonds it is even larger. QET assumes that dissociation occurs only after several vibrations therefore molecular ion will "forget" at which site the ionization took place, i.e. the rate of bond cleavage is independent of original ionization process. 2) The dissociation rate is small compared with that of redistribution of original excitation energy (which is gained from attacking electron) through all degrees of freedom in molecular ion. There is enough time for excess energy, inherited by molecular ion at EI or DRC processes, to be redistributed before fragmentation in fast radiationless transitions among electronic, vibrational and rotational states. These transitions are accompanied by corresponding variation of bond lengths and angles.

Chapter 2 - The effect of atmospheric aerosols on the Earth's climate is unambiguous. Atmospheric aerosols formed via nucleation and subsequent condensational growth, which contribute significantly to the concentration of cloud condensation nuclei, influence the Earth radiation balance indirectly by modifying cloud properties and precipitation. A clear understanding of atmospheric nucleation is critically important for assessing the lifecycle of atmospheric particles and predicting the global climate changes associated with the aerosol radiative forcing. The global scale phenomena are closely connected to processes of gas to particle conversion during nucleation. Nucleation rates are very sensitive to the thermochemistry of initial steps of the aerosol particle formation where matter in a state, neither liquid nor gaseous, but in indeterminate state of molecular cluster is growing by the attachment of vapor monomers. The small pre-nucleation clusters are essentially quantum in nature, and must be treated using quantum methods. Recently, it has been shown that the application of quantum methods to the thermochemistry of clusters composed of atmospheric precursors is certainly beneficial, and allows solving numerous problems that could not be adequately treated using the continuum/bulk liquid classical nucleation theory. This paper describes the recent developments in the thermochemistry of pre-nucleation clusters composed of common atmospheric nucleation precursors and discusses the atmospheric implications of the new thermochemical data. The importance of the cluster thermochemistry is outlined and future developments towards the new atmospheric nucleation theory, which will be based on the thermochemistry obtained from "first principles", is discussed. The nucleation models constrained by thermochemical data obtained using quantum methods can be incorporated into global models to study new particle formation in planetary atmosphere and associated implications to climate change.

Chapter 3 - Development of fire and heat shield materials with reduced combustibility consists of several stages.

Heat energy from an external source, delivered to the material surface, can increase its temperature so that degradation of material components or chemical reaction between them begins. Combustible degradation products formed are capable of igniting and combust independently of flame spreading. The problem of creating fire and heat shield materials is closely connected with the questions of material combustibility decrease.

Under normal conditions, heat transfer proceeds from higher heated (hot) bodies to less heated (cold) ones by means of heat conductivity or convection. Under fire conditions, irradiation is the main type of heat transfer from flame or incandescent material surface. The radiant flux falling on the body surface consists of three components [1 - 3].

The radiant flux permeates through the cover to the protected object surface and heats it up, which is unwanted for fire and heat shield materials. Heat flux absorbed by the material is accumulated in it or transforms into different types of energy [4].

Chapter 4 - Plasma surface treatments are widely used in industrial applications. Plasma assisted surface treatments improve the wear, corrosion resistance and fatigue strength by the change at the properties of surface and near-surface region of material. The variety of surface treatment is too large for this chapter to cover all of them in detail. The scope of this chapter is limited on thermochemical processes assisted by plasma which have been widely applied by manufacturing sector in Europa, North America and Japan during the last two decades. These processes involve heating into the ferrite or austenite regions of a materials part in an environment that allows elements such as C, N and B to be diffused through surface of materials. Firstly, plasma state and glow discharge are given in this chapter and, then thermochemical processes assisted by plasma are explained in detail. This chapter is especially focused on compound layer formed during thermochemical process.

Chapter 5 - The constant-volume combustion energy, $\Delta_c U$ (LLM-105, s, 298.15 K), and kinetic behavior of the decomposition reaction of the title compound (LLM-105) are determined by a precise rotating bomb calorimeter, TG/DTG and DSC, respectively. Its standard enthalpy of combustion, $\Delta_c H_m^\theta$(LLM-105, s, 298.15 K), standard enthalpy of formation, $\Delta_f H_m^\theta$(LLM-105, s, 298.15 K) and kinetic parameters of the exothermic main decomposition reaction in a temperature-programmed mode [the apparent activation energy (E_a) and pre-exponential factor (A)] are calculated. The values of $\Delta_c U$ (LLM-105, s, 298.15 K), $\Delta_c H_m^\theta$(LLM-105, s, 298.15 K), and $\Delta_f H_m^\theta$(LLM-105, s, 298.15K) of LLM-105 are (-9729.89±3.88)$J \cdot g^{-1}$, (-2091.48±0.84)$kJ \cdot mol^{-1}$ and (-54.22±0.99) $kJ \cdot mol^{-1}$, respectively. The kinetic model function in differential form and the value of E_a and A of the decomposition reaction of LLM-105 are $(1-\alpha)^{0.71}$, 154.41$kJ \cdot mol^{-1}$ and $10^{11.22}$ s^{-1}, respectively. The critical temperature of thermal explosion of LLM-105 is 328.19°C. The values of $\Delta S^{\neq}$, $\Delta H^{\neq}$ and $\Delta G^{\neq}$ of this reaction are -35.65$J \cdot (mol \cdot K)^{-1}$,149.58$kJ \cdot mol^{-1}$ and 170.28$kJ \cdot mol^{-1}$, respectively. The compatibility of LLM-105 with components in propellant is investigated by vacuum stability test. The experimental data indicate that the binary systems LLM-105/RDX (cyclotrimethylenetrinitramine), LLM-105/DINA (N-nitro-dihydroxyethylamine-dinitrate), LLM-105/AP (ammonium perchlorate), LLM-105/β-Pb salt (lead 2,4-dihydrxybenzoate), LLM-105/(NC/NG mixture), LLM-105/DEP (diethylphthalate) have good compatibility. The theoretical investigation on LLM-105 as structure unit is carried out by DFT-B3LYP/6-311++G method, and the atomic net charges and the population analysis are discussed.

Chapter 6 - The authors applied computational electronic energy methods, G3MP2B3 and CCSD(T)-CBS (W1U), for which we have the MAD (Mean Absolute Deviation) between calculated and experimental enthalpies of formation, $\Delta_f H^o(298.15)$, for sets of well known radicals. For G3MP2B3, the MAD is 3.9 kJ mol^{-1}, and for CCSD(T)-CBS (W1U), it is 2.3 kJ mol^{-1}. The former method can be applied to systems with up to 20 C-atoms or more; but for the latter, it is limited to systems up to 4 – 5 C-atoms. The systems $^{\bullet}C_2H$, C_2H_2, $^{\bullet}C_3H_3$, C_3H_4, n-$^{\bullet}C_4H_3$, i-$^{\bullet}C_4H_3$, and C_4H_4 were calculated for both computational methods. The systems $^{\bullet}C_6H_5$ (phenyl), C_6H_6, $^{\bullet}C_8H_7$ (styryl), $^{\bullet}C_{10}H_9$ (4-phenyl-1,3-butadien-1-yl and 2-(1,3-butadien-1-yl)phenyl) transition structures and intermediates, and $C_{10}H_8$ (naphthalene) were calculated only by G3MP2B3. The properties are energy minimum structures as well as transition structures for the title reactions, harmonic vibrational wave numbers ν, moments of inertia I_A, I_B, and I_C, heat capacities $C^o_p(T)$, standard entropies $S^o(T)$, thermal energy contents $H^o(T) - H^o(0)$, standard enthalpies of formation $\Delta_f H^o(T)$, and atomic spin densities ρ. The temperature T is extended from 0 to 3000 K. Parent molecules of radicals were calculated to obtain bond dissociation energies $D^o(298.15)$.

Chapter 7 - The exergetic analysis of ideal processes gives parameters to be optimised to design processes converting or transforming inlet exergy as completely as possible. Thermochemical dipoles consist of two linked physico-chemical processes which produce or consume a working gas. These dipoles allow cold or heat production through the management of the physico-chemical processes involved. New couplings of such dipoles permits production of cold at low temperature (-22°C for deep-freezing purpose) from low grade heat (at 70°C, coming from thermal wastes or produced by solar collectors for example). In such systems, inlet exergy is completely used and no excess cold or energy is produced.

Chapter 8 - This review includes one hundred and thirty heteroocta – and heterooligonuclear iron clusters of known crystal structures. The heterometal atoms include both transition and non – transition metals, including both hard (class A) and soft (class B) examples with one example of actinide (uranium). The clusters range from heterooctametallic of up to sixty five metal atoms.

The shortest Fe–M (non – transition) and Fe–M (transition) bonds are 2.471(2) Å for Fe–Ge and 2.540(4) Å for Fe-Pt. The shortest M-M and M'-M bonds are 2.456(1) Å for Hg-Hg and 2.346(2) (4)Å for Co-Ge. The shortest Fe-Fe bond is 2.517(3) Å. There are some relationships between the structural parameters which are discussed.

Chapter9 - Research and development of nanoscale luminescent and laser materials are part of the rapidly advancing nanoscience and nanotechnology. Besides well-known quantum dots, lanthanide-doped oxide nanocrystals form a new promising class of nanophosphors in view of various application because of unique spectroscopic properties related to two types of electronic transitions: 4f-4f and 4f-5d transitions. The subjects of this chapter are the synthesis of rare earth ions doped nanocrystals and their spectral properties. The authors begin this chapter with some introductory comments on the nature of spectroscopy of rare earth ions, and then discuss an important issue that also needs to be addressed is whether the rare earth ions are really being incorporated in the host lattice in the synthesis of rare earth ions doped nanocrystals. A crucial test for the incorporation is an excitation spectrum of the rare earth emission. Citing examples suggest that some of the difficulties encountered in nanodoping are often attributed to "self-purification". Finally, they will discuss the effect of nanostructure on the optical properties of rare earth ions. Specifically, the various mechanisms of spectral properties of rare earth ions from bulk to nanocrystals are detailed. The luminescence properties of rare earth doped nanocrystals differ from corresponding bulk materials in several ways: (i) spectroscopic changes, as a broadening and shift(related to the 5d energy level change) of the luminescence bands in nanocrystals, resulting from a structural disorder intrinsic to the small size; (ii) a lower luminescence quantum yield related to the presence of chemical species adsorbed at the surface, especially the OH groups; (iii) a shift towards higher values of the optimum doping concentration, due to the alteration of energy transfers rate from luminescence centers to quenching centers; (iv) new luminescent peak occurs due to the new site occupation of rare earth ion on the surface or another phase; (v) the change of photoluminescence decay time due to surface-bound centers. All the results are emphasis on the photoluminescence from the rare earth-doped oxide nanocrystals in the last 15 years.

Chapter 10 - Template synthesis processes proceeding for metal ion – ligand synthon A – ligand synthon B triple systems into solutions and biopolymer-immobilized matrix, have been and analyzed with thermodynamical point of view. It has been shown that final results of these processes should be distinguished extremely considerably. The distinction predicted has been displayed by examples of template synthesis for some M(II)-,M(III) – (N,S)-,(N,O,S)-ambidentate ligand synthon – carbonyl-containing ligand synthon in water-ethanol solutions and metalhexaceanjferrate gelatin-immobilized matrix implantates

Short Communcation - The authors describe an extremely efficient approach based on the non-iterative orthogonal tight-binding scheme to calculate the structure and energetics of large hydrocarbon and carbon systems. The results are reported for an extensive set of molecules and radicals including linear and branched alkanes, polycyclic aromatic compounds, polyene and polyyne systems, and fullerenes. The calculated values of heats of formation are close to the experimental data with the mean absolute error of 2.26 kcal/mol obtained for 113 comparisons. The internal consistency of the calculated data allows for reliable prediction of reaction enthalpies for bond dissociation, isomerization, and cyclization reactions. The method provides accurate structural geometries and good estimates of vibrational frequencies. They conclude that the proposed computational scheme is very promising for atomic calculations and dynamics simulations of hydrocarbons, fullerenes and carbon nanostructures.

Short Communcation - Standard enthalpies of formation at 298 K for the FSO_x (x=1 to 5) radical series are determined at the B3LYP, B3PW91, mPW0, B98, B97-2 and PBE0 levels of density functional theory and at the *ab initio* Gaussian-3 models G3MP2B3, G3B3, BAC-G3MP2B3 and BAC-G3B3. Enthalpies of formation were calculated from total atomization energies and from isodesmic reaction energies. The B98, B97-2 and PBE0 hybrids lead to the most reliable results of the tested functionals. However, the Gaussian-3 and mainly the computations with bond additivity corrections (BAC) combined with the 6-311+G(3df) basis set provide the more accurate results. The best values obtained for FSO, FSO_2, FSO_3, $FS(O_2)OO$ and $FS(O_2)OOO$ radicals are -66.6, -94.0, -120.4, -109.8 and -126.3 kcal mol^{-1}, with a stated error limit of ±2 kcal mol^{-1}. A comparison with previous values is reported.

Commentary - The critical importance of clear understanding and insight of the thermochemistry of nucleating clusters/particles for quantitatively assessing the climate-related, health and environmental impacts of atmospheric aerosols is commonly accepted. Traditionally, the thermochemistry of nucleation clusters has been studied using the bulk luquid/ capillarity approximation, which is a cornerstone of the classical nucleation theory (CNT). The simplicity and convenience of the CNT is largely offset by moderately poor performance and excessive sensitivity to input parameters, whose experimental determination is quite difficult. At the present time, there are two classes of methods that are imposed as an alternative to the classical theory: molecular based methods and computational quantum methods. Molecular-based models are more advanced theoretically than the CNT and they compute nucleation kinetically. However, interactions between vapour monomers and *i*-mers in the existing molecular studies are usually approximated using empirical (e.g. TIP4P, TIP5P, Dang–Chang) potentials parameterized to the bulk liquid properties. The aforementioned empirical potentials "were never intended to treat small clusters" and their applicability to the pre-existing and nucleating clusters is questionable. In our view, the existing molecular-based nucleation approaches should be classified as semi-classical because they implicitly treat molecular clusters as liquid-like droplets. The application of inappropriate potentials has lead to the failure of molecular-based methods to predict the sign preference in aqueous systems and dimerization free energy for water. It is important to note that molecular-based methods using empirical potentials are generally not invalid or defective. However, the further improvement of the quality of their model predictions will require more realistic potentials that could be obtained using e.g. quantum methods or experiments.

In: Thermochemistry and Advances in Chemistry Research ISBN: 978-1-60692-376-4
Ed: Anton G. Mirskiy

Chapter 1

DISPUTABLE PROBLEMS OF ORGANIC MASS SPECTROMETRY

Dmitry Ponomarev[1]* and Viatcheslav Takhistov[2]
[1] St.Petersburg Forest Technical Academy, 194021 St.Petersburg, Institutski per.5,
[2] Center of Ecological Safety RAS, 197042 St.Petersburg, Korpusnaja ul. 18,

1. QUASI-EQUILIBRIUM THEORY

In both processes of positive (electron impact - EI) and negative ions formation (dissociative resonance capture – DRC) the molecular ion $M\cdot^{+}$ or $M\cdot^{-}$, respectively, appears in electronically excited state. This happens because at EI the energy of bombarding electron 70 eV is essentially higher than the ionization energy of *any* molecule (7-15 eV) while at DRC process, since this event is exothermic and resonance one, the *whole* energy of attaching electron is transferred to a molecule and then to a molecular ion.

What happens then with a molecular ion being in electronically excited state irrespective whether it is registered or not? The answer could be found in quasi-equilibrium theory (QET) of Eyring, the architect of statistical thermodynamics [1]. The essence of QET in a brief form, in addition to Ref.1, could be also found in Ref. 2 (negative ions) and Refs 3 and 4 (positive ions) (the literature is given in Chapters 1 and 3). The essence of QET is following. 1) The time necessary for dissociation of initially formed molecular ion is *large* compared with the time interaction of electron with molecule the former being necessary for formation and excitation of molecular ion. This time is ~10^{-16} s. The shortest time being observable for dissociation (this is a period of a single vibration) is about 10^{-14} s for C-H bond, for other bonds it is even larger. QET assumes that dissociation occurs only after several vibrations therefore molecular ion will "forget" at which site the ionization took place, i.e. the rate of bond cleavage is independent of original ionization process. 2) The dissociation rate is small compared with that of redistribution of original excitation energy (which is gained from attacking electron) through all degrees of freedom in molecular ion. There is enough time for

* E-mail: juniper@mailbox.alkor.ru and tahistov@vt10515.spb.edu

excess energy, inherited by molecular ion at EI or DRC processes, to be redistributed before fragmentation in fast radiationless transitions among electronic, vibrational and rotational states. These transitions are accompanied by corresponding variation of bond lengths and angles.

In a result, the ions with smaller electronic and larger vibrational energy content are obtained. This excess (as compared with the ground state of molecular ion) vibrational energy can concentrate at any bond of a molecular ion (but with highest probability at the weakest one) which will break resulting in formation of a fragment or a rearrangement ion; in the latter case a new bond may be formed. Some of the fragment ions might still possess an excess energy sufficient for cleavage of its weakest bond or for isomerization, and the redistribution of its excess energy will occur similar to that which happened in the molecular ion, giving rise to secondary then tertiary etc. fragment ions. Hence 3) fragmentation and isomerization products are produced in the series of consecutive and competing monomolecular reactions; 4) the ions which are formed in mass spectrometer represent the isolated systems being in the state of "internal" equilibrium [3]. Therefore, the prefix "quasi" appears in the term "quasi-equilibrium" theory (QET).

QET unites the processes occurring in mass spectrometer with those taking place in solution [5].

Now we stress some specific features of QET application to negative ions formation. Here, *all* processes of DRC, leading to fragmentation as well as the processes of elastic and non-elastic scattering of electrons by molecular systems, proceed through quasi-stationary states irrespective of their capability to form or not the bound stationary systems [2]. The specificity of application of QET to negative ions has been known since the time when the efficiency curves of their formation were measured experimentally in the electron energy range of 0-15 eV. This specificity is reflected in formation of fragment ions with a given m/z not in all of this energy range but only in several (often only in one) more or less narrow energy domains. This immediately manifests the important difference in the processes of formation of negative from that of positive ions: the symmetry of electronic state of the daughter (the fragment one or that of isomeric to initial one) ion has to correlate with the symmetry of electronic state of the parent, e.g. the molecular ion. This correlation might be taken into account to explain the DRC processes when the set of fragment ions, generated in the chain of consecutive fragmentation steps from molecular ion is observed only in the narrow range of energies rather than in the whole interval of energies being measured 0-15 eV [2].

In mass spectrometry of positive ions the set of crossing excited states makes it useless the application of symmetry rules for description of fragmentation processes. Naturally, in case of superposition of two or more resonance states, being close in their energies, similar situation might occur in mass spectrometry of negative ions, especially in the case of complex organic molecules.

Earlier we proposed the model of QET application to rearrangement processes [6-8]. It was suggested and then established that for such processes the QET might not be referred to the *summary* process beginning from the initial structure and finishing with the fragment, but rather to the *each* individual rearrangement step (which could be several ones) of fragmentation.

One more specification might be introduced. When the QET appeared in 1952 [1] and later on the early stages of its development the mass spectrometrists were studying relatively small molecules. Nowadays, the molecules with masses of several thousand or several hundred thousands dalton are subjected to fragmentation by electrons, photons and by other particles [5]. If in such case we leave intact the QET postulate about complete redistribution of excess electronic and, consequently, of vibrational energy by all degrees of freedom then in such large molecules the so small part of energy might be shared by each degree of freedom that no fragmentation takes place in either positive or negative molecular ion. However, since in these large molecules the fragmentation is always observed, this signifies that redistribution of excess energy has its geometrical and, consequently, its time limits (both limits are unknown) both being well achieved while the fragmentation might take place without complete redistribution of energy. This means that a certain part of ions can fragment from electronically excited states. However, this part is small one and does not define the mass spectra of both positive and negative ions in its essence. According to our observations, coming from study of mass spectra of positive ions at several electron ionization energies for more than 300 compounds [4, 6, 8, 9] the portion of such processes occurring from excited states exhibits not more than 2-3%. For negative ions, as one might suggest, this portion is also small and can be referred to the ions with small life times (several microseconds) in case when the measured lifetimes of molecular (and fragment) ions are tens and hundreds of milliseconds.

The major part of known macromolecules is of regular structure, i.e. with systematically repeated structurally identical fragments in the chain and/or in the space like natural or synthetic polymers, polypeptides, proteins, carbohydrates, nucleic acids, dendrimers and others. That means that in a given fragment (being common in a gross molecule) the complete redistribution of excess electronic to vibrational and rotational energies takes place and, consequently, the isomerization and fragmentation in consensus with the most important principle of QET, i.e. the occurrence of the most energetically favorable process(es), will take place.

The other theories of mass spectra are known. One of them, Rice-Rampsberger-Kassel-Markus (RRKM) theory with all necessary original sources is given in the works of Lorquet [10] and Waite [11]. The author [10], criticizing both theories, QET and RRKM, represents the quantum-chemical version of the monomolecular decomposition of the molecules possessing large excess energies [10]. The author considers other possible mechanisms of energy randomization [10]. Maccoll proposes one more quantum-chemical model of mass spectrum formation, which according to the author's opinion appears to be an "ideal one" for calculation of mass spectra and explanation of its variation with electron energy and temperature [12].

The key moment in so called "violation" of QET is the suggestion for *certain* fragmentation processes and for *certain* compounds that these processes originate from molecular ion with unrandomized excess energy or from isolated electronically excited states, i.e. the fragmentation proceeds by energetically unfavorable route called "non-ergodic" [13]. Thus, Turecek suggests such non-ergodic process for *one* of the fragmentation routes of acetone [13] but why only for a single one? Ho and others arrive at violation of QET in fragmentation of Me_3CPh molecule [14]. Traeger finds "countrathermodynamic" process in mass spectra of ethers [15].

Maccoll for explanation of the absence of the peaks of molecular ions in two compounds [16] attracts the suggestion of Lindholm [17] that fragmentation can proceed from an "optical metastable state" of a molecule rather than from a molecular ion. The idea of fragmentation from an electronically excited molecular ion whose peak is either registered or not is often attracted in explanation of fragmentation of negative ions (see later).

In our description of large amount of isomerization and fragmentation processses for both positive and negative ions [4, 6, 8, 9, 18] we have nowhere attracted the notions of electronically excited ions or isolated states. Instead, we explained the occurrence of "unusual" processes by isomerization of molecular and fragment ions having enough time to perform this isomerization to give the rearranged ions instead of those which are expected from initial, unrearranged ions.

Just at the possibility of rearrangement ions formation it is necessary to pay attention even if their structures seem to be unusual or, moreover, unexpected. Burgers et al, analyzing the suggestions of some workers on violation of QET on the example of isomeric $C_2H_3O^+$ ions, are especially correct in their statement that until *all* stable isomers (including such "unusual" as $CH_3O(^+)$=C:) are not generated and are not involved in analysis the hypothesis about incomplete randomization of energy might be delayed [19]. Weber et al also point out that the conception of isolated electronic states, earlier being attracted to explain some unexpected effects in organic mass spectrometry, was later rejected in favor of different interpretation [20]. Applying the methods of kinetics at field ionization it is shown that isomerization of alkenes molecular ion with double bond shift takes place at the time of $<10^{-10}$ s being sufficient for H-randomization. 1-Octene decomposes in specific way at the time $<2 \times 10^{-11}$ s, i.e. when the excess energy has no time to be randomized. However, beginning from $\tau = 10^{-10}$ s 1-octene decomposes already non-specifically, i.e. the double bond (and H-atoms) is shifted [21]. Hence, it is evident that at the time scale of mass spectrometer (10^{-6} s) the molecular ion has enough time (4 orders in the example quoted [21]) for performing the rearrangement at long distances along the carbon chain and for randomization of energy not only in the initial molecular ion but also in its isomers, and in fragment ions.

Let us consider fragmentation of $C_6F_5SiMe_3$ molecule: $M^{\cdot+}$ (40), $[M\text{-}Me]^+$ (19), $[M\text{-}MeSiF_3]^{\cdot+}$ (42), $MeSiF_2^+$ (18), $Me_2Si(^+)F$ (100) [22]. Its molecular ion has several very simple and energetically favorable fragmentation routes typical for $RSiMe_3$ molecules (see Ref. 4.

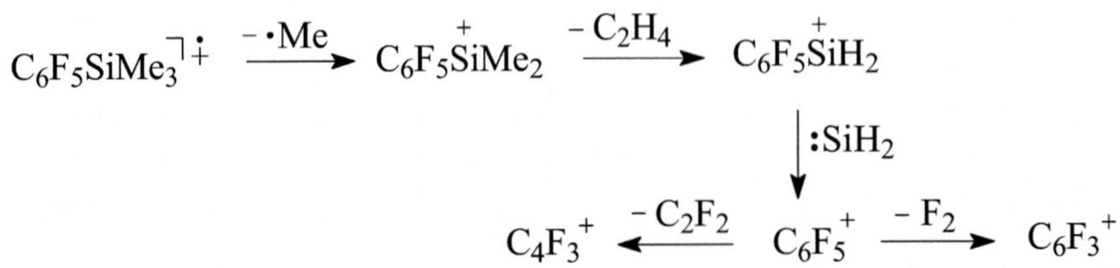

However, fragmentation proceeds completely in a different way. The molecular ion ventures many tens of structures until F-atom(s) appear at Si-atom since it is extremely favorable energetically: not the formation of the ions themselves (C- and Si-fluorinated ions are of similar thermodynamic stabilities) but rather of the ions' skeletons since the F-affinity of Si-atom is extremely large (see Ref. 4). The conclusion is straightforward: the system has sufficient time to realize the cleavage of many bonds (including those in benzene ring) and the formation of many new bonds in favor of production of very stable isomeric molecular ion(s) giving rise to thermodynamically very stable fragments.

Mass spectrum of [3] catenane is of special interest owing to the formation of rearrangement ion

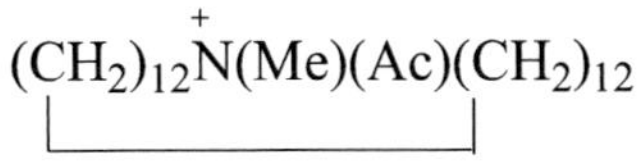

which is produced in result of migration of Me-group from either MeO-group at benzene ring or from MeCON-group of one macrocycle to N-atom of another macrocycle [23]. To perform such migration it is necessary that Me-group appears in close proximity to N-atom. The specificity of gross macromolecule is such that their rings are not linked with each other. Moreover, the approach to tertiary N-atom is sterically hampered. All this means that gross macromolecular ion has to try many hundreds of conformations and positions of rings and groups in the space to let the skeletal rearrangement occur. Hence, there is enough time for the whole system to redistribute the excess energy introduced by bombarding electron, in this case, from one of the rings to another one. In a result, a very stable fragment ion >N(:)C($^+$)Me is formed.

Similar "intermolecular" H-atom migration takes place in [2] catenane. This event, in addition, is interesting in another aspect – the randomization of energy, in agreement with QET, occurs at a such low energy as 14 eV [24].

Finally, if in such very complex and specific cases (as in these three examples being discussed) the complete randomization of excess energy happens and there is no any indication on the involvement of electronically excited states of molecular ion in "non-ergodic" (see above) processes, then appeal to these notions in such simple molecules as PhCMe3 [14], n-pentanol [16], acetone [13] or aliphatic ethers [15] is hardly justified.

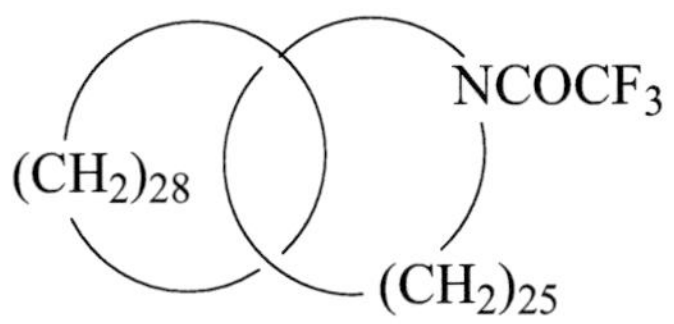

[2] catenane

14 eV: $M^{+\bullet}$ (100), $[M-CF_3]^+$ (100)

$>\ddot{N}C(OH)CF_3^{\,+}$ (20)

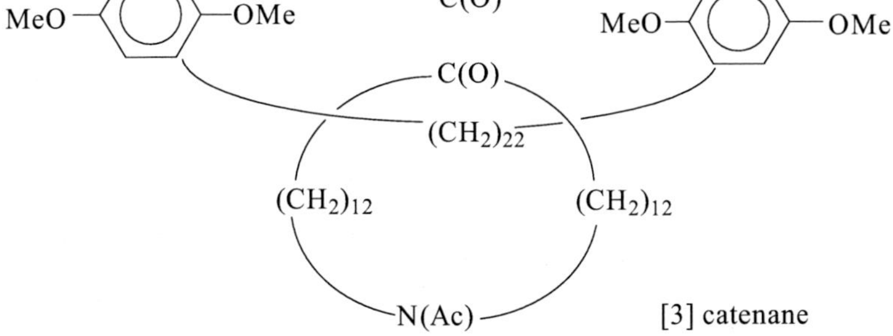

[3] catenane

2. Fragmentation Rules for Positive Ions

Analysis and modern formulation of QET together with our large experience in empirical calculation of the large massives of the enthalpies of formation for ions [4, 6 –9, 25-34], free radicals [4, 9, 27, 35-43] and molecules [4, 8, 9, 27-32, 36-39, 41, 44, 45] permitted us to propose general fragmentation rules for simple bond cleavage, hydrogen and skeletal rearrangements of organic and elementorganic compounds under electron impact. We stress that in previous years we created the separate methodologies of estimating unknown, checking or correcting the known molecules enthalpies of formation (based on the enthalpies shift procedure [4, 39]), another methodology for cations (based on the series of isodesmic reactions [4, 6]) and one more methodology for free radicals and biradicals [4, 35, 42, 43] (also based on the series of isodesmic reactions) – all these species taking part in fragmentation and isomerization of molecules proceeding in a mass spectrometer.

Up to now about 21 million (~20 million in 2003 [46]) of organic and elementorganic compounds are known and for several hundred thousands of them the mass spectra were registered. It is impossible to describe those by *classes* as it was done earlier. Mass spectrometry has to repeat (and it is really to) the previous stages of organic chemistry. At the first stage of its development, the chemical properties of organic compounds were described by the *classes* and later by the specific *types* of chemical reactions, each of the types being characterized by its own specific mechanism, i.e. the definite sequence of bond ruptures and new bond formations based on definite thermodynamic, kinetic and spectroscopic grounds. The fragmentation types were originally introduced to mass spectrometry by K.Biemann [47], then tried by W.Benz [48] and later by us [4, 18, 49-51] but unfortunately have never been used by practicing mass spectrometrists.

The principal difference between presentation of fragmentation and isomerization types performed in earlier works and the recent ones [4, 50] is that in the latter systematic *thermochemical* approach is introduced in description of fragmentation and isomerization mechanisms and schemes. A large amount of ions appearance energies (AEs), critical energies (E_{cr}), enthalpies of formation of ions and free radicals illustrates the preference of a given type in fragmentation or isomerization of a particular molecular or fragment ion, the competition between formation of fragment ions, the structural pecularities as the driving force(s) for fragmentation. The introduction of those types permitted to represent very complex schemes including dozens of ions without additional application of other techniques (usually expensive and thus not accessible to major part of MS laboratories) and to attract the mass spectra of nearly two thousand compounds of a wide set of classes of organic and elementorganic compounds [4, 18, 49-51].

We briefly describe those types, which – mechanically – are well known to the readers. Simple bond cleavage types: A_σ-type (the cleavage of α-bond in the absence of resonance stabilization); A_π-type (the cleavage of β-bond to double or triple bond); $A_{\pi\pi}$-type (the cleavage of β-bond to the system of conjugated multiple bonds including aromatic systems); A_{ar}-type (the cleavage of β-bond to π-system in the cycle leading to formation of non-classical aromatic ions (tropylium etc.); A_R-type (the formation of a stable free radical as the essential part of driving force for formation of a cation with intense peak); B_1-type (the cleavage of β-bond to unshared electron in radical cation or to radical centers); B_2-type (the version of B_1-type with double bond with heteroatom); C-type [the cleavage of α (C_1) or β

(C_2)-bonds to even-electron cationic centers with formation of carbene (:CH_2)-like (C_1) species or neutral molecules (C_2); D-type (elimination of a diatomic fragment in a cycle); D_1-type [the version of D-type in six-membered ring with a double bond (Retro Diels-Alder fragmentation)].

For each proposed isomerization or fragmentation type we can give only a few (usually a single) example while many hunderds of those are given in our previous publications [4, 18, 49-51] [ΔH_f^0 – in kcal mol^{-1}; E_{cr} = AE – IE; A_R ($A_{\pi\pi}$) – means stabilization by several types, here by A_R and by $A_{\pi\pi}$; the ion abandunces are given either in percents of base ion (100%) or in percents of the total ion current (6.2% Σ)]

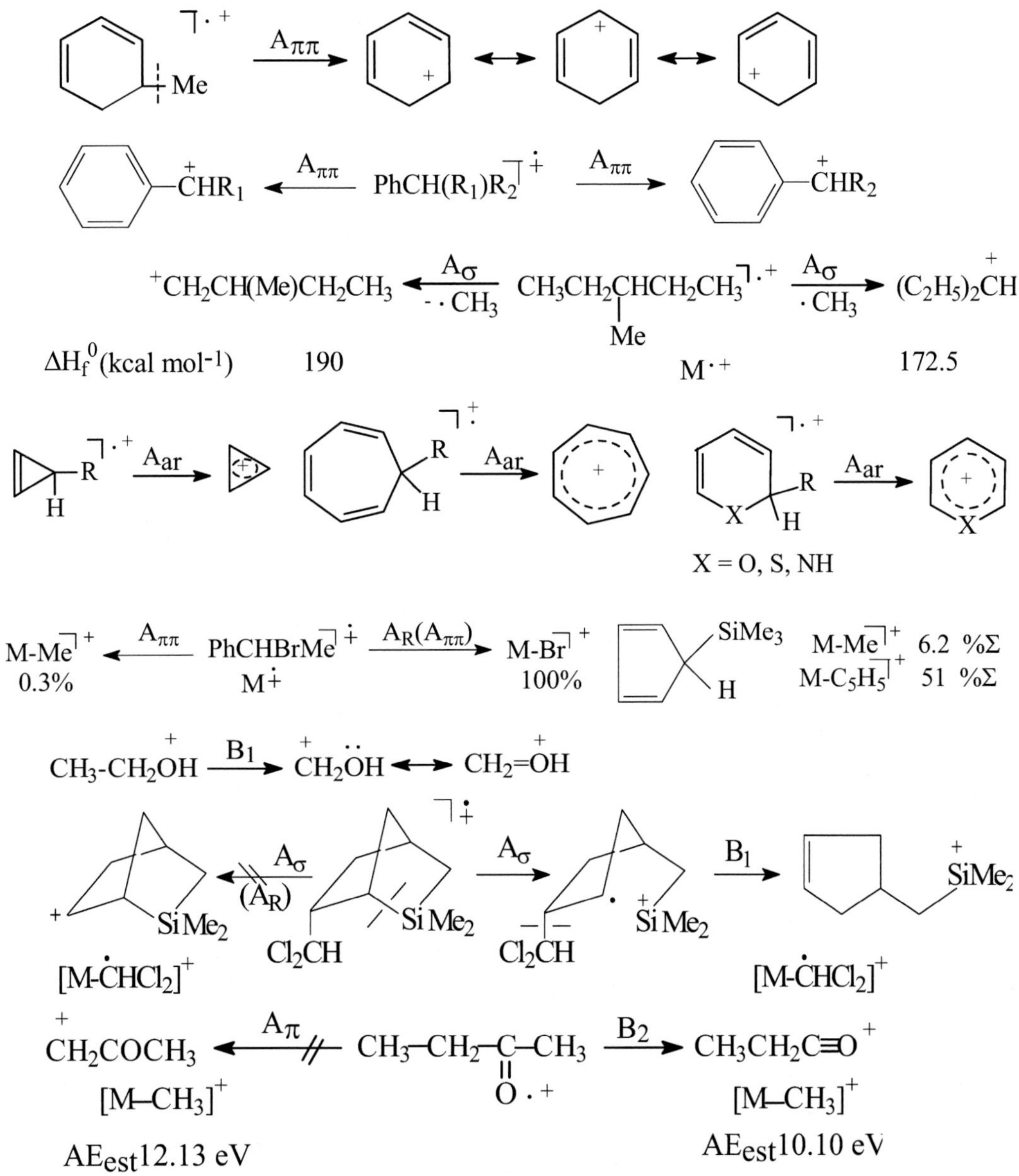

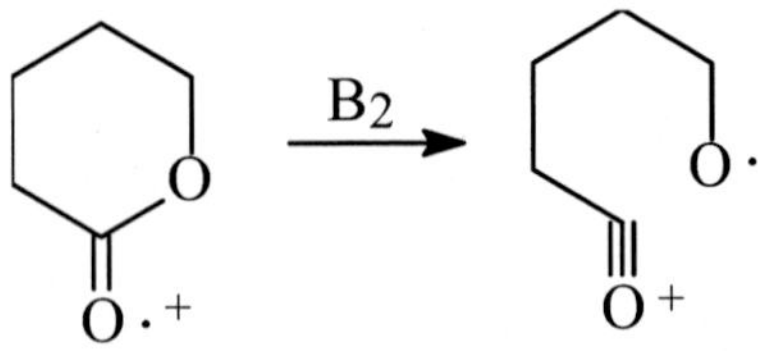

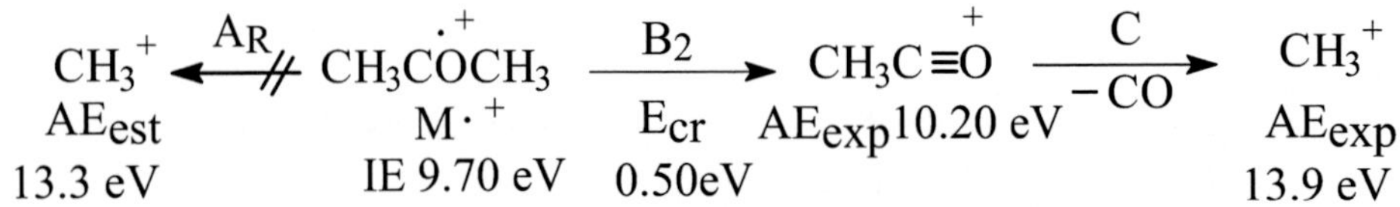

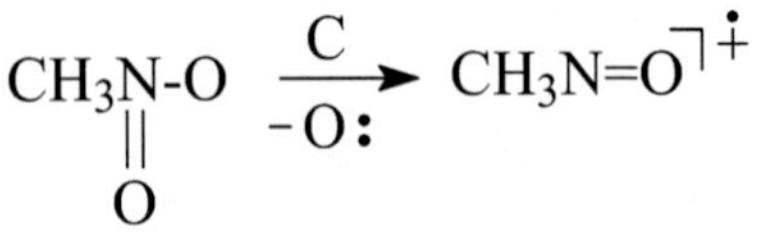

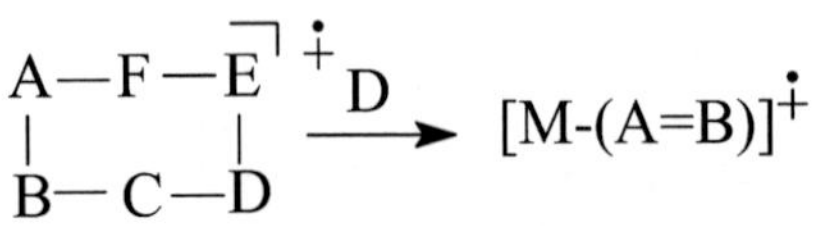

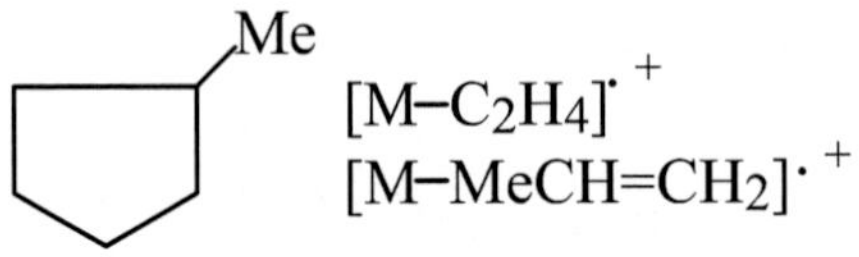

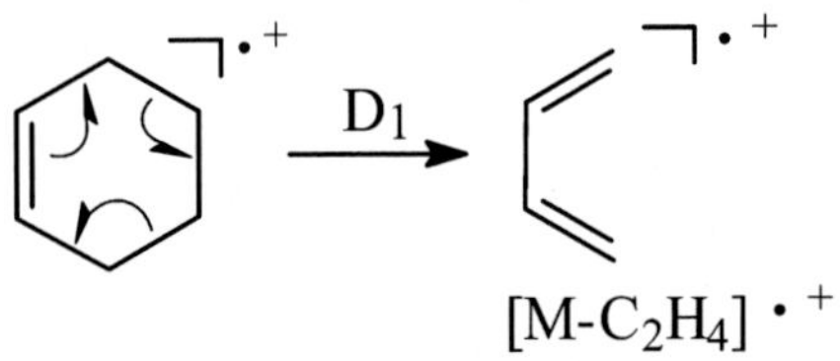

Now, H-shift isomerization and fragmentation types: K-type (1,2-H-migration); L-type (1,3-shift to cationic, radical cationic or free radical centers – through 4-membered transition state with concerted α-bond cleavage; M-type (H-shift to ionized double or triple bond through six-membered transition state with concerted β-bond cleavage; for carbonyl compounds – McLafferty rearrangement); N-type represents an isomerization of a cyclic radical ion; it comprises two steps: a) cleavage of a cyclic bond leading to formation of isomeric distonic ion, i.e. with separated charged and free radical centers, e.g. $\cdot OCH_2CH_2^+$ followed then by b) H-shift from the position adjacent to the formal free radical center to formal cationic center (or vice versa); O-type (migration of two H-atoms to the same function); P-type (neutral molecule HX elimination from both odd- and even-electron ions); R-type [non-specific H-migrations – including remote shift and randomization (scrambling) of H-atoms] [R+B_1(C)] – means two consecutive processes while B_1(C) signifies involvement of two types B_1 and C]; fishhook signifies one electron shift, while usual arrow means two electron shift.

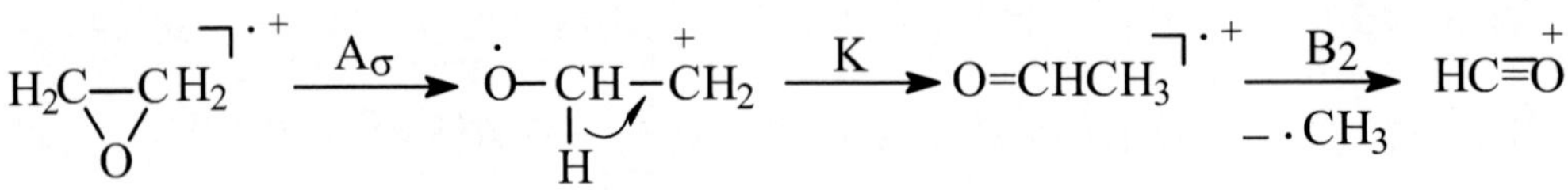

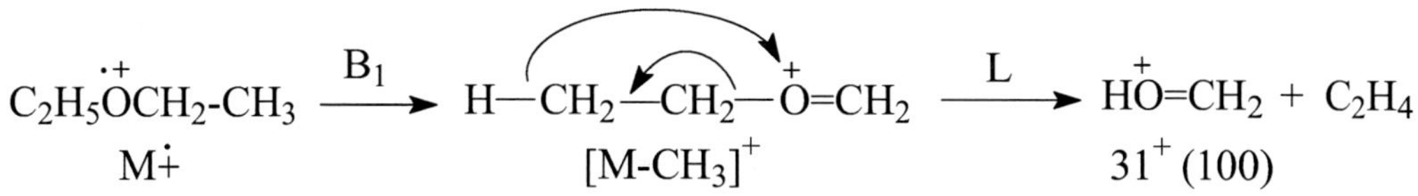

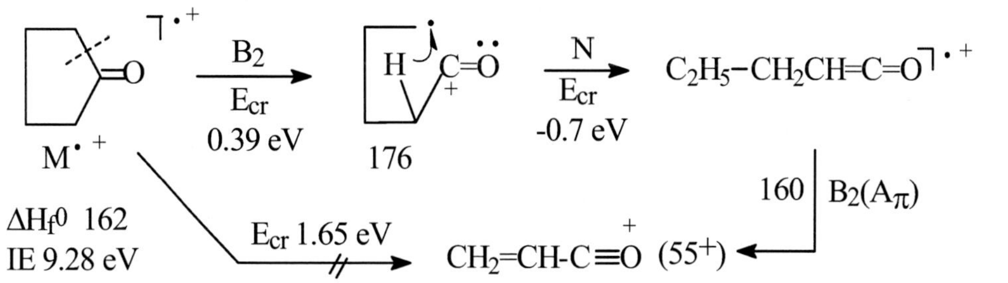

Now come skeletal rearrangement types: S_c-type (migration of atom or group to cationic center); S_R-type (migration to or interaction with free radical center); S_π-type (interaction of two π-systems); S_s-type (anchimeric assistance to bond rupture from π-system or remote atoms and groups, "internal solvation"); S-type rearrangements (skeletal rearrangements with unidentified or very complex mechanisms) ([S_c + C] signifies two processes occurring in concerted mode)

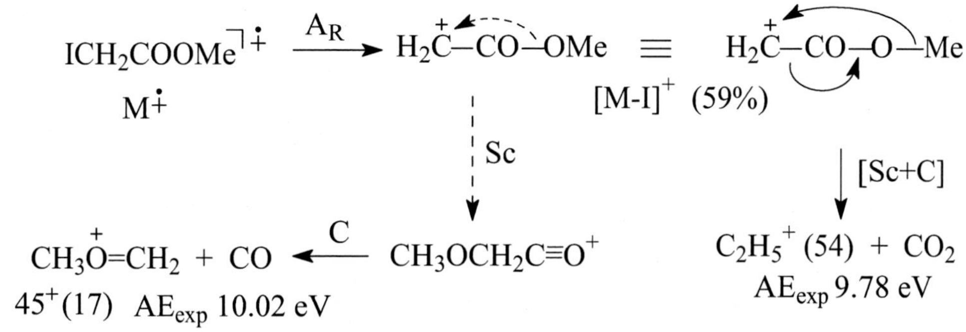

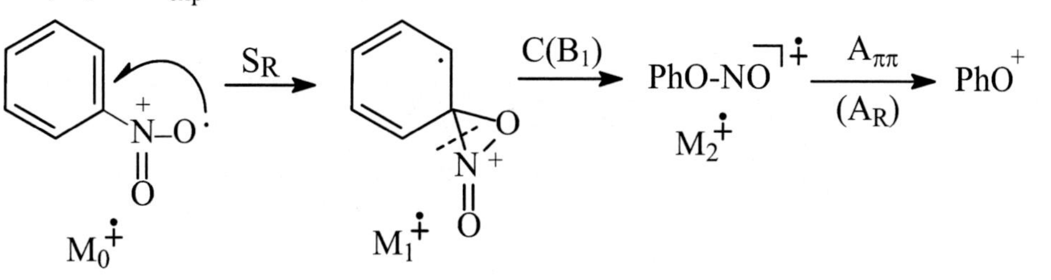

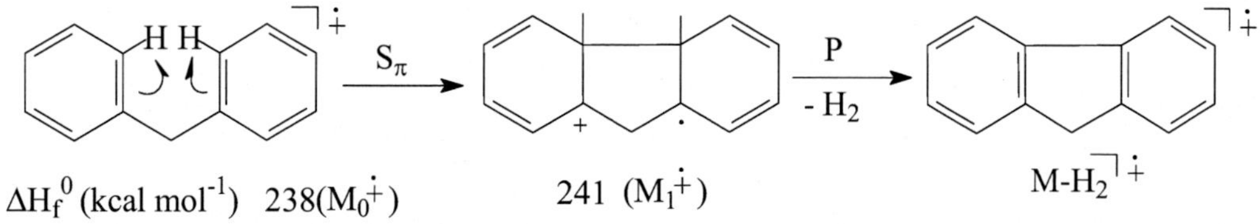

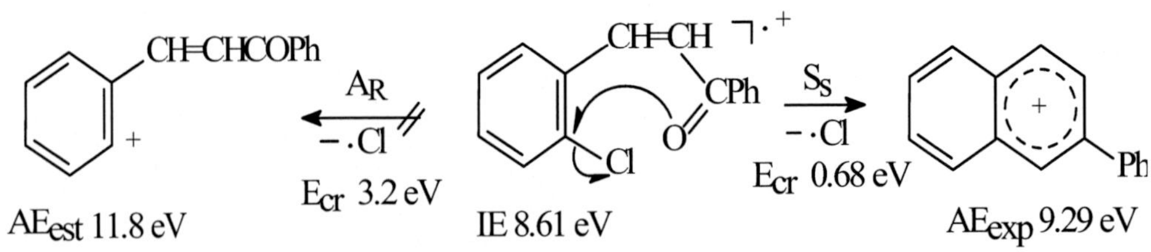

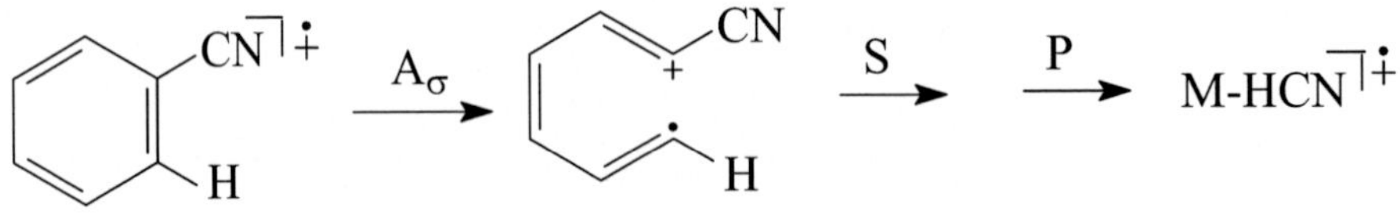

$$Me_3SiC{\equiv}CCF_3^{\cdot+} \xrightarrow[*]{S} C_4H_3F^{\cdot+}$$

$M^{\cdot+}$ $\quad$ M- $Me_2SiF_2^{\cdot+}$ (54%)

Many examples of isomerization and fragmentation types with thermochemical approach are presented in our previous works [4, 18, 49-51].

3. Specific Features of Isomerization and Fragmentation of Negative Ions

We propose the following processes for simple bond cleavage in negatively charged ions: A_1 [σ-bond cleavage in α-position to an atom or group with high electron affinity (EA) value]; A_2 (σ-bond cleavage in β-position to an atom with lone pairs); A_3 (σ-bond cleavage in γ-position to an atom with high EA value); A_p (σ-bond cleavage with stabilization of the negative charge through polarizability effect); A_{ar} (ring-substituent bond cleavage with formation of aromatic systems, e.g. cyclopentadienyl anion); A_R-type reveals the process in which stabilization of the free radical center is the driving force of the bond cleavage; B_1 (β-bond cleavage to free radical center in distonic ions); C-type (α- or β-bonds cleavage to anionic center); D-type (elimination of two-atomic fragment of the ring system); D_1-type (mass spectrometric analogue of retro Diels-Alder fragmentation, e.g. in ionized cyclohexene). The H-shift types (K, L, M, N, O, P, R) are common with those for positive ions. Skeletal rearrangements S_R, S_π, S_s and S are also common with those for positive ions while S_c-type is replaced by S_A, i.e. migration of an atom (except H) or a group to anionic center or interaction with this center. The values in brackets on the scheme signify the position (in eV) of a peak's maximum; while its intensity is measured in this maximum. M^{*-} signifies $[M\text{-}H]^-$ - quasi-molecular ion obtained by diverse methods of deprotonation techniques [33]. We also introduced a novel A_p-type which signifies the participation of polarizability (PAZ) effect in stabilization of negative charge. For example, the electronegativities EN of –O- and –S- groups are 8.0 and 5.75 units, respectively (for F – 10) [39], however, if electron affinity (EA) < 0 for $\cdot CH_2OMe$ free radical for $\cdot CH_2SMe$ it exhibits 0.87 eV [33]. Similarly, in XCH_2COOMe molecules owing to high PAZ of large I atom the ICH_2^- peak is 11% (0.7 eV) whereas XCH_2^- peaks are absent with X = F, Cl, Br [9].

$$^{-}X \xleftarrow{A_1} XCH_2COOCH_3^{\bullet -} \xrightarrow{A_1} CH_3O^{-}$$

Cl^{-} 99.2%Σ (0.3 eV) $M^{\bullet -}$ (0) 0.38%Σ (6.6 eV)

Br^{-} 92.4%Σ (0 eV) 0.5% Σ (5.5 eV)

$$XCH_2COOCH_3^{\bullet -} \not\rightarrow XCH_2COO^{-}$$

$$MeCOCH_2{-}Et^{\bullet -} \xrightarrow{A_3} MeCOCH_2^{-}$$

$M^{\bullet -}$ (0) M-Et^{-} (100)

$$^{-}Cl \xleftarrow{A_1} Cl_2CHCOOCH_3^{\bullet -} \xrightarrow[(A_p)]{A_2} Cl_2CH^{-}$$

100 (0 eV) $M^{\bullet -}$ (no) 0.28 (1.2 eV)

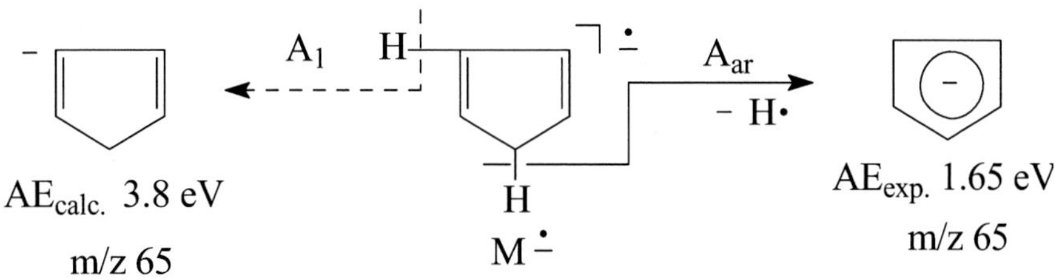

$$CH_3\dot{C}HCD(Et)COO^{-} \xleftarrow[-\bullet H]{A_R} Et_2CDCOO^{-} \xrightarrow[-\bullet D]{A_R} Et_2\dot{C}COO^{-}$$

45% M* 55%

$$Et_2CDCOO^{-} \xrightarrow[A_R]{-\bullet C_2H_5} CH_3CH_2\dot{C}(D)COO^{-}$$

100%

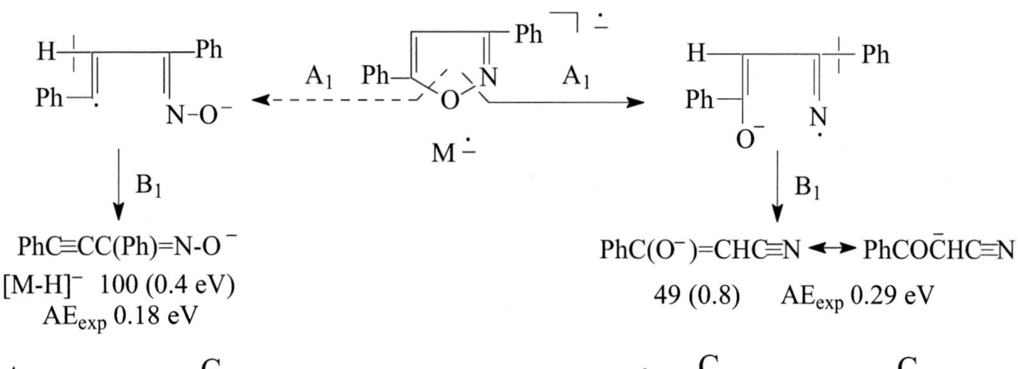

$$Ph{-}CPh_2{-}O^{-} \xrightarrow{C_2} Ph^{-} + Ph_2C{=}O \qquad CO_2^{\bullet -} \xleftarrow[-\bullet R]{C} R{-}COO^{-} \xrightarrow[-CO_2]{C} R^{-}$$

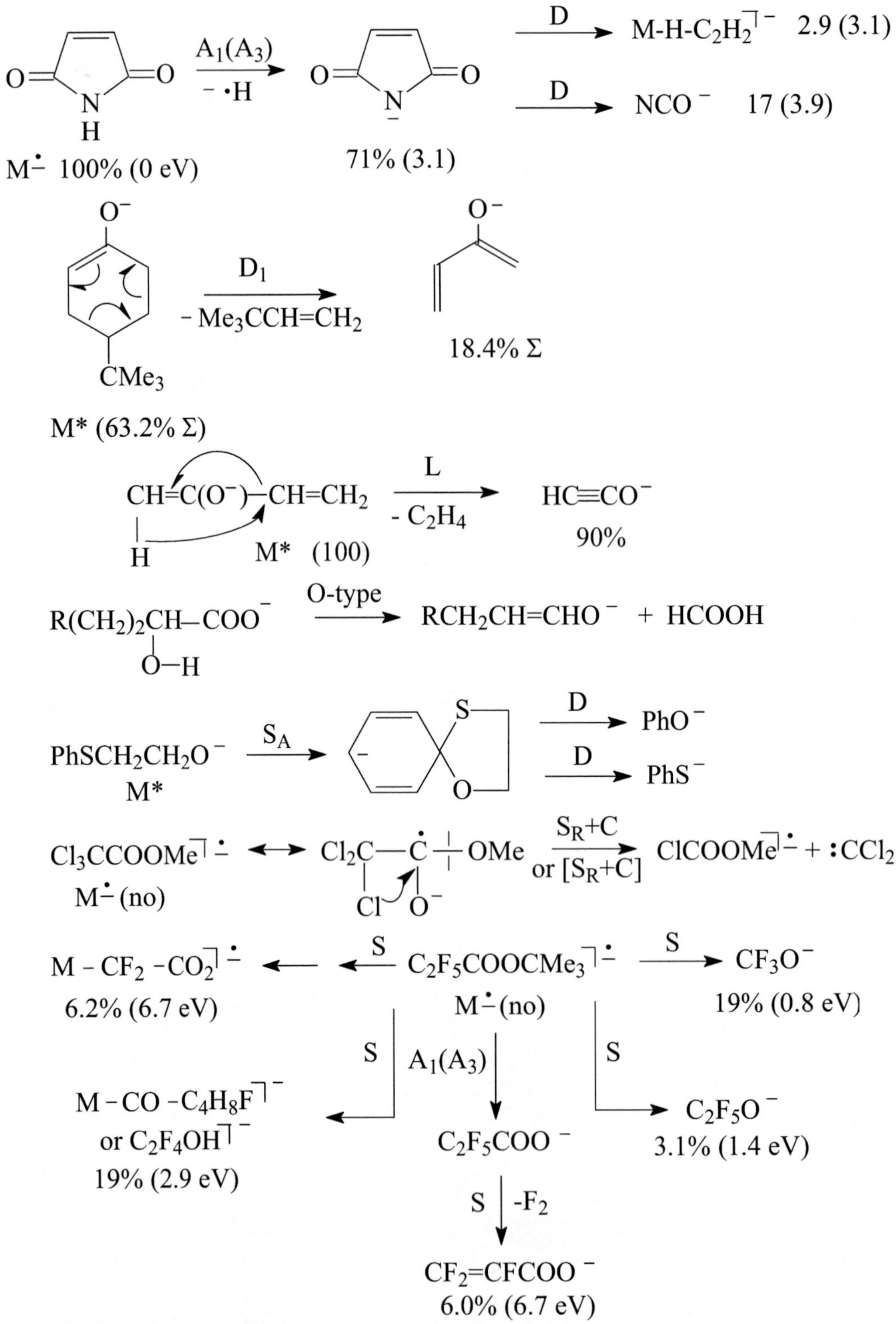

We give only a few examples of fragmentation types for negative ions since most of them, especially of H-rearrangements and skeletal rearrangements are common with those types for positive ions. However, many examples can be found in Refs 33 and 34. Here we concentrate readers attention on some specific features negative ions isomerization and fragmentation which are disputable or novel ones for those mass spectrometrists working in this domain. But in presentation of our ideas we shall lean against fragmentation types.

We remind to the reader that fragment negative ions are formed not in the whole range of energies applied to register mass spectra, say, at 0-15 eV but rather in a few (or even in a single) rather narrow domains of energies [33, 34, 52]. This happens (unlike with positive ions) because the process of negative ions formation is a resonance one and the stability of resonance state(s) at a particular energy obviously dictates the thermochemistry of fragmentation (see Section 1). In compound I for formation of ions Ia (m/z 225) there are two such domains with maxima at 1.2 and 7.2 eV. We insist that thermochemistry of its formation at these two energies might differ substantially while attraction of excited states or "non-ergodic" processes with involvement of identical stoicheometry or thermochemistry (as it is suggested in Ref. 53) to explain the existence of the ion with the same structure in two energetically remote domains is in obvious contradiction with Quasi-Equilibrium Theory. Thus, we suggest formation of *several* neutrals to let the system fit the stable resonance state at 7.2 eV while at 1.2 eV the routine simple bond cleavage takes place. At 7.2 eV a set of fragmentation types might be involved in a concerted process of *several* bonds cleavage. We calculated the ΔH_f^0 values of 15 species which in summary constitute $[C_8H_{17}\cdot]$ formulae to give AEs 225^+ from 0.8 to 8.45 eV finally choosing AE 6.65 (experimental peak maximum 7.2 eV) and suggested the Scheme 1.

Scheme 1

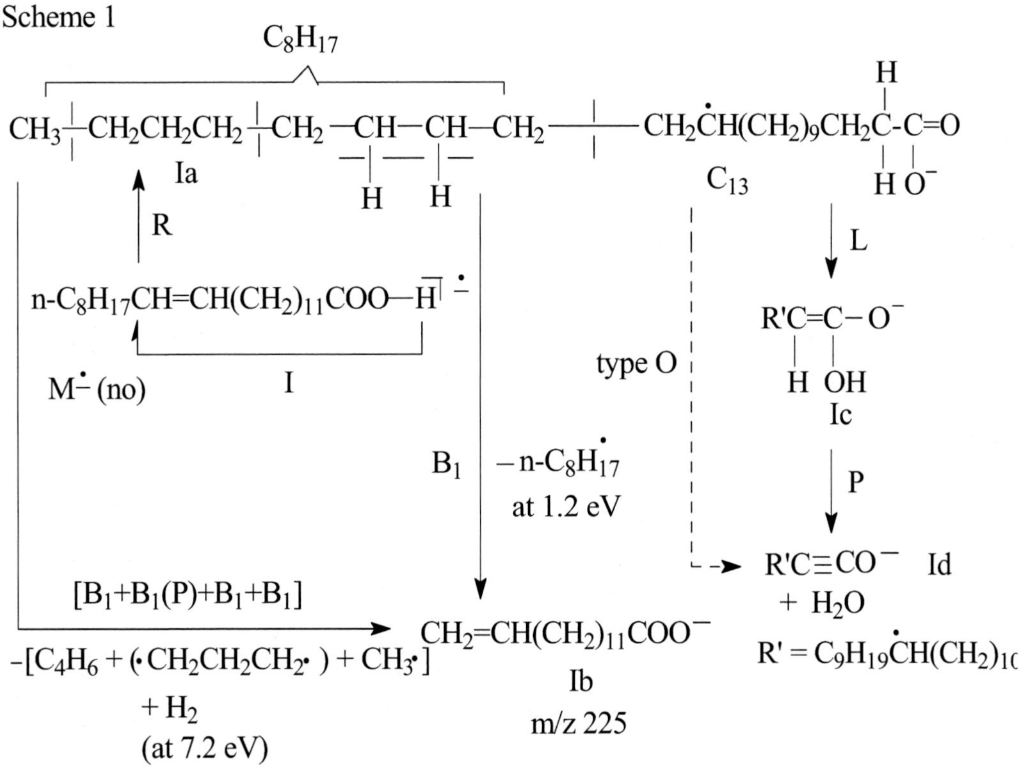

Formation of Ia ion through R-type H-migration to double bond in I and the creation of free radical center at C-13 then provokes an avalange (a cascade) of concerted β-bond cleavages (with addition of H_2 molecule elimination by P-type) to produce Ib ion at 7.2 eV while at 1.2 eV a routine cleavage of a single bond being β to free radical center occurs.

We suggest similar scheme for fragmentation of quasi-molecular ion II with formation of abundant ion IIa (m/z 115). It is obvious that direct formation of this ion by $A_1(A_3)$ process (route "a") requires transfer of negative charge from $RCOO^-$ moiety to the bond being cleaved (such scenario is suggested by the authors [54]) which is really doubtful. The possibility of formation of several neutrals in one step is not accepted in the literature (see Ref. 33).

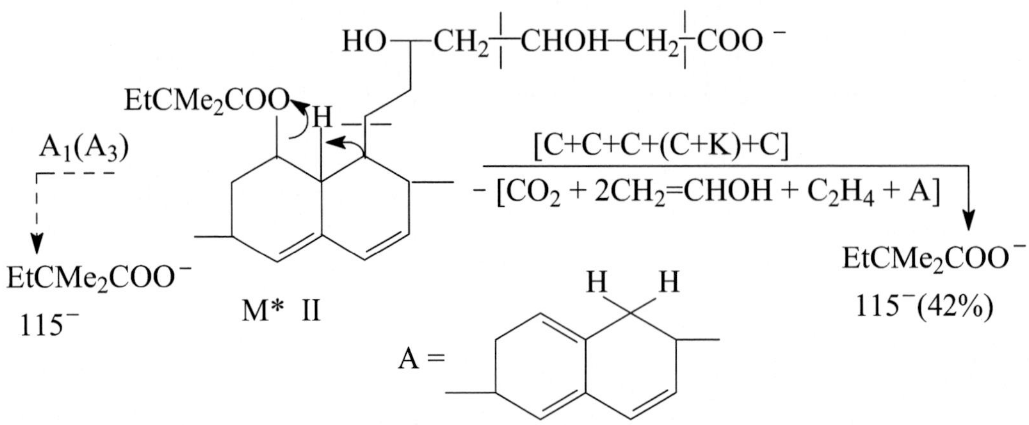

Several disputable problems do exist in relation to negatively charged molecular ion $M^{\cdot-}$. The first one is the absence of its peak in mass spectra of most compounds studied. The lack of $M^{\cdot-}$ happens much more often than of the positively charged analogue $M^{\cdot+}$. The absence of $M^{\cdot-}$ in compounds having electron affinities (EAs) < 0 is clear – these molecules cannot form stable species but rather these in temporary bound states. But how to explain the absence of the $M^{\cdot-}$ peak in molecules possessing EAs > 0? Carette et al claim [55] that for large polyatomic molecules, the excess of internal energy (gained from reacting electron) in molecular anion can be absorbed by a great number of vibrational modes resulting in ions with long lifetime and thus being registered. Thus, the larger is the molecule the higher is the stability of $M^{\cdot-}$ ion. Hence, most of the "small" molecules do not give the peak of $M^{\cdot-}$ ion [55]. Similar conclusion were declared in Refs 56 and 57. We do not agree with this position. Firstly, the peaks of $M^{\cdot-}$ have been registered in very small molecules MeCOCOMe, $(HOOC)_2CH_2$, $MeNO_2$ (seven atoms), $(NC)_2C=C(CN)_2$, $CF_3C\equiv CCF_3$ (98%Σ!) [2] but are absent in some very large molecules [2, 33, 34]. We think that the reason for absence of negatively charged molecules (like with the positively charged ones [4]) is their instability in relation to fragmentation when the ground state of a certain [fragment ion + neutral] system lies *below* that of molecular ion, i.e. when the latter being produced it immediately decomposes. This statement is supported by mass spectra of the compounds p-RC_6H_4COOMe IIIa-d [34].

R	n-C_3F_7 IIIa	$(CF_3)_2CF$ IIIb	$(CF_3)_3C$ IIIc	$(CF_3)_3CCH_2$ IIId
$M^{\cdot-}$	100 (0) 53.3%Σ	2.5 x 10^{-3} (0)	7.5 x 10^{-5} (0)	0
base peak	M^{-}	$[M\text{-}HF]^{\cdot-}$ 94.6% Σ	$[M\text{-}CF_3]^{-}$ 99.3% Σ	$(CF_3)_3C^{-}$ 98.2% Σ

All these compounds possess EAs > 0. If compound IIIa produces highly stable $M^{\cdot-}$ its isomer IIIb, on the contrary, gives a very feeble signal. The reason is that in the latter the energetically profitable fragmentation route appears with formation of very stable ion IIIe and neutral molecule HF. Similarly, compound IIIc gives $M^{\cdot-}$ ion of extremely low abundance owing to formation of highly stable ion IIIf while the ion $(CF_3)_3C^-$ (IIIg) in compound IIId coupled with stable radical IIIh makes $M^{\cdot-}$ so unstable that its peak is not registered.

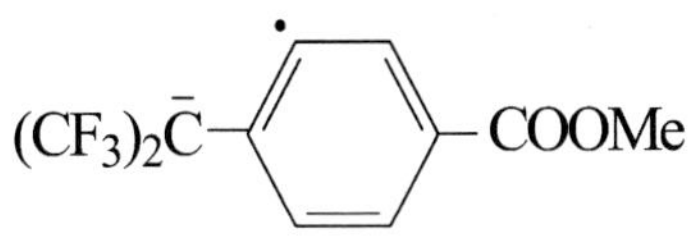

IIIe

$(CF_3)_2\bar{C}PhCOOMe$ IIIf $(CF_3)_3\bar{C}$ IIIg $\cdot CH_2PhCOOMe$ IIIh

Many examples of this kind are given in Refs 33 and 34.

In relation to Quasi-Equilibrium Theory (QET) (Section 1) it was assumed that not more than 2-3% of positively charged fragment ions could be (but not necessarily) formed with violation of QET i.e. produced not from the $M^{\cdot+}$ion with completely randomized excess energy but rather from $M^{\cdot+}$ ions being in electronically excited state. Similar situation can be assumed for negatively charged $M^{\cdot-}$ ions, i.e. that a certain, very small, part of these ions may be produced in electronically excited state, namely decomposing at the initial stage of their formation. Thus, probably, registration of such extremely weak $M^{\cdot-}$ peaks as in compounds IIIb and IIIc happens due to the presence of such electronically (but not vibrationally) excited ions.

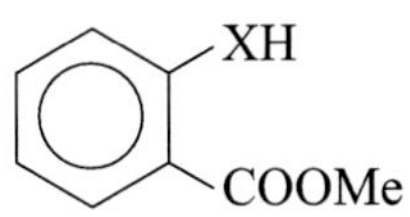

IV (X=O) $M^{\cdot-}$ (2%)
V (X=S) $M^{\cdot-}$ (98%Σ)

p-RPhCOCOPh
VI
R = H, Me, NMe_2
$M^{\cdot-}$>99%Σ

p-$O_2NPhNHAc$
VII
$M^{\cdot-}$
98%Σ

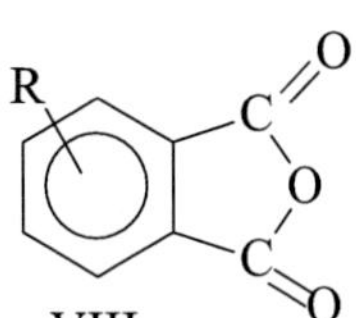

VIII
R = 3-NHAc
$M^{\cdot-}$ 98%Σ
R = H, 3-NH_2
$M^{\cdot-}$ 100%Σ

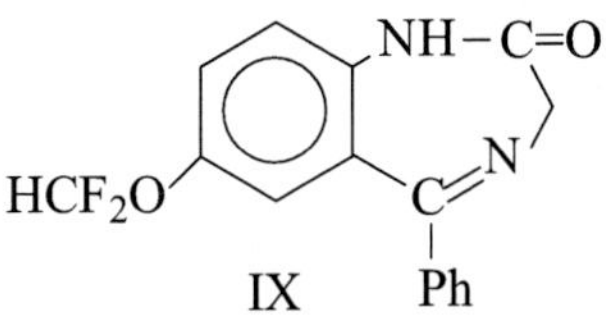

IX

$M^{\cdot-}$ 100% (0 eV)
or 99.9%Σ

X
$M^{\cdot-}$
100%Σ

$CF_3C{\equiv}CCF_3$
XI
$M^{\cdot-}$ 98%Σ

Another specifity arises with molecular negative ions being considered is their unusually high abundance, e.g. in compounds IV-XI. Some molecules do not fragment at all (from Ref. 34). This does not happen in MS of positive ions.

Since, to our knowledge, in the literature there is no explanation of this event we dare to suggest our own one. We think that the $M^{\cdot -}$ ions possessing very high excess energies (we stress that electron capture is a resonance process thus the whole energy of attacking electron is transferred to the $M^{\cdot -}$ ion) is spend on the "internal" fragmentation, i.e. the cleavage of the bonds in the rings – compounds V-X or the π-bond cleavages in compound XI. One should realize how much energy might be spent to "open" triple bond (XIa-b)

$$CF_3C{\equiv}CCF_3^{\rceil \dot{-}} \qquad CF_3\bar{C}{=}\dot{C}CF_3 \qquad CF_3\dot{\bar{C}}{-}\ddot{C}CF_3$$

XI$^{\rceil \dot{-}}$ XIa XIb

A specific problem related to molecular anions is how to explain the appearance of *several* molecular ion on energetical scale in rather a wide range of energies 0-3 eV or even higher [2, 34]. Tobita et al give the following explanation of this fact [58]. The resonances which are observed in the parent anion mass spectra at energies higher than zero are attributed to processes involving concomitant electronic excitation of molecule M by a 'fast' electronic excitation and capture of the 'thermolized' electron by so-produced electronically excited molecule M*, viz. Eqn 1.

$$e^- \text{(fast)} + M \rightarrow M^* + e^- \text{(thermal)} \rightarrow M^{*-} \quad (1)$$

This is so called two-electron process of negative $M^{\cdot -}$ formation explaining, according to the authors [58], the appearance of two (in addition to that at zero energy) domains of existence of $M^{\cdot -}$ for acridine XII, of three domains for tetracene XIII and more than one for perylene XIV. Actually, in the latter the large width of the parent ion peak indicates contribution of more than one resonance (0-1.5 eV) [58]. Deconvolution technique by a nonlinear algorithm, PicFit, unfortunately being applied very rare, permits to find additional domains of existence of molecular and fragment ions [58-61].

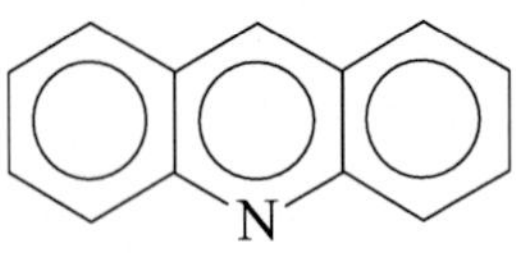

XII

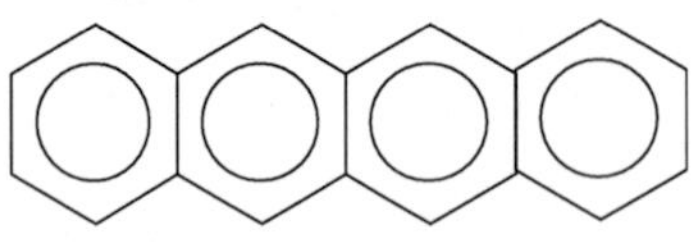

XIII

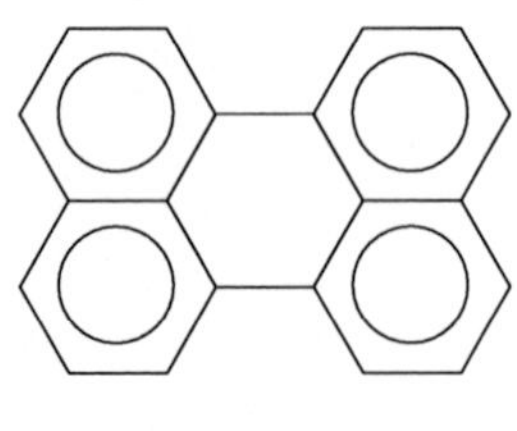

XIV

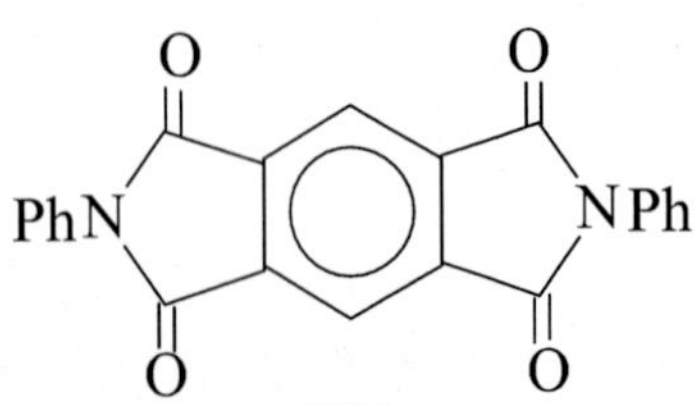

XV

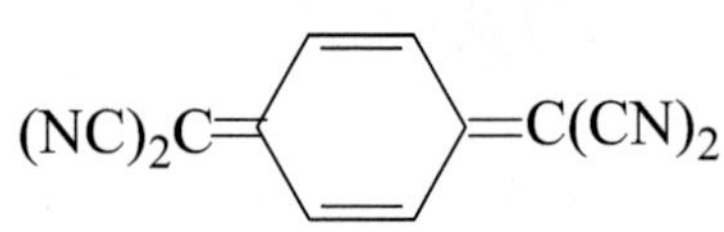

XVI

We have a different opinion on this occasion. Firstly, why these events take place only in selected molecules producing 2-4 M·⁻ ions while the most of the compounds produce a single ion giving very narrow peak. Secondly, we paid attention, analyzing data in Refs 2, 4, 33 and 34, that all the compounds giving M·⁻ peak in two or more regions possess the cyclic structure, e.g. XII-XVI. Compounds XV and XVI produce four peaks of M·⁻ in the range 0-4.2 (XV) and 0-6 eV (XVI). To our knowledge, all acyclic compounds (with a single, to our knowledge, exclusion – see later) have a molecular anion registered only in one domain.

Thus, we suggest that the bond(s) cleavage in ring system performs formation of isomeric molecular anions. We performed analysis of formation of isomeric M·⁻ ions from antraquinone XVII, which produces them in three energetical regions [59]. We suggest formation of three isomers XVIIa,b,c species (Eqn 2)

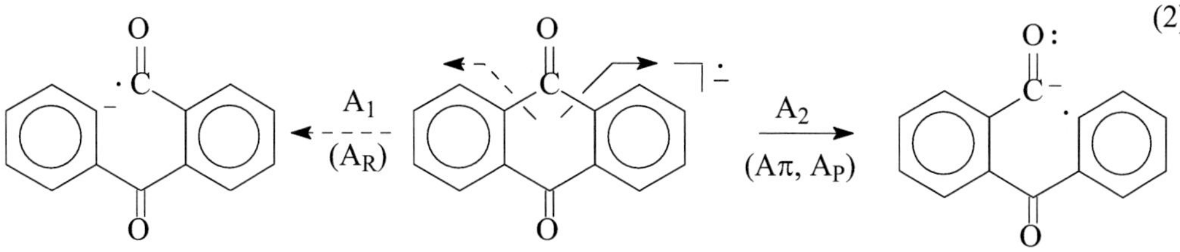

(2)

XVIIc XVIIa XVIIb

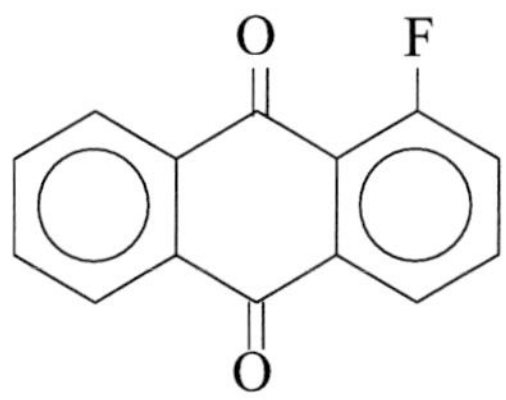

XVIII

Our calculation of the AEs for XVIIa,b,c isomers (for details see Ref. 34) gave the following values: AE_{calc} XVIIa (-1.71), AE_{exp} 0 eV; AE_{calc} XVIIb 1.38, E_{max} (exp) 1.7 eV and AE_{calc} XVIIc 0.01 eV, E_{max} (exp) 0.44 eV. We performed similar analysis for compound XVIII. Here, 4-5 domains of M·⁻ anions existence are realized [61]. We found 8 isomers which can be formed at CO-C_β bond cleavage owing to the asymmetric structure of XVIII molecule. Some of them possess close ΔH_f^0 values thus giving common unresolved peak.

We found a single non-cyclic compound which also gives molecular anion in two energetical regions, this is Me_2C=NOC(OH)=NMe pesticide [60] while 14 other pesticides (all cyclic) give M·⁻ anions in two domains [60].

Our suggestion is that this compound could exist in two tautomeric forms each giving its own molecular ion at different energies because these forms possess different enthalpies of formation and EA values.

We performed similar analysis [34] of experimental data on measured molecular anions lifetimes [2, 57]. Since, most often the range of lifetime values is very large, the mean value is presented as the final characteristics. In experiment with Rydberg K-atom collisions Suess et al [57] found that C_6F_6 molecules (XIX) form their molecular anions with a broad range of lifetimes that extend from 1 ns to 50 μs but produce them in three quite clearly separated time domains. The assumed lifetimes used in this fit of experimental data are 5, 35 and 245 μs with relative weightings 81, 14 and 5%, respectively, indicating that electron capture leads to the population of a number of different negative ions states [57]. We interpret these data and the fact of broad range of lifetimes for molecular anions by formation of several structures of molecular ions rather than by formation of a single initial structure of molecular anion but being in different electronic states as suggested by the authors [57]. Such several structures for C_6F_6 (XIX) are easily predicted (XIXa-e):

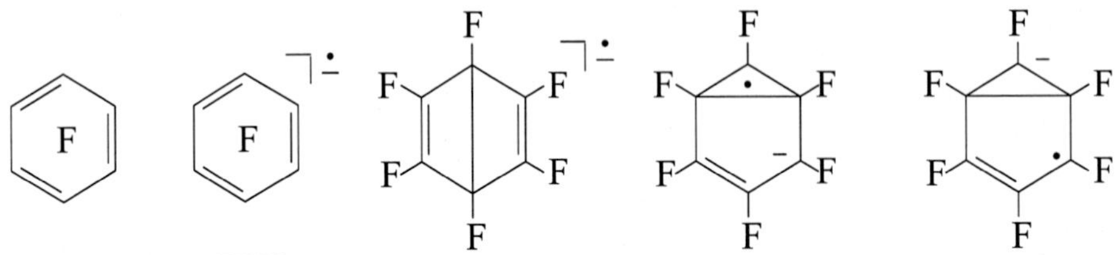

XIX XIXa XIXb XIXc XIXd

$^{-}$CF=CFCF=CFCF=CF$^{\bullet}$

XIXe

Some other specific features of negative ions formation, isomerization and fragmentation can be found in Refs 33 and 34.

4. Charge Localization. Remote-Charge Fragmentation

The conception of charge localization in molecular and fragment ions is widely circulated in organic mass spectrometry (see, e.g. Refs 49, 62-64). The representation of radical cations in monofunctional aliphatic compounds, like RNH($\cdot^{+}$)R’, RS($\cdot^{+}$)R”, RC(=O$\cdot^{+}$)R’ or RC≡N$\cdot^{+}$, has no objections since the character of the highest occupied molecular orbital (HOMO) well correlates with such structures. In general, the character of HOMO is safely defined in experiment by photoelectron spectroscopy (PES) (see [3]) although the “mixing” (i.e. interaction) of the different orbitals can arise leading to certain difficulties in assigning such character. In polyfunctional compounds (benzene ring might be also considered as a function) the site of charge localization is not always can be defined. For example, in benzaldehyde the PES bands, corresponding to π-electrons of benzene ring and CO-group are unresolved and thus the “site” of localization of charge becomes uncertain. If the strong electron withdrawing, e.g. p-CF_3-group, is introduced in benzene ring this makes MO of π-electrons more stable energetically (i.e. increases IE), while the position of MO for oxygen lone pair in CO-group changes negligibly the latter MO becoming now HOMO. H → p-MeO replacement, on the contrary, will destabilize MO of benzene ring (i.e. will essentially decrease IE of the molecule) exerting only minor effect on MO of CO-group. However, all three molecules p-XPhCHO (X=H, CF_3, MeO) fragment identically giving the most abundant XPhC≡O^{+}, i.e. [M-H]$^{+}$ primary ions.

Thus we see that the idea of charge localization becomes useless. Moreover, one might not ascribe to the localized charge the function of the *driving* force for fragmentation like it is done by some authors, e.g. formation of $[M\text{-}Me]^+$ ion is ascribed to participation of Ph-O($\cdot^+$)Me ion with charge localization at O-atom [5].

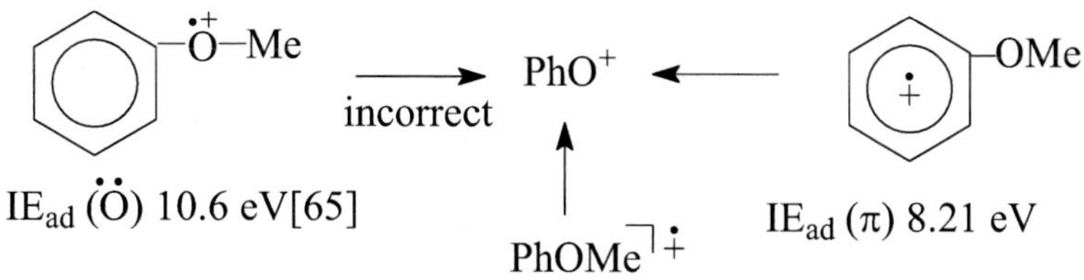

Therefore, when representing the structure of molecular ion – with rare exclusion of some monofunctional aliphatic compounds with low IEs – we bring the charge out of the brackets. We also do not see much sense in representation of the molecule's structure with fixed center of "future" ionization in ion-molecule reactions, say, in molecules' protonation since this can cause rude errors. For example, IE_1 and IE_2 exhibit for the molecule $Me_2N(:)C{\equiv}N$ and $Me_2NC{\equiv}N:$ 8.86 and 14.58 eV, respectively [65], however, protonation occurs at ≡N: group since it gives thermodynamically essentially more stable ion $Me_2N(:)\text{-}C(^+){=}N(:)H \leftrightarrow Me_2N(:)C{\equiv}N(^+)H$ compared with its isomer $Me_2N(^+)(H)C{\equiv}N$ [65].

In representation of radical-anions we recommend to use only the general structures $[AB]^{\cdot -}$ since the data on the character of lowest unoccupied molecular orbitals (LUMO) are very scarce while the general trends in structure/LUMO character (as compared with those for HOMO) are unknown.

At the same time at representation of fragmentation schemes we systematically use the structures of distonic ions, i.e. the ions with spatially separated cationic (anionic) and free radical centers either conditionally (like in resonance structures) (XXa-d) or existing in reality, e.g. at ring opening in molecular ions of cyclic compounds (XXI)

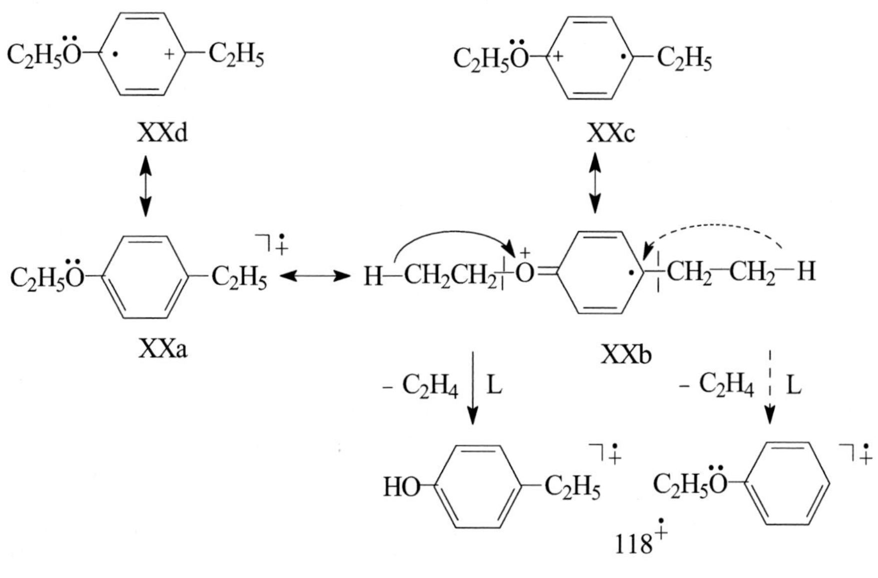

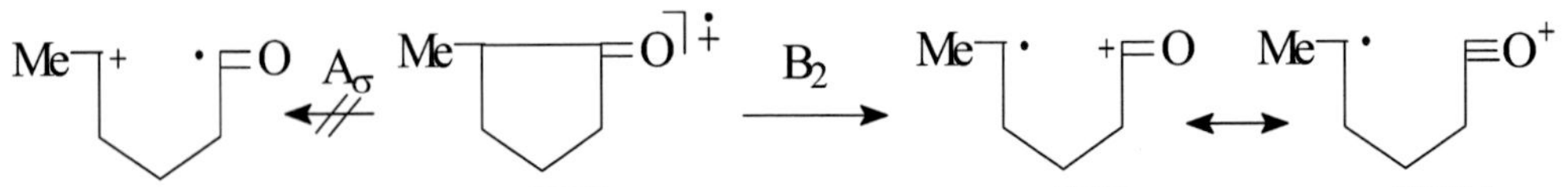

XXIb XXIa XXIc XXId

In representing distonic ions and resonance structures one might give preference to those with better stabilized charge rather than the free radical center (site). Thus, in XXb ↔ XXc ion the charge is better stabilized by oxygen lone pairs while in the structure XXd – it is not. Similarly, the ions XXIc ↔ XXId are essentially more stable than the ion XXIb. This comes out from the isodesmic reaction for model ions

	Me_2CH^+ +	MeCH=O →	Me_2CH_2 +	$MeC{\equiv}O^+$ +	Q
ΔH_f^0 (kcal mol^{-1})	189	-39.7	-25	149.5	24.8

in which $MeC{\equiv}O^+$ is more stable than Me_2CH^+ ion by 24.8 kcal mol^{-1}. In presenting resonance structures it is possible to give only one of them, say XXb or XXc, XXIc or XXId. In general, one might lean upon common sense. For example, if acetyl cation may be pictured by any of resonance structures either $CH_3C(^+)$=O: or $CH_3C{\equiv}O^+$, with electron withdrawing CF_3-group – it is better to take $CF_3C{\equiv}O^+$ resonance isomer since here the positive charge appears in remote position from destabilizing CF_3-group. The cations of vinylic type it is better not to include in resonance partners, e.g.

$$HC{\equiv}C\overset{+}{C}H_2 \nleftrightarrow H\overset{+}{C}{=}C{=}CH_2$$

XXIIa XXIIb

since the contribution of XXIIb structure is negligible while stabilization in XXIIa occurs through polarizability effect of π-system under the action of the powerful acceptor, the positive charge, rather than through resonance as it takes place in allyl cation.

Much (we think too much) attention in the literature is paid to fragmentation of even electron ions with fixed positive charge (for fragmentation of those with negative charge see Section 3). These ions, generated at EI, like $R\text{-}N(^+)H_3$ or $RC(^+)(OH)_2$ as well as at other methods (FAB or CID), e.g. $[M+Li]^+$, $[M+H]^+$ etc very often decompose without direct participation of the charged moiety of the ion, e.g. eliminating the molecules of H_2O or H_2 from the part of the ion which does not carry the charge. Many authors classify such processes belonging to the separate (specific) fragmentation type – "remote-charge fragmentation". A large set of examples of charge-remote fragmentation is collected by Gross [66]. But we do not see any specificity in these processes since as it is stressed above we do not assign to the positive charge the driving force for fragmentation. If, say, C-type fragmentation occurs with cleavage of α- or β-bonds to the cationic center (Section 2) this only manifests that such process is favorable energetically rather than the cationic center "provokes" fragmentation. Similarly, β-bond cleavage relative to free radical center by B_1-type (Section 2) takes place because it is energetically more profitable (smaller bond dissociation energy – BDE) compared with α- or γ-bond cleavages with essentially larger BDE) but not because free radical center provokes fragmentation. By the same reasons elimination of stable neutral molecules from "non-charged" part of the ion is thermodynamically profitable exactly as it is profitable in case of elimination from (or close to) the charged part of the ion. We do not see any necessity to ascribe participation of the charge in fragmentation through space as it is done by some authors [66] (but one might not mix up it with S_S-type fragmentation – anchimeric assistance to the bond cleavage through space – Section 2). The remote-charge fragmentation occurs in the systems where through-space interaction is not possible (in meta-disubstituted bensenes, 1,3-disubstituted adamantanes and others). Finally, organic chemistry knows a lot of similar with mass spectrometry processes proceeding in *neutral* molecules, say, in pyrolytic reactions and this does not cause any surprise as it is correctly underlined by Gross [66]. From the other side, we would like to stress that the remote-charge fragmentation once again demonstrates the basic principle of quasi-equilibrium theory (see Section 1) about complete randomization of excess energy gained from colliding electron (or attached – for negative ions) and with further fragmentation following the most energetically favorable route.

We notice that the idea of charge localization was originally suggested in 1940 [67].

5. Ion-Dipole (Complexes Ion/Neutral Complexes; Bridged Ions)

Some authors insist on participation in rearrangement processes of intermediates with non-classical structures like ion/neutral species, "bridged" ions etc. (see Reviews [68-70]). Thus Schwarz et al for the process of skeletal rearrangement $[CF_3COOCH_2CHMe_2]^{+\cdot} \rightarrow Me_3C^+ + CF_3COO\cdot$ suggest participation of three possible intermediates [71]

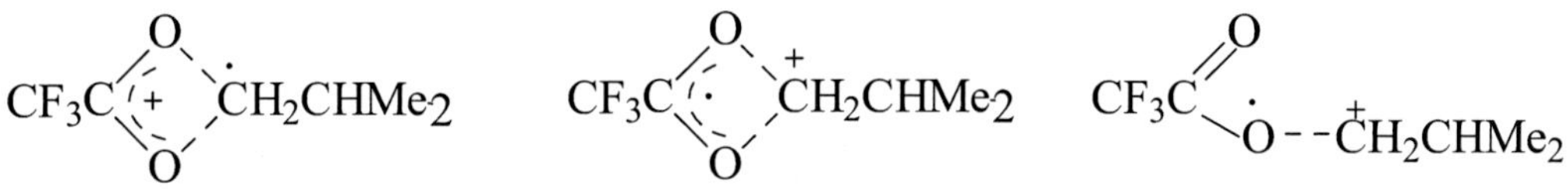

There is no doubt that such structures do really exist, e.g. it is evident that H-migration through K-type proceeds without fail with intermediate of the bridged ion

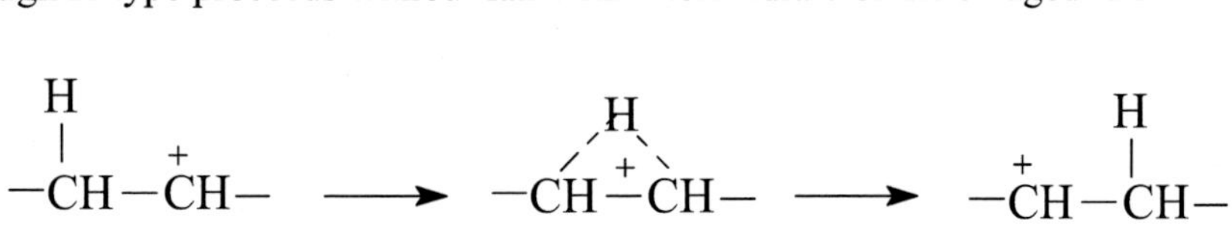

However, we stay against including such transient species in the fragmentation schemes since they make them more complex without introducing any positive moment in understanding the mass spectrometric process. On the contrary, we propose to follow strictly the isomerization and fragmentation types, elaborated in our works (see Section 2) which stand on firm thermochemical approach and representation of intelligent schemes in which the stabilization of cationic (and free radical) centers is systematically and insistently taken into account, e.g.

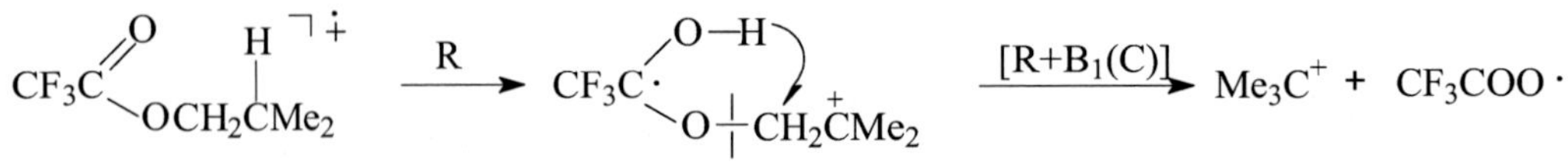

Some authors, instead of multistep rearrangement processes in which each step is considered from the position of energetical accessibility and stability of each isomer, suggest the formation of a single complex, ion/neutral species of unknown structure which gives birth again through unknown mechanism to *all* fragment ions, e.g. [70]

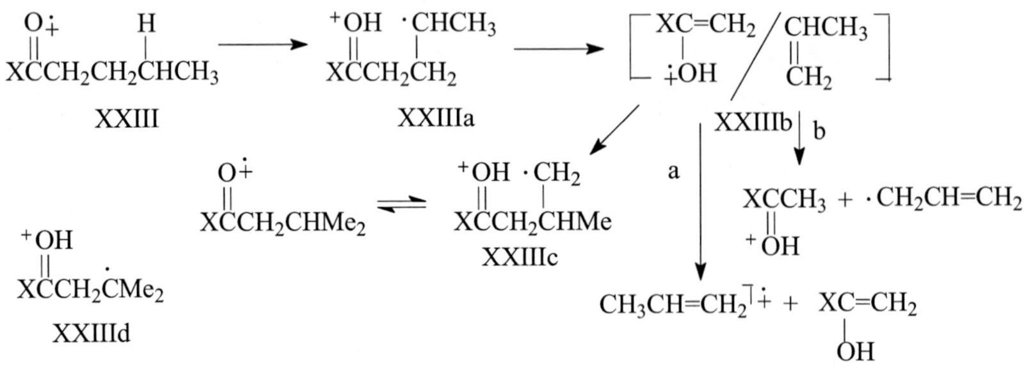

The participation of complexes of XXIIIb-type gained the special name of "cage effect" [70]. In mass spectrometry of negative ions representation of such complexes, ion/neutral species, is systematically carried out by Bowie (Refs 33 and 34). We do not see any sense in including such complexes with rather vague "cage effects" which "govern" the rearrangement process but suggest, instead, to introduce the isomerization types until formation of the final products. From pedagogical point of view, the acception of "cage effect" makes certain harm helping the authors to escape thinking about real mechanism of isomerization and fragmentation.

In the scheme given above (from Ref. 70) the route "a" is doubtful since in consensus with Stevenson-Audier rule (Refs 6-9, 49-51) the charge will reside on the most stable partner, here $[XC(OH)CH_2]$ (i.e. $[XC(OH){=}CH_2]^{\cdot +}$ having essentially lower IE compared with its partner $MeCH{=}CH_2$. The formation of primary free radical center in the ion XXIIIc is also doubtful whereas his isomer, distonic ion XXIIId, is more stable. Thus, introduction into scheme of complex ion/neutral species XXIIIb in no way explains the formation of the fragments and the isomerization of the hydrocarbon chain. We propose the different isomerization and fragmentation scheme for compound XXIII

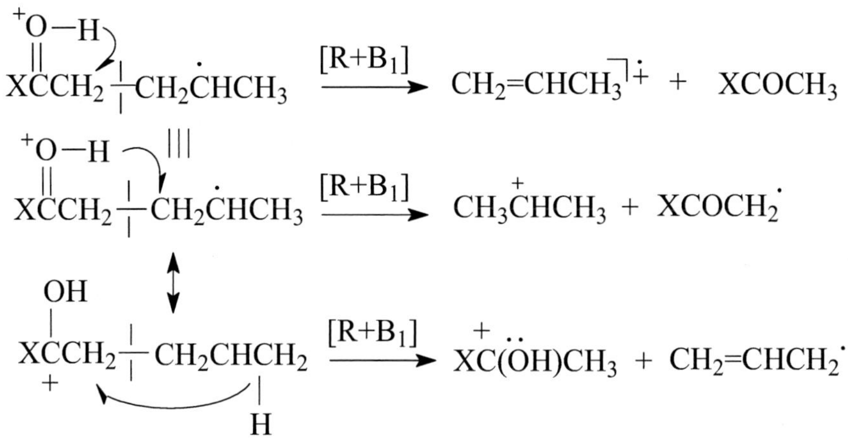

XXIIIa

Cacace et al showed that cyclohexyl cation experimentally obtained at radioactive decay of tritium with a rate of 10^{-10} s rearranges to 1-Me-cyclopentyl cation [72]. We suggest the following scheme of formation of the latter from ionized substituted cyclohexanes

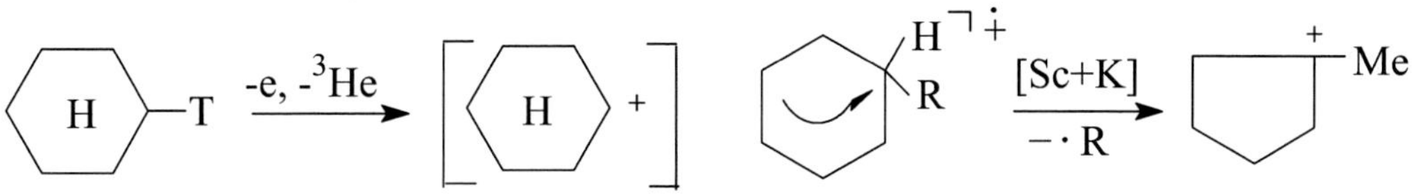

Although the ions with five-coordinated carbon are well known, e.g. the ion CH_5^+, generated and applied in chemical ionization MS, however the acceptance of hypervalent carbon in no way is justified in fragmentation schemes, e.g. in isomerization of cyclohexyl cation as it is suggested in Ref. 73.

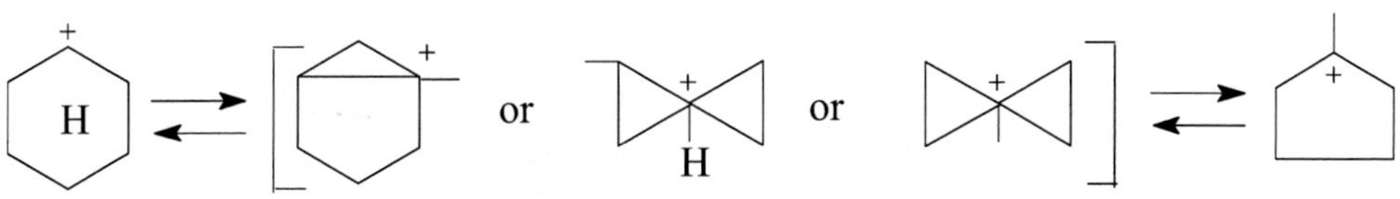

At the same time the structures of phenylated and alkylated cyclopropane ions used for explanation of fragmentation of Me_3CPh and Et_3CPh molecules (as well as of protonated cyclopropane – $C_3H_7^+$ isomer), suggested by Meyerson as early as in 1956 [74], appeared to be quite viable since such structures were later obtained in solution (see Ref. 75)

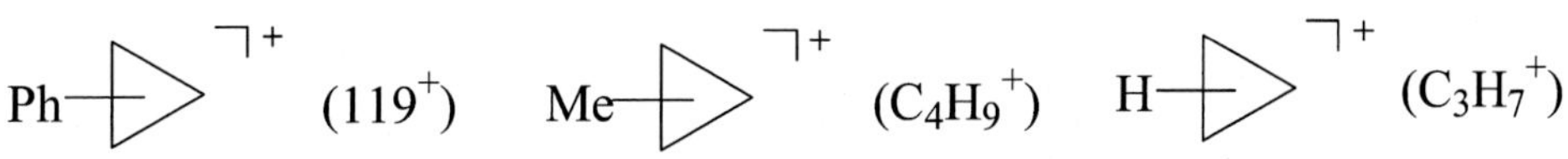

The proposal of Brauman to replace the especially unfavorable energetically three-membered intermediate XXIVa in the process of nitro-nitrite rearrangement by ion/neutral complex XXIVb [76] also deserves attention.

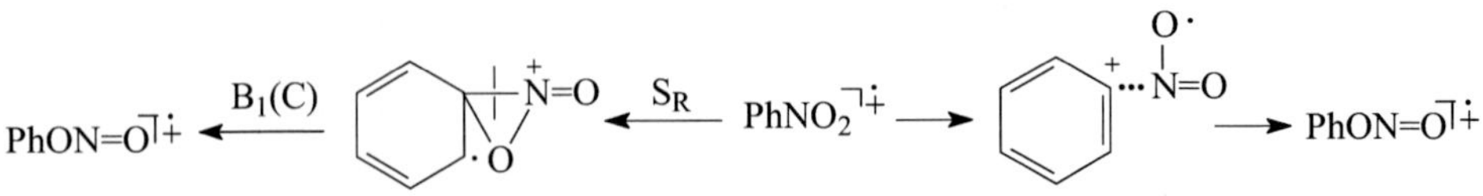

XXIVa XXIVb

Now we consider some other non-classical ions. According to Beauchamp et al [77] the linear 2-substituted ethyl cations (XXVa)

$\ddot{X}CH_2CH_2^+$ (X = F, Cl, OH, NH_2) XXVa

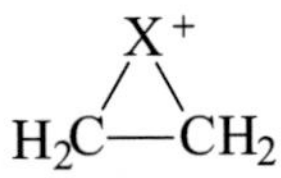

XXVb

$CH_3\overset{+}{C}H\ddot{X}$ (X = F, Cl, OH, NH_2) XXVc

are more stable than their cyclic isomers (XXVb) and only with X=Br and SH the latter by 1-3 kcal mol^{-1} are more stable. However, this difference is too small to confirm their existence. Moreover, the interference of the third, the most stable isomer XXVc is possible. The radical-cations with halogen and oxygen in hypervalent states were observed in the works [78, 79]. The first appeared to be less while the second more stable compared with their classical isomers (recalculated for ΔH_f^0 Me· 35.1 kcal mol^{-1} from [78]); ΔH_f^0 – in kcal mol^{-1}

$Me\overset{+}{Cl}CH_2\cdot$	$Me\dot{C}H\overset{+}{Cl}H$	$Me\overset{+}{Br}CH_2\cdot$	$Me\dot{C}H\overset{+}{Br}H$	$\dot{C}H_2CH_2\overset{+}{O}H_2$	$\dot{C}H_2\overset{+}{O}(H)Me$
242	232	253.5	237.5	181	176

$MeCH_2Cl$	$MeCH_2Br$	$MeCH_2OH$	MeOMe
226.6	222.5	182.5	186.5

In nucleophylic substitution reactions the analogous structures were obtained for aromatic compounds [80]

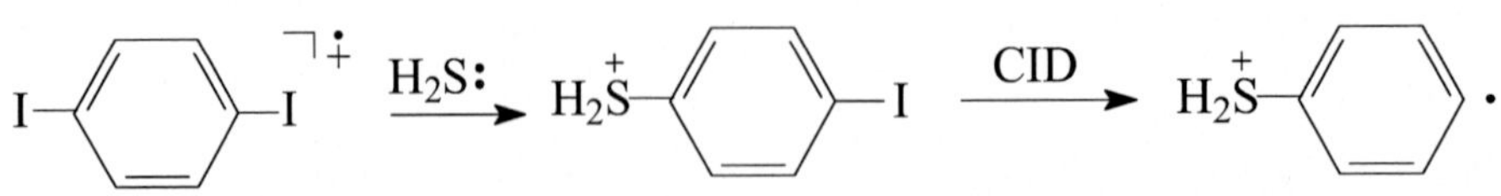

6. Tautomers. Stereochemical Effects

Mass spectra of diketones $PhCOCH_2COR$ (R = alkyl, c-C_6H_{11}, c-C_6F_{11}, perfluoroalkyl) [81] whose molecules are completely enolized in gas phase exhibit the peak of the $[M\text{-}R]^+$

ion, i.e. PhC(OH)=CHC≡O^+ ion which is essentially more intense than the peak of $[M-Ph]^+$ or RC(OH)=CHC≡O^+ ion. This means, that irrespective of the content of each of enol forms in neutral diketones (for some of them the tautomer PhCOCH=C(OH)R is even preferable in solution), the decomposing molecular ion has already the structure of the enol which gives more stable cation PhC(OH)=CHC≡O^+, i.e. 1,5-H-shift and isomerization of initial $M\cdot^+$ occurs. It is interesting that similar process takes place in fixed isomers $CF_3C(OCH_3)$=$CHCOCH_3$ and CF_3COCH=$C(OCH_3)CH_3$ since AE of $[M-CH_3]^+$ (as well as $[M-CF_3]^+$) for both structural isomers) appeared to be identical. The conclusion is straightforward: if CH_3-group so easily migrates then 1,5-H-shift is expected to occur in any similar system.

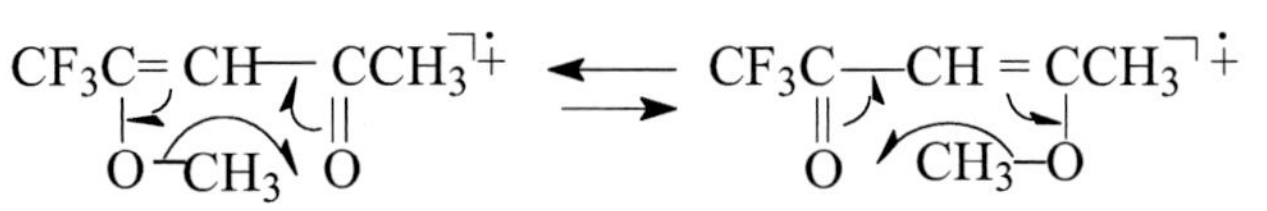

Hence, it is expected that initial molecular ion of 1,3-dicarbonyl compound (diketone, β-ketoester, diester) will convert to the enol form which gives the more stable fragment. Is it possible that $M\cdot^+$ of enol form isomerizes to keto form? From AEs (exp.) and derived ΔH_f^0 values of $[M-R]^+$ ions from PhCOCH=C(OH)R, $[M-R_F]^+$ ion from $CH_3COC(OH)$=$CHCOR_F$ (R_F = perfluoroalkyl), $[M-CMe_3]^+$ ion from $Me_3CC(OH)$=$CHCOCMe_3$, $[M-CF_3]^+$ ion from p-XC_6H_4COCH=$C(OH)CF_3$ (X = H, F, CH_3) molecules it is concluded that all of them give fragment ions in enol form of the PhC(OH)=CHC≡O^+ type. Unexpectedly, it appears that the ions $[M-R_F]^+$ from p-$CH_3OC_6H_4COCH$=$C(OH)R_F$ (R_F = CF_3 and n-C_6F_{13}) at the threshold of their formation correspond to keto form $CH_3OC_6H_4COCH_2C$≡O^+ although it by ~25.5 kcal mol^{-1} is less stable than $CH_3OPhC(OH)$=CHC≡O^+ tautomer. That follows from corresponding isodesmic reaction. Such large difference in stabilities is quite expected if their structures are compared.

So, why the less stable ion is obtained? In order to transfer the molecular ion $M\cdot^+$ (enol) to $M\cdot^+$ (diketo) it is necessary to spend 0.6-0.8 eV for aliphatic [82] or 1.0-1.2 eV for aromatic compounds [81]. Since E_{cr} (equivalent of activation energy E_a for monomolecular reactions) for formation of RC(OH)=CHC≡O^+ ions (R = Ph, XPh, Me, CF_3 and others) is *smaller* compared with isomerization energy $M\cdot^+$ (enol) → $M\cdot^+$ (keto), then such isomerization does not happen. However, when the powerful electron releasing CH_3O-group is introduced in benzene ring the ionization energy (IE) of a molecule drastically falls [e.g. from 9.3 to 8.23 eV in XC_6H_4COCH=$C(OH)CF_3$ at H → p-OCH_3 replacement] thus increasing E_{cr} for the process $M\cdot^+$ → $[M-CF_3]^+$ up to ~2.6 eV instead of 1.0-1.2 eV [81]. Therefore the $M\cdot^+$ (enol) gets the possibility to isomerize (instead of fragmentation) to $M\cdot^+$ (keto), and the latter fragments on its own most energetically favorable way, giving $CH_3OC_6H_4COCH_2C$≡O^+ instead of expected $CH_3OC_6H_4C(OH)$=CHC≡O^+ isomer [81]. High AE_{exp} 11.73 eV for $[M-Ph]^+$ ion from PhC(OH)=CHCOPh molecule manifests formation of an ion of unknown structure with ΔH_f^0 by 25.5 or 17 kcal mol^{-1} higher than PhC(OH)=CHC≡O^+ and $PhCOCH_2C$≡O^+ isomers, respectively. Isomerization of initial $M\cdot^+$ to $M\cdot^+$ of unknown structure occurs because E_{cr} = 2.17 eV is also very high for energetically the lowest process of PhC(OH)=CHC≡O^+ formation.

Elimination of H_2O molecule in the following example in no way proves that the molecule exists chiefly in dienol form but rather that the process of enolization through L-type (see Section 2) in molecular ion increases with decrease of electron energy

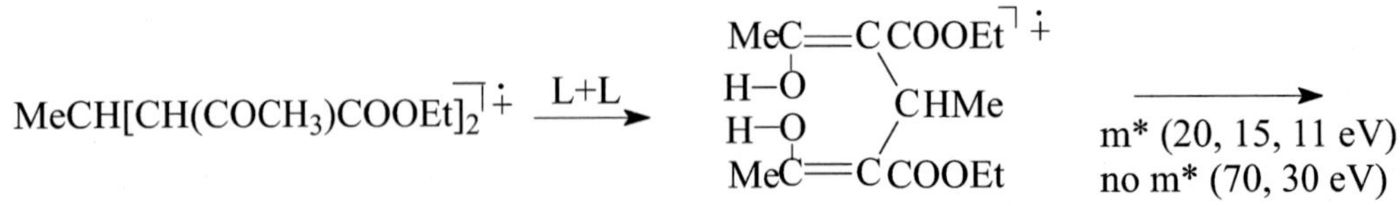

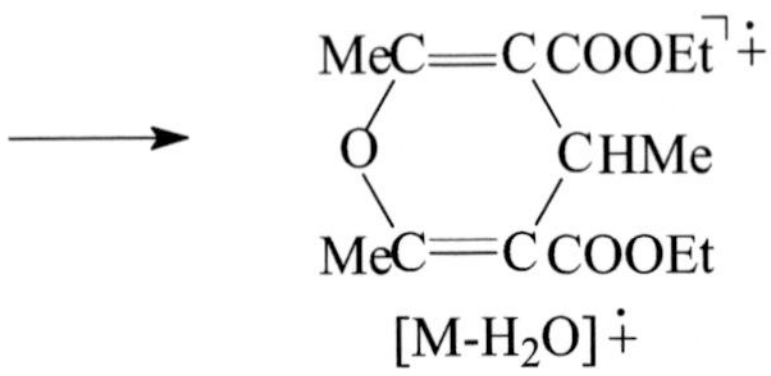

eV: 70 30 20 15 11

%Σ: 0.2 0.4 1.4 2.1 13.6

Thus we think that determination of tautomers ratio using relative abundances of selected ions (as it is claimed in some works) belonging only to a given tautomer – is unreliable. Furthermore, the tautomers ratio varies at sample evaporation and very sensitive to any forms of interaction with metallic and nonmetallic surfaces of inlet system and ion source.

However, mass spectrometry with the negative ions formation might be recommended for distinguishing of tautomers since L-type isomerization for $[\text{keto}]^{\cdot-} \leftrightarrow [\text{enol}]^{\cdot-}$ does not occur [33].

The application of organic mass spectrometry to solving the stereochemical problems has been widely discussed and is continued to be discussed in the literature. First of all we mention the large series of the works of Djerassi published in 60-80 years of the last century (more than 300 works) dedicated to the solution of structural and *stereochemical* problems by MS methods.

The important domain is the distinction of isomers and stereoisomers by methods of chemical ionization, ion-molecule reactions and collision activation. As to the distinction of the stereoisomers by EI one should be very cautious. If the mass spectra of several stereoisomers with known structures are available then their interpretation helps to reveal their specific features, however, mainly not in the fragmentation routes but rather in the differences in common ions' abundances those ions mainly being primary ones.

However, if the stereoisomers with the definite structures (determined by IR, NMR or from synthesis) are not available then it appears problematic to ascribe the fragmentation process from experimental mass spectrum to a particular stereoisomer. We remind to the reader, that even structural isomers, e.g. C_7H_8 isomers, p- and m-disubstituted benzenes, isomeric alkenes and cycloalkenes, alkene derivatives, say unsaturated acids or esters, although give the different mass spectra but their interpretation remains uncertain for distinction of isomers. Double bond shift in aliphatic chain, benzene ring disclosure, H- and halogen walking over the benzene ring appear to be a routine event (see Refs 4, 18, 49-51). Therefore, ring opening in saturated carbo- and heterocyclic mono- and polycyclic compounds inspiring the stereochemical changes compared with initially ionized molecule makes these changes even more inevitable at their fragmentation.

7. Determination of Ions' Structure from Mass Spectra of Collisional Activation (CA)

The CA method in the regime of MS/MS and MS_n is an extremely effective mass spectrometric method for structural analysis. Introduction of this method into practice performed a real revolution in structural analysis of isolated individual compounds and individual compounds as the components of complex mixtures. MS/MS method can be considered as a specific "chromatograph" which through consecutive fragmentation of either daughter ions or of molecular ones followed then by decomposition of fragment ions in complex mixture yet more and more individualizes the compound being analyzed, thus distinguishing it from other compounds, including isomers. The method and its application are described in detail in Ref. 5 and Refs therein. In this section we critically consider only one aspect of CA application, i.e. determination of ions structures with small masses.

The collision chamber is usually arranged in the field-free space where monomolecular dissociation takes place. This means that the ions with a given mass, selected for study, are the ions generated in the ion source, which possess the energies being insufficient to decompose at the time scale of existence of metastable ions, i.e. microseconds. In general, the time scale of events going on in mass spectrometer with CA chamber is given below [83].

Time of event (s)	
Ionization, EI	10^{-16}
Bonds vibrations (the start of chemical process)	10^{-14}—10^{-12}
Rotation	10^{-8}
Ions leave the ion source	10^{-6}
Ions pass the first free space	3×10^{-6}
Ions enter the collision chamber (the second field-free space)	1.3×10^{-5}
Excitation by collision	10^{-15}—10^{-14}
Time of crossing of collision chamber by an ion	10^{-7}
Ions enter the collector	2.3×10^{-5}

The undecomposed ions, selected for introduction to collision chamber, possess the wide range of internal energies. A priori it is assumed that the ions with a given mass A^+ *retain* its structure irrespective of the content of internal energy in the ion. This internal energy is left either from the precursor ion (if it is a fragment ion) or from bombarding electron if it is a molecular ion $M^{\cdot +}$. Only at such postulate (retaining the structure) CA spectrum could be treated as structurally specific. To fulfil this condition it is highly desirable to generate the ion A^+ with the content of internal energy being as low as it is possible. Although the ways of decreasing this energy and exploiting only the ions with low content of energy are well known [83] but in practice are not always applied by mass spectrometrists working with CA methods. But even when such procedure is performed this does not yet secure the occurrence of ion A^+ with a single structure. If saying about molecular ion there are several classes of organic compounds which are *exothermally* transferred into isomers, e.g. ionized acetylenes to cyclopropenes or ionized carboxylic acids and esters to $RC(^+)(OH)OR'$ ions (Refs 4 and 50). Naturally, there are much more examples when isomerization requires very small energy and, what is worse, in such energetically low processes not a single but rather several isomers are obtained with close enthalpies of formation.

There exists another serious problem of application of CA spectra for ions structures elucidation. The assignment of structure to a particular isomer is based on the comparison of CA spectra of several isomers each of them, naturally, possessing a specific, unique structure. This "specifity" is hinged over the selection of such initial molecules which, according to the mass spectrometrist opinion, will give at dissociation strictly the isomer of the planned structure. Here the uncertainty arises whether it is possible to "guarantee" such structural "specifity" (if to take into account the well known possibilities of molecular ions' isomerizations demonstrated in this edition). Moreover, one should be confident that CA spectrum does not correspond to some unknown isomer, which is not taken into account [83]. This means that *all* structural isomers have to be generated that is hardly possible even for simple ions. For example, for such simple ions as $[C_2H_3O]^+$ one suggests 8! quite reasonable structures

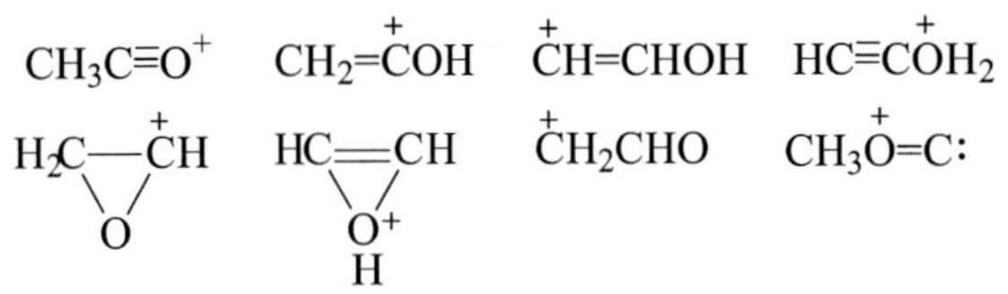

But how to generate each of them in a "pure" form and run its CA spectrum?

Therefore, when some workers quite confidently suggest the structure of a particular ion originating from a compound under investigation owing to identity (or closeness) of its spectrum with that obtained for a model compound without consideration of many other possible isomers, such identification cannot be treated as the safe one.

CA spectra could be close, e.g. for $CH_3C{\equiv}O^+$ and $CH_2{=}C{=}O(^+)H$ isomers [84]. The existing large differences in experimental conditions of obtaining CA spectra in diverse laboratories add the difficulties in comparison of CA mass spectra and identification of isomers, especially in case of relatively large molecules like ionized phenol and its isomers [37].

However, the most serious problem which is encountered in application of CA spectra for structural analysis is that initially obtained isomers being structurally pure could very fast isomerize at the moment of or immediately after collision with reagent gas (usually He) and then decompose [85]. Such isomerization processes are well documented: $HC{\equiv}O^+ \rightarrow HO(^+){=}C:$ [86], diverse $[C_2H_3S]^+$ isomers $\rightarrow CH_3C{\equiv}S^+$ [87], the skeletons of R-substituents isomerize in pyridinium ions $C_5H_5N(^+)\text{-}R \rightarrow C_5H_5N + R_1^+$ [88, 89]. Other examples are given in Ref. 85. Contrary to the extremely high barrier for isomerization $[HC{\equiv}COH]^{\cdot+} \rightarrow CH_2{=}C{=}O^{\cdot+}$ 79 kcal mol^{-1} CA spectrum of ionized ketene contains the rearrangement ion 29^+ $HOC:^+/HC{\equiv}O^+$ whereas CA spectrum of its isomer, ionized ethynol, gives the signal of rearrangement ion $CH_2^{\cdot+}$ [90]. Energetical barrier for $Me_3C^+ \rightarrow MeCH(^+)Et$ transfer exhibits 32 kcal mol^{-1} however both ions give indistinguishable CA mass spectra [91]. Besides high barriers for izomerisation of some ions, these processes themselves might require high energetical expenses due to transformation of well stabilized ions to less thermodynamically stable isomers like $Me_3C^+ \rightarrow MeCH(+)Et$ [91] or $HC{\equiv}O^+ \rightarrow HOC{:}^+$ [90]. Still these processes take place.

We interpret all these facts in such a way that overcoming of such high barriers and transformation of stable isomers to less stable ones could be achieved only with an aid of reagent gas which is far from being "neutral" carrier of energy but takes an active part in chemical reaction similar to the vessels' walls taking part in chemical kinetics' reactions.

Finally, we stress that the data gained from CA spectra are insecure to be used in structural analysis and especially in quantitative thermochemical investigations. For example, from measurement of AE $[M\text{-}CH_2{=}C{=}O]^{\cdot +}$ in $[PhOCOCH_3]^{\cdot +} \rightarrow [C_6H_5OH]^{\cdot +} + CH_2{=}C{=}O$ process and comparison of CA spectra of phenol $[PhOH]^{\cdot +}$ and $[M\text{-}CH_2{=}C{=}O]^{\cdot +}$ from $[PhOCOCH_3]^{\cdot +}$ the structure of $[M\text{-}CH_2{=}C{=}O]^{\cdot +}$ was erroneously assigned to ionized phenol $PhOH^{\cdot +}$ in AE experiments [92] thus leading to a wrong value for enthalpy of formation for ketene $CH_2{=}C{=}O$ molecule [92] (for details see Ref. 37).

8. Kinetic Shift

According to Dunbar [93] kinetic shift is the supplementary energy above thermochemical threshold at the time scale of mass spectrometer. The criteria for kinetic shift determination have been formulated by Dunbar [93-95] and then repeated in Review of Lifshitz [96]. The advocates of existence of kinetic shift specify the *conventional* kinetic shift which corresponds to observation of metastable ions in conventional mass spectrometer with additional energy which is necessary for achievement of dissociation rate 10^{-3} s^{-1} (or peaks intensity of 1%) and the *intrinsic* kinetic shift – this is the excess energy which is necessary to achieve the dissociation rate 10^{-1} s^{-1} to compete with radiation relaxation [93, 96].

In this definition we do not find any physical picture which necessitates the increase of energy compared with calculated thermodynamic threshold but rather see the problem of *observability* or non-observability of an ion which are directly linked with an instrument's possibility to fix the beginning (onset) of registration of an ion i.e. with sensitivity of registrating system. This has no any relation to thermodynamics of the decomposition process. To our opinion, if AE_{calc} for a certain ion is, say, 10. 00000 eV then at energy of 10.00001 eV this ion would be formed without fail (without any "kinetic shift") but whether this ion would be registered or not completely depends only on *sensitivity* of registrating system.

The supporters of kinetic shift consider as the most important argument in favor of its existence the compulsory decrease of AE value with increase of τ, i.e. the time of ion's residence in ionization chamber. According to Ref. 97 ΔH_f^0 Ph^+ from PhI molecule exhibits 274.1 (τ 11.2 μs) or 263.9 (τ 2 μs) kcal mol^{-1}. From IE Ph· 8.32±0.04 eV (photoelectron spectroscopy) [98] ΔH_f^0 Ph^+ is 270.4 kcal mol^{-1} if ΔH_f^0 Ph· 78.5 kcal mol^{-1} is taken [4]. Applying photoion-photoelectron coincidence spectroscopy Dannacher et al obtained AE Ph^+ (from PhI) 11.20 (τ 1 μs), 11.015 (5.9 μs), 11.020 (21 μs) and 11.015 (57 μs), i.e. demonstrated the routine fluctuation in measurement of experimental quantities while no any contribution of "kinetic shift" was observed [99]. ΔH_f^0 Ph^+ in this work 267.6 kcal mol^{-1} was found. Similar discrepancies in AE Ph^+ values in Ref. 100 (Lifshitz) and Ref. 101 (Sergeev) were found originating from PhBr molecule. AE Ph^+ exhibits 12.05 (6 ms), 11.8 (500 μs) and

11.708 (2 ms) [100] while Sergeev gives 11.75 eV [101]. The data of Sergeev on ΔH_f^0 Ph^+/PhI 268.4 kcal mol^{-1} correspond to data of other authors with exception of those of Lifshitz. In all works quoted the method of photoionization mass spectrometry was applied, however the real experimental conditions and the instruments were different.

The decrease of AE Ph^+ with the increase of its residence time τ in ionization chamber, observed in Refs 97 and 100, we explain by the increase of the amount of the ions being registered due to low (compared with the works of other authors) sensitivity of the instrument in ordinary conditions of running mass spectrum. In addition, from data given in Refs 97 and 100 it follows that experimental errors are higher in these works compared with other authors. Their data on ΔH_f^0 Ph^+ (from PhI) 262.9 kcal mol^{-1} (2 μs) are in obvious contradiction with the data from other studies (see above).

The enthalpy of formation for $C_6H_5^+$ ion from PhX molecules exhibits 285.7, 272.7, 269.4 and 268.4 kcal mol^{-1} with X = H, Cl, Br and I, respectively [101]. The augmentation of ΔH_f^0 $C_6H_5^+$ from benzene by 17.7 kcal mol^{-1} compared with that from PhI molecule evidences not the participation of kinetic shift (why not with halobenzenes?) but rather that the higher AE_{exp} $C_6H_5^+$ corresponds to the structure other than Ph^+ ($phenyl^+$). The reason lies in larger energetical preference of benzene ring opening compared with C-H bond cleavage, the former taking place before the latter process occurs. In the moment of ring disclosure in this and other cases the isomerization of molecular ion can occur with H-shift which leads to stabilization of the system compared with that formed at the exact moment of ring opening. Such ring opening does not occur in case of PhBr and PhI molecules before elimination of Br and I atoms, respectively, because E_{cr} for those processes is lower at the threshold of their formation compared with E_{cr} for bond cleavage in the ring. Thus the ion $C_6H_5^+$ has the structure of Ph^+ ion while the small increase in ΔH_f^0 $C_6H_5^+$ (from PhCl) evidences the beginning of isomerization of molecular ion and formation of the mixture of structures for $C_6H_5^+$ ion one of them being Ph^+ cation.

This situation is a routine one. A lot of similar examples is given in Ref. 4 but we have to come back to this question in connection with kinetic shift problem. The authors [102] *a priori* assign the structure for $[C_5H_6]^{\cdot+}$ or $[M\text{-}CO]^{\cdot+}$ ion originating from phenol PhOH molecule to be ionized cyclopentadiene and since AE_{exp} $C_5H_6^{\cdot+}$ is by 1 eV surpasses that AE_{calc} for ionized cyclopentadiene the authors ascribe this difference to involvement of kinetic shift. It is clear that CO molecule can be expelled only after ring opening in the $[PhOH]^{\cdot+}$ molecular ion and H-migration. Thus, in any way the rearrangement of molecular ion with initial structure of phenol $[PhOH]^{\cdot+}$ will happen. Besides cyclopentadiene we numbered 17 (!) stable neutral molecules with the formula C_5H_6. No any arguments in favor of ionized cyclopentadiene isomer are presented [102]. Even if to attract the data on CA spectra this would not help in assignment the structure of $[M\text{-}CO]^{\cdot+}$ ion from phenol PhOH at the threshold of its formation. Similarly, for $[M\text{-}HNC]^{\cdot+}$ or $C_5H_6^{\cdot+}$ ion obtained from aniline $PhNH_2$ molecule the authors without any arguments accept the structure of ionized cyclopentadiene [100]. And again the difference in $\Delta\Delta[AE_{exp} - AE_{calc}]$ (for $[cyclopentadiene]^{\cdot+}$) now being 2 eV is ascribed to involvement of kinetic shift. In this case, in addition to 17 possible $C_5H_6^{\cdot+}$ isomers the variation in neutral HCN or HNC molecules becomes possible. Thus the total number of fragmentation versions for [HNC] loss approaches 35! Which structure of an ion and a neutral (or their mixtures) corresponds to AE_{exp} is obviously unknown.

It is also important to keep in mind that variation in residence time τ in the ionization chamber may change the ion's structure or to change the ratio of the structures if there are several ones and this is not surprising. We give a single example. At low τ values (microseconds) the molecular ion $M^{\cdot+}$ from HC≡CCH$_2$CH$_2$C≡CH ($C_6H_6^{\cdot+}$) molecule was observed. However, at higher τ values (milliseconds) initial molecular ion isomerizes to molecular ion of unknown structure followed by H-elimination giving rise to an ion of Ph^+ structure and since its AE appears to be smaller than the IE of initial molecule HC≡CCH$_2$CH$_2$C≡CH the peak of its $M^{\cdot+}$ is not registered [103] (see Ref. 4).

Now we give one more example of fragmentation for which the authors bring into play the notion of kinetic shift [104]. The process of elimination of CH_2O molecule from ionized anisole $[PhOCH_3]^{\cdot+}$ was studied with suggestion of formation of an ionized benzene molecule $[PhH]^{\cdot+}$ as a fragment. AE_{calc} $PhH^{\cdot+}$ is 9.72 eV while AEs_{exp} $[M\text{-}CH_2O]^{\cdot+}$ are 11.59 (τ 6 μs), 11.28 (32 μs), 11.31 (46 μs), 11.16 (500 μs) and 11.06 eV (2000 μs). Quoting the private communication of Dutui and Baer, at even larger values of τ residence times in ion chamber, AE may approach the value 9.88 eV, i.e. AE_{exp} 9.72 eV for $[PhH]^{\cdot+}$ ion.

Let us consider the possible routes for isomerization of $M^{\cdot+}$ ion of anisole $PhOCH_3$ molecule. IE ($PhOCH_3$) is 8.20 eV [104]. AE $[M\text{-}H]^+$ which is energetically the most favorable process of *simple* bond cleavage (i.e. without rearrangement) is 10.83 eV (using ΔH_f^0 $PhOCH_2^+$ 182.4 kcal mol^{-1} [106]). Hence E_{cr} for this process is 10.83-8.20 = 2.63 eV. For other two processes of simple bond ruptures – CH_3 elimination and benzene ring opening at the bond adjacent to CH_3O-group E_{cr} values are estimated as ~3.15 and 3.7 eV, respectively. Thus, in the limit of E_{cr} 2.63 eV *any* rearrangement (isomerization) processes might take place (see Section 6). The first such process is obvious – this is H-atom migration from CH_3-group to ionized benzene ring requiring only ~0.77 eV (E_{cr} and ΔH_f^0 – in kcal mol^{-1})

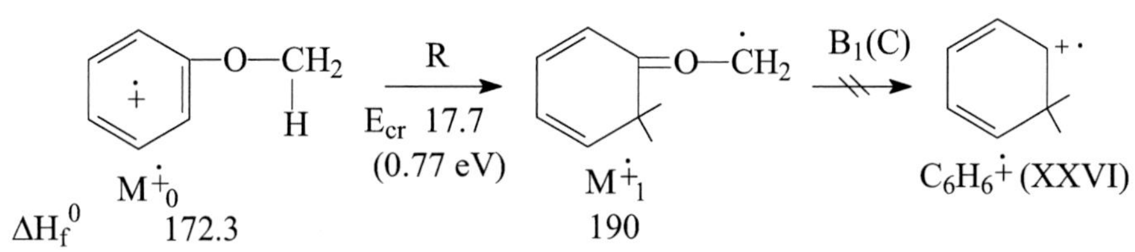

This means that isomerization will occur without fail and the newly born molecular ion $M_1^{\cdot+}$ (isomeric with initial one $M_0^{\cdot+}$) will further isomerize and/or fragment independently of the structure of initial $M_0^{\cdot+}$ ion. The further isomerization will naturally take place since CH_2O elimination from $M_1^{\cdot+}$ isomer might give energetically unfavorable ion $C_6H_6^{\cdot+}$ (XXVI) having the structure of ionized carbene. We can number not less than 40! neutral C_6H_6 isomers, each of them can produce stable $C_6H_6^{\cdot+}$ radical cations. Many of them can give the high values of AE $[M\text{-}CH_2O]^{\cdot+}$ ions corresponding to AE_{exp} found in the work [104]. At such large number of possible $C_6H_6^{\cdot+}$ isomers their structure(s) can vary with increase of τ value.

By this example, considered in detail, we demonstrated the absence of necessity to attract the notion of "kinetic shift" whereas the raising of AE_{exp} compared with "expected" AE is explained by formation of rearranged (isomerized) ion(s). The authors who accept kinetic shift notion in interpretation of thermochemical data some time run to extremes. Butler et al [107] for the process $[n\text{-}PrCOOH]^{\cdot+} \rightarrow Me_2CH^+ + \cdot COOH$ the value of ΔAE only 0.11 eV (AE_{calc} 10.85; AE_{exp} 10.96 eV) or 2.6 kcal mol^{-1} ascribed to interference of kinetic shift. It is clear that this "deviation" does not come out of the experiment's accuracy. In addition, one must take in account the accuracies of ΔH_f^0 values of all participants of the process under discussion and discrepancies in literature values of the enthalpies of formation for Me_2CH^+ and $\cdot COOH$ species. We recalculated the data used in Ref. 107 taking the following ΔH_f^0 values for n-PrCOOH (-113.7±1.0 [108], Me_2CH^+ (189) [4, 50] and $\cdot COOH$ (-47) [4]) and obtained AE_{calc} Me_2CH^+ 11.07 eV while AE_{exp} is 10.97 eV [107].

Another extremity in relation to kinetic shift is demonstrated due to our analysis of the data in Ref. 109. For ion $[M\text{-}C_2]^{\cdot +}$ from fullerene C_{60} the authors suggested the kinetic shift value of 30! eV (or about 700! kcal mol^{-1}). The detailed analysis of this work is given in our communication [7] the essence of which is coming in the search of isomer $[M\text{-}C_2]^{\cdot +}$ whose structure differs from one proposed by the authors [109]. They suggested that in result of $\cdot C{\equiv}C\cdot$ expulsion to form $C_{58}^{\cdot +}$ ion two bonds are ruptured while we proposed the cleavage of more bonds, each cleavage creating two additional, energetically unfavorable, free radical centers of vinylic type. This might be accompanied by very large increase of the ion's enthalpy of formation and, consequently, by increase of $AE[M\text{-}C_2]^{\cdot +}$, i.e. formation of other isomer of $C_{58}^{\cdot +}$ compared with that suggested in Ref. 109.

By direct experiments on AEs measurements of fragment ions with variation of residence time τ of ions in ion chamber it was quite definitely demonstrated the absence of kinetic shift for many processes of fragment ions formation: $[M\text{-}I]^+$ (from PhI) [99], $[M\text{-}I]^+$ (CF_3I) [110], $MeC{\equiv}O^+$ + $\cdot CH_2OH$ (MeCOOMe) [111], $MeC{\equiv}O^+$, $CH_2{=}O(^+)H$, $C_2H_5^+$ and $C_2H_4^{\cdot +}$ ($MeCOCH_2OH$) [112]. The enthalpies of formation for cyclopropenyl $C_3H_3^+$ ion, gained from 8 precursors are identical although some of the $C_3H_3^+$ are obtained in result of complex rearrangements [113, 114]. All this unequivocally demonstrates the absence of kinetic shift.

The direct denial of kinetic shift comes out from the results on ΔH_f^0 determination for free radicals from AEs of positive ions as the partners of the former [28, 115-119] since these ΔH_f^0 values are identical or very close to the values gained from chemical kinetics [46, 120, 121] and IR-spectroscopy [40, 41, 122]. It is curious that in mass spectrometry of negative ions the notion "kinetic shift" is not used although the time scale (10^{-6} s) is the same for both ionization modes. Apparently, owing to specifity of negative ions registration in energetcal experiments the mass spectrometrists manage to register fragment ions down to 10^{-7}% of base ion abundance and to determine their energetical characteristics (see Refs 33, 34 and Section 3).

The conception of kinetic shift does certain harm to organic mass spectrometry since it casts doubts on thermochemical characteristics of ions, free radicals and molecules gained from data on measured ions' AEs. Without the latter (measured or estimated) the description of fragmentation processes under EI is hardly possible. Here we give a single example in support of our statement. The authors [123] measured AEs for $[M\text{-}CH_3]^+$ ions from 32 C_6H_{10} isomers, and obtained the values of enthalpies of formation for this ion in the range of 198.2-226.2 kcal mol^{-1}. They assigned to *all* those 32 $C_5H_7^+$ ions the *single* structure of 3-cyclopentenyl cation, since its ΔH_f^0 is the lowest one among all isomers 198.2 kcal mol^{-1}. The higher values of ΔH_f^0, calculated from AEs exp being higher than 198.2 kcal mol^{-1} the authors ascribed to interference of kinetic shift [123]. Analyzing the data of this work [123] we as always, attributed the variations in enthalpies of formation to variations of structures of $C_5H_7^+$ isomers rather than to kinetic shift and extracted from the quoted work [123] the yet unknown enthalpies of formation for several ions [4].

9. Competition Shift

According to this artificial notion, closely related to kinetic shift, if the AEs of two competing ions are measured one of which having the low AE value while the other – the higher one, then the latter "has to" accumulate the amount of energy to be sufficient for measurement against the background of the more energetically favorable ion. Thus, the AE of less favorable ion might be shifted to higher values compared with its thermochemical threshold [4, 50]. To our opinion since the AE_{calc} of any ion does not depend on AEs of other ions the problem of registration of the beginning for appearance (i.e. its threshold) of less favorable and less abundant ion is determined only by *sensitivity* of the instrument. If its sensitivity is high enough, the AEs of competing ions even being very different (by several electronvolts) might be successfully measured. The random examples taken from Reference Book [105] support this simple idea, e.g. AEs $[M\text{-}F]^+$ and $[M\text{-}Cl]^+$ (from CF_2Cl_2) are 14.15 and 12.10 eV, respectively, or 13.25 and 11.57 eV ($CFCl_3$), respectively, $[M\text{-}H]^+$ and $[M\text{-}OH]^+$ (CH_3OH) – 11.55 and 13.82 eV, respectively, while AEs of less energetically favorable ions exactly correspond to their calculated thermodynamic thresholds.

Even better the absence of competition shift could be demonstrated at AEs measurements for negative ions since they are determined in essentially larger range of energies (0-15 eV) compared with ΔAE 0-2(3) eV (see Section 3) for positive ions.

10. Activation Energy Of Reversed Reaction

Some authors suppose that augmentation of AEs for fragment ions one can explain by essential value of activation energy E_a 0.3-2.0 eV for reversed reaction of reunion of an ion with a free radical. From kinetics of free radical reactions one very useful conception appeared which was supported experimentally that the reaction of two free radicals, i.e. their reunion into a molecule does not require any energy [124]. We think, however, that there exist two special types of such reactions when E_a for reunion of two free radicals may disagree with this conception, i.e. $E_a > 0$. One of them is construction of a molecule A-B from A· and B· free radicals if both comprise strong electron-withdrawing substituents thus electrostatic repulsion may create a certain barrier for such process, e.g. $2CF_3\cdot \rightarrow CF_3\text{-}CF_3$. Another type is reunion of two bulky groups which have to overcome certain spatial repulsion like in reaction $2Me_3C\cdot \rightarrow Me_3C\text{-}CMe_3$ in which the strained molecule has to be created.

The activation energy for gas phase reaction between an ion and a free radical (or molecule) is negligible and it is suggested that it is close to zero [125-127]. If we consider the reaction of HX elimination from ionized ethylenes when the structure of cyclic intermediates "a" and "b" may give rise to suspicion that the reversed reaction of reunion of ionized acetylene and HX molecule is energetically unfavorable, i.e. requires certain barrier to be overcome in reality it appeared that such barrier is close to zero since AE $[HC{\equiv}CH]^{\cdot +}$ ions corresponds exactly to calculated thermochemical threshold of their formation [6].

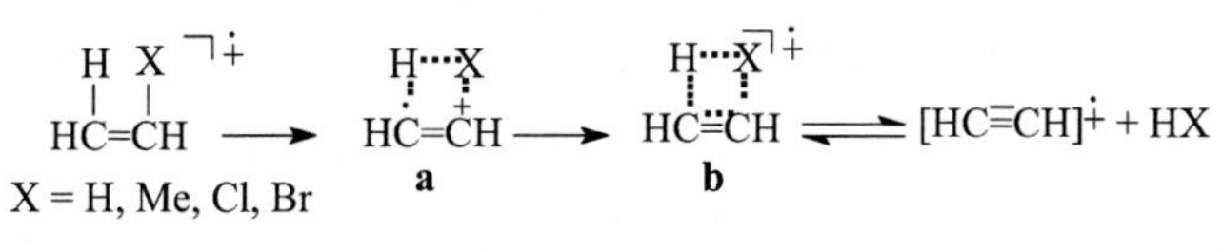

If in this case, when intermediate has the structure of four-membered cycle with double bond, E_a for reversed reaction appeared to be zero then in other cases it might be definitely close to zero. If we consider the different mechanism of HX elimination through H- to X-atom migration with formation of, to our opinion, rather "strange" intermediates like "c" (as it is suggested in Refs 6 and 128) still it is not possible to avoid the intermediance of species "a" and "b"

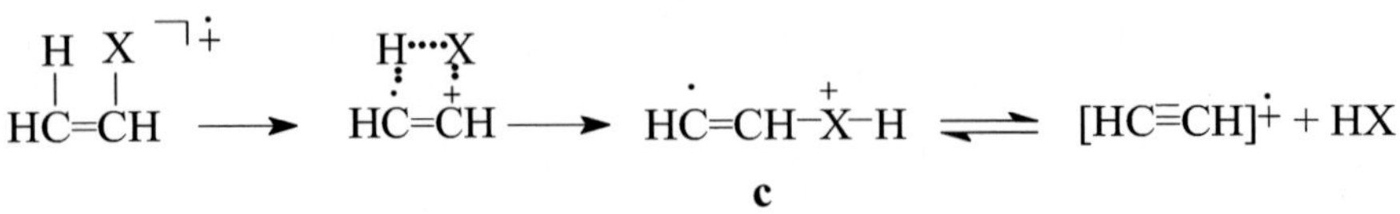

Finally, if to spread the mechanism suggested for parent ethylene molecule in Ref. 128, to substituted ethylenes, then the three-membered cycle "d" does not create a noticeable E_a for the reaction of interaction of HX molecule with C-atom.

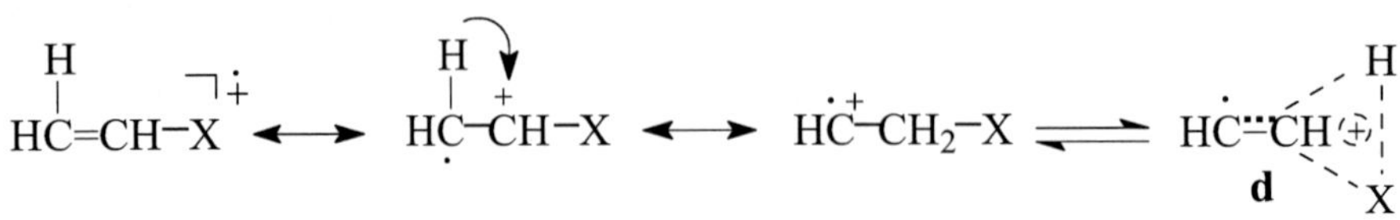

Hence, the high values of E_a for reversed reactions of reunion of ions and free radicals (or molecules) in rearrangement processes, which are suggested in some works appear to be far from reality. Some other problems related to E_a of reversed reactions, are considered in the next Section.

11. METASTABLE IONS

Metastable ions render a very important assistance to a mass spectrometrist in interpretation of mass spectra. Here we go into detail with two problems related to formation of metastable ions. Let us consider four reactions, which are accompanied by formation of intense peaks of metastable ions [129]

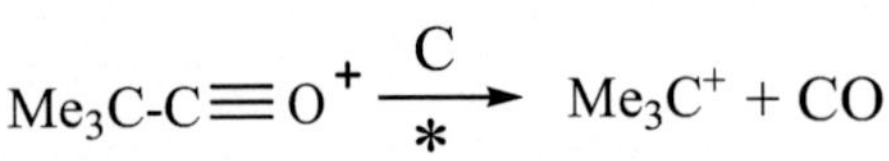 (3)

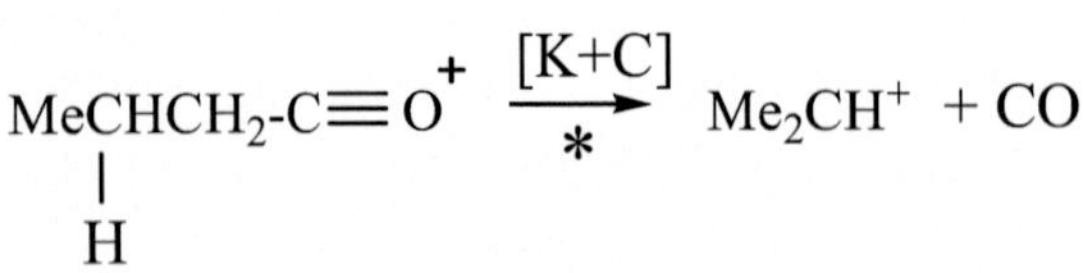 (4)

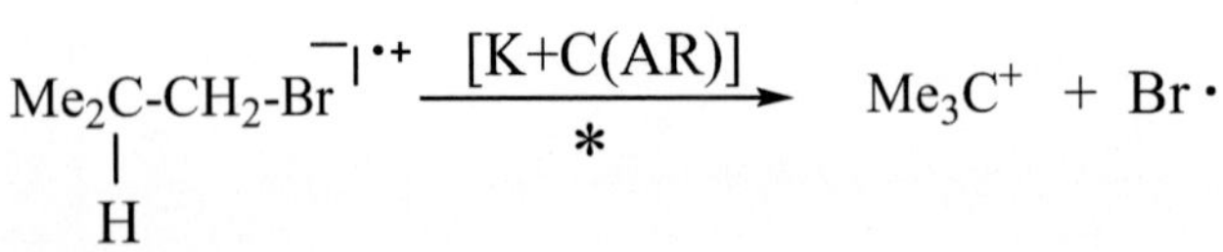 (5)

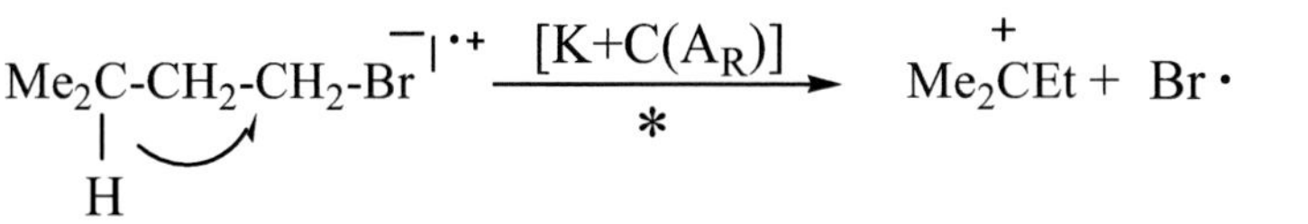

(6)

For reaction (3) a narrow peak of metastable ions is observed which is typical for simple bond cleavage. For reactions 4-6 the wide (flat or dishform) peaks are registetred which evidence the occurrence of rearrangement processes. From the measured peaks' width the activation energies E_a of reversed energies for bimolecular reaction of ions and free radicals are calculated the E_a values being rather large [125]. However, as it was considered in previous section the activation energies E_a of reversed reactions of Me_2CH^+ + CO (Eqn 4), Me_3C^+ + Br· (Eqn 5) or $Me_2C(^+)Et$ + Br· (Eqn 6) are close to zero. Thus, the E_a energies calculated from the width of the peaks of metastable ions correspond not to E_a for these "reversed" reactions mentioned but, rather, to the energies of rearrangement "expenses" to form less stable ions from more stable ones, i.e. $Me_2CH^+ \rightarrow MeCH_2CH_2^+$, $Me_3C^+ \rightarrow Me_2CHCH_2^+$ or $Me_2C(^+)Et \rightarrow Me_2CHCH_2CH_2^+$ reversed reactions in Eqns 4, 5 and 6, respectively. These data could be applied for calculation of the enthalpies of formation for energetically less stable isomers which are hardly could be gained from AE measurements.

The authors of the work [130] (e.g. pp. 69, 74 and 107) give examples of carbene $:CH_2$ formation which is supported by peaks of metastable ions, like $c\text{-}C_7H_7^+ \rightarrow C_6H_5^+ + :CH_2$. Such process is extremely unfavorable energetically due to very high enthalpies of formation for $:CH_2$ carbene 79 (triplet) and 88 kcal mol^{-1} (singlet) [4]. ΔH for this reaction being 143.5! kcal mol^{-1} or 6.2! eV. Thus, we doubt in occurrence of such process since tropylium $c\text{-}C_7H_7^+$ ion has much more energetically favorable channels for fragmentation, these are H_2 and C_2H_2 elimination which are also supported by corresponding peaks of metastable ions. Therefore, we think that CH_2 elimination does not occur in monomolecular reaction but either in a result of some bimolecular reaction or with participation of some surfaces in a mass spectrometer. Even in molecule XXVII in which CH_2 elimination gives a stable species of ionized benzene the peak of $[M\text{-}CH_2]^{\cdot+}$ ion exhibits only 0.7% of the base peak [4] although the presence of catalysis or ion-molecule reaction cannot be excluded in this case as well.

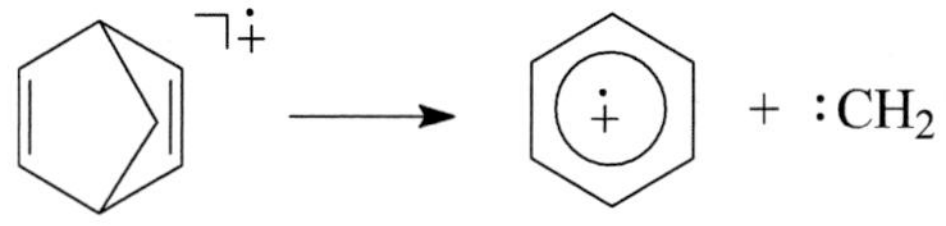

XXVII

Even yet less energetically favorable process, supported by peak of metastable ion m*, was represented in Ref. 131.

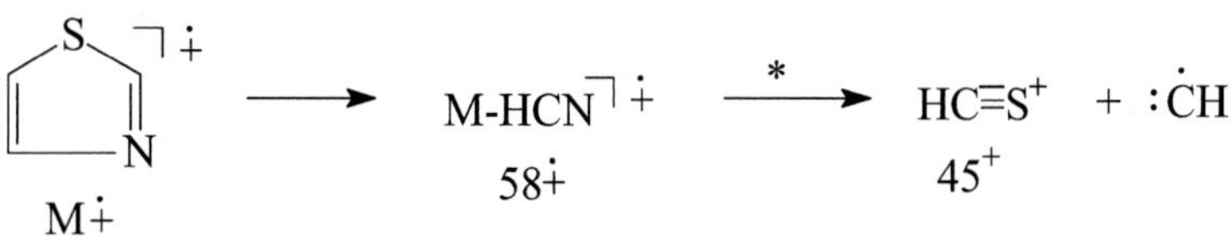

The ion 58·$^+$, irrespective which structure it possesses ·CH=CHS$^+$, [HC≡CSH]·$^+$ or [CH_2=C=S]·$^+$ has the much more energetically favorable fragmentation route of H$^-$elimination with formation of HC≡CS$^+$ ↔ HC($^+$)=C=S ion whereas the elimination of ·:CH species with very high ΔH_f^0 142 kcal mol^{-1}[50] in no way is justified. Apparently, we again encounter the ion-molecule reaction or CH-species is formed with the "aid" of certain surfaces in a mass spectrometer.

12. Elimination of Several Particles in a Single Process

Traditionally it is believed that elimination of two, even more of three etc particles in one step is rarely encountered in fragmentation process [66]. It was shown in Refs 4 and 50 that elimination of entire fragment AB is usually more energetically favorable than of its parts A + B, since the enthalpy of formation for AB species usually is smaller than of A + B pair. However, there are many exclusions from this "rule" (see Ref. 50). We add other examples: elimination of C_nH_{2n+2}, more exact [C_nH_{2n} + H_2] from aliphatic chain in the salts of carboxylic acids [66, 132]; formation of such pairs at fragmentation of halogen- and nitroderivatives of 1,2,4-triazoles [133] [:CCl_2 + N_2], [HCN + N_2], [ClCN + N_2], [HCN + HCNO], [HCN + N_2O], [Cl + N_2], elimination in one step of [C_2H_3 + CO_2] and [H_2O + H_2O] from o-NO_2PhC($^+$)Me_2 (or [M-Me]$^+$) [4, 50]; formation of 2 CH_2=S molecules expelled from cyclic polysulfides cyclo-[CH_2-S-]$_5$ [4]. The interaction of two ortho-substituents in diphenylamine results in expulsion of two species at the final step of rearrangement process being supported by the peaks of metastable ions [134]

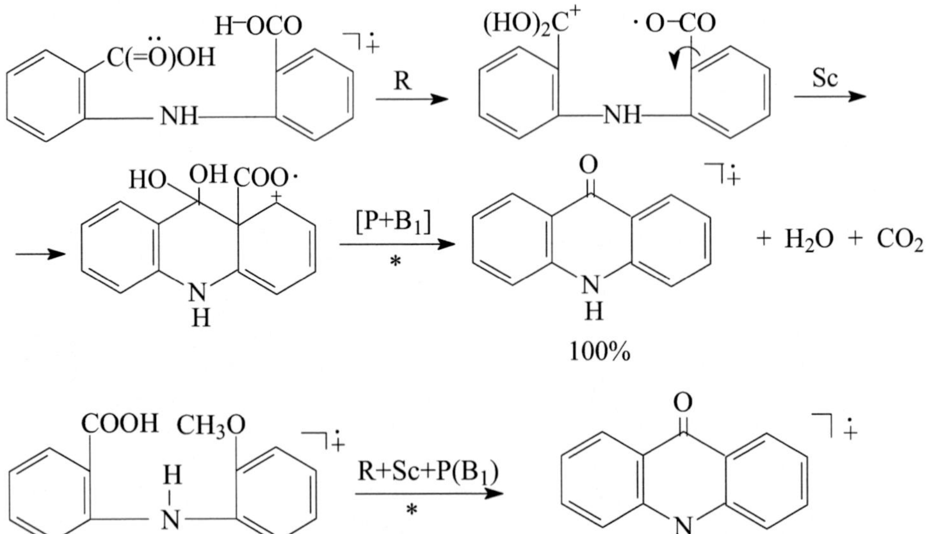

[M-(H_2O + CH_2O)]$^{+\cdot}$ (36)

The formation of neutral fragment with mass 58 in methyl esters could proceed only through elimination of two molecules [9]

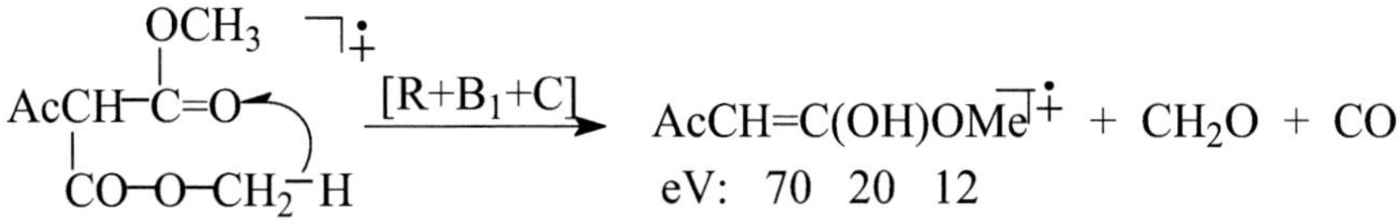

eV: 70 20 12

%Σ 1.7 2.6 14.7

Many similar examples are given in Ref. 4. The fragmentation of negative ions can proceed through formation of several and even of many neutral species which is governed by position of resonance state on energetical scale (Section 3).

13. Ion Pair

Some workers suggest the formation of fragment ions not through the usual intermediate molecular ion in the process $AB + e \rightarrow AB\cdot^{+} + 2e \rightarrow A^{+} + B\cdot$ but rather by direct formation of ion pair $AB + e \rightarrow A^{+} + B^{-} + e$. Let us consider this process from thermochemical position. We compare two fragmentation processes which proceed by heterolytic $AB \rightarrow A^{+} + B^{-}$ and by homolytic mechanism $AB \rightarrow A\cdot + B\cdot$. The simple calculations evidence that even in inorganic molecules fragmentation by heterolytic (ionic) mechanism requires more energy compared with homolytic process (from Refs 105 and 135)

	NaCl	Na^{+} Cl^{-}	$Na^{\cdot}$ $Cl^{\cdot}$
ΔH_f^0(kcal mol^{-1})	-43.4	Σ [144.2-54.3] = 89.9	Σ[25.7+29] = 54.7

	CsCl	Cs^{+} Cl^{-}	$Cs^{\cdot}$ $Cl^{\cdot}$
ΔH_f^0(kcal mol^{-1})	-57.9	Σ [108.0-54.3] = 53.7	Σ[18.3+29] = 47.3

Cl-atom is chosen as a partner since it possesses the highest among all halogenides EA values [135]. Now, if we consider organic compound $CF_3COOSiMe_3$ expected to be one of the most profitable for formation of ion pair due to high stability of both CF_3COO^{-} and $SiMe_3^{+}$ ions the result will be the same: the formation of two free radicals requires less energy compared with ion pair formation (EA $CF_3COO\cdot$ 4.49 eV and ΔH_f^0 $CF_3COO\cdot$ 192.4 kcal mol^{-1} [4]):

	Me_3Si^{+} CF_3COO^{-}	$Me_3Si\cdot$ $CF_3COO\cdot$
ΔH_f^0(kcal mol^{-1})	[145 - 295.9] = -150.9	[-3.0 -192.4] = -195.4

The situation changes if we come from gas to condensed phase. The ionicity of inorganic salts is well known for crystalline state, melts, solutions and for complexes like $NaAlCl_4$ ($Na^{+}AlCl_4^{-}$). It turned out that some organic compounds can be ionic in condensed phase. For example, if the compound $CF_3COOSiMe_3$ is a liquid with its boiling point 86-89^{0}C [136], its close analogue $CF_3COOSnMe_3$ with estimated boiling point about 115-120^{0}C appears to be a solid one with melting point 122-124^{0}C [137] which we explain only by the *ionic* structure for the latter. Another example is the Me_2NCH_2Cl molecule whose boiling point we estimate to be about 70^0C (from data for Cl-alkanes). However, it appears that all compounds Me_2NCH_2X (X = Cl, Br, I) exist as solids with high melting points [138] which we again explain only by their ionic character $Me_2N(^+)=CH_2/X^-$. Many similar examples could be found in Refs 139-141. They are dedicated to those organic molecules which might dissociate (in electrochemical cell or at photoexcitation) either giving two radicals AB → A· + B· or two ions AB → A^+ + B^-. For A^+ ion the most stable species were selected as tropylium or triphenylcyclopropenylium c-$Ph_3C_3^+$ cations while for B^- ion also the most known stable as fluorenyl anions were taken as well. It was found that in isolated molecule the homolytic bond cleavage is *always* energetically more favorable compared with heterolytic process. However, in solution the solvent may promote the formation of ion pair A^+ and B^- through their solvation and sometimes such processes become more favorable compared with homolytic bond cleavage [139-141].

Thus, monomolecular fragmentation with ion pair formation cannot occur in gas phase while observation in photoionization studies of A^+ fragment with its AE being essentially lower than the thermodynamic threshold for $AB\cdot^+ \rightarrow A^+ + B\cdot$ process can be assigned to the occurrence of certain form of chemical ionization or of participation of the surface in a mass spectrometer.

Since the pair of free radicals is always thermodynamically more stable compared with ion pair then, naturally, the parent molecule is expected to be thermodynamically more stable compared with its fragments. Therefore the quantum-chemical calculations giving the opposite result [142] look strange when fragment ion pair was found to be essentially more stable than its parent molecule ($\Delta\Delta H_f^0$ – in kcal mol^{-1}).

	$(N\equiv C)_3C\text{-}C(NH_2)_3$	$(N\equiv C)_3C^- + {}^+C(NH_2)_3$	$(N\equiv C)_3C\cdot + \cdot C(NH_2)_3$
[142]	0.0	-60.2	-
our estimate	0.0	+96	+77

14. Reorganization Energy of Ions and Radicals in Fragmentation Process

There exists rather common opinion that at the bond cleavage either in ions or in neutral molecules (with formation of two radicals) the reorganization of fragments' structures takes place to such an extent that it is necessary to "correct" the enthalpy of formation of those fragments to take in account this reorganization energy [143, 144]. The detailed analysis of this conception and its critics is given in our previous works [6, 9]. Here we briefly repeat some of our arguments. In the processes of decomposition of $CH_3C\equiv O^+ \rightarrow CH_3^+ + CO$ and $[RO\text{-}NO]\cdot^+ \rightarrow RO^+ + NO$ one expects the highest possible reorganization of the neutral part of the ions, i.e. of CO and NO species, respectively, which when being the parts of the ions, represent the biradical :CO and classical ·N=O free radical, respectively. In the process of fragmentation they are "transferred" into nonclasical, really existing carbon and nitrogen monooxides, respectively, with their ΔH_f^0(exp.) –26.2 and 21.8 kcal mol^{-1} [124], respectively, whereas ΔH_f^0(est.) of their "unrearranged" classical "isomers" appears to be higher by 65! and 40! kcal mol^{-1}, respectively, with $\Delta\Delta H_f^0$ shift +39 and +62 kcal mol^{-1}, respectively [6, 40]. However, ΔH_f^0(exp.) of ions CH_3^+ and RO^+, obtained at fragmentation of $CH_3C{\equiv}O^+$ and $[RO\text{-}NO]^{\cdot +}$ ions, respectively, appeared to be identical with their ΔH_f^0 values gained from photoionization studies of $CH_3\cdot$ and RO· radicals, or from AEs of CH_3^+ (from CH_3X) and RO^+ (from RO-R' molecules) where no "reorganization" is expected. The calculation of such "reorganization" energies and description of this conception, in general, is given, besides Refs 6 and 40, also in other works [121, 143, 144].

15. Electron Impact (EI) and Photoionization (PI) Mass Spectra

Some differences in EI and PI mass spectra were noticed long ago. For example, AE_{exp} of $ClCH_2^+$ ion (from CH_3Cl) corresponds exactly to the calculated AE at EI while at PI this ion is not observed at the energies of 1 eV higher its threshold. Eland et al [145] proposed the explanation of this fact by occurrence of autoionization states at EI which are absent at PI. According to their opinion, at PI process only one electron transitions have the essential probabilities while at EI, especially at low energies, the two electron transitions also become possible. In $Me_3CC{\equiv}CH$ and $Me_3CC{\equiv}CCOCMe_3$ molecules the opposite effect was observed: the peak of $M^{\cdot +}$ ion was not fixed at low energies at EI while at PI it was possible to measure their IEs [9]. Butler et al [107] in their study of n-PrCOOH molecule found that at PI its molecular ion does not isomerize while at EI a certain part of $M^{\cdot +}$ ion isomerizes to low-energetical isomer $[C_2H_5CH{=}C(OH)_2]^{\cdot +}$.

The mass spectra of many compounds were run at low energies at EI and at 10.2 and 12.1 eV at PI [9]. Here we give the partial mass spectra of $Me_3EC{\equiv}CI$ [9] (peaks' intensities – in $\%\Sigma_{29}$)

E	EI				PI			
	eV:	$M^{\cdot +}$	$[M\text{-}Me]^+$	$[M\text{-}I]^+$	eV:	$M^{\cdot +}$	$[M\text{-}Me]^+$	$[M\text{-}I]^+$
C	70	7.2	7.0	8.7				
	20	7.2	5.9	21.2				
	12	24.4	4.1	51.2	12.1	36	25	39
					10.2	72	0.1	2.9
Si	70	9.0	40.3	6.7				
	20	19.5	56.4	16.7				
	12	32.3	39.3	28.0	10.2	25.4	64.1	10.6
					12.1	65.3	2.1	32.5
Ge	70	8.8	40.5	2.4				
	20	23.9	53.6	8.1	10.2	22.3	74.7	2.8
	12	29.6	45.7	11.0	12.1	67.4	24.0	8.0

It is evident that both fragmentation routes and ratios of peaks' intensities are close for both methods in both simple bond cleavages ($[M\text{-}Me]^+$) and in rearrangement processes ($[M\text{-}I]^+$). Since PI process is of resonance character its depth is somewhat larger compared with

EI. We estimate very roughly that PI mass spectrum at 10.2 eV (α-Lyman line of H-lamp) corresponds to EI mass spectra at 15-18 eV.

16. Theoretical Calculations

In recent years, giving the readers a wide set of enthalpies of formations for molecules, free radicals, positive and negative ions, obtained by empirical procedures, we – for comparison – always quote the theoretical values computed by semi-empirical, ab initio, DFT (density functional theory) methods. In this connection we have analyzed several hundreds publications with rather unfavorable conclusion – all of them are not trustworthy when applied to thermochemical problems [37, 39, 40, 42, 43, 45, 146]. The most detailed critics of theoretical methods is given in Ref. 146. The main results extracted from that work is the following. Firstly, the number of known quantum-chemical methods surpasses any reasonable limits. For example, in Ref. 147 more than 1300! methods are presented. Only in a single issue of J.Chem.Phys. 120 (2004) N 13 in 13 publications (pp. 5875-6004) 340 novel quantum-chemical methods are given. In our home collection we keep about 2100(!) methods. Of course, our list of these methods is far to be complete but continues to be replenished. Thus, in first issues of J.Chem.Phys. (vol. 122) and J.Phys.Chem. (A109) for 2005 in four publications it is announced about creation of 80! novel methods related to thermochemical problems [147-150].

Secondly, the interval of thermochemical values computed by diverse methods is very wide even for very simple (including diatomic) molecules, ions and free radicals. Thirdly, deviation of computed enthalpies of formation from experimental ones for some "uncomfortable" species like aromatic molecules, cumulenes (e.g. CO_2), molecules with elements in hypervalent states (e.g. HNO_3 or H_2SO_4), polyhalogeneted molecules (e.g. C_2Cl_4, SiF_4, $AlCl_3$) exhibits tens of kilokalories to both positive or negative sides.

In this section we briefly analyse only some works related to thermochemistry of positive ions. For comparison, we quote either experimental data taken from Reference Book [135] or those being checked (or derived) in the series of isodesmic reactions compared with structurally close model cations and being published in our previous works [4, 6, 8, 9, 40, 50, 146].

Analysis of the data collected in Tables 1-3 demonstrates the unreliability of theoretical data. If some of them are very close to experimental ones however other ones manifest large deviations either from experiment or from values estimated by simple additive scheme with correction on stabilization or destabilization of a target compared with a model ion. If those deviations from experiment were of systematic character then one might introduce the corrections however those deviations are random rather than systematic. In addition, essential differences in the values for the same species are observed when several theoretical methods are applied in the same work or a single method but with application of several working reactions [152]). In Ref. 153 the $\Delta\Delta H_f^0$ values for many fragments relative to $[H_2NCH_2COOH]^{\cdot +}$ ions are computed and are recalculated in this edition to gain the enthalpy of formation for particular species (Table 2). Similarly, in Ref. 154 the $\Delta\Delta H_f^0$ values for $C_3H_3^+$ isomers are given relative to cyclo-$C_3H_3^+$ ion whose ΔH_f^0 256 kcal mol^{-1} [135] is used to calculate ΔH_f^0 of other isomers (Table 3). If essential deviations from experiment are observed for such simple ions as $^+NH_2{=}CH_2$ or $HC{\equiv}NH^+$ [153] how can we trust to the computed ΔH_f^0 values for a set of ten other fragments or for about fifty intermediates, performed by CCSD(T)//B3LYP with 6-31+G(d,p) basis set [153].

Table 1. Enthalpies of formation, ΔH^f_0 (g) (298 K), for some fluorinated ions (in kcal mol^{-1})

	$F_2CH(^+)CHF$	FCH_2^+	F_2CH^+	CF_3^+	$MeCHF^+$	$MeCF_2^+$
MNDO [151]	97.3	182.7	132.3	100.7	164.7	116.5
literature data	92 [26]	199 [4]	142 [4]	90.3 [4]	163 [26]	110.5 [4]
	$CF_3CH_2^+$	$FCH_2CH_2^+$	$FCH_2CH(^+)F$	$CF_3\cdot$	CF_3^-	
MNDO [151]	121.0	187.3	130.8	-137.2	-178.9	
literature data	105 [26]	185 [26]	133.6 [26]	-113.5 [40]	-155.5 [40, 135]	

Table 2. Enthalpies of formation, ΔH^f_0 (g) (298 K), for some N-comprising ions (in kcal mol^{-1})

	$CH_2{=}C(^+)CH{=}NH$	H_2N-c-$C_3H_2^+$	$MeC(^+)HCN$	$^+CH_2CH_2CN$
MINDO/3 [152]	221	239	259	275
MNDO [152]	221	244	251	282
literature data	221.4 [135]	215 [146]	255	260**
	$^+CH_2CH_2NC$	$^+CH{=}CHCH{=}NH$	$^+NH_2{=}CH_2$*	$HC{\equiv}NH^+$
MINDO/3 [119]	273	287	172.3 or 158.6 [153]	207.3 or 208.6 [153]
MNDO [119]	296	283		
literature data	285**	285**	178 [135]	226 [135]

*) CCSD(T)//B3LYP [154]
**) our estimate

Table 3. Enthalpies of formation, ΔH^f_0 (g) (298 K), for $C_3H_3^+$ isomers (in kcal mol^{-1})

[154]	$^+CH_2C{\equiv}CH$	$CH_3C{\equiv}C^+$	$^+CH_2CH{=}C$:
HF/4-31G	270.2	363	341.4
HF/6-31G	289.9	374.3	356.1
MP2/6-31G*	287.6	395.7	372.0
MP2/6-31G**	287.9	394.5	371.3
literature data	281 [135]	375*	320*

*) our estimate

Another series of works [155, 156] is devoted to theoretical calculations of Q values for isodesmic reaction $CH_3^+ + CH_3X \rightarrow CH_4 + {}^+CH_2X + Q$ which we used to derive the ΔH_f^0 of $^+CH_2X$ ions (ΔH_f^0 – in kcal mol^{-1}) (Table 4)

Table 4. Enthalpies of formation, ΔH^f_0 (g) (298 K) of $^+CH_2X$ ions (in kcal mol^{-1})

	$^+CH_2F$	$^+CH_2OH$	$^+CH_2NH_2$	$^+CH_2CHO$	$^+CH_2CN$	$^+CH_2CF_3$
	210 [155]	177.3 [155]	187 [155]	245.2 [156]	306.7 [156]	136.6 [156]
	199 [155]	166.3 [155]	174.7 [155]			
Literature data	199 [135]	168 [135]	178 [135]	204*	286*	105*

*See Ref. 4

The essential overestimation of ΔH_f^0 values for ions with traditionally treated as electron-withdrawing groups CHO, CN, CF_3 (Table 4) reveals that theoretical methods are not sensitive to possibilities of these groups to display the polarizability effect which is expected to be large notably in the case of cations.

Another example of low sensitivity of theoretical methods to structural changes can be demonstrated by analysis of data of Ref. 157. The authors computed ΔH_f^0 values for about 60 hydrocarbon ions which, in general, are close to the experimental values selected by the authors. However, no consensus is observed between the values for structurally close ions. Thus, 3- and 4-Me cyclohexyl cations appeared to be by 4.5 and 2.6 kcal mol^{-1} *less stable* (i.e. CH_3-group is acceptor?), respectively, in the framework of isodesmic reactions compared with its model cyclohexylcations while a minor, if any, *stabilization* (0.5-1 kcal mol^{-1}) is expected since Me-group is inserted to remote position to the cation center. Similarly, the stabilities of cyclohexyl- and cyclopentyl cations are expected to be identical for the same reasons while the authors computed the large, 3.5 kcal mol^{-1} stabilization for the former [158]. The consecutive introduction of electron-donating Me-group to the stable benzyl cation might give the essential attenuation of the stabilization effect. Thus, taking Q_1 9.9 for $PhCH_2^+ \rightarrow PhCH(^+)Me$ we expect $Q_2 \sim$ 5-5.5 kcal mol^{-1} for $PhCH(^+)Me \rightarrow PhC(^+)Me_2$ isodesmic reaction while theoretical result is 8.4 kcal mol^{-1} [157]. The detailed analysis of the theoretical work [157] is given in Ref. 32.

In some cases the theory gives the enthalpies of formation for the ions which do not exist at the time scale of mass spectrometer isomerizing or decomposing to more stable systems. For example, cyclohexyl cation cannot be caught in a mass spectrometer due to its isomerization to 1-Me-cyclopentyl ion, it can be fixed only at radioactive decay of tritium in c-$C_6H_{11}T$ molecule (see Section 5). Another example is the computation of the ΔH_f^0 values for $H_2NCH_2^-$, $HOCH_2^-$ or $MeOCH_2^-$ anions [158] which do not exist since the EA<0 for corresponding free radicals or of the $FCH_2CH_2^-$ anion which without barrier transforms to F^- + C_2H_4 pair [4, 33, 34].

The examples given above for structurally rather simple ions cast doubts on theoretical calculations for more complex ions. Thus, one hardly can trust to the computation of the enthalpies of formation for 16(!) $[CH_3O_2]^+$ isomers among which only $(HO)_2CH^+$, $^+CH_2OOH$ and CH_3OO^+ have reasonable structures [159] or to the computation of ΔH_f^0 for ten $C_6H_6^{3+}$ and fifteen $C_5H_3^{\cdot ++}$ isomers [160] or of ΔE_{rel} [MP4/6-311//G(d,p)] for "isomers" given below [161] (kcal mol^{-1}):

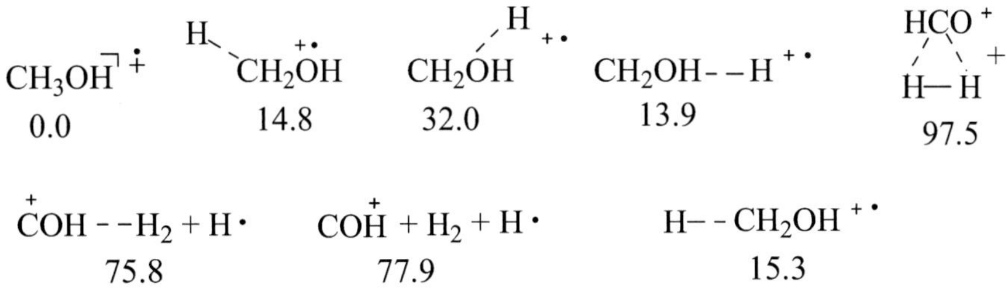

We made an attempt to apply the known enthalpies of formation of cations and radical cations to compute ΔH_f^0 for HC≡C· free radical [39]. From 21 isodesmic (working) reactions applying three ab initio and three semiempirical methods (PM3, AM1, MNDO) we obtained 126 values in the range of 50.6-177.3 kcal mol^{-1} whereas the real value extracted from very detailed analysis of experiment and empirical observations exhibited the value for ΔH_f^0 HC≡C· 123±2 kcal mol^{-1} [39]. Among those twenty one working reactions ten of them included ions (Table 5).

Table 5. Calculated values of ΔH_f^0 HC≡C· (kcal mol^{-1}) (from Ref. 39)

Method Working reaction		PM3	AM1	MNDO	ab initio		
					a	b	c
1	$HC{\equiv}CH + H_2O^{+\cdot} \rightarrow HC{\equiv}C\cdot + H_3O^+$	170.2	161.1	162.1	134.8	144.0	134.0
2	$MeC{\equiv}CH + H_2O^{+\cdot} \rightarrow HC{\equiv}C\cdot + MeO(^+)H_2$	172.7	111.8	174.8	135.0	135.8	126.8
3	$HC{\equiv}CH + HO^+ \rightarrow HC{\equiv}C\cdot + H_2O^{+\cdot}$	137.6	128.5	131.6	50.6	56.0	115.1
4	$MeC{\equiv}CH + HO^+ \rightarrow HC{\equiv}C\cdot + MeOH^{+\cdot}$	135.8	128.1	150.6	60.7	60.7	117.4
5	$HC{\equiv}CH + HS^+ \rightarrow HC{\equiv}C\cdot + H_2S^{+\cdot}$	123.5	127.0	146.0	76.0	77.6	92.3
6	$MeC{\equiv}CH + HS^+ \rightarrow HC{\equiv}C\cdot + MeSH^{+\cdot}$	131.8	129.8	156.8	79.5	79.3	88.7
7	$HC{\equiv}CH + NH_2^+ \rightarrow HC{\equiv}C\cdot + NH_3^{+\cdot}$	149.8	132.4	139.5	81.4	81.6	104.5
8	$MeC{\equiv}CH + NH_2^+ \rightarrow HC{\equiv}C\cdot + MeNH_2^{+\cdot}$	157.4	137.1	151.0	85.5	85.3	108.3
9	$HC{\equiv}CH + NH_3^{+\cdot} \rightarrow HC{\equiv}C\cdot + NH_4^+$	168.4	156.9	161.7	131.9	133.3	130.8
10	$MeC{\equiv}CH + NH_3^{+\cdot} \rightarrow HC{\equiv}C\cdot + MeNH_3^+$	173.1	160.8	169.6	131.4	130.0	127.6

* a) extended, b) medium, c) rough

Our results coming out of the Table 5 appeared to be unsatisfactory as with data of other authors (see Ref. 39). The expected value of 123 kcal mol^{-1} with deviation of, say, 5 kcal mol^{-1} was observed only for four values from sixty one. No any regularity in these values is noticed. We specially mention the reaction 9 (Table 5) whose results are also disappointing. This reaction was performed experimentally in a mass spectrometer [162] and it was shown to be nearly thermoneutral thus giving ΔH_f^0 HC≡C· 126.7 kcal mol^{-1} (ΔH_r here was not measured exactly) which is rather close to our empirical value 123±2 kcal mol^{-1} (see Ref. 39).

An important element of theoretical calculations appears to be the choice of experimental ΔH_f^0 values for a molecule or a fragment to be compared. If the set of the latter is very wide, like in case of HC≡C· free radical, then such choice becomes rather random if the thorough analysis of experimental values is not made. When a single experimental value is available its trustworthiness is not always made by theoreticians which can lead to serious errors. Here we speak about molecules whose experimental determination of their enthalpies of formation meets with certain difficulties. Thus, recently we drastically changed ΔH_f^0(g) for ketene CH_2=C=O molecule from –11.4 accepted in Reference Books [108, 135] to –24 kcal mol^{-1} (see Refs 37 and 163). In the latter work it was demonstrated that besides a very wide range of theoretical values in the range of +19.0 to –24, in the last years most theoreticians adjust their results to the value –11.4 rather than to the real value of –24 kcal mol^{-1} [37]. Such adjustment to the desired value we illustrate by one more example.

By EI method, without monochromator, the AEs of CF^+ and CH^+ ions originating from FC≡CH molecule were measured and the ΔH_f^0 of the latter 25.5 kcal mol^{-1} was derived [164]. It was really difficult to select less appropriate molecules for measurement of AEs for CF^+ and CH^+ ions since the probability to get them (and their neutral fragments) in excited states at cleavage of a triple (!) bond is very high. And really, as we found [39], the ΔH_f^0 FC≡CH molecule appeared to be too high. This can be easily established by applying the routine additity schemes. Thus, taking the safe value for ΔH_f^0 PhF molecule we gain ΔH_f^0 [FC≡CH] = ΔH_f^0 [PhF] - ΔH_f^0 [C_6H_6] + ΔH_f^0 [HC≡CH] = -27.7 [108] – 19.7 [108] + 54.5 [108] = 7.1 kcal mol^{-1}. Applying another additive scheme we have ΔH_f^0 [FC≡CH] = ΔH_f^0 [HC≡COH] - ΔH_f^0 (RF→ROH) = 10 [37] – (4-6) = 4-6 kcal mol^{-1}. The enthalpic shift $\Delta\Delta H_f^0$ for ROH→RF was originally noticed by Benson [165] and later was verified many times [4, 26, 39]. And, finally, according to the third additive scheme ΔH_f^0 [FC≡CH] = ΔH_f^0 [FCH=CH_2] - ΔH_f^0 [CH_2=CH_2] + ΔH_f^0 [HC≡CH] we have ΔH_f^0 [FC≡CH] 8.8 or 7.5 kcal mol^{-1} using ΔH_f^0 [FCH=CH_2] –33.2 [108] or –34.5 [38] kcal mol^{-1}, respectively.

All this means that ΔH_f^0 FC≡CH molecule is <9 kcal mol^{-1}. However, quantum-chemical calculations adjust their results to a priori doubtful experimental value of 25.5 kcal mol^{-1} [164]. In our previous work [39] the following theoretical ΔH_f^0 values for FC≡CH molecule taken from 13 works are quoted: 24.6, 25, 25.3±0.3, 25.5, 26, 26.4, 28.2, 30, 30.1, 31.2 kcal mol^{-1}.

REFERENCES

[1] Rosenstock HM, Wallenstein MB, Wahrhaftig AL, Eyring H. Absolute rate theory for isolated systems and the mass spectra of polyatomic molecules. *Proc.Natl.Acad.Sci.,* 1952, 38, 667-678.

[2] Khvostenko VI. Mass spectrometry of negative ions in organic chemistry (in Russ.), Moscow: Nauka (Science), 1981.

[3] Levsen K. Fundamental aspects of organic mass spectrometry. Weinheim: Verlag Chemie, 1978.

[4] Takhistov VV, Ponomarev DA. Organic mass spectrometry (in Russ.). BBM: St.Petersburg, 2005.

[5] Lebedev AT. Mass spectrometry in organic chemistry (in Russ.). Binom: Moscow, 2003.

[6] Takhistov VV, Ponomarev DA. Isodesmic reactions and thermochemistry of ions. *Org.Mass Spectrom.* 1994, 29, N 8, 395-412.

[7] Takhistov VV, Ponomarev DA. A Reply to the comment on :"Isodesmic reactions and thermochemistry of ions". *Eur.Mass Spectrom.* 1995, 1, N 1, 107-110.

[8] Ponomarev DA, PhD Thesis, Leningrad, 1974.

[9] Takhistov VV. DrSci Thesis, Leningrad, 1989.

[10] Lorquet JC. Basic questions in mass spectrometry. *Org.Mass Spectrom.* 1981, 16, N 11, 469-482.

[11] Waite BA, Miller WH. Model studies of mode specifity in unimolecular reaction dynamics. *J.Chem.Phys.* 1980, 73, N 8, 3713-3721.

[12] Maccoll A. Low-energy, low-temperature mass spectra. 14. A simple model for the rate coefficient. *Org.Mass Spectrom.* 1991, 26, N 4, 235-240.

[13] Tureček F, McLafferty FW. Non-ergodic behavior in aceton-enol ion dissociation. *J.Am.Chem.*Soc. 1984, 106, N 9, 2525-2528.

[14] Cho YS, Choe JC, Kim MS. Non-Rice-Ramsperger-Kassel-Marcus quasi-equilibrium theory dissociation of tert.butylbenzene molecular ion. *J. Phys. Chem.* 1995, 99, N 21, 8645-8651.

[15] Traeger JC, Morton TH. Activation barriers for addition of methyl radicals to oxygen-stabilized carbocation. *J. Phys. Chem.* 2005, 109A, N 45, 10467-10473.

[16] Bowen RD., Maccoll A. Low energy, low temperature mass spectra of some small saturated alcohols and ethers. *Org.Mass Spectrom.* 1984, 19, N 8, 379-384.

[17] Gustaffson E, Lindholm E. Ionization and dissociation of H_2, N_2 and CO in charge exchange collision with positive ions. *Arkiv Fys.* 1960, 18, N 13, 219-239.

[18] Ponomarev DA, Takhistov VV. General fragmentation rules of organic and elementoorganic compounds under electron impact. In: Atta-ur-Rahman M., editor. Recent advances in analytical techniques. Amsterdam: Elsevier, 1998; 369-492.

[19] Burgers PC, Holmes JL, Szulejko JE, Mommers AA, Terlouw JK. The gas phase ion chemistry of the acetyl cation and isomeric $[C_2H_3O]^+$ ions. On the structure of $[C_2H_3O]^+$ daughter ions generated from the enol acetone radical cation. *Org.Mass Spectrom.* 1983, 18, N 6, 254-262.

[20] Weber R, Levsen K, Burgers PC, Terlouw JK. Collision induced dissociation of ionized methyl iodide: evidence for the existence of an isolated electronic states. *Org.Mass Spectrom.* 1981, 16, N 11, 514-515.

[21] Borchers F., Levsen K., Schwarz H., Wesdemiotis C., Winkler HU. Isomeization of linear octene cations in the gas phase. *J.Am.Chem.Soc.* 1977, 99, N 19, 6359-6365.

[22] Miller JM. Mass spectral studies of some perhalogenoaromatic compounds. *J.Chem.Soc. (A)* 1967, N 5, 828-837.

[23] Schill G., Risser K., Fritz H., Vetter W. Synthesis, isolation, and identification of [3] catenane translation isomers. *Angew.Chem.* 1981, 93, N 2, 197-201.

[24] Vetter W., Logemann E., Schill G. Mass spectrometric investigation of some catenanes and macrocycles.. *Org.Mass Spectrom.* 1977, 12, N 6, 351-369.

[25] Takhistov VV, Misharev AD, Orlov VM., Ponomarev DA, Voss J. Thermochemistry and mass spectrometry of aromatic compounds with tertiary alkyl groups. *Eur.J.Mass Spectrom.* 1995, 1, N 4, 353-361.

[26] Takhistov VV, Pashina TA, Ismagilov NG, Orlov VM, Rodin AA, Barabanov VG. Experimental and theoretical determination of the enthalpies of formation of molecules, cations, anions, and radicals containing fluoroalkyl and other fluorinated groups. *Russ. J.Org.Chem. (English Transl.)*. 1995, 31, N 12, 1582-1588.

[27] Muftakhov MV, Vasil'ev YV, Mazunov VA, Takhistov VV, Ponomarev DA. Rearrangement processes of negative ions in the gas phase: $[M-Hal]^-$ ions in halogenated azobenzenes. Resonance stabilization of negative ions. *J.Mass Spectrom.* 1995, 30, N 2, 275-281.

[28] Domnin IN, Takhistov VV, Ponomarev DA. Thermochemistry of organic and heteroorganic species. Part IV. Photoionization studies of isomerisation and fragmentation of polysubstituted cyclopropenes: 3-methyl-3-ethynyl- and tetrachlorocyclopropene. *Eur.Mass Spectrom.* 1998, 4, N 3, 151-156.

[29] Domnin IN, Takhistov VV, Ponomarev DA. Thermochemistry of organic and heteroorganic species. Part V. Photoionization studies of isomerisation and fragmentation of polysubstituted cyclopropenes: 3-methyl-3-acetyl- and related compounds. *Eur.Mass Spectrom.* 1999, 5, N 3, 169-182.

[30] Domnin IN, Ponomarev DA, Takhistov VV, Pihlaja K. Calculation of the enthalpies of formation of substituted cyclopropenes. *Russ.J.Org.Chem. (Engl. Transl.).* 1999, 35, N 1, 28-33.

[31] Muftakhov MV, Vasil'ev YA, Khatymov RV, Mazunov VA, Takhistov VV, Travkin OV, Yakovleva EV. Thermochemistry of negatively charged ions. II. Energetics of formation of negative ions from acridanone and some of its derivatives. *Rapid Commun.Mass Spectrom.* 1999, 13, N 10, 912-923.

[32] Takhistov VV, Domnin IN, Ponomarev DA. Thermochemistry of organic and heteroorganic species. Part IX. Photoionization studies of isomerization and fragmentation of polysubstituted cyclopropanes: phenyl-substituted cyclopropanes. *Eur.Mass Spectrom.* 2000, 6, N 1, 53-64.

[33] Ponomarev D, Takhistov V. Thermochemistry of organic and heteroorganic species. Part XVII. Structural aspects and thermochemical approach to isomerization and fragmentation of negative ions. Simple bond cleavage and hydrogen rearrangement processes. *J.Mol.Struct.* 2000, 784, N 1, 177-197.

[34] Ponomarev D, Takhistov V. Thermochemistry of organic and heteroorganic species. Part XIX. Structural aspects and thermochemical approach to isomerization and fragmentation of negative ions: skeletal rearrangements, molecular ion and artifacts. *J.Mol.Struct.* 2006, 784, N 1, 198-213.

[35] Ponomarev D, Takhistov V. Some regularities in the relative thermodynamic stabilities of free radicals. *J.Mol.Struct.* 1997, 435, N 3, 259-274.

[36] Ponomarev D, Takhistov V. What are isodesmic reactions? *J.Chem.Educ.* 1997, 74, N 2, 201-203.

[37] Golovin AV, Ponomarev DA, Takhistov VV. Thermochemistry of organic and heteroorganic species. Part VIII. Ketene and structurlly related species. *J.Mol.Struct.* 2000, 524, N 3, 259-281.

[38] Orlov YD, Bouchoux G, Takhistov VV, Ponomarev DA. Thermochemistry of organic and heteroorganic species. Part XI. A new estimation scheme of the enthalpies of formation for $RCH_2\cdot$ free radicals. *J.Mol.Struct.* 2002, 608, N 1, 109-122.

[39] Golovin AV, Takhistov VV. Thermochemistry of organic and heteroorganic species. Part XII. Mono- and disubstituted acetylenes and ethynyl free radicals. New electronegativity scale. *J.Mol.Struct.* 2004, 701, N 1, 57-91.

[40] Golovin AV, Takhistov VV. Thermochemistry of organic and heteroorganic species. Part XIII. Interrelationship for ν_{CH} values in R<u>H</u> molecules and enthalpy of formation of R· free radicals in aliphatic compounds. *J.Mol.Struct.* 2005, 734, N 1-3, 219-233.
[41] Ponomarev D, Takhistov V. Thermochemistry of organic and heteroorganic species. Part XVI. Application of IR spectra of unsaturated aliphatic molecules to the thermochemistry of vinylic and allylic free radicals. *Internet Electron J.Mol.Design.* 2005, 4, N 6, 367-389.
[42] Ponomarev D, Takhistov V, Slayden S, Liebman J. Thermochemistry of organic, elementorganic and inorganic species. Part XX. Enthalpies of formation for free radicals of main group elements' halogenides. *J.Mol.Struct.* 2008, 876, N 1, 15-33.
[43] Ponomarev D, Takhistov V, Slayden S, Liebman J. Thermochemistry of organic, elementorganic and inorganic species. Part XXI. Enthalpies of formation for bi- and triradicals of main group elements' halogenides. *J.Mol.Struct.* 2008, 876, N 1, 34-65.
[44] Ponomarev D, Takhistov V. Thermochemistry of organic and elementoorganic species. Part III. The enthalpies of formation of the hydrides of main group elements. *J.Mol.Struct.* 1999, 477, N 1, 91-103.
[45] Takhistov VV, Golovin AV. Thermochemistry of organic, heteroorganic and inorganic species. Part XV. The enthalpies of formation for halogenides of main group elements. *J.Mol.Struct.* 2006, 784, N 1, 47-68.
[46] Luo YR. Handbook of bond dissociation energies. Boca Raton: CRC Press; 2003.
[47] Biemann K. Mass spectrometry. Organic chemical applications. New York: McGraw-Hill Book Co.; 1962.
[48] Benz W. Massenspektrometrie Organischer Verbindungen. Frankfurt/Main: Akademische Verlagsgesellschaft: 1969.
[49] Takhistov VV. Practical mass spectrometry (in Russ.). Leningrad: Univ.Press; 1977.
[50] Taskhistov VV. Organic mass spectromery (in Russ.). Leningrad: Nauka (Science), 1990.
[51] Ponomarev DA, Golovin AV, Takhistov VV. Free radicals in mass spectrometry. Part I. General fragmentation rules for organic and organometallic compounds under electron impact. *Eur.J.Mass Spectrom.* 2002, 8, N 6, 409-434.
[52] Muftakhov MV, Khatymov RV, Mazunov VA, Ponomarev DA, Takhistov VV, Vatlina LP. Thermochemical determination of the structure of negative ions on the basis of data from resonance electron capture mass spectrometry. Phenol, its chlorinated derivatives and thioanalogue. *Rapid Commun.Mass Spectrom.* 2000, 14, N 16, 1482-1484.
[53] Voinov VG, Clayes M. Charge-remote fragmentation characteristics of monosaturated fatty acids in resonance electron capture: differentiation between cis and trans isomers. *Int.J.Mass Spectrom.* 2001, 205, N 1, 57-64.
[54] Qin XZ. Collision-induced dissociation of the negative ions of simvastatin hydroxy acid and related species. *J.Mass Spectrom.* 2003, 38, N 6, 677-686.
[55] Carette M, Zerega Y, Perrier P, Andre J, March RE. Rydberg electron-capture mass-spectrometry of 1,2,3,4-tetrachlorodibenzo-p-dioxin. *Eur.J.Mass Spectrom.* 2000, 6, N 5, 405-408.
[56] Naft WT, Kooper CD, Compton RN. Transient negative-ion states in alicyclic and aromatic fluorocarbon molecules. *J.Chem.Phys.* 1968, 49, N 6, 2784-2788.

[57] Suess L, Parthasaraphy R, Dunning FB. Non-dissociative low-energy attachment to SF_6, C_6F_6, $C_{10}F_8$, and c-C_7F_{14}: negative ion lifetimes. *J.Chem.Phys.* 2002, 117, N 24, 11222-11227.

[58] Tobita S, Meinke M, Illenberger E, Christophorou LG, Baumgärtel S. Polycyclic aromatic hydrocarbons: negative ions formation following low energy (0-15 eV) electron impact. *Chem.Phys.* 1992, 161, N 3, 501-508.

[59] Asfandiarov NL, Fal'ko VS, Lukin VG, Nafikova EP, Pshenichnuyk SA, Fokin AI, Lomakin GS, Chizhov YV. Violation of frozen shell approximation in dissociative electron capture by halogenated antraquinones. *Rapid Commun.Mass Spectrom.* 2001, 15, N 19, 1869-1878.

[60] Laramée JA, Mazurkiewicz P, Berkout V, Deinzer ML. Electron monochromator-mass spectrometry instrument for negative ion analysis of electronegative compounds. *Mass Spectrom.Rev* 1996, 15, N 1, 15-42.

[61] Asfandiarov NL, Fokin AI, Lukin VG, Nafikova EP, Lomakin GS, Fal'ko VS, Chizhov YV. Shape resonances in slow electron scattering by aromatic molecules. I. Antraquinone derivatives. *Rapid Commun.Mass Spectrom.* 1999, 13, N 12, 1116-1123.

[62] Wagner PJ. Mass spectral fragmentation of aminoketones in which charge is not localized exclusively on nitrogen. *Org.Mass Spectrom.* 1970, 3, N 11, 1307-1315.

[63] Budzikiewicz H, Djerassi C, Williams DH. Mass spectrometry of organic compounds, San Francisco: Holden Day; 1967.

[64] McLafferty FW, Turecek F. Interpretation of mass spectra, 4th Ed. Mill Valley: University Science Books, 1993.

[65] Mölder UX. PhD Thesis, Univ.Tartu, 1984.

[66] Adams JA, Gross M. Charge-remote fragmentation of closed-shell ions. A thermolytical analogy. *J.Am.Chem.Soc.* 1989, 111, N 2, 435-440.

[67] Cummings CS, Bleakney W. Products of ionization by ET in MeOH and EtOH. *Phys.Rev.* 1940, 58, N 6, 781.

[68] Heinrich N, Schwarz H. On the role of ion/dipole complexes in the isomerization/dissociation reactions of ionized acetic acid and its enol in the gas phase. *Int.J.Mass Spectrom.Ion Proc.* 1987, 79, N 3, 295-310.

[69] Longevialle P. Ion-neutral complexes in the unimolecular reactivity of organic cations in the gas phase. *Mass Spectrom. Rev.* 1992, 11, N 3, 157-192.

[70] McAdoo DJ, Morton TH. Gas-phase analogues of cage effects. *Acc.Chem.Res.* 1993, 26, N 6, 295-302.

[71] Schwarz H, Stahl D. Unimolecular water loss from protonated alcohols in the gas phase: the effect of ion/dipole interactions on the isomerization of incipient carbocations. *Int. J.Mass Spectrom.Ion Proc.* 1982, 36, N 3, 285-289.

[72] Cacace F. The radiolytic approach to gas-phase ion chemistry. *Acc.Chem.Res.* 1988, 21, N 6, 215-222.

[73] Gäumann T, Houriet R, Stahl D, Tabet J-C, Heinrich N, Schwarz H. Further examples of skeletal rearrangements of the Wagner-Meerwein type in chemical ionization mass spectrometry: the case of $[C_6H_9]^+$ ions. *Org.Mass Spectrom.* 18, 1983, N 5, 215-218.

[74] Rylander PN, Meyerson S. Rearrangement of 3-chloro-3-ethylpentane during acid-catalyzed alkylation. *J.Am.Chem.Soc.* 1956 , 78, N 16, 6683-6684.

[75] Meyerson S. Cationated cyclopropanes as reaction intermediates in mass spectra: an earlier incarnation of ion-neutral complexes. *Org.Mass Spectrom.* 1989, 24, N 4, 267-270.

[76] Osterheld TH, Baer T, Brauman JT. IR multiple-photon dissociation of the nitrobenzene radical cation. A paradigm for competitive reaction. *J.Am.Chem.Soc.* 1993, 115, N 4, 6284-6289.

[77] Berman DW, Anicich V, Beauchamp JL. Stabilities of isomeric halonium ions C_2H_4X+ (X=Cl, Br) by photoionization mass spectrometry and ICR spectroscopy. General consideration of the relative stabilities of cyclic and acyclic isomeric onium ions. *J.Am.Chem.Soc.* 1979, 101, N 5, 1239-1248.

[78] Blanchette MC, Holmes JL, Lossing FP. The ethyl halides: stable neutral and radical cation isomers C2, H5, X] where X=F, Cl, Br, I. *Org.Mass Spectrom.* 1987, 22, N 11, 701-709.

[79] Holmes JL, Lossing FP, Terlouw JK, Burgers PC. The radical cation $[CH_2OH_2]^{\cdot+}$ and related stable gas-phase ion-dipole complexes. *J.Am.Chem.Soc.* 1982, 104, N 10, 2931-2932.

[80] Thoen KT, Smith RL, Nousiainen JJ, Nelson ED, Kenttämaa HI. Charged phenyl radicals. *J.Am.Chem.Soc.* 1996, 118, N 36, 8669-8676.

[81] Orlov VM, Rashkes YV, Siretskaya TV, Takhistov VV. Thermochemical decription of tautomeric processes in ionized 1,3-diketones. *Russ.J.Gen.Chem.* 1988, 58, N 2, 429-438.

[82] Schweig A, Vermeer H, Weidner U. A photoelectron spectroscopic study of keto-enol tautomerism in acetylacetones – a new application of photoelectron spectroscopy. *Chem.Phys.L*ett. 1974, 26, N 2, 229-233.

[83] Holmes JL. Assigning structures to ions in gas phase. *Org.Mass Spectrom.* 1985, 20, N 4, 169-183.

[84] Burgers PC, Holmes JL, Szuleiko JE, Mommers A, Terlouw JK. The gas phase ion chemistry of the acetyl cation and isomeric $[C_2H_3O]^+$ ions. On the structure of the daughter $[C_2H_3O]^+$ ions generated from the enol of acetone radical cation. *Org.Mass Spectrom.*1983, 18, N 6, 254-262.

[85] Cooks RG, ed. Collisional Spectroscopy, New York: Plenum Press. 1978.

[86] Burgers PC, Holmes JL, Mommers AA. Laboratory experiments on the interstellar species $[HCO]^+$ and $[COH]^+$. *J.Am.Chem.Soc.* 1985, 107, N 5, 1099-1101.

[87] Eberlin NN, Majumdar TK, Cooks RG. Structures and mechanisms of reactions of isomeric $C_2H_3O^+$ and $C_2H_3S^+$ ions revealed through ion/molecule reactions in conjunction with 2D and 3D mass spectrometry. *J.Am.Chem.Soc.* 114, N 8, 2884-2896.

[88] Katritzky AR, Watson CH, Dega-Szafran Z, Eyler JR. CAD of N-alkylpyridinium cations to pyridine and alkyl cations in the gas phase. *J.Am.Chem.Soc.* 1990, 112, N 7, 2471-2478.

[89] Katritzky AR, Watson CH, Dega-Szafran Z, Eyler JR. CAD of N-alkylpyridinium cations and olefins in the gas phase. *J.Am.Chem.Soc.* 1990, 112, N 7, 2479-2484.

[90] Van Baar B, Weiske T, Terlouw JK, Schwarz H. Hydroxyacetylene: generation and characterization of the neutral molecule, radical cation and dication in the gas phase. *Angew.Chem.Int.Ed.* 1986, 25, N 3, 282-284.

[91] Wagner-Redeker W, Illies AJ, Kemper PR, Bowers MT. Structure and reactivity of gas-phase ions: $C_4H_4^{\cdot+}$. *J.Am.Chem.Soc.* 1983, 105, N 18, 5719-5724.

[92] Aubry C, Holmes JL, Terlouw JK. Effect of methyl substitution on the thermochemistry of ketene. *J.Phys.Chem.* 1997, 101, N 33, 5958-5961.

[93] Faulk D, Dunbar RC. Time-resolved photodissociation of three tert.butyl benzene ions. *J.Phys.Chem.* 1991, 95, N 18, 6932-6936.

[94] Dunbar RC. What are slowest metastable-ion decompositions. *Int.J.Mass Spectrom. Ion Proc.* 1983, 54, N 1-2, 109-118.

[95] Huang FS, Dunbar RC. Time-resolved photodissociation of methylnaphtalene ion. An illustration of kinetic shifts in large-ion dissociation. *J.Am.Chem.Soc.* 1990, 112, N 22, 8167-8169.

[96] Lifshitz C. Tropylium ion formation from toluene: solution of an old problem in organic mass spectrometry. *Acc.Chem.Res.* 1994, 27, N 5, 138-144.

[97] Malinovich Y, Lifshitz C. Time-dependent mass spectra and breakdown graphs. 7. Time-resolved PI MS of iodobenzene. The heat of formation of $C_6H_5^+$. *J.Phys.Chem.* 1986, 90, N 10, 2200-2203.

[98] Butcher V, Costa ML, Dyke JM, Ellis AR, Morris A. A study of the phenyl radical by vacuum UV PE spectroscopy. *Chem.Phys.* 1987, 115, N 2, 227-237.

[99] Dannacher J, Rosenstock HM, Buff R, Parr AC, Stockbauer RL, Bombach R, Stadelmann J-P. Benchmark measurement of iodobenzene ion fragmentation route. *Chem.Phys.* 1983, 75, N 1, 23-25.

[100] Lifshitz C, Gotohiguian P, Roller R. Time dependent mass spectra and breakdown graphs. The kinetic shift in aniline. *Chem.Phys.Lett.* 1983, 95, N 1, 106-108.

[101] Sergeev YL, Akopyan ME, Vilesov FI. Photoionization of phenyl halides. *Opt.Spectrosc.(Russ.)* 1970, 29, N 1, 119-123.

[102] Lifshitz C, Gefen S. Time-dependent mass spectra and breakdown graphs. The kinetic shift in phenol. *Org.Mass Spectrom.* 1984, 19, N 4, 197-198.

[103] Lifshitz C, Gefen S. Time-dependent mass spectra and breakdown graphs. I. Hexadiyne. *Int.J.Mass Spectrom. Ion Proc.* 1980, 35, N 1/2, 31-37.

[104] Ziesel JP, Lifshitz C. Time-dependent mass-spectra and breakdown graphs. 10. Dissociative photoionization of anisole. *Chem.Phys.* 1987, 117, N 2, 227-237.

[105] Levin RD, Lias SG. Ionization Potential and Appearance Potential Measurements, 1971—1981, Washington, DC, 1982.

[106] Ponomarev DA, Takhistov VV, Misharev AD, Orlov VM. Energetical characteristics of ionization and fragmentation of ethers of substituted phenols in the gas phase. *Russ.J.Gen.Chem.* 1993, 64, N 6, 1006-1009.

[107] Butler JJ, Fraser-Monteiro ML, Fraser-Monteiro L, Baer T, Hass JR.. Thermochemistry and dissociation dynamics of state-selected $C_4H_8O_2^+$ ions. 2. Butanoic acid. *J.Phys.Chem.* 1982, 86, N 5, 747-752.

[108] Pedley JB, Naylor RD, Kirby SP. Thermochemical data of organic compounds, 2nd ed., Chapman and Hall, London, 1986.

[109] Yeo RK, Ruscic B, Berkowitz J. Vacuum UV photoionization mass spectrometric study of C_{60}. *J.Chem.Phys.* 1992, 90, N 2, 911-918.

[110] Berman W, Beachamp JL. ICR and PI investigations of the thermochemistry and reactions of ions derived from CF_3I. *Int.J.Mass Spectrom.Ion Phys.* 1981, 39, N 1, 47-54.

[111] Osamura Y, Takeuchi T, Nishimoto K. An MO study of fragmentation mechanism of ethane and propane radical cations. *Bull.Chem.Soc.Jpn.* 1988, 61, N 10, 3387-3391.

[112] Arakawa R.. Mass spectral study of ionized hydroxyacetone dissociation. *Bull.Chem.Soc.Jpn.* 1991, 63, N 3, 1022-1024.

[113] Lossing FP. Free radicals in mass spectrometry. XLV. Ionization potentials and heats of formation of C_3H_3, C_3H_5, and C_4H_7 radicals. *Can.J.Chem.* 1972, 50, N 24, 3973-3981.

[114] Werner AS, Baer T. Absolute unimolecular decay rates of energy selected $C_4H_6^+$ metastable ions. *J.Chem.Phys.* 1975, 62, N 7, 2900-2910.

[115] Burkey TJ, Castelhano AL, Griller D, Lossing FP. Heats of formation and ionization potentials of some aminoalkyl radicals. *J.Am.Chem.Soc.* 1983, 105, N 14, 4701-4703.

[116] Holmes JL, Lossing FP, Maccoll A. Heats of formation of alkyl free radicals from appearance energies. *J.Am.Chem.Soc.* 1988, 110, N 22, 7339-7342.

[117] Holmes JL, Lossing FP. Heats of formation and bond dissociation energies in halogen substituted methyl and ethyl radicals. *J.Am.Chem.Soc.* 1988, 110, N 22, 7343-7345.

[118] Holmes JL, Lossing FP, Mayer PM. Heats of formation of oxygen-containing organic free radicals from AP measurements. *J.Am.Chem.Soc.* 1991, 113, N 26, 9723-9728.

[119] McMillen DF, Golden DM. Hydrocarbon bond dissociation energies. *Ann.Rev.Phys.Chem.* 1982, 33, 493-532.

[120] Siretskaya TV, Takhistov VV, Pimenova SM, Orlov VM, Pustobaev VN. Formation enthalpies of acetylene ketones fragments in gas phase. *React.Kinet.Catal.Lett.* 1986, 31, N 1, 15-19.

[121] Orlov YD., Lebedev YA., Saifullin IS. Thermochemistry of organic free radicals (in Russ.), Moscow: Nauka (Science), 2001.

[122] McKean DC. CH bond dissociation energies, isolated stretching frequencies, and radical stabilization energies. *Int.J.Chem.Kinet.* 1989, 21, N 6, 445-469.

[123] Wolkoff P, Holmes JL, Lossing FP. On the formation of cyclopentenium cations from all $C_6H_{10}^{\cdot +}$ molecular ion structures. *Can.J.Chem.* 1980, 58, N 3, 251-257.

[124] Benson SW. Thermochemical Kinetics, New York: Wiley, 1976.

[125] Cooks RG, Beynon JH, Caprioli R.M, Lester GR. Metastable ions, Amsterdam: Elsevier, 1973.

[126] Lehman TA, Bursey MM. ICR spectrometry, New York: Wiley, 1976.

[127] Kalman JR, Fairwater RB, Hvistendahl G, Williams DH. Methyl radical expulsions occuring with kinetic energy release. *J.Chem.Soc.Chem.Commun.* 1976, N 15, 604-605.

[128] Williams DH, Hvistendhal G. Kinetic energy release in relation to symmetry forbidden reactions. *J.Am.Chem.Soc.* 1974, 96, N 21, 6753-6755.

[129] Williams DH, Stapleton J, Bowen RD. The direct observation of energy release in carbonium ion rearrangements. *Tetrahedron Lett.* 1978, 19, N 32, 2919-2922.

[130] Polyakova AA, Khmel'nitsky RA. Introduction to mass spectrometry of organic compounds (in Russ.), Moscow: Khimia, 1966.

[131] Salmona G, Vincent EJ. Isomerization of thiazole and isothiazole prior to fragmentation. *Org.Mass Spectrom.* 1978, 13, N 2, 119-120.

[132] Adams J. Charge-remote fragmentations: analytical applications and fundamental studies. *Mass Spectrom. Rev.* 1990, 9, N 2, 141-186.

[133] Pevsner MS, Takhistov VV. Fragmentation of halogen- and nitroderivatives of 1,2,4-triazole under electron impact. *Russ.J.Org.Chem.* 1999, 35, N 12, 1852-1856.

[134] Ramana DV, Srinivas R, Mahalakshmi P. Unusual neutral and radical losses in 2'-substituted N-phenylanthranilic acid on electron impact. *Org.Mass Spectrom.* 1991, 26, N 4, 305-310.

[135] Lias SG, Bartmess JE, Liebman JF, Holmes JL, Levin RD, Mallard WG. *J.Phys.Chem.Ref.Data,* 1988, 17 (Suppl 1).

[136] Pike RM. Chemistry of silylamines. III. Condensation of silylamines with silylesters. *Rec.Travaux Chim.Pay-Bas* 1961, 80, N 8, 819-824.

[137] Kuivila HG, Rahman W, Fish RH. Free radical addition of trimethyl tin hydride to alkenes. *J.Am.Chem.Soc.* 1965, 87, N 13, 2835-2840.

[138] Böhme H, Lehners W, Keitzer G. Uber die Spaltung von Diaminmethan-Derivatives mit Halogenwasserstofen. *Chem.Ber.* 1958, 91, N 2, 340-345.

[139] Maslak P, Narvaez JN. Mesolytic cleavage of C-C bonds. Comparison with homolytic and heterolytic processes in the same substrate. *Angew.Chem.Int.Ed.* 1990, 29, N 3, 283-285.

[140] Camaioni DM. Calculation of structures and bond dissociation energies of radical cations : the importance of through-bond delocalization in bibenzylic systems. *J.Am.Chem.Soc.* 1990, 112, N 26, 9475-9483.

[141] Arnett EM, Flowers RA. Bond cleavage energies for molecules and their associated radical ions. *Chem.Soc.Rev.* 1993, 22, N 1, 9-15.

[142] Cioslowski J, Mixon ST. Unusual bonding in the 1,1,1-triamino-2,2,2-tricyanoethane molecule. *Chem.Phys.Lett.* 1990, 170, N 2/3, 297-300.

[143] Simoes JAM, Beauchamp JL. Transition metal-hydrogen and metal-carbon strength. The ways to catalysis. *Chem.Rev.* 1990, 90, N 4, 629-688.

[144] Exner K, Schleyer PR.. Theoretical bond energies. A critical evaluation. *J.Phys.Chem.* 2001, A105, N 13, 3407-3416.

[145] Eland JH, Frey R, Schulte H, Brehm B.Unimolecular dissociations and internal conversions of methyl halide ions. *Int.J.Mass Spectrom.Ion Proc.* 1976, 22, 155-170.

[146] Takhistov VV, Ponomarev DA. Physics and chemistry: differentiation of the responsibility on two examples. In Collection of Works: Regional school-conference on mathematics and physiscs, vol.III (in Russ.), Ufa: Bashkir Univ., 2004, p. 27-45.

[147] Lynch BJ, Zhao Y, Truhlar DG. The 6-31B(d) basis set and the BMC-QCISD and BMC-CCSD multicoefficient correlation methods. *J.Phys.Chem.* 2005, 109A, N 8, 1643-1648.

[148] Hesselmann A, Jansen G, Schulz M. Density functional theory-symmetry-adapted intermolecular perturbation theory with density fitting: a new efficient method to study intermolecular interaction energies. *J.Chem.Phys.* 2005, 122, 014103 (14 p.).

[149] Stampfuss P, Wenzel W. Improved implementation and application of the individually selecting configuration interaction method. *J.Chem.Phys.* 2005, 122, 024110 (5 p.).

[150] Fouqueau A, Casida ME, Daku LML, Hauser A, Neese F. Comparison of density functionals for energy and structural differences between the high [$^5T_{2g}$:$(t_{2g})^4(eg)^2$] and low [$^1A_{1g}$:$(t_{2g})(eg)^0$] spin states of iron (II) coordination compounds. II. More functionals and the hexaminoferrous cation $[Fe(NH_3)_6]^{2+}$. *J.Chem.Phys.* 2005, 122, 044110 (13 p.).

[151] Dewar MJS, Rzepa HS. Ground states of molecules. 40. MNDO results for molecules containing fluorine. *J.Am.Chem.Soc.* 1978, 100, N 1, 58-67.

[152] Heerma W, Sarneel MM, Dijkstra G. Structure and formation of gaseous $[C_3H_4N]^+$ ions. *Org.Mass Spectrom.* 1986, 21, N 10, 681-687.

[153] 153.Lu HF, Li FY, Lin SH. Theoretical study of the fragments generated from a glycine radical cation. *J.Phys.Chem.* 2004, 108A, N 42, 9233-9243.

[154] Li WK, Riggs NV. An ab initio molecular orbital study of the $C_3H_3^+$ potential energy surface. *J.Mol.Struct.(THEOCHEM)* 1992, 257, 189-204.

[155] Hopkinson AC, Lien MH. Substituent effects in carbocations CX^+, CHX^+, and $CH_2X^{\cdot +}$, and in singlet and triplet carbenes. Proton affinities of singlet carbenes. *Can.J.Chem.* 1985, 63, N 12, 3582-3586.

[156] Paddon-Row MN, Houk KN, Tidwell TT. The magnitude of destabilization of stabilized cations by cyano and trifluoromethyl groups. *Tetrahedron* 1982, 23, N 4, 383-386.

[157] Reindl B, Clark T, Schleyer PR. A new method for empirical forcefield calculations on localized and delocalized carbocations. *J.Comput.Chem.* 1996, 17, N 12, 1406-1430.

[158] Tupytsyn IF, Cane AA, Puzanov YV, Shibaev AY. Proton affinity and specifity of mechanism of carbanions stabilization. I. Comparative study of accessibility of diverse quantum-chemical methods. *Russ.J.Gen.Chem.* 1989, 59, N 5, 1141-1145.

[159] Ha TK, Nguen MT. Hypothesized structures for the $CH_3O_2^+$ cation isomers in several ion-neutral reactions. An ab initio self-consistent field study. *J.Phys.Chem.* 1984, 88, N 19, 4295-4298.

[160] Koch W, Schwarz H. Structures and stabilities of $C_6H_6^{3+}$ isomers. An ab initio molecular study (using split valence 3-21G basis set). *Chem.Phys.Lett.* 1985, 113, N 2, 145-150.

[161] Ma NL, Smith BJ, Pople JA, Radom L. Rearrangement and dissociative reactions of the methanol radical cation $[CH_3OH]^{\cdot +}$: a comparison of theory and experiment. *J.Am.Chem.Soc.* 1991, 113, N 21, 7903-7912.

[162] Iraqi M, Petrank A, Peres M, Lifshitz C. Proton transfer reactions of $C_2H_2^{+}$: the bond energy $D_0(C_2H\text{-}H)$. *Int.J.Mass Spectrosc.Ion Proc.* 1990, 100, 679-691.

[163] Takhistov VV, Ponomarev DA, Golovin AV. Thermochemistry of ketene molecule. *Rus.J.Gen.Chem.* 2003, 73, N 11, 1873-1875.

[164] Kloster-Jensen E, Pascual C, Vogt J. Mass spectrometric studies of mono- and dihaloacetylenes. *Helv.Chim.Acta* 1970, 53, N 8, 2109-2118.

[165] Benson SW. Thermochemistry and kinetics of sulfur-containing molecules and radicals. *Chem.Rev.* 1978, 78, N 1, 23-36.

In: Thermochemistry and Advances in Chemistry Research ISBN: 978-1-60692-376-4
Ed: Anton G. Mirskiy

Chapter 2

THERMOCHEMISTRY OF NUCLEATING CLUSTERS: A BRIDGE BETWEEN THE QUANTUM WORLD AND PLANETARY ATMOSPHERE

Alexey B. Nadykto and Fangqun Yu*

Atmospheric Sciences Research Center, State University of New York at Albany, 251 Fuller Rd., Albany 12203, NY, USA

ABSTRACT

The effect of atmospheric aerosols on the Earth's climate is unambiguous. Atmospheric aerosols formed via nucleation and subsequent condensational growth, which contribute significantly to the concentration of cloud condensation nuclei, influence the Earth radiation balance indirectly by modifying cloud properties and precipitation. A clear understanding of atmospheric nucleation is critically important for assessing the lifecycle of atmospheric particles and predicting the global climate changes associated with the aerosol radiative forcing. The global scale phenomena are closely connected to processes of gas to particle conversion during nucleation. Nucleation rates are very sensitive to the thermochemistry of initial steps of the aerosol particle formation where matter in a state, neither liquid nor gaseous, but in indeterminate state of molecular cluster is growing by the attachment of vapor monomers. The small pre-nucleation clusters are essentially quantum in nature, and must be treated using quantum methods. Recently, it has been shown that the application of quantum methods to the thermochemistry of clusters composed of atmospheric precursors is certainly beneficial, and allows solving numerous problems that could not be adequately treated using the continuum/bulk liquid classical nucleation theory. This paper describes the recent developments in the thermochemistry of pre-nucleation clusters composed of common atmospheric nucleation precursors and discusses the atmospheric implications of the new thermochemical data. The importance of the cluster thermochemistry is outlined and future developments towards the new atmospheric nucleation theory, which will be based on the thermochemistry obtained from "first principles", is discussed. The nucleation models constrained by thermochemical data obtained using quantum methods can be

* Emails: alexn@asrc.cestm.albany.edu; yfq@asrc.cestm.albany.edu

incorporated into global models to study new particle formation in planetary atmosphere and associated implications to climate change.

1. INTRODUCTION

Atmospheric particles influence the Earth climate directly, by absorbing and scattering solar radiation, or indirectly by modifying cloud properties [1-2]. The most of uncertainties in current assessment of global climate changes are associated with the aerosol indirect effect. New particle formation via nucleation and subsequent condensation is an important source of atmospheric aerosols [3]. Although the contribution of secondary particles to the total particulate matter mass is quite small, they usually dominate the particle number concentration of atmospheric aerosols and cloud condensation nuclei (CCN) [4]. CCN concentration is a key factor determining the cloud properties and thus aerosol indirect radiative forcing. On the other hand, high concentrations of secondary ultrafine particles associated with traffic [5-7] are found to lead to adverse public health effects [8-11]. To address both issues, a clear understanding of the formation mechanism of new viable embryonic droplets (i.e., nucleation) is critically important and yet to be achieved.

Most of the problems in the atmospheric nucleation theories are related to a complex, essentially multicomponent, nature of atmospheric nucleation and the application of the conventional classical nucleation theory, whose thermochemistry is based on the capillarity approximation treating molecular clusters as liquid droplets. Atmospheric nucleation involves a number of different species including e.g. sulfuric acid, ammonia, ions and organics (Fig. 1). At the present time, there are several candidate mechanisms that could possibly be responsible for atmospheric nucleation:

1. Binary homogeneous nucleation (BHN) of H_2SO_4 and H_2O [12-15];
2. Ternary homogeneous nucleation (THN) of H_2SO_4-H_2O-NH_3 [16-18];
3. Ion-mediated nucleation (IMN) of H_2SO_4-H_2O-Ion [19-23];
4. Organics-enhanced nucleation H_2SO_4-H_2O-organics [24-25].

Despite decades of intensive studies and impressive progress achieved in both theory and instrumentation within the last decade, the mechanisms controlling the burst of new particles in the Earth atmosphere remain unclear. BHN was long considered as the dominant nucleation mechanism in the atmosphere. However, it was unable to explain recently observed nucleation events in the lower troposphere. Classical model of THN [17] developed in the response to the failure of BHN predicts NH_3 at ppt level to enhance the binary nucleation by up to ~30 orders of magnitude and has been used to explain many observed nucleation events. However, Yu [16] pointed out that the existing laboratory experiments [18, 26-29] do not support the predictions of classical THN and concluded that, based on a kinetic THN model constrained by experimental data, the contribution of THN to atmospheric particle formation is likely to be small. The widely used classical THN model [17] has been recently corrected [30], leading to a significant reduction in predicted nucleation rates, and the revised THN model became unable to explain the formation of new particles in the boundary layer. The presence of ions, which essentially work as a catalyst, helps to overcome the nucleation barrier and leads to significant enhancement in nucleation rates (Fig. 2). IMN

[19-21] has been shown to be likely one of the most important particle formation mechanisms in the atmosphere [23].

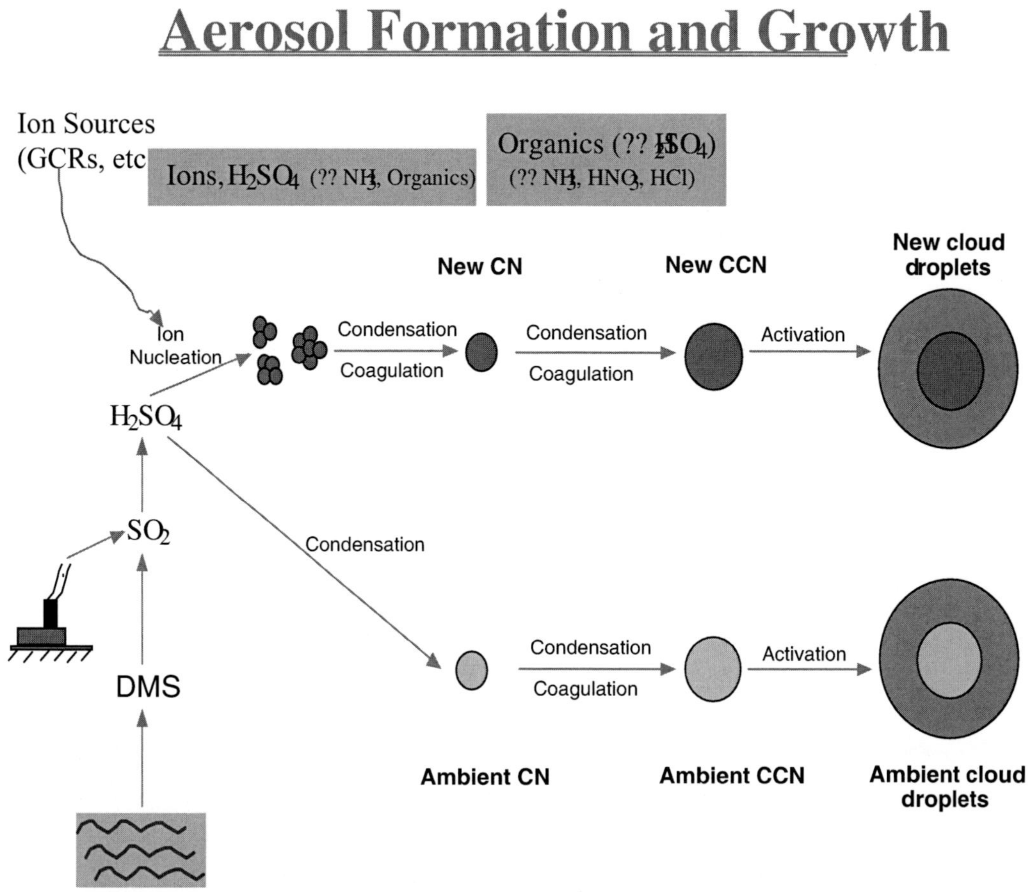

Figure 1. Schematic representation of the key processes and species involved in the formation of new condensation nuclei (CN) and their growth into cloud condensation nuclei (CCN). The possible dominating species involved in the different phases of CN formation and growth processes are also indicated. The organics species are known to play an important role in growing the CN into the size of CCN. Sulfuric acid is known to be involved in most of nucleation events observed in the atmosphere, while the exact role of ammonia and organics in atmospheric nucleation is unclear.

The predictions of the second generation IMN model [21], which was built upon on the substantial progress in understanding the molecular nature and thermodynamics of small charged and neutral clusters achieved in the last few years [31-34] and a number of recently available laboratory data [35-36], suggest that IMN can lead to significant nucleation in the lower atmosphere [21,23]. While the involvement of ions in most (90%) of observed nucleation events has been corroborated from the field measurements obtained in boreal forests [37-38], the relative importance of IMN and homogeneous nucleation remains a subject of ongoing debates [39-42].

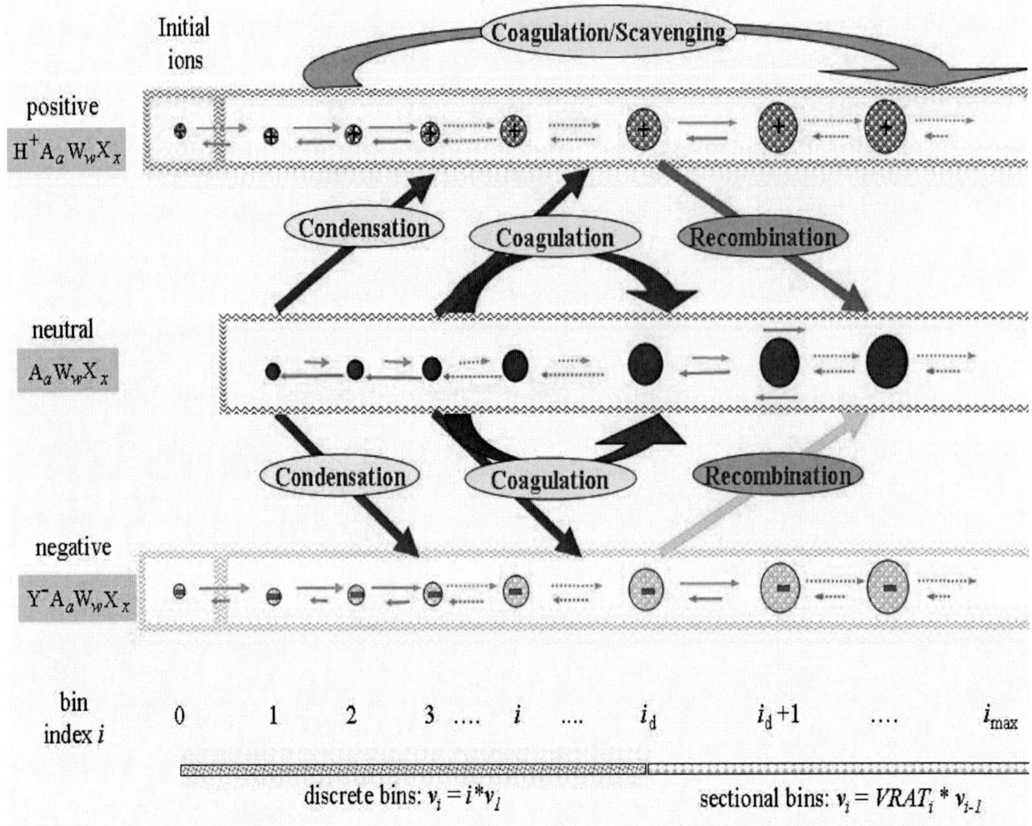

Figure 2. Schematic representation of particle formation processes involving positively charged, neutral, and negatively charged clusters in the atmosphere (from Yu, 2006 [21]). A, W, and X represent the sulfuric acid (H_2SO_4), water (H_2O), and other possible species (for example, organics, NH_3, etc.), respectively, while a, w, and x refer to the number of A, W, and X molecules in the clusters/droplets, respectively. Y^- is an abbreviation for the negative core ion. The application of the computational quantum methods can help to identify the key species involved in the atmospheric nucleation and provide important thermochemistry information needed to simulate the dynamics of cluster formation and evolution.

The experimental work of Zhang et al. [24] has shown that the presence of organic acids can enhance the nucleation of H2SO4 and H2O. More recently, Nadykto and Yu [25] pointed out that ammonia is certainly not an exclusive stabilizer of H2SO4-H2O clusters in the atmosphere because low-molecular formic and acetic acids, which are among the most abundant organic acids in the atmosphere, are as efficient in stabilizing the binary clusters as NH3. A substantial amount of further work is needed in order to determine the organic species that can enhance the atmospheric nucleation and to quantify their effect on atmospheric nucleation rates.

As we mentioned earlier, most of the difficulties in the theoretical description of different nucleation mechanisms arise from the application of the conventional classical nucleation theory (CNT) [43-44], which is based on the liquid droplet formalism introduced by Lord

Kelvin in 19th century (see Fig. 3). The well known problems of the CNT include the inability to account for the internal structure and quantum nature of molecular pre-nucleation clusters and the excessive sensitivity of CNT nucleation rates to macroscopic/bulk input parameters of multicomponent mixtures (density, surface tension, saturated vapor pressure), which are often poorly determined or absent.

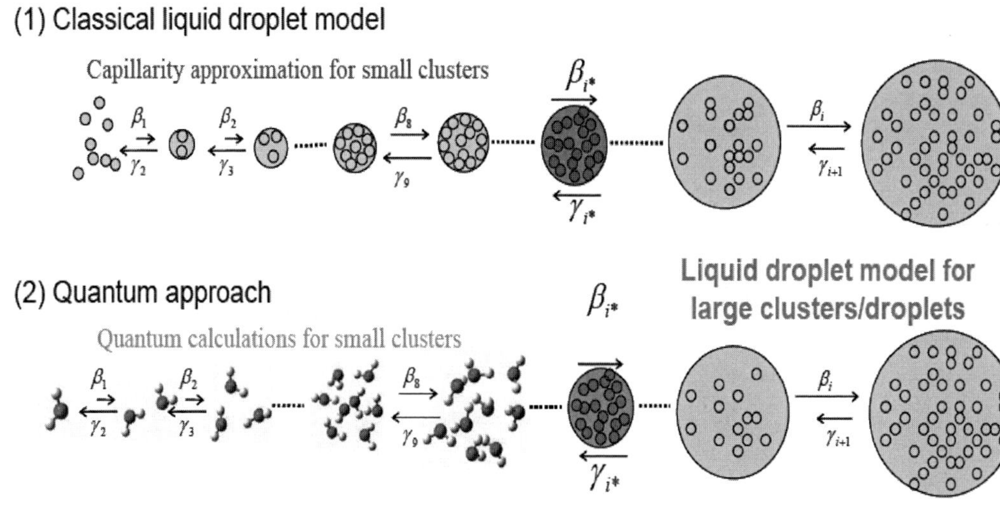

Figure 3. A comparison of the nucleation calculations based on the conventional bulk liquid approximation and microscopic/quantum approach.

The major weakness of the CNT is that the treatment of the thermodynamics of initial cluster growth steps with the bulk liquid approximation, which works well for large clusters with less distinct internal structure, is no longer valid. In contrast to continuum CNT, quantum theory accounts for the actual chemical identites controlled by the wave function, and is able to predict the cluster thermochemistry with quite high degree of confidence (Fig. 3). Molecular-based methods [45-50] that are also often imposed as a viable alternative to the CNT do not explicitly use the properties of bulk liquid to compute nucleation rates. However, they are semi-classical in nature because interactions between vapor monomers and *i*-mers are usually approximated using empirical (e.g. TIP4P, TIP5P, Dang–Chang) potentials parameterized in such a way as to reproduce the properties of bulk liquid. The application of empirical potentials parameterized to bulk liquid leads to significant problems in describing the initial cluster formation steps and resulting nucleation rates.

The application of the computational quantum methods to the thermochemistry of clusters composed of atmospheric species has been shown to be very promising. The hydration level of sulfuric acid monomers has a strong effect on the stability or evaporation coefficient of neutral H_2SO_4-H_2O clusters [15]. The hydration thermodynamics of sulfuric acid monomers obtained from density functional theory (DFT) with several different methods and basis sets [25, 51] has been used to constrain and improve the existing binary as well as ion-mediated nucleation models [15, 42]. The quantum approach has also been applied to the ion-induced nucleation in aqueous systems and the classical problem of the sign preference established back 1897 in pioneering Wilson's experiments in the cloud chamber has been solved [22,52]. Du et al. [53] applied the quantum-mechanical correction to nucleation in water vapors and achieved a significant progress (see Fig.4) in predicting nucleation rates at whole range of experimental conditions.

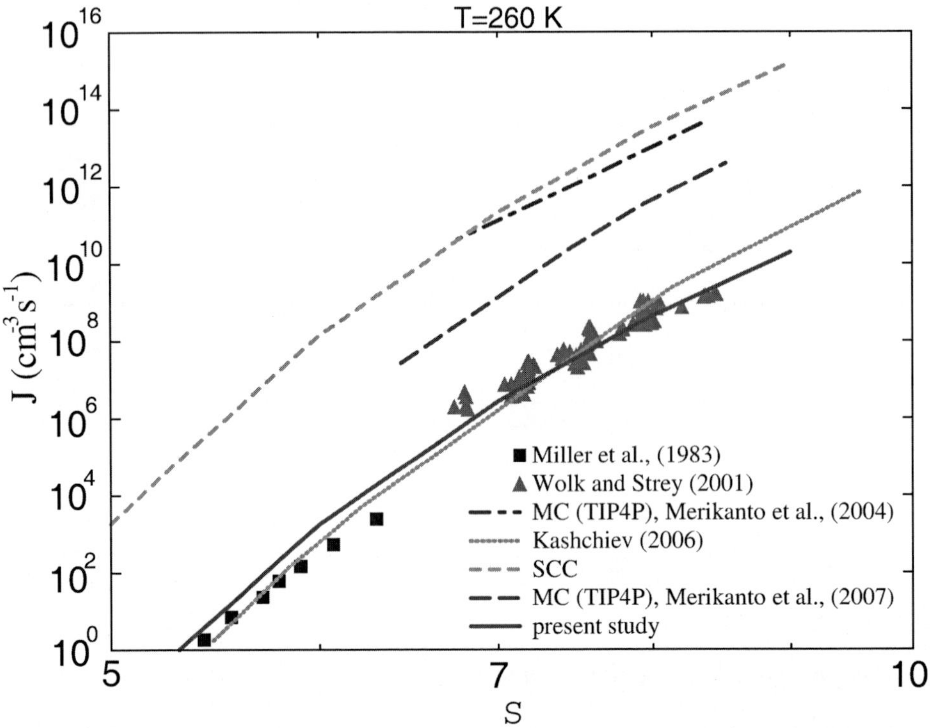

Figure 4. Comparison of nucleation rates obtained using the CNT constrained by quantum calculations with experiments and other theoretical studies. (From *Du et al.* [53])

The quantum chemical calculations of cluster free energies needed to derive/constrain the nucleation thermodynamics are based on the optimized molecule/cluster geometries and calculated vibrational spectrum. The present paper is focused on the following aspects of the thermochemistry of clusters consisting of atmospheric species:

a) comparison of calculated and experimental optimized geometries and vibrational spectrums,
b) formation free energies for different classes of atmospheric pre-nucleation clusters,
c) impacts of the cluster thermochemistry on atmospheric nucleation.

In this work, we review the existing studies and report new ones, discuss the thermochemical data for different atmospheric cluster obtained using quantum methods and their implications, and outline the future developments towards new atmospheric nucleation theory, which will be based on the thermochemistry obtained from "first principles". Kinetic nucleation models taking advantage of clustering thermochemistry and their applications in the atmosphere are also discussed.

2. Overview of Computational Quantum Methods

The most common computational quantum methods [54-55] can be divided into two distinct categories: (a) *ab initio*; (b) density functional theory (DFT). In principle, all knowledge about a system can be obtained from the wave function, which is obtained by solving the Schrödinger equation for a complete multi-electron system. The term *ab initio* literally means "from first principles"; however, it does not imply that the Schrödinger equation is solved exactly. The main idea of *ab initio* is to select a method that can lead to a reasonable approximation to the solution of the Schrödinger equation and then to choose a basis set that will implement this method properly. The original Hartree-Fock method is incapable of giving a correct solution to the Schrödinger equation even if a very large and flexible basis set is employed (Hartree–Fock limit) because in the reality electrons are not paired up in the way that the original Hartree-Fock theory suggests. The difference in energy between the exact result and Hartree-Fock limit is called the "correlation energy". There are two classes of methods dealing with the "correlation problem" – variation methods (e.g. CID, CISD, full CI) and perturbation methods (e.g. MP*n*, QCISD). At the present time, the Møller and Plesset second order method (MP2) is the most commonly used *ab initio* method. In addition to pure *ab initio* methods, there exist compound *ab initio* methods (e.g. CBS-APNO, G2, G3, G2MP2, G3MP2) involving a variety of basis sets and advanced theoretical corrections to provide good energy estimates. The main idea of the DFT is to describe an interacting electronic system via its density - a one-body quantity - instead of its many-body wave function, providing in this way accurate results comparable in quality with *ab initio* calculations at lower computational costs. DFT is neither a Hartree-Fock method nor a post-Hartree-Fock method because it is constructed in a different way and the resulting orbitals are often referred to as "Kohn-Sham" orbitals. The most common density functionals have been derived by careful comparison with experiments. At the present time, numbers of atoms in a complex/cluster that can be treated by *ab initio* (MP2) and DFT are~ 40 and ~150. In order to extend the size range, hybrid (e.g. quantum mechanics-molecular mechanics (QM/MM)) methods [56-58] can be applied. At the present time DFT is the most common method used to study atmospheric clusters/complexes due to its reasonable predictivity and low computational costs.

3. Comparison of Experimental and Theoretical Geometries of Atmospheric Molecules and Clusters

3.1. Molecular Geometries

The determination of equilibrium geometries is essential part of free energy computations. Geometries of most common atmospheric molecules are relatively well studied. Both theoretical calculations and experimental data are available for water H_2O, ammonia NH_3, sulfuric acid H_2SO_4, formic acid CH_2O_2 and acetic acid $C_2H_4O_2$. Fig. 5 and Tables 1-4 present theoretical geometries of the aforementioned species. Geometries calculated in the present study at *ab initio* MP2/6-311++G(3df,3pd) and DFT B3LYP/6-

311++G(3df,3pd) and PW91PW91/6-311++G(3df,3pd) levels of theory have been compared with experimental data.

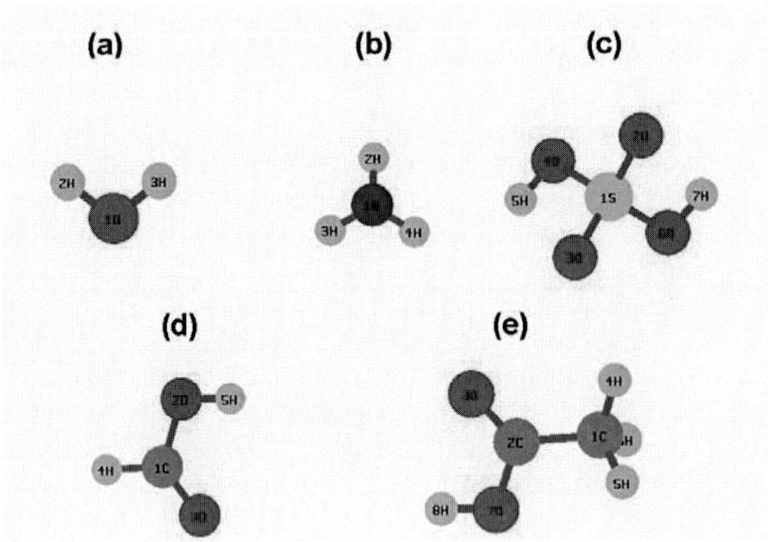

Figure 5. Geometries of (a) water H2O; (b) ammonia NH3; (c) sulphuric acid H2SO4; (d) formic acid CH2O2; and (e) acetic acid C2H4O2.

Table 1. Comparison of theoretical geometries of water and ammonia with experimental data.

parameter		H_2O		
	B3LYP	PW91	MP2	exp.[a]
R(1,2)	0.961	0.9685	0.958	0.958±0.001
A(1,2,3)	105.0485	104.233	104.18	104.477±0.5
parameter		NH_3		exp.[b]
R(1,2)	1.013	1.021	1.011	1.012
A(1,2,3)	107.12	106.34	106.95	106.67

a Reference [59]. b Reference [60].

Table 2. Comparison of theoretical geometries of the formic acid with experimental data

	formic acid			
parameter	exp[a]	MP2	PW91	B3LYP
rCO	1.202	1.202	1.205	1.196
rCO	1.343	1.343	1.352	1.343
rCH	1.097	1.092	1.104	1.096
rOH	0.972	0.968	0.979	0.97
aOCO	124.9	125.11	125.25	125.21
aHCO	124.1	125.16	125.37	125.12
aHOC	106.3	106.47	106.7	107.9

a Reference [60].

Table 3 . Comparison of theoretical geometries of the acetic acid with experimental data.

		acetic acid		
parameter	exp.[a]	PW91	B3LYP	MP2
rCC	1.517	1.503	1.502	1.498
rCO	1.212	1.212	1.202	1.207
rCO	1.361	1.365	1.356	1.354
rCH	1.1	1.1	1.085; 1.09	1.083
aCCO	126.6	126.2	126.05	126.24
aCCO	110.6	111.4	111.62	111.21
aOCO	123	122.4	122.33	122.54

a Reference [61].

Table 4 . Comparison of theoretical H2SO4 geometries with experimental data.

parameter	MP2	MP2/TZPa	B3LYP	B3LYPb	B3LYPc	B3LYPd	PW91	exp.e
R(O1-H2)	0.968	0.97	0.968	0.975	0.975	0.968	0.976	0.97±0.01
R(O2-S)	1.5835	1.627	1.592	1.634	1.643;1.621	1.609	1.612	1.574±0.01
R(O4-2S)	1.419	1.446	1.421	1.456	1.464	1.429	1.433	1.422±0.01
(H1-O2-S)	108.32	107.59	109.48			108.3	107.93	108.5±1.5
(O2-S-O4)	108.82	105.18	105.56				105.16	106.4±0.5
(O3-S-O4)	108.20	108.5	108.78				108.8	108.6±0.5
(O4-S-O5)	124.33	125.62	123.96				124.47	123.3±1
(O2-S-O3)	102.05	101.33	102.15				102.18	101.3±1

a Reference [62] b-c Reference [63]. d Reference [64]. e Reference [65].

As may be seen from Tables 1-4, both ab initio MP2 and DFT PW91PW91 and B3LYP methods accurately predict molecular geometries of common atmospheric species. MP2 predictions agree best with the experimental data; however, the performance of both DFT methods is very close to that of MP2. Both MP2 and DFT methods predict bonding lengths and angles with accuracy of < 0.5 % on average, which is close to typical experimental uncertainties.

3.2. Geometries of clusters consisting of atmospheric nucleation precursors

While geometries of molecules of common atmospheric species have been studied in detail in the past, the experimental geometrical properties are available only for a limited number of clusters relevant to atmospheric nucleation. Tables 5-7 present the comparison of theoretical and experimental data for the water dimer $(H_2O)_2$, water trimer $(H_2O)_3$ and monohydrate of the sulfuric acid $(H_2SO_4)(H_2O)$.

Table 5. Comparison of theoretical bonding lengths (angstroms) and angles (degrees) for water dimer $(H_2O)_2$ with experimental data.

	OO	OH	OHO
MP2/aug-cc-pvDz[a]	2.917	1.951	171.4
MP2/aug-cc-pvTz[a]	2.907	1.946	171.7
MP2/aug-cc-pvQz[a]	2.903	1.943	171.8
CCSD(T)/TZ2P[b]	2.914		
CCSD(T)(Full)/TZ2P(d,f) + diff[c]	2.909		
CCSD(T)(Full)/(IO275→∞)[d]	2.912		
PBEPBE/aug-cc-pvTz	2.896	1.922	171.93
PBEPBE/6-311++G(3df,3pd)	2.895	1.92	171.92
PW91PW91/aug-cc-pvTz	2.882	1.9	171.41
PW91PW91/6-311++G(3df,3pd)	2.878	1.905	171.97
Expt.	2.952[e], 2.946[f]		

Reference [66]; [b]Reference [67]; [c] Reference [68] ; [d].Reference [69]; [e] Reference [70]; [f] Reference [71]; [g] Reference [72]; [h] Reference [73].

Table 6. Comparison of theoretical bonding lengths (angstroms) and angles (degrees) for water trimer $(H_2O)_3$ with experimental data.

	$r(O_1\text{-}O_4)$	$r(O_4H_9)$	$A(H_5O_4H_6)$	$A(H_9O_7O_4)$
MP2/aug-cc-pvTz[a]	2.785	1.8925	105.28	19.04
MP2/ aug-cc-pvQz[a]	2.779	1.8902	105.45	19.18
PBE/ aug-cc-pvTz	2.765	1.868	105.48	19.36
PBEPBE/6-311++G(3df,3pd)	2.764	1.86	105.41	19.41
PW91PW91/ aug-cc-pvTz	2.751	1.85	105.68	19.28
PW91PW91/6-311++G(3df,3pd)	2.753	1.85	105.56	18.93
expt. [b]	2.845			

[a] Reference [74] ; [b] Reference [75]

As may be seen from Tables 5 and 6, the comparison of experimental and theoretical geometries of water dimer and water trimer, the predictions of ab initio and DFT methods are very close. However, neither ab initio nor DFT perfectly predict the few available experimental bonding lengths. Further experimental work is needed for the through assessment of the performance of computational quantum methods in the case of water clusters. The comparison of the structural data given in Table 7 shows that PW91PW91, B3LYP and MP2 with 6-311++G(3df,3pd) basis set accurately predict the geometry of the sulfuric acid monohydrate. A considerable difference between B3LYP and MP2 predictions and experimental data is observed only in the case of H9-O4 bond controlling the cluster stability. While B3LYP and MP2 over-estimate the H9-O4 bond length by ~ 5-6 %, PW91PW91 predicts this bonding length in excellent agreement with the experiment.

Table 7. A comparison of (H2SO4)(H2O) geometry obtained using ab initio and DFT methods with experiments. Abbreviations MP2/6-311++G(3df,3pd) and PW91/6-311++G(3df,3pd), B3LYP/D95++, B3LYP/6-311++G(2d,2p), PW91/DNP, PW91/TZP and exp refer to the Nadykto and Yu (2007) [25]; Re et al., 1999 [63], Bandy and Ianni (1998) [64] , Ding et al., (2003) [76] , Al Natsheh et al. (2004) [77] and Fiacco et al. (2003) [78] respectively.

	MP2/ (3df,3pd)	B3LYP/ D95++	B3LYP/ (2d,2p)	PW91/ DNP	PW91/ TZP	PW91/ (3df,3pd)	exp
R(H1-O2)	0.967	0.975		0.978	0.98	0.976	0.95
R(O2-S)	1.579	1.636	1.611	1.634	1.63	1.614	1.578
R(O3-S)	1.557	1.603		1.604	1.59	1.576	1.567
R(O4-S)	1.43	1.466	1.439	1.463	1.46	1.446	1.464
R(O5-S)	1.42	1.458	1.43	1.453	1.44	1.435	1.41
R(H6-O3)	0.997	1.009	0.999	1.023	1.03	1.022	1.04
R(H6-O8)	1.654	1.651	1.682	1.621	1.61	1.61	1.645
R(H9-O8)	0.976					0.98	0.98
R(H10-O8)	0.96					0.97	0.98
R(H9-O4)	2.15	2.23	2.207	2.141		2.05	2.04
A(H1-O2-S)	108.23					107.82	108.5
A(O2-S-O4)	104.37					104.15	104.71
A(O3-S-O4)	109.18					109.18	106.71
A(O3-S-O5)	107.33					107.8	106.7
A(H6-O3-S)	108.23					108.15	108.6
A(O4-S-O5)	122.7					122.37	123.3
A(O2-S-O3)	102.93					103.2	101.8
A(H9-O8-H10)	105.74					105.73	107

5. Comparison of Experimental and Theoretical Vibrational Spectrums of Atmospheric Molecules and Clusters

Accurate calculations of the vibrational spectra are critically important for the free energy calculations. The vibrational spectra can be computed using either the most common Rigid Rotor Harmonic Approximation (RRHOA), or different anharmonic approximations. Rigorous theoretical treatment of anharmonic effects is important indeed. However, drawing a meaningful conclusion about the fidelity of anharmonic approximations is impossible without referring to a specific electronic structure method and comparing results produced by this method with reliable spectroscopic experiments.

In the present study, the vibrational spectra calculated using DFT PW91PW91/6-311++G(3df,3pd) and B3LYP/6-311++G(3df,3pd) in standard harmonic (RRHOA) approximation have been compared to experimental data and anharmonic theoretical studies.

4.1. Vibrational spectrums of atmospheric molecules

Tables 8-10 present the comparison of theoretical spectrums of ammonia, water, formic acid and acetic acids calculated using DFT at B3LYP/6-311++G(3df,3pd) and PW91PW91/6-311++G(3df,3pd) levels. As may be seen from Tables 8-10, both DFT methods agree well with the experimental data. The degree of agreement of PW91PW91/6-311++G(3df,3pd) with the experiments is a bit higher than that of B3LYP/6-311++G(3df,3pd). The vibrational spectrums of simplest atmospheric molecules have been studied by a number of methods, and the complete survey on their spectrums can be found in NIST WebBook [79].

Table 8. Comparison of theoretical vibrational spectrums of ammonia and water with experimental data.

	NH_3				H_2O	
mode	B3LYP	PW91	exp.[a]	B3LYP	PW91	exp.[a]
1	3594	3512.5	3444	3914.8	3820.1	3756
2	3475	3391.1	3337	3815.7	3718.5	3657
3	1661	1619.2	1627	1626.0	1593.6	1595
4	1028	1021.1	950			

a Reference s [80], [59].

Table 9. Comparison of theoretical vibrational spectrum of formic acid with experimental data.

	formic	**acid**	
Symmetry	**exp.**[a]	**PW91**	**B3LYP**
A'	3570	3612	3732
A'	2943	2984	3052
A'	1770	1766	1819
A'	1387	1353	1403
A'	1229	1260	1299
A'	1105	1083	1125
A'	625	678	631
A''	1033	1003	1053
A''	638	607	677

a Reference [60].

Table 11 presents the frequency data for sulfuric acid monomer (H_2SO_4). The spectroscopy of (H_2SO_4), the key nucleation precursor, has been discussed in the literature. Miller et al. [62] noticed that DFT PW91PW91/TZP [77] agree well with the experimental data and anharmonic CC-VSCF MP2/TZP results. The comparison of the DFT results presented in Table 11 shows that the basis set dependency of spectra predicted by PW91 PW91 is small, especially in the low-frequencies region. Low–lying frequencies of (H_2SO_4) predicted by $PW91_{xc}$ and anharmonic MP2/aug-cc-pvdz preditcions are in better agreement with the experimental data than those given by CC-VSCF MP2/TZP.

Table 10. Comparison of theoretical vibrational spectrum of acetic acid with experimental data.

			acetic acid	
Mode Number	**Symmetry**	**exp.[a]**	**PW91**	**B3LYP**
1	A'	3583	3639	3756
2	A'	3051	3118	3164
3	A'	2944	3001	3054
4	A'	1788	1765	1820
5	A'	1430	1424	1473
6	A'	1382	1355	1409
7	A'	1264	1294	1337
8	A'	1182	1161	1203
9	A'	989	960	1000
10	A'	847	841	859
11	A'	657	567	586
12	A'		412	425
13	A"	2996	3064	3112
14	A"	1430	1428	1478
15	A"	1048	1026	1070
16	A"	642	665	665.4
17	A"	534	532	547
18	A"	93	60	77

a Reference [80].

Table 11. Experimental and theoretical frequencies of (H_2SO_4) (cm^{-1})

	CC-VSCF MP2/ TZP[a]	PW91/TZP[b]	PW91/a-vtz[c]	PW91/ (3df,3pd)[c]	exp.1[a]	exp.2[b]
1	3590	3632	3646	3663	3609	3567
2	3500	3627	3642	3659		3563
3	1434	1426	1387	1423	1465	1452
4	1182	1160	1141	1176	1220	1216
5	1128	1153	1125	1126	1157	1157
6	1118	1138	1114	1110	1138	1136
7	821	801	793	816	891	882
8	779	745	737	767	834	831
9	525	513	500	521	568	558
10	519	499	490	508	550	548
11	480	463	450	467		506
12	489	413	401	415	422	422
13	379	341	338	351	281	379
14	459	321	300	320		288
15	380	221	221	238	215	224

[a] anharmonic CC-VSCF MP2/TZP calculations and compilation of experimental data by *Milller et al.* [62]; [b] harmonic calculations and compilation of experimental data by Al *Natsheh et al.* [77] , [c]*Nadykto et al.*[81].

Tables 12 and 13 present a comparison of the computed fundamentals of (H_3O^+) and ($H_5O_2^+$) clusters with the experimental data. While CC-VSCF MP2/TZP accurately predicts free OH-stretching of (H_3O^+), MP2(full)/aug-cc-pvtz predicts accurately free OH-stretching in both (H_3O^+) and ($H_5O_2^+$) spectra. Unlike VSCF and CC-VSCF, both MP2(full)/aug-cc-pvtz and PW91 PW91 predicts the degeneration of vibrational frequencies in both (H_3O^+) and ($H_5O_2^+$) spectra. The agreement of frequencies obtained at different levels of theory with experimental data for (H_3O^+) is reasonable, with the exception for mode 6, for which the deviation of PW91 PW91, CC-VSCF –MP2/TZP and MP2(full)/aug-cc-pvtz from the experimental data is excessively large. In the case of ($H_5O_2^+$), PW91PW91 predicts OH-stretching better than both CC-VSCF MP2/TZP and VSCF MP2/TZP. CC-VSCF MP2/TZP and $PW91_{xc}$ spectra of ($H_5O_2^+$) are in good agreement, with the exception for low-lying modes 14 and 15. These modes predicted by PW91 PW91 are much closer to MP2(full)/aug-cc-pvtz results than those given by CC-VSCF MP2/TZP.

Table 12. Experimental and theoretical frequencies of (H3O+) (cm-1)

	VSCF[a]	CC-VSCF[a]	**MP2(full) /a-vtz**[b]	PW91*	PW91**	Exp[a]
1	3582	3509	3517	3568	3575	3519
2	3532	3478	3517	3568	3575	3519
3	3535	3454	3431	3483	3488	3390
4	1631	1621	1630	1641	1639	1626
5	1627	1618	1630	1641	1639	1626
6	573	432	717	849	832	526

[a] VSCF and CC-VSCF calculations at MP2/TZP level and compilation of experimental data [82] ;
[b]anhamonic results (present study).

Table 13. Experimental and theoretical frequencies of ($H_5O_2^+$) (cm^{-1})

	VSCF[a]	CC-VSCF[a]	**MP2(full)/ a-vtz**[b]	PW91*	PW91**	exp[a]
1	3634	3579	3671	3719	3730	3684
2	3632	3577	3671	3719	3730	3684
3	3669	3593	3595	3639	3650	3609
4	3623	3518	3587	3631	3642	3609
5	1802	1853	1835	1717	1716	
6	1659	1662	1458	1650	1648	
7	1494	1494	1409	1488	1486	
8	1451	1442	1333	1417	1424	
9	1230	1209	1416	1122	1111	
10	625	599	241	595	594	
11	579	559	456	532	527	
12	568	529	311	522	518	
13	542	494	343	446	442	
14	612	501	342	416	411	
15	428	383	148	161	167	

[a] VSCF and CC-VSCF calculations at MP2/TZP level and compilation of experimental data [82] [b]anharmonic result (present study)

Vibrational fundamentals of HNO_x (x=2,3,4) and nitric acid hydrate $(HNO_3)(H_2O)$ are shown in Tables 14-16. The very accurate calculations of the vibrational spectrum of HONO have been presented by Richter et al. [83]. The level of the approximations used to treat vibrations in [83] is so high that their results are nearly numerically exact for the potential function used. PW91 PW91 averages small (~2%) deviation from the experimental data and gives scaling factors weighted to high and low frequencies of 0.988 and 1.009, respectively. Low-lying modes 5 and 6 predicted by PW91 PW91 are in excellent agreement with the experiment within 11 and 4 cm^{-1}„ respectively. As seen from Table 14 and 15, which show the fundamental transitions of HNO_3 and HNO_4, the only advantage of anharmonic DPT2 /MP2/TZP, CC-VSCF MP2/TZP and MP2/TZP over PW91 PW91 is the prediction of the OH-stretching. Contrast to DPT2 /MP2/TZP, CC-VSCF that are unable to predict NOH torsion (mode 9) of HNO_3 and HOON torsion (mode 10) of HNO_4 (Table 16), PW91 PW91 reproduces these modes in agreement with the experimental data. PW91PW91 also predicts more accurate NO_2 asymmetric stretch (mode 2) than all the aforementioned methods. While OH-stretch of HNO_3 (mode 3) given by PW91 PW91 agree with the experiment within 49 cm^{-1}, the deviation of DPT2 /MP2/TZP, CC-VSCF MP2/TZP and MP2/TZP from the experimental data exceeds 100 cm^{-1}.

Table 14. Experimental and theoretical frequencies of *trans* (HONO) (cm^{-1})

	MCTDH/FD[a]	PW91*	PW91**	exp
1	3587	3643	3656	3591
2	1698	1723	1732	1700
3	1269	1223	1229	1263
4	796	761	765	790
5	601	584	589	595
6	538	538	545	543

[a] theoretical results and compilation of experimental data [83]

Table 15. Experimental and theoretical frequencies of (HNO_3) (cm^{-1})

	DPT2 /MP2/TZP[a]	CC-VSCF /MP2/TZP [a]	MP2/TZP [b]	PW91*	PW91**	Exp[a]
1	3547	3547	3597	3609	3621	3550
2	1840	1840	1841	1705	1763	1710
3	1263	1261	1323	1293	1330	1326
4	1296	1296	1296	1255	1270	1304
5	876	876	867	838	844	879
6	782	782	751	746	743	763
7	642	643	641	590	604	647
8	572	573	579	548	553	580

[a] anharmonic calculations at DPT2 /MP2/TZP and CC-VSCF/MP2/TZP level and compilation of experimental data [82]; [b] [84]

Table 16. Experimental and theoretical frequencies of (HNO_4) (cm^{-1})

	CC-VSCF/MP2/TZP	DPT2-VSCF/MP2/TZP	PW91*	PW91**	exp
1	3545	3545	3580	3590	3540
2	1915	1915	1760	1774	1728
3	1342	1343	1375	1382	1397
4	1303	1303	1310	1319	1304
5	928	928	951	957	945
6	787	788	779	779	803
7	702	699	698	703	722
8	622	626	616	620	654
9	392	388	390	391	483
10			369	377	340
11	310	299	274	279	
12	152	127	155	158	145

[a] anharmonic calculations at DPT2 and CC-VSCF/MP2/TZP level and compilation of experimental data [82]

4.2. Vibrational spectrums of molecular clusters

The vibrational spectra of atmospheric $(H_2O)_2$, $(H_2O)_3$, $(H_2SO_4)(H_2O)$, $(HNO_3)(H_2O)$, water-methanol and methanol-water complexes are presented in Tables 17-22.

As seen from Tables 17 and 18, the difference between spectra of $(H_2O)_2$ obtained at different levels of theory is quite large. While the difference in high (> ~3000 cm^{-1}) frequencies is small, it increases dramatically as the wavenumber decreases. In the case of low-lying modes 11 and 12, the deviation between the theoretical predictions (400 and ~ 500 cm^{-1}, respectively) is exorbitant. Ab initio anharmonic mean-field grid-based methods and methods based on higher energy derivatives produce slightly better values of the OH-stretching modes than PW91PW91. However, PW91PW91 obviously over-performs VSCF and CC-VSCF methods in the critically important low- frequencies (< 200 cm^{-1}) region. While the difference between PW91PW91 results and experiments does not exceed 50-60 cm^{-1}, both VSCF MP2 and CC-VSCF MP2 deviate from experimental data > 300 cm^{-1}. The large deviation in low-lying frequencies controlling the thermochemical stability of the cluster may cause significant uncertainties in the dimerization enthalpies, entropies and free energies. The anharmnonic frequencies for $(H_2O)_2$ obtained at MP2(full)/a-vtz and MP2(full)/a-vqz levels agree better with the experimental data than those given by other methods. However, their predictions in the low frequencies region are affected by a pronounced basis set dependency.

In the case of water trimer, the agreement among model predictions obtained at different levels of theory is very good. Mode 11 is the only exception. While MP2(full)/aug-cc-pvqz perfectly matches the experiment, PW91PW91, VSCF and CC-VSCF-MP2/TZP predictions deviate from the experimental value by 159, 280 and 272 cm^{-1} , respectively. No experimental data for low-lying frequencies of water trimer is currently available, and, thus, drawing a meaningful conclusion about the performance of different methods in the low-frequencies region is impossible at the present time. Another important detail is that the difference between experimental fundamentals of both water dimer and trimer is quite large. This is not surprising because experimentally determined frequencies are sensitive to matrix effects, and only few data points obtained in gas-phase experiments have been reported.

Table 17. Experimental and theoretical frequencies of $(H_2O)_2$ (cm^{-1})

	VPT2 /MP2/a-vtz[a]	MP2(full)/a-vqz [b]	VSCF MP2/TZP[c]	CC-VSCF/MP2/TZP[c]	Exp[c]	Exp[b]
1	3753	3776	3763	3724	3745	3745
2	3722	3769	3733	3745	3730	3735
3	3615	3671	3768	3647	3600	3660
4	3554	3597	3560	3565	3530	3601
5	1592	1600	1612	1605	1601	1616
6	1580	1590	1565	1564	1599	1599
7	502	517	769	732		523
8	310	308	550	521		311
9	138	144	451	409		143
10	114	108	259	147	150	108
11	113	113	414	309		103
12	60	65	545	419		>88

[a] VPT2 calculations at MP2/aug-cc-pvtz level [85]; [b] anharmonic calculations at MP2/aug-cc-pvqz and compilation of experimental data [86]; [c] VSCF and CC-VSCF anahrmonic calculations at MP2/TZP level and compilation of experimental data [87]

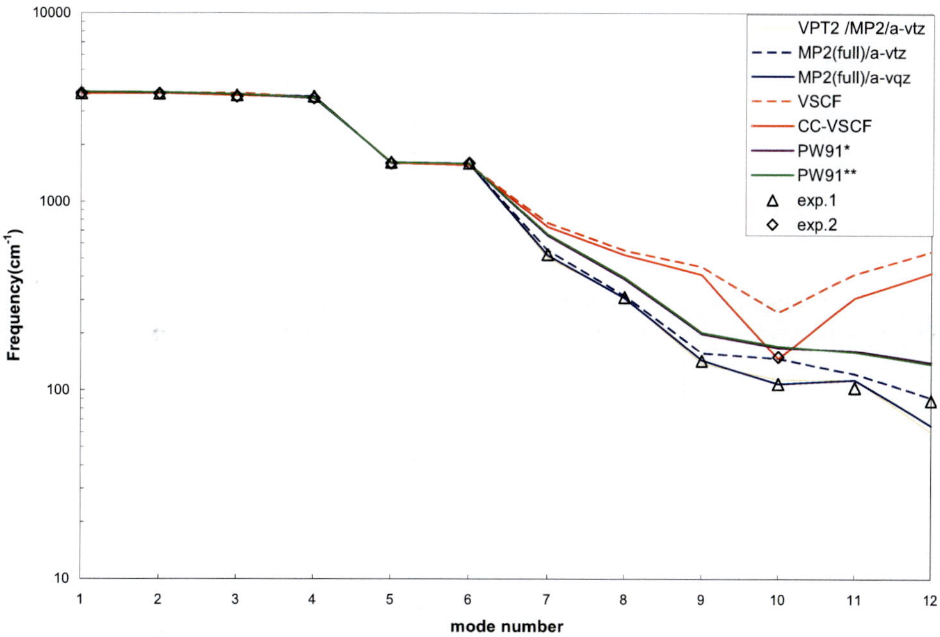

Figure 6. Comparison of theoretical results for the vibrational spectrum of $(H_2O)_2$ with experimental data.

Table 18. Experimental and theoretical frequencies of $(H_2O)_3$ (cm^{-1})

	VSCF MP2/TZP[a]	CC-VSCF/MP2/TZP[a]	**MP2(full)/a-vtz[b]**	PW91	PW91	exp.1[a]	exp.2[b]
1	3714	3725	3732	3772	3781	3703	
2	3709	3723	3730	3770	3779		3717, 3726
4	3469	3429	3500	3402	3406	3400	3530, 3533
6	3319	3283	3422	3296	3301	3340	3518
7	1633	1623	1593	1624	1626	1632	
8	1601	1595	1595	1603	1607	1620	1608
9	1599	1592	1595	1601	1603	1602	
11	849	841	569	721	728		569

[a]; anharmonic VSCF and CC-VSCF calculations at MP2/TZP level and compilation of experimental data [87] [b]anharmonic calculations at MP2/aug-cc-pvtz and compilation of experimental data [86]

Table 19. Experimental and theoretical frequencies of $(H_2SO_4)(H_2O)$ (cm^{-1})

	CC-VSCF MP2/TZP[a]	MP2/TZP[a]	PW91/TZP[b]	PW91/a-vtz[c]	PW91/ (3d,3pd)[c]	exp.[a]
1	3738	3986	3730	3757	3770	3745
2	3593	3841	3638	3650	3668	3640
3	3628	3822	3508	3562	3576	3573
5	1573	1619	1607	1593	1594	1600
6	1436	1468	1464	1437	1444	1500
8	1172	1187	1140	1128	1155	1205
10	868	884	917	882	893	887
11	834	820	856	844	871	834
19	529	349	349	337	349	554

[a] anharmonic CC-VSCF MP2/TZP calculations and compilation of experimental data *Millller et al.,,* (2005) [62] ; (2005)[c] harmonic calculations and compilation of experimental data by Al *Natsheh et al.* (2004) [77] , [c]*Nadykto et al.,* (2007)[81].

Table 20. Experimental and theoretical frequencies of $(HNO_3)(H_2O)$ (cm^{-1})

	DPT2 /MP2/TZP[a]	CC-VSCF /MP2/TZP [a]	**MP2/TZP** [b]	PW91*	PW91**	Exp[a]
3	3146	3146	3270	2972	2976	3025
4	1809	1809	1811	1686	1696	1694
6	1466	1466	1443	1442	1451	1412
7	1309	1309	1305	1261	1271	1308
8	932	927	924	910	923	931
10	764	763	759	760	763	770

[a] anharmonic calculations at DPT2 and CC-VSCF/MP2/TZP level and compilation of experimental data [82];

[b] Reference [84]

As seen from Table 20, both harmonic DFT and anharmonic *ab initio* results are consistent with the experimental data. The only difference is that NO_2 asymmetric stretch (mode 4) of $(HNO_3)(H_2O)$ has been predicted by PW91PW91 is much close to the experimental value that those given by DPT2 /MP2/TZP, CC-VSCF MP2/TZP and MP2/TZP.

Table 21. Experimental and theoretical frequencies (cm^{-1}) and IR intensities (km/mol) for methanol-water complex

	frequencies						intensitie s				
	VSCF a	CC-VSCF[a]	PW91 *	PW91 **	PW91 /TZP	exp	VSCF[a]	PW91*	PW91* *	PW91 /TZP	exp
1	3705	3661	3798	3811	3774	3714	78	73	70	80	44
2	3689	3645	3697	3712	3682	3627	4	6	6	12	6
3	3572	3595	3537	3551	3523	3536	347	497	496	501	100
4	3020	3007	3032	3040	3034	2982	42	38	38	33	8
6	2930	2891	2912	2916	2909	2835	68	85	87	75	12
7	1617	1625	1595	1594	1604	1601	77	68	69	78	23
8	1524	1523	1461	1462	1465	1475	2	3	3	3	1
9	1515	1516	1446	1445	1449	1464	1	2	2	2	2
10	1497	1495	1423	1427	1427	1448	12	3	3	5	1
11	1428	1420	1388	1392	1397	1380	32	35	34	35	7
13	1135	1128	1083	1086	1085	1103	21	4	4	6	3
14	1067	1064	1034	1040	1031	1048	94	104	104	102	30

[a] VSCF and CC-VSCF calculations at MP2/TZP level and compilation of experimental data [87]

Table 22. Experimental and theoretical frequencies (cm^{-1}) and IR intensities (km/mol) for water - methanol complex

	frequencies						intensities				
	VSCF[a]	CC-VSCF[a]	PW91*	PW91**	PW91 /TZP	exp	VSCF[a]	PW91*	PW91**	PW91/ TZP	exp
1	3742	3792	3775	3789	3763	3704	97	65	61	69	40
2	3683	3713	3722	3737	3701	3663	56	36	36	43	17
3	3530	3501	3489	3509	3480	3539	321	512	500	535	180
4	2975	2935	3011	3017	3069	3018	59	34	34	11	13
6	2955	2907	2953	2956	2950	2974	54	54	55	46	28
7	1593	1582	1612	1614	1621	1614	75	38	36	44	45
10	1503	1500	1427	1431	1427	1429	8	3	3	3	
13	1104	1093	1048	1049	1049	1078	30	5	4	7	4
14	1039	1037	1002	1007	1003	1032	148	119	122	122	100

[a] VSCF and CC-VSCF calculations at MP2/TZP level and compilation of experimental data [87]

Tables 21 and 22 present the comparison of theory and experiments for methanol-water and water–methanol complexes, respectively. The comparison shows a similar performance of CC-VSCF MP2/TZP and PW91 PW91methods. While CC-VSCF MP2/TZP predicts OH-stretching of the water-methanol complex better than PW91 PW91, PW91 PW91 over-performs CC-VSCF in the region of low frequencies. The average deviation of the CC-VSCF and PW91 PW91 methods from the experimental data for methanol-water complex is 2.1 and 1.53 %, respectively that indicates good agreement of both methods with the experiments. In the case of water-methanol complex, the CC-VSCF MP2/TZP and PW91 PW91 predictions of OH stretching are close. Both methods predict the vibrational spectrum of the water-methanol in agreement with the experimental data within 20-80 cm^{-1}. The average deviation of the CC-VSCF MP2/TZP and PW91 PW91 methods from the experimental data is nearly identical. Both methods predict the lower frequencies of water-methanol and methanol-water complexes in a good agreement with the experimental data. In addition, both VSCF and PW91 PW91 IR intensities are in qualitative agreement with the experiment.

4.3. Analysis

The comparison of the theoretical harmonic spectrums computed using the DFT PW91PW91, which is the most common method used in atmospheric studies, with experimental data and anharmonic *ab initio* studies shows that the quality of the harmonic DFT spectrums is close to that of more demanding *ab intio* method with anharmonic corrections included. This is a valuable argument in favour of the application of low-cost harmonic approximation for DFT calculations of free energies. Due to excessively large computational requirements imposed by anharmonic calculations, the amount of information needed for a thorough evaluation of different anharmonic methods is very limited. Another important detail is that the deviation in model predictions obtained using different anharmonic corrections is excessively large. In the case of water dimer, which is the simplest hydrogen-bonded complex, the deviation in low-laying vibrational frequencies predicted using different anharmonic approximations exceeds [52] 400 cm^{-1} or 500%. In addition, there is strong method and considerable basis set dependences of the *ab initio* anharmonic results whose costs are prohibitively high, when large basis sets are applied [86]. Considerable uncertainties in the few available experimental low-laying frequencies responsible for the vibrational contribution to the Gibbs free energies, sensitivity of the experimental spectra to the matrix and absence of a complete list of gas-phase frequencies for simplest hydrogen bonded systems such as e.g. water dimer further complicate this issue [86] This suggests that a substantial amount of further work is needed to draw a meaningful about applicability of different anharmonic methods to thermochemistry of atmospheric pre-nucleation clusters/complexes.

Application Of Quantum Methods To Study The Thermochemistry Of Atmospheric Nucleation Species

At the present time, quantum computations are quite demanding, and, thus, the size of a system that can be treated using computational quantum methods is limited. However, the nucleation investigation can benefit from using the thermochemical data obtained from "first principles" due to : (a) High sensitivity of nucleation rates to free energies of the formation of initial (or pre-critical) clusters; (b) Relatively small (~10) number of H_2SO_4 molecules in critical clusters formed during typical atmospheric nucleation events; (c) Possibility of identifying new species involved in nucleation processes; (d) Possibility of direct utilization of obtained results for explicitly kinetic nucleation modeling.

Neutral water clusters

Table 23 present the comparison of the thermochemical data computed by DFT and model chemical methods with the results obtained using the conventional CNT. The comparison of the experimental and theoretical enthalpies and Gibbs free energies shows that predictions made by all quantum-chemical methods are consistent with experiments. As may be seen from Table 23, the three out of four model chemistries (CBS-APNO, G3 and G2) give very close results. Both DFT methods are in good agreement within 0.5-0.7 kcal/mole/step with CBS-APNO, G3 and G2. Another important detail is that the predictions of the conventional CNT systematically deviate from the thermochemical data obtained from "first principles". In addition to systematic deviation from quantum theory, the CNT erroneously assign the non-zeros value to $0 \rightarrow 1$ transition.

Neutral binary H_2SO_4 -H_2O clusters

The modern era of quantum studies of multicomponent atmospheric clusters [25,53, 63,64,76,77,81,92-105] began with the pioneering systematic investigation of sulfuric acid hydrates by Bandy and Ianni [64]. This is not surprising because the hydration of the monomeric sulfuric acid in the atmosphere is a fundamental phenomenon whose description is included in the most of up-to–date nucleation models [15]. Molecular hydrate complexes have been detected in sulfate aerosols and their stability has been corroborated in experiments. Unfortunately, the first systematic study [64] performed using the B3LYP method was unable to predict the hydration free energies in agreement with latter experiments [106]. Nadykto and Yu [25] and, independently, Kurten et al. [97] pointed out that the disagreement with experimental data was caused by problems of B3LYP in the description of the hydrogen bonding. At the present time, PW91PW91 is the most common density functional used in the quantum-mechanical studies of atmospheric species. Although it would be an exaggeration to present the PW91PW91 as an ultimate development in the quantum chemistry of hydrogen bonded systems, it provides good geometries, excellent vibrational frequencies and quite accurate cluster free energies [22, 25, 81, 96-99, 102-104].

Table 23. A comparison of the stepwise ΔH^0 and ΔG^0 for $(H_2O)_{n-1} + H_2O \rightarrow (H_2O)_n$ reaction calculated using DFT, model chemistries, MP2, and CNT with experiments.

n	Model chemistries[a]				DFT methods				
	QB3	APNO	G3	G2	PBE aug-cc-pvtz	PW91 aug-cc-pvtz	MP2 6-311++G(2d,2p)[b]	Expt.[c]	CNT[d]
					ΔH^0(kcal/mol)				
2	-3.39	-3.43	-3.51	-3.43	-3.5	-3.82	-3.06, -3.15, -3.08	-3.59±0.5	
3	-7.84	-8.53	-8.52	-8.25	-9.33	-9.9			
4	-9.5	-10.1	-10.46	-10.2	-11.58	-12.18			
5	-6.6	-7.28	-7.43	-7.15	-8.09	-8.65			
6	-6.41	-7.03	-6.96	-6.66	-7.54	-8.06			
7					-5.75	-6.25			
8					-17.58	-18.28			
9					-8.87	-9.44			
10					-10.17	-10.82			
					ΔG^0(kcal/mol)				
1									2.77
2	1.74	2.08	1.94	2.01	2.78	2.55	2.74, 2.76, 2.84	1.95±0.9	0.78
3	4.31	1.43	1.83	1.97	1.49	0.93			0.32
4	0.04	-1.07	-1.66	-1.29	-1.65	-2.39			0.07
5	1.38	-0.28	-0.46	-0.11	0.24	-0.03			-0.1
6	2.21	1.92	1.77	1.98	0.86	0.27			-0.23
7					1.89	1.21			-0.33
8					0.14	-0.24			-0.41
9					0.5	-0.15			-0.48
10					0.47	-0.03			-0.53

a Reference [88]; b Reference [89]; c Reference [90]; d. Reference [44]

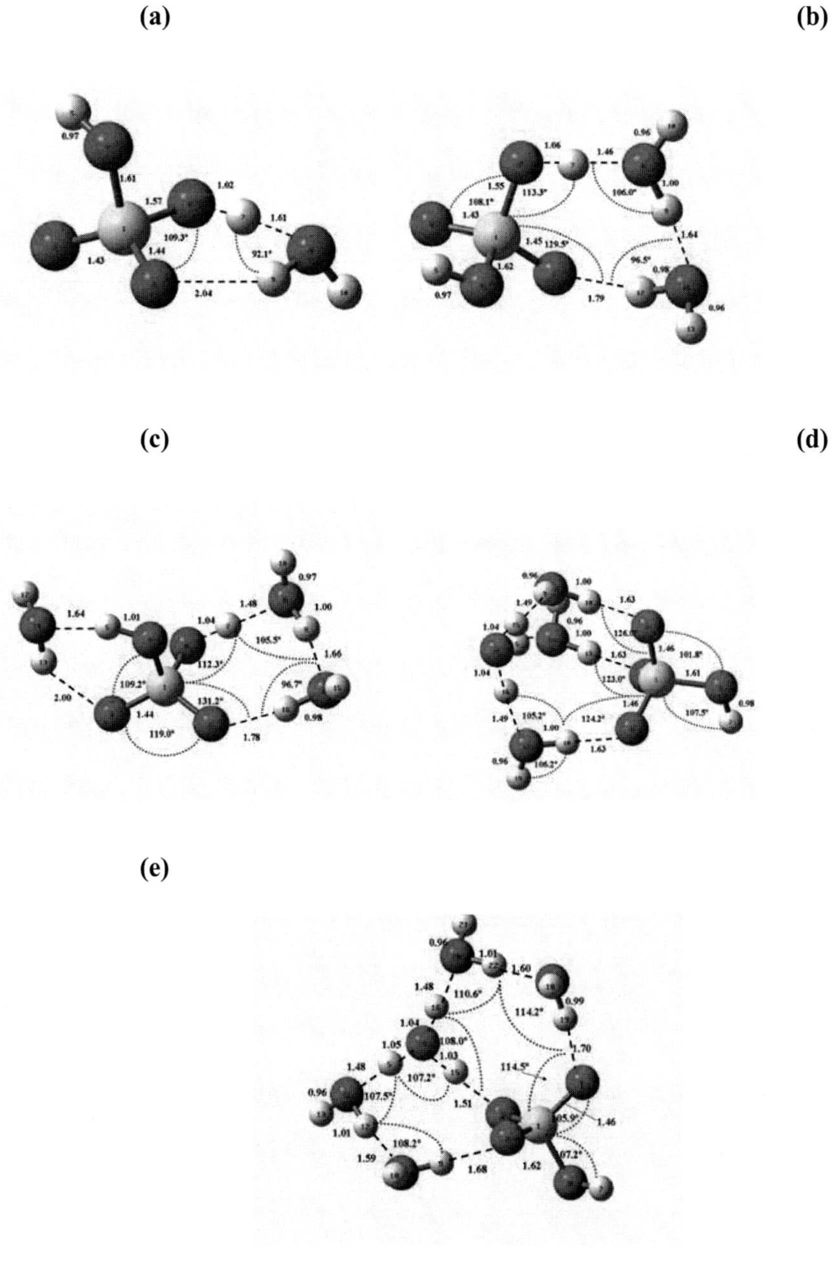

Figure7. Most stable isomers of (a) H_2SO_4 (H_2O); (b) H_2SO_4 $(H_2O)_2$; (c) H_2SO_4 $(H_2O)_3$; (d) H_2SO_4 $(H_2O)_4$; (e) H_2SO_4 $(H_2O)_5$ calculated at PW91PW91/6-311++G(3df,3pd) level of theory.

The computational quantum methods express molecule/complex/cluster energies using the optimized geometry and calculated vibrational frequencies. Fig.7 presents the equilibrium geometries of most stable isomers of H2SO4 (H2O)n (n=1-5). Tables 23-26 give the enthalpy, entropy, and Gibbs free energy changes of various clustering reactions calculated based on different methods and comparisons with available experimental data.

Table 24. Enthalpies, entropies, and Gibbs free energy of hydration calculated at T=298.15K and P=101.3 KPa. Abbreviations PW91/(3df,3pd), PW91/DNP, B3LYP/(2d,2p), and exp. Refer m to Nadykto and Yu, [25] and [107], Ding et al. [76], Bandy and Ianni [64] and Hanson and Eisile [106] , respectively.

	PW91/(3df,3pd)			PW91/ DNP	**B3LYP/(2d,2p),**	**exp**
reaction	ΔH	ΔS	ΔG	ΔG	ΔG	ΔG
$H_2SO_4+H_2O \Leftrightarrow (H_2SO_4)(H_2O)$	-11.76	-31.80	-2.28	-2.5	-0.6	-3.6±1
$(H_2SO_4)(H_2O)+H_2O \Leftrightarrow (H_2SO_4)(H_2O)_2$	-12.57	-32.08	-3.00	-1.8	-0.1	2.6± 0.3
$(H_2SO_4)(H_2O)_2+H_2O \Leftrightarrow (H_2SO_4)(H_2O)_3$	-11.34	-31.71	-1.89	-2.2	-0.5	
$(H_2SO_4)(H_2O)_3+H_2O \Leftrightarrow (H_2SO_4)(H_2O)_4$	-14.49	-37.15	-3.42	-3.7		
$(H_2SO_4)(H_2O)_4+H_2O \Leftrightarrow (H_2SO_4)(H_2O)_5$	-9.99	-29.42	-1.22	-2.4		

Table 25. Enthalpies, entropies, and Gibbs free energy of hydration for sulfuric acid dimer calculated at T=298.15K and P=101.3 KPa. Abbreviations PW91/(3df,3pd), PW91/DNP, B3LYP refer to the present study, Ding et al. [76] and Ianni and Bandy [93], respectively.

	PW91/ 6 311++G(3df,3pd)			**PW91/DNP**	**B3LYP**
reaction	ΔH	ΔS	ΔG	ΔG	ΔG
$(H_2SO_4)_2+H_2O \Leftrightarrow (H_2SO_4)_2(H_2O)$	-14.44	-36.99	-3.41	-2.6	0.2
$(H_2SO_4)_2(H_2O)+H_2O \Leftrightarrow (H_2SO_4)_2(H_2O)_2$	-12.85	-33.12	-2.98	-2.6	-0.3
$(H_2SO_4)_2(H_2O)_2+H_2O \Leftrightarrow (H_2SO_4)_2(H_2O)_3$	-14.56	-37.59	-3.36	-4.2	-0.2
$(H_2SO_4)_2(H_2O)_3+H_2O \Leftrightarrow (H_2SO_4)_2(H_2O)_4$	-14.81	-36.95	-3.80	-5.1	-0.5
$(H_2SO_4)_2(H_2O)_{24}+H_2O \Leftrightarrow (H_2SO_4)_2(H_2O)_5$	-11.95	-32.00	-2.41	-4.3	-0.3

Table 26. Enthalpies, entropies, and Gibbs free energy changes of associated with formation of (H2SO4)2(H2O)n by attachment of sulfuric acid. T=298.15K and P=101.3 KPa. Abbreviations PW91/(3df,3pd),PW91/DNP and B3LYP refer to the present study, Ding et al. [76] and Ianni and Bandy [23] respectively.

	PW91/ (3df,3pd)			**PW91/DNP**	**B3LYP**
reaction	ΔH	ΔS	ΔG	ΔG	ΔG
$H_2SO_4+H_2SO_4 \Leftrightarrow (H_2SO_4)_2$	-16.16	-35.46	-5.59	-3.1	-2.5
$(H_2SO_4)(H_2O)+(H_2SO_4) \Leftrightarrow (H_2SO_4)_2(H_2O)$	-18.84	-40.65	-6.72	-3.3	-1.7
$(H_2SO_4)(H_2O)_2+(H_2SO_4) \Leftrightarrow (H_2SO_4)_2(H_2O)_2$	-19.12	-41.69	-6.69	-4.0	-1.9
$(H_2SO_4)(H_2O)_3+(H_2SO_4) \Leftrightarrow (H_2SO_4)_2(H_2O)_3$	-22.34	-47.57	-8.16	-6.6	-1.2
$(H_2SO_4)(H_2O)_4+(H_2SO_4) \Leftrightarrow (H_2SO_4)_2(H_2O)_4$	-22.66	-47.37	-8.54	-8.0	-2.0
$(H_2SO_4)(H_2O)_5+(H_2SO_4) \Leftrightarrow (H_2SO_4)_2(H_2O)_5$	-24.62	-49.96	-9.72	-9.9	-5.1

Neutral ternary H_2SO_4-H_2O-NH_3 clusters

The structure and thermochemical properties of H_2SO_4-H_2O-NH_3 clusters have been studied theoretically and reported in number of publications [25,93,97,99]. However, neither structural nor spectroscopic experimental data are available at the present time. Optimized geometries of the most stable isomers of $(NH_3)(H_2SO_4)(H_2O)_n$ (n=1-5) are presented in Fig. 8. The thermochemical data for $(NH_3)(H_2SO_4)(H_2O)_n$ and $(NH_3)(H_2SO_4)_2(H_2O)_n$ are presented in Figs. 8 and 9 and Tables 26 - 31.

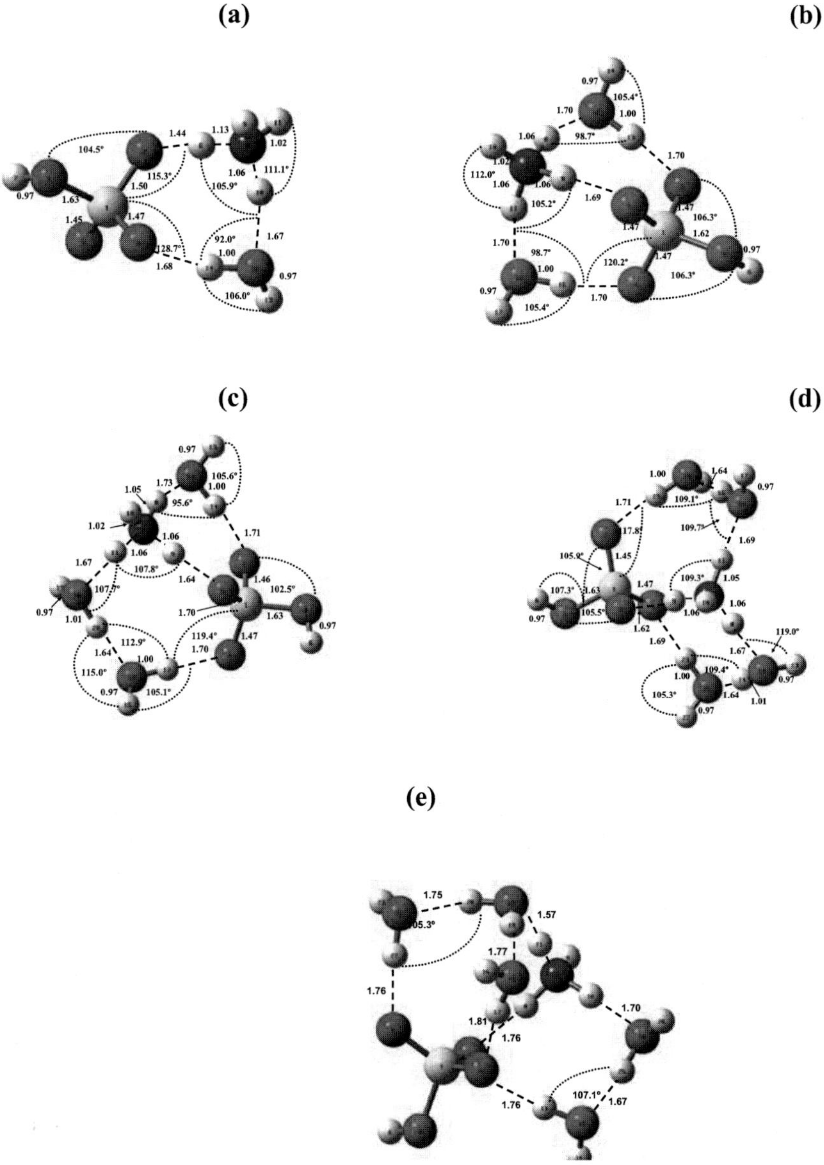

Figure 8. Most stable isomers of (a) $(NH_3)(H_2SO_4)(H_2O)$; (b) $(NH_3)(H_2SO_4)(H_2O)_2$; (c) $(NH_3)(H_2SO_4)(H_2O)_3$; (d) $(NH_3)(H_2SO_4)(H_2O)_4$; (e) $(NH_3)(H_2SO_4)(H_2O)_5$ calculated at PW91PW91/6-311++G(3df,3pd) level of theory [107].

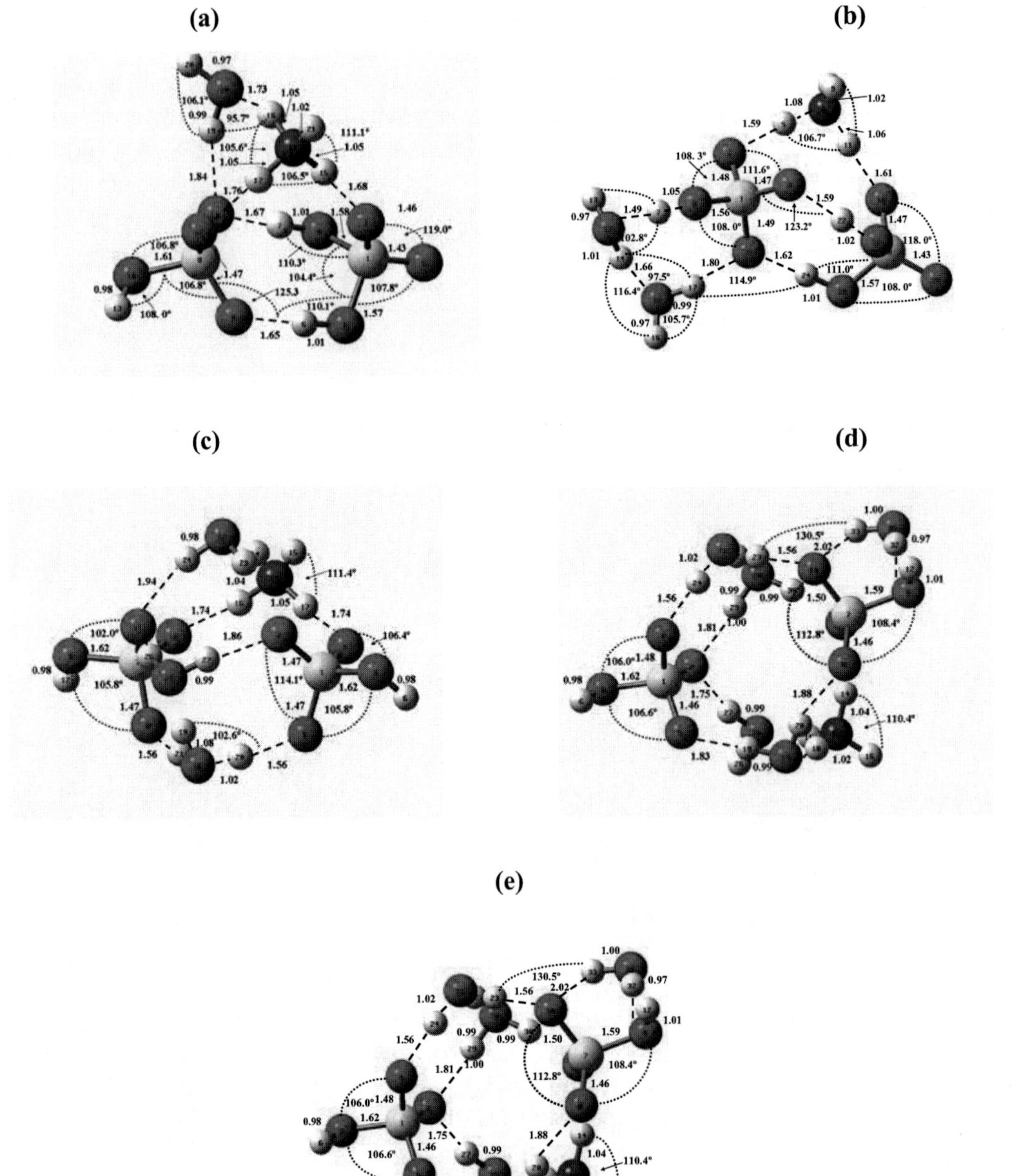

Figure 9. Most stable isomers of (a) $(NH_3)(H_2SO_4)_2(H_2O)$; (b) $(NH_3)(H_2SO_4)_2(H_2O)_2$; (c) $(NH_3)(H_2SO_4)_2(H_2O)_3$; (d) $(NH_3)(H_2SO_4)_2(H_2O)_4$; (e) $(NH_3)(H_2SO_4)_2(H_2O)_5$ calculated at PW91PW91/6-311++G(3df,3pd) level of theory [107].

Table 27. Enthalpy, entropy, and Gibbs free energy changes associated with the formation of (H2SO4)(H2O)n(NH3) by attachment of (NH3) [107]. T=298.15K and P=101.3 KPa. Abbreviation PW91/DNP refers to [97].

	PW91/ 6-311++G(3df,3pd)			PW91/ (3df,3d) 265 K	PW91/DNP 265K
reaction	ΔH	ΔS	ΔG	ΔG	ΔG
H_2SO_4+NH_3 ⇔ (H_2SO_4)(NH_3)	-16.72	-30.01	-7.77	-8.77	-8.70
(H_2SO_4)(H_2O) + NH_3⇔ (H_2SO_4)(H_2O)(NH_3)	-15.91	-30.23	-6.90	-7.90	-7.60
(H_2SO_4)(H_2O)$_2$+ NH_3⇔ (H_2SO_4)(H_2O)$_2$(NH_3)	-17.42	-35.14	-6.94	-8.11	-9.10
(H_2SO_4)(H_2O)$_3$+ NH_3⇔ (H_2SO_4)(H_2O)$_3$(NH_3)	-17.68	-32.40	-8.01	-9.09	-7.80
(H_2SO_4)(H_2O)$_4$+ NH_3⇔ (H_2SO_4)(H_2O)$_4$(NH_3)	-13.96	-23.74	-6.88	-7.67	-7.50
(H_2SO_4)(H_2O)$_5$+ NH_3⇔ (H_2SO_4)(H_2O)$_5$(NH_3)	-14.66	-30.18	-5.66	-6.66	-6.80

Table 28. Enthalpy, entropy, and Gibbs free energy changes associated with the formation of (H2SO4)(H2O)n(NH3) by attachment of (H2O) [107]. T=298.15K and P=101.3 KPa. Abbreviation PW91/DNP refers to [97].

	PW91/ 6-311++G(3df,3pd)			PW91 * 265 K	PW91/DNP 265K
reaction	ΔH	ΔS	ΔG	ΔG	ΔG
(H_2SO_4)(NH_3)+H_2O⇔ (H_2SO_4)(NH_3)(H_2O)	-10.96	-32.03	-1.41	-2.47	-2.00
(H_2SO_4)(NH_3)(H_2O)+H_2O⇔ (H_2SO_4)(NH_3)(H_2O)$_2$	-14.07	-36.98	-3.05	-4.27	-4.13
(H_2SO_4)(NH_3)(H_2O)$_2$+H_2O⇔ (H_2SO_4)(NH_3)(H_2O)$_3$	-11.60	-36.98	-2.96	-1.80	-3.00
(H_2SO_4)(NH_3)(H_2O)$_3$+H_2O⇔ (H_2SO_4)(NH_3)(H_2O)$_4$	-10.77	-28.49	-2.28	-3.22	-2.42
(H_2SO_4)(NH_3)(H_2O)$_4$+H_2O⇔ (H_2SO_4)(NH_3)(H_2O)$_5$	-10.70	-35.85	-0.01	-1.20	-2.54

Table 29. Enthalpies, entropies, and Gibbs free energy changes associated with the formation of (H2SO4)2(H2O)n(NH3) by attachment of (NH3) [107]. T=298.15K and P=101.3 KPa. Abbreviation PW91/DNP refers to [97]

	PW91/ 6-311++G(3df,3pd)			PW91 * 265 K	PW91/DNP 265K
reaction	ΔH	ΔS	ΔG	ΔG	ΔG
(H_2SO_4)$_2$+NH_3 ⇔ (H_2SO_4)$_2$(NH_3)	-25.67	-39.68	-13.84	-15.15	-15.10
(H_2SO_4)$_2$(H_2O) + NH_3⇔ (H_2SO_4)$_2$(H_2O)(NH_3)	-22.91	-34.01	-12.77	-13.89	-15.90
(H_2SO_4)$_2$(H_2O)$_2$+ NH_3⇔ (H_2SO_4)$_2$(H_2O)$_2$(NH_3)	-21.46	-35.14	-10.98	-13.80	-16.70
(H_2SO_4)$_2$(H_2O)$_3$+ NH_3⇔ (H_2SO_4)$_2$(H_2O)$_3$(NH_3)	-22.34	-31.87	-12.84	-13.89	-15.00
(H_2SO_4)$_2$(H_2O)$_4$+ NH_3⇔ (H_2SO_4)$_2$(H_2O)$_4$(NH_3)	-17.33	-26.72	-9.36	-10.25	-9.80
(H_2SO_4)$_2$(H_2O)$_5$+ NH_3⇔ (H_2SO_4)$_2$(H_2O)$_5$(NH_3)	-17.55	-25.48	-9.95	-10.80	-12.10

Table 30. Enthalpy, entropy, and Gibbs free energy changes associated with the formation of (H2SO4)2(H2O)n (NH3) by attachment of H2SO4 [107]. T=298.15K and P=101.3 KPa.

	PW91/ 6-311++G(3df,3pd)		
reaction	ΔH	ΔS	ΔG
$(H_2SO_4)(NH_3)+H_2SO_4 \Leftrightarrow (H_2SO_4)_2(NH_3)$	-25.11	-45.13	-11.65
$(H_2SO_4)(H_2O)(NH_3)+(H_2SO_4) \Leftrightarrow (H_2SO_4)_2(H_2O)(NH_3)$	-25.83	-44.42	-12.59
$(H_2SO_4)(H_2O)_2(NH_3)+(H_2SO_4) \Leftrightarrow (H_2SO_4)_2(H_2O)_2(NH_3)$	-23.16	-41.69	-10.73
$(H_2SO_4)(H_2O)_3(NH_3)+(H_2SO_4) \Leftrightarrow (H_2SO_4)_2(H_2O)_3(NH_3)$	-27.01	-47.03	-12.98
$(H_2SO_4)(H_2O)_4(NH_3)+(H_2SO_4) \Leftrightarrow (H_2SO_4)_2(H_2O)_4(NH_3)$	-26.04	-50.35	-11.02
$(H_2SO_4)(H_2O)_5(NH_3)+(H_2SO_4) \Leftrightarrow (H_2SO_4)_2(H_2O)_5(NH_3)$	-27.51	-45.26	-14.01

Table 31. Enthalpies, entropies, and Gibbs free energy changes associated with the formation of (H2 SO4)2(H2O)n (NH3) by attachment of H2 O [107] . T=298.15K and P=101.3 KPa.

	PW91/ 6-311++G(3df,3pd)		
reaction	ΔH	ΔS	ΔG
$(H_2SO_4)_2(NH_3)+H_2O \Leftrightarrow (H_2SO_4)_2(H_2O)(NH_3)$	-11.68	-31.32	-2.35
$(H_2SO_4)_2(H_2O)(NH_3)+H_2O \Leftrightarrow (H_2SO_4)_2(H_2O)_2(NH_3)$	-11.40	-34.25	-1.19
$(H_2SO_4)_2(H_2O)_2(NH_3)+H_2O \Leftrightarrow (H_2SO_4)_2(H_2O)_3(NH_3)$	-15.44	-34.31	-5.21
$(H_2SO_4)_2(H_2O)_3(NH_3)+H_2O \Leftrightarrow (H_2SO_4)_2(H_2O)_4(NH_3)$	-9.80	-31.81	-0.32
$(H_2SO_4)_2(H_2O)_4(NH_3)+H_2O \Leftrightarrow (H_2SO_4)_2(H_2O)_5(NH_3)$	-12.17	-30.76	-3.00

Nadykto and Yu [25] and, independently, *Kurten et al.* [97] have reached the following conclusions about the thermochemistry of ternary clusters:

a) The stabilizing effect of ammonia on the formation of small H_2SO_4-H_2O clusters is likely to increase with the number of sulfuric acid molecules in the cluster.
b) Thermochemistry of H_2SO_4 and NH_3 in small H_2SO_4-H_2O-NH_3 clusters is either virtually independent of or depends weakly on the water content. This suggests that the stabilizing effect of ammonia at initial steps of the cluster growth is associated mainly with the sulfuric acid.

Stabilization of H_2SO_4 -H_2O complexes by low-molecular organic acids

Ammonia has been considered as an exclusive principle stabilizer of H_2SO_4-H_2O clusters in the atmosphere since 1990s. However, recent study of *Nadykto and Yu* [25] have shown that common low molecular organic acid such as formic and acetic acids can also efficiently stabilize H_2SO_4-H_2O clusters. Interaction of formic acid and acetic acids, whose concentrations in the atmosphere are as high or higher than ammonia concentrations, with sulfuric acid and water leads to the formation of stable hydrogen-bonded complexes, whose stability is close to the stability of complexes of sulfuric acid with ammonia. In addition to the formation of strong hydrogen-bonded complexes with sulfuric acid and water, formic and acetic acids form quite thermodynamically stable complexes with ammonia. This indicates that ammonia is certainly not an exclusive stabilizer of H_2SO_4 -H_2O clusters in the atmosphere and that the involvement of common organic species, alongside with or without ammonia, in clustering and subsequent nucleation of sulfuric acid and water should be studied in detail further.

Neutral gas-phase hydrates of HNO_3 in the stratosphere

Theoretical studies of the coexistence of gas-phase hydrates of HNO_3 under the stratospheric conditions have been carried out by Al Natsheh with co-authors [94-95]. The family of nitric acid trihydrates (NAT), which dominate the composition of the Ia PSCs, are considered to be thermodynamically stable. However, their formation mechanisms and thermodynamic properties were poorly understood, and the formation pathways for different conformer of NAD and NAT, as well as the NAD–NAT transformations, remained puzzling. Arnold [108] suggests that physical chemistry of stratospheric clouds may involve a formation of metastable nitric acid hydrates. According to Arnold, such hydrates are less stable than the NAT and they differ from NAT in both catalytic and optical properties and HCl solubility. A clear understanding of the role of the metastable NAD in the stratosphere is important because NAD particles grow much faster than NAT crystals. Another important detail is that difference in the chemical and physical properties of NAD and NAT can affect heterogeneous surface chemistry of PSCs, especially the transformation of chlorine reservoir species, such as $ClONO_2$ and HCl, into active chlorine forms that readily photolyze in sunlight.

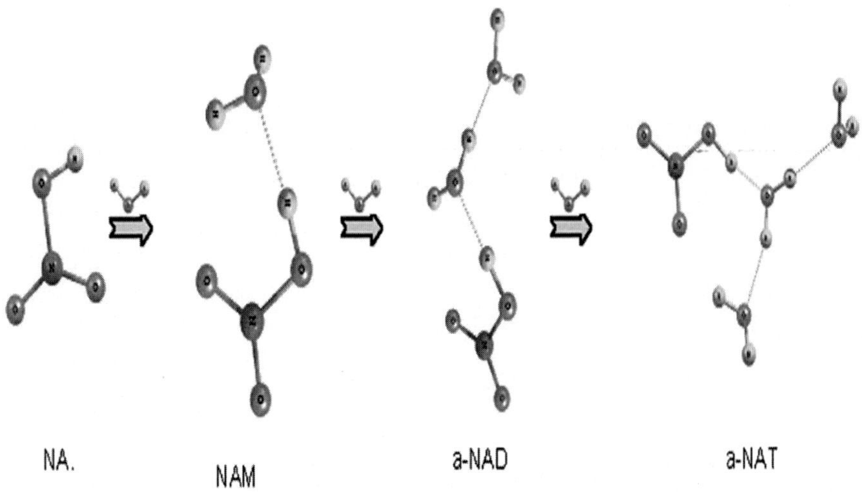

Figure 10. HNO_3-NAM-NAD-NAT transformations

NAD has been investigated in a number of experiments. In the past, it was assumed that there exists one isomer of NAD only. However, recent spectroscopic studies have hinted at the presence of a second NAD conformation. In the recent study quantum-mechanical study [94], three equilibrium structures of the gas-phase NAD, including two new structures not found in the earlier theoretical earlier, have been identified.

NAD-a (□)

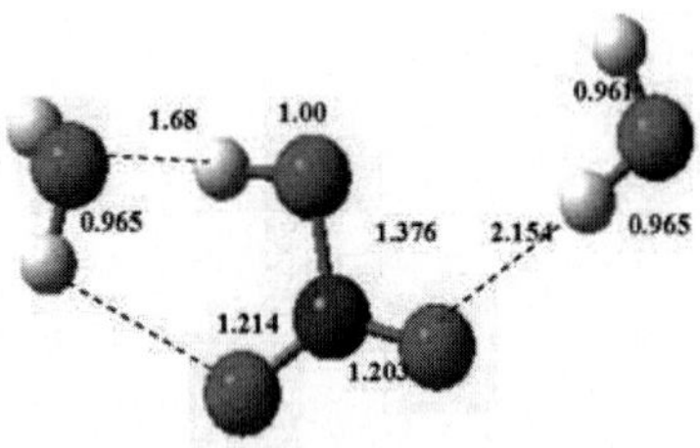

NAD-b (α)

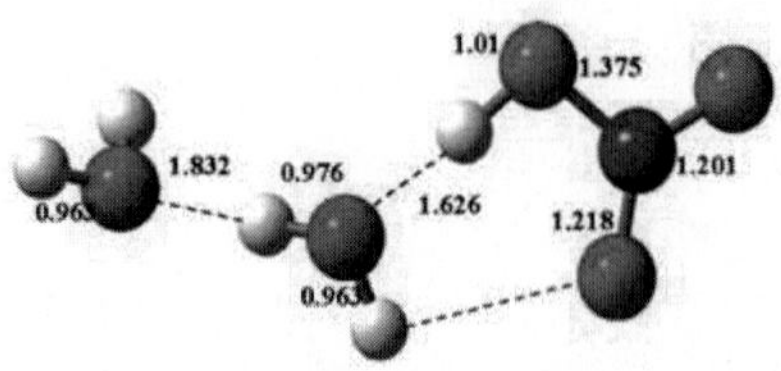

NAD-c (γ)

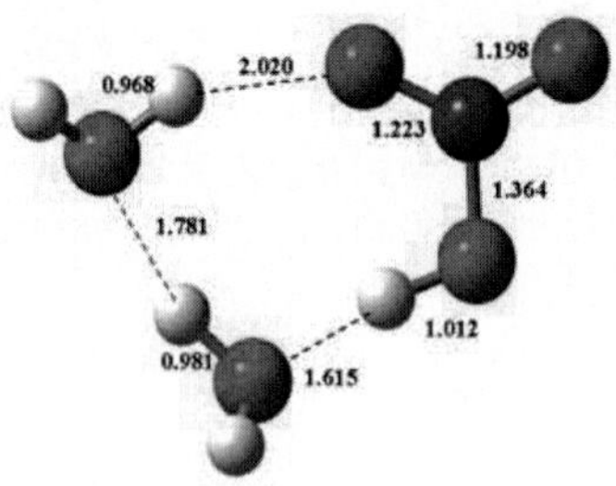

Figure 11. Equilibrium conformations of NAD.

Al Natsheh et al. [94-95] found that although the formation of the gas-phase β-NAD is thermodynamically favorable under stratospheric conditions, the difference in the formation free energy between β-NAD and α-NAD is small. Therefore, both forms are expected to be abundant. The comparison of the thermodynamic data for α- and β-NAD and α- and β-NAT suggests that the formation of NAT from NAD is thermodynamically favorable only if the α-NAD ⇒ α-NAT transformation occurs. The formation of NAT from the nitric acid and NAM is more favorable thermodynamically than the NAD ⇒ NAT transformation; however, the direct nitric acid ⇒ NAD, nitric acid ⇒ NAT, and NAM ⇒ NAT processes under stratospheric conditions are likely kinetically limited. The small differences in the formation free energies of NAM, α-NAD, β-NAD, α-NAT and β-NAT suggest that hydrated nitric acid in the stratosphere is a mixture composed mainly of NAM, α-NAD, β-NAD, α-NAT and β-NAT. On the other hand, the availability of water and strongly hygroscopic nature of nitric acid may drive the acquisition of more water molecules into clusters that reduces the NAM fraction. NAM grows into α-NAD and β-NAD, which transforms into α-NAT and β-NAT, respectively. These considerations suggest that nitric acid crystals under stratospheric conditions are formed mainly from the gas-phase β-NAD and β-NAT. Further research is needed to determine the extend to which the structure of gas-phase hydrates of nitric acid changes during the crystals formation.

. Ionic $H_3O^+(H_2O)_n$ clusters

The importance of the ambient ionization on the cluster formation is well established. The thermochemistry of the hydration of simple and common atmospheric ion H3O+ has been studied using the computational quantum methods in the past. Table 32 presents the comparison of calculated hydration enthalpies and entropies for H3O+(H2O)n-1 + H2O → H3O+(H2O)n reaction with the recent experiments [110]. As may be seen from Table 32, the computational quantum methods (both ab initio and DFT) agree well with each other and recent experiments of Froyd and Lovejoy [110]. Moreover, they were able to predict the abrupt drop in the enthalpy change at n=4 related to the formation of the second hydration/solvation shell in good agreement with recent experiments [110].

Table 32. Comparison of hydration enthalpies for $H_3O^+(H_2O)_{n-1} + H_2O \rightarrow H_3O^+(H_2O)_n$ reaction calculated using DFT PW91PW91/6-311++G(3df,3pd) in *Nadykto et al.* (2006) [22] with available *ab initio* studies (MP2/cc-pVTZ (MP2), MCCM-UT-CCSD//MP2/cc-pVTZ (MCCM)) [109] and experimental data of Froyd and Lovejoy [110]. The entropies obtained using the DFT are also compared with experimental values (entropies were not reported in the *ab initio* study [109]). The data are given at ambient temperature of 298.15 K and ambient pressure of 101.3 KPa.

ΔH				ΔS	
DFT	MP2	MCCM	exp.	DFT	exp.
-37.87	-37.21	-35.3		-30.05	
-22.75	-22.98	-21.72	-20.9	-27.35	-24.1
-18.12	-19.35	-18.24	-18	-26.05	-27.9
-13.65	-13.71	-12.66	-12.8	-27.22	-23.8
-12.07	-12.4	-11.42	-12.2	-26.53	-26.9
-11.16			-10.6	-25.72	-25.1
-9.78			-10	-24.58	-25.8

Ionic $HSO_4^-(H_2O)_n$ and $(H_3O)^+(H_2SO_4)(H_2O)_n$ clusters

Fig.12 and 13 present the equilibrium structures of most stable isomers of $HSO_4^-(H_2O)_n$ and $(H_3O)^+(H_2SO_4)(H_2O)_n$ optimized at PW91PW91/6-311++G(3df,3pd) level of theory.

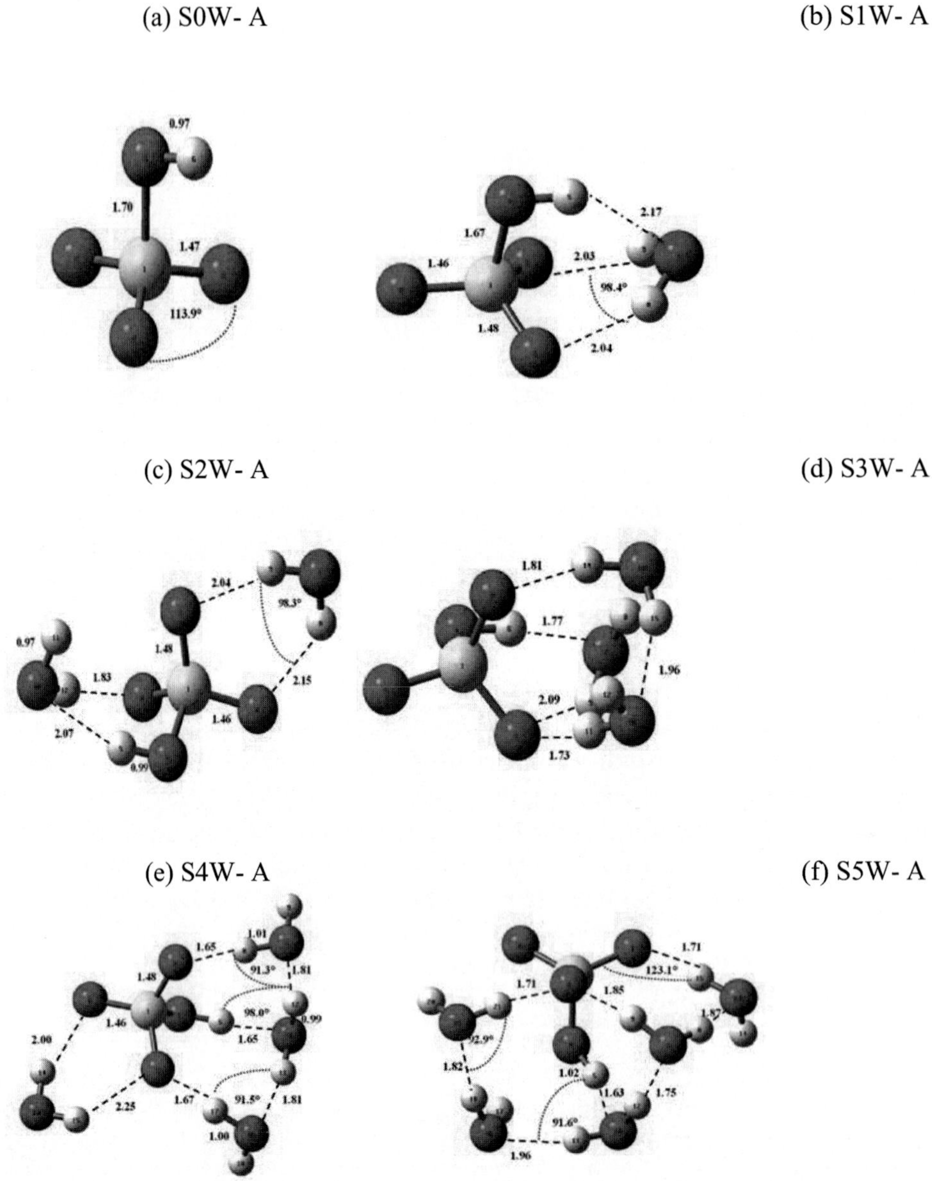

Figure 12. Most stable isomers of (a) HSO_4^- (b) $HSO_4^-(H_2O)$; (c) $HSO_4^-(H_2O)_2$; (d) $HSO_4^-(H_2O)_3$; (e) $HSO_4^-(H_2O)_4$; (f) $HSO_4^-(H_2O)_5$ calculated at PW91PW91/6-311++G(3df,3pd) level of theory.

Tables 33 and 34 present the comparison of the hydration enthalpies, entropies and Gibbs free energies with the experimental data [110-111] and other theoretical studies [98]. Fig. 14 presents the comparison of hydration free energies for aqueous clusters formed over different cores relevant to atmospheric nucleation: H2SO4, hydrogensulfate ion HSO4- and (H3O)+(H2SO4). The comparison of the hydration free energies given in Tables 33 and 34 shows that the hydration of the cation (H3O)+(H2SO4) is much stronger than that of anion HSO4-. This finding is consistent with the recent studies on the sign preference for water [22]. The difference in the stepwise hydration free energy is very large (~10 kcal mole-1);

however, it decreases quickly when the cluster is growing. Another important observation is that the difference in the hydration free energies for clusters formed over different cores also decreases with the hydration number. This gives us an important indication of the fast convergence of the stepwise Gibbs free energy to the bulk value. These considerations lead to a logical conclusion that quantitative assessment of the relative stability of hydrates clusters formed over different core species can be done by studying the smallest clusters only.

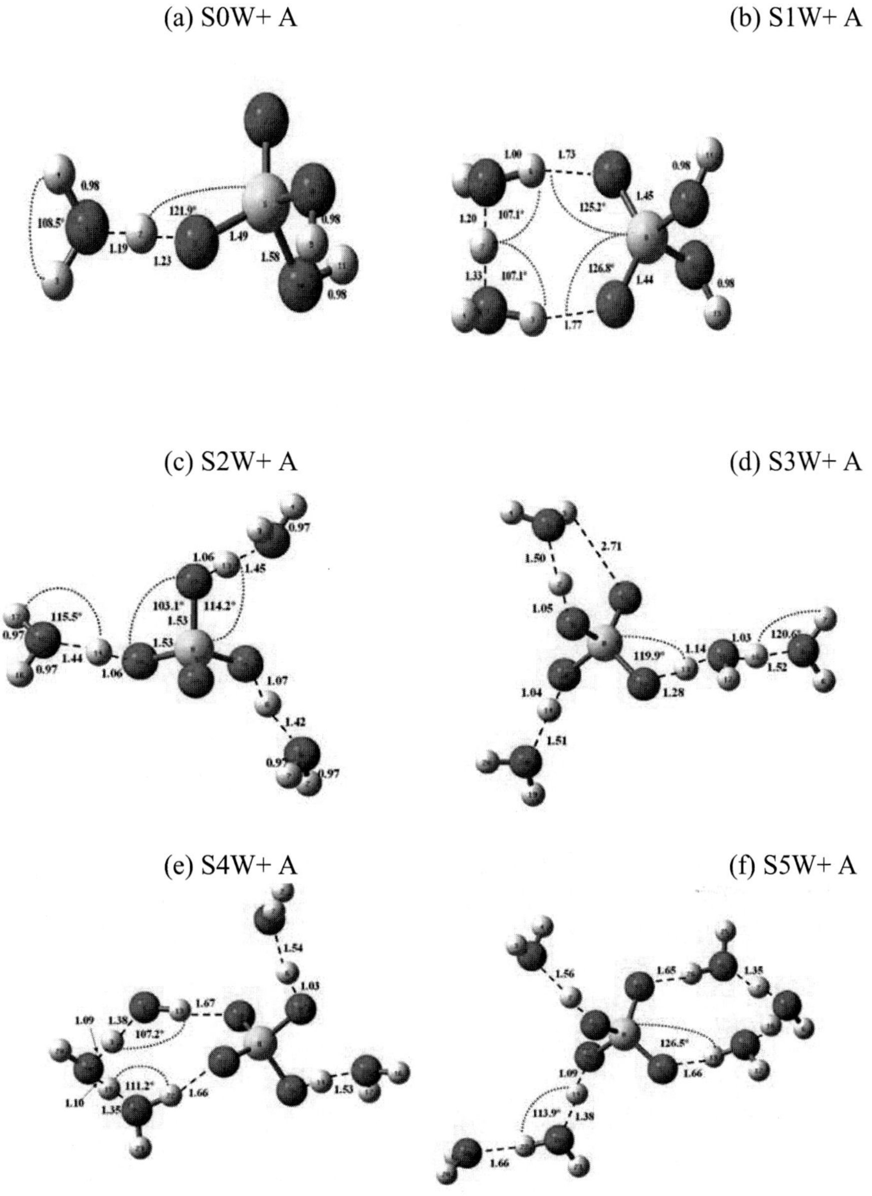

Figure 13. Most stable isomers of (a) $(H_3O)^+(H_2SO_4)$, (b) $(H_3O)^+(H_2SO_4)(H_2O)$; (c) $(H_3O)^+(H_2SO_4)(H_2O)_2$; (d) $(H_3O)^+(H_2SO_4)(H_2O)_3$; (e) $(H_3O)^+(H_2SO_4)(H_2O)_4$; (f) $(H_3O)^+$ (H_2SO_4) $(H_2O)_5$ calculated at PW91PW91/6-311++G(3df,3pd) level of theory (from [103]).

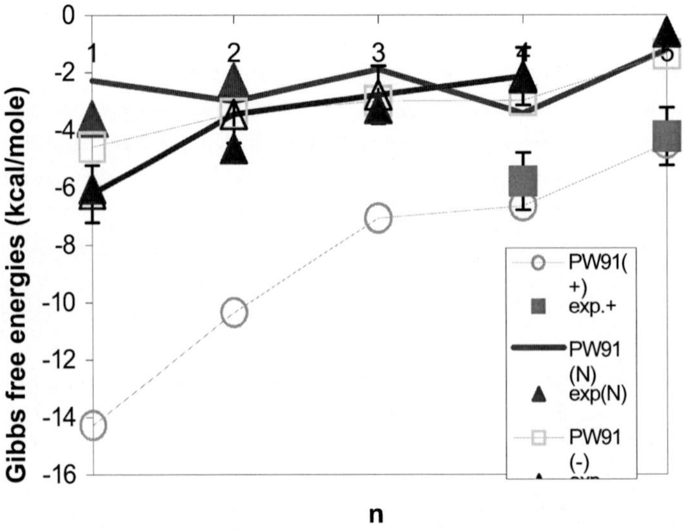

Fig.14. The comparison of the PW91PW91/6-311++G(3df,3pd) results with the theoretical [25, 107] and experimental values [106] (neutrals), [110-111] (ions) of the step-wise hydration free energies associated with the formation of $H_2SO_4(H_2O)_n$ (N) , $HSO_4^-(H_2O)_n$ (-) and $(H_3O)^+(H_2SO_4)(H_2O)_n$ (+).

Table 33. Enthalpies, entropies, and Gibbs free energy of HSO4-(H2O)n hydration calculated at T=298.15K and P=101.3 KPa. Superscripts a, b, and c and refer to the present study, MP2/aug-cc-pv(d+D)Z study of Kurten et al. (2007) [98] and experimental results of Froyd and Lovejoy (2003) [111], respectively.

	ΔH	ΔS	ΔG
$HSO_4^- + H_2O \Leftrightarrow (HSO_4^-)(H_2O)$	-14.13^a	-35.07^a	-4.58^a
	-12.95^b	-23.00^b	-6.22^b
	-12.90^c	-23.00^c	-6.04 ± 0.7^c
$(HSO_4^-)(H_2O) + H_2O \Leftrightarrow (HSO_4^-)(H_2O)_2$	-14.59^a	-34.44^a	-3.42^a
	-13.39^b	-33.35^b	-3.45^b
	-11.20^c	-22.00^c	-4.64 ± 0.7^c
$(HSO_4^-)(H_2O)_2 + H_2O \Leftrightarrow (HSO_4^-)(H_2O)_3$	-12.21^a	-31.15^a	-2.99^a
	-14.46^b	-38.95^b	-2.78^b
	-12.40^c	-30.70^c	-3.25^c
$(HSO_4^-)(H_2O)_3 + H_2O \Leftrightarrow (HSO_4^-)(H_2O)_4$	-10.91^a	-26.40^a	-2.98^a
	-9.22^b	-23.26^b	-2.14^b
	-13.20^c	-36.90^c	-2.20 ± 0.4^c
$(HSO_4^-)(H_2O)_4 + H_2O \Leftrightarrow (HSO_4^-)(H_2O)_5$	-13.07^a	-39.22^a	-1.38^a
	-11.70^c	-37.20^c	-0.61 ± 1.5^c

The comparison of the computed free energies with the experimental data shows that the DFT results are in good agreement with the experimental data. The average difference of the present DFT results for HSO4-(H2O)n from the experimental data does not exceed 0.28 kcal mole-1 per step that is smaller than the typical experimental uncertainty.

Table 34. Enthalpies, entropies, and Gibbs free energy of (H3O)+(H2SO4)(H2O)n hydration calculated at T=298.15K and P=101.3 KPa. Superscript e refers to the experimental data [111].

n	ΔH	ΔS	ΔG
$(H_3O)^+(H_2SO_4)(H_2O)_0 + H_2O \Leftrightarrow (H_3O)^+(H_2SO_4)(H_2O)_1$	-24.45	-34.14	-14.27
$(H_3O)^+(H_2SO_4)(H_2O)_1 + H_2O \Leftrightarrow (H_3O)^+(H_2SO_4)(H_2O)_2$	-16.84	-21.78	-10.35
$(H_3O)^+(H_2SO_4)(H_2O)_2 + H_2O \Leftrightarrow (H_3O)^+(H_2SO_4)(H_2O)_3$	-18.63	-38.79	-7.07
$(H_3O)^+(H_2SO_4)(H_2O)_3 + H_2O \Leftrightarrow (H_3O)^+(H_2SO_4)(H_2O)_4$	-14.84	-27.45	-6.65
	-13.6[e]	-26.2[e]	-5.79±0.7[e]
$(H_3O)^+(H_2SO_4)(H_2O)_4 + H_2O \Leftrightarrow (H_3O)^+(H_2SO_4)(H_2O)_5$	-12.48	-26.85	-4.47
	-11.9[e]	-25.7[e]	-4.23±1.1[e]

e ref.[111]

Ionic $HSO_4^-(NH_3)(H_2O)_n$ clusters

Fig.15 presents equilibrium geometries of (HSO4-)(NH3)(H2O)n. Tables 35 and 36 present calculated hydration enthalpies, entropies and Gibbs free energies associated with the formation of (HSO4-)(NH3)(H2O)n by addition of water and ammonia, respectively.

Table 35. Enthalpies, entropies, and Gibbs free energy of (HSO4-)(NH3)(H2O) n hydration calculated at T=298.15K and P=101.3 KPa. Superscripts a refers to MP2/aug-cc-pv(d+D)Z study [97].

	ΔH	ΔS	ΔG
$(HSO_4^-)(NH_3) + H_2O$ □ $(HSO_4^-)(NH_3)(H_2O)$	-15.79	-35.53	-5.20
	-13.07[a]	-30.00[a]	-4.23[a]
$(HSO_4^-)(NH_3)(H_2O) + H_2O$ □ $(HSO_4^-)(NH_3)(H_2O)_2$	-15.77	-38.51	-4.29
$(HSO_4^-)(NH_3)(H_2O)_2 + H_2O$ □ $(HSO_4^-)(NH_3)(H_2O)_3$	-11.04	-28.53	-2.53
$(HSO_4^-)(NH_3)(H_2O)_3 + H_2O$ □ $(HSO_4^-)(NH_3)(H_2O)_4$	-10.90	-29.01	-2.25
$(HSO_4^-)(NH_3)(H_2O)_4 + H_2O$ □ $(HSO_4^-)(NH_3)(H_2O)_5$	-12.69	-34.17	-0.28

As seen from Tables 33 and 35, the presence of ammonia does not lead to a noticeable enhancement in the hydration strength. The hydration free energies of (HSO4-)(H2O)n and (HSO4-)(NH3)(H2O)n are close, although the hydration of (HSO4-) (H2O)n is somewhat systematically stronger (~ 0.5 kcal.mol) than that of (HSO4-)(NH3)(H2O)n. The Gibbs free energies of the initial (n=1,2) steps of hydration of (HSO4-)(NH3) is more negative than those of (H2SO4)(NH3); however, the hydration free energies of latter steps are nearly identical. PW91PW91/6-311++G(3df,3pd) and MP2/aug-cc-pv(d+D)z [97] hydration free energies for (HSO4-)(NH3) + H2O ⇔ (HSO4-)(NH3)(H2O) reaction agree within ~1 kcal mole-1.

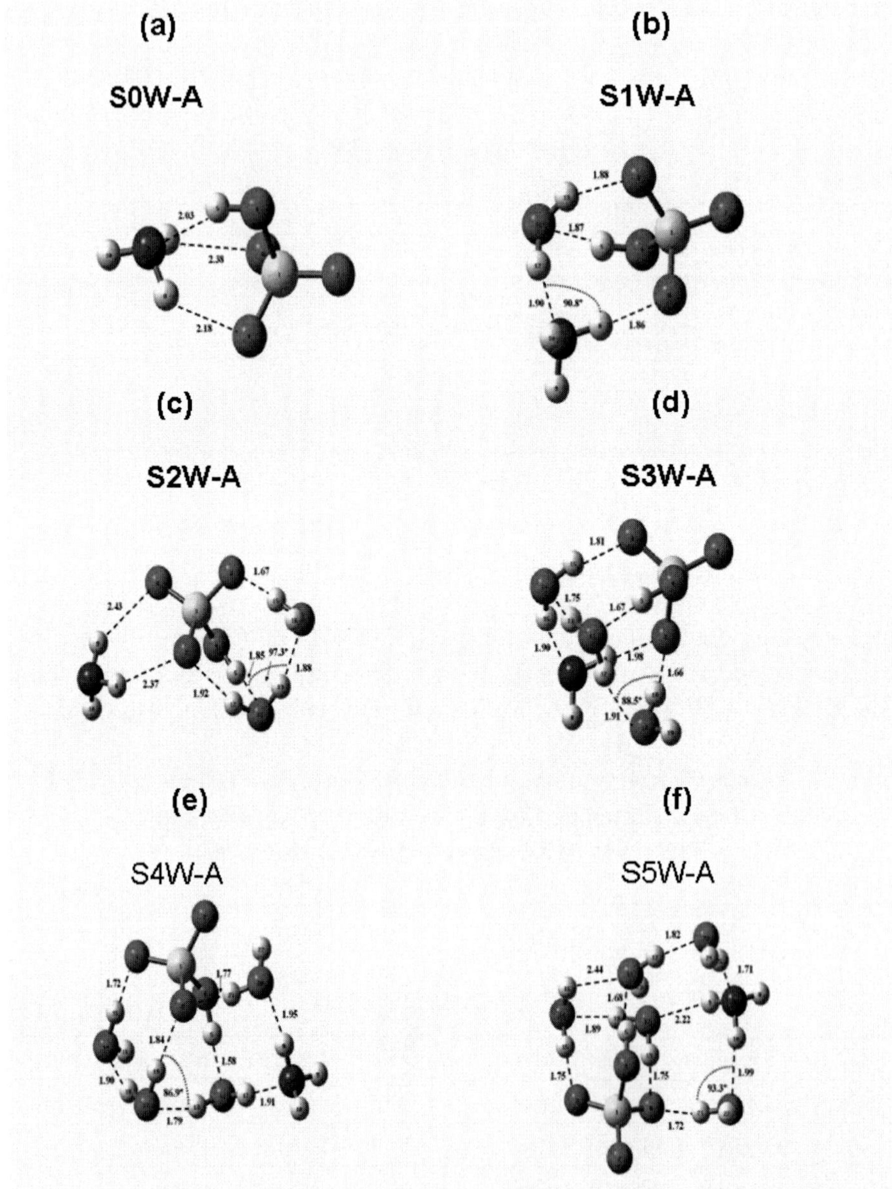

Figure 15. Most stable isomers of (a) $(HSO_4^-)(NH_3)$; (b) $(HSO_4^-)(NH_3)(H_2O)$; (c) $(HSO_4^-)(NH_3)(H_2O)_2$; (d) $(HSO_4^-)(NH_3)(H_2O)_3$; (e) $(HSO_4^-)(NH_3)(H_2O)_4$; (f) $(HSO_4^-)(NH_3)(H_2O)_5$ obtained at PW91PW91/6-311++G(3df,3pd) level of theory.

Table 36. Enthalpies, entropies, and Gibbs free energy of (HSO4-)(NH3)(H2O)n formation by addition of ammonia. T=298.15K and P=101.3 KPa. Superscript a refers to MP2/aug-cc-pv(d+D)Z study [97].

	ΔH	ΔS	ΔG
$(HSO_4^-) + (NH_3) \Leftrightarrow (H_2SO_4)(NH_3)$	-7.22	-27.90	1.10
	-9.24[a]	-32.37[a]	0.69[a]
$(HSO_4^-)(H_2O) + (NH_3) \Leftrightarrow (HSO_4^-)(NH_3)(H_2O)$	-9.44	-32.41	1.12
$(HSO_4^-)(H_2O)_2 + (NH_3) \Leftrightarrow (HSO_4^-)(NH_3)(H_2O)_2$	-5.90	-26.51	2.01
$(HSO_4^-)(H_2O)_3 + (NH_3) \Leftrightarrow (HSO_4^-)(NH_3)(H_2O)_3$	-9.46	-33.86	0.71
$(HSO_4^-)(H_2O)_4 + (NH_3) \Leftrightarrow (HSO_4^-)(NH_3)(H_2O)_4$	-9.44	-36.47	1.43
$(HSO_4^-)(H_2O)_5 + (NH_3) \Leftrightarrow (HSO_4^-)(NH_3)(H_2O)_5$	-9.07	-31.42	2.52

The results of the quantum-chemical study of negatively charged binary and ternary clusters lead us to the following conclusions:

The presence of NH3 does not enhance the thermochemical stability of HSO4-(H2O)n and ammonia is unlikely involved in the gas-phase hydration of hydrogensulfate ion under the atmospheric conditions.

a) The total free energy changes associated with the formation of charged $(HSO_4^-)(NH_3)(H_2O)_n$ is less negative than that associated with the formation of neutral $(H_2SO_4)(NH_3)(H_2O)_n$ due to the positive affinity of NH_3 to (HSO_4^-). This leads us to conclude that the assessment of charged clusters in the classical nucleation theory as presumably more stable thermodynamically compared to neutrals is not applicable to pre-nucleation negatively charged ternary clusters, or generally multicomponent molecular clusters. This is a clear indication that not only quantitative but also qualitative assessment of the thermodynamic stability of atmospheric ternary clusters is impossible without the quantum-mechanical treatment.

Ionic $(NH_4^+)(H_2SO_4)(H_2O)_n$ clusters

Fig.16 presents the equilibrium geometries of most stable isomers of $(NH_4^+)(H_2SO_4)(H_2O)_n$. obtained at PW91PW91/6-311++G(3df,3pd) level of theory. Table 37 and Fig.17 presents comparison of hydration enthalpies and Gibbs free energies associated with the formation of $(NH_4^+)(H_2SO_4)(H_2O)_n$, $(NH_4^+)(H_2O)_n$,, $(NH_3)(H_2SO_4)(H_2O)_n$, $(H_3O^+)(H_2SO_4)(H_2O)_n$ and $(H_3O^+)(H_2O)_n$. As seen from Table 37 and Fig.17, the protonation of ammonia has a profound effect on the stability of hydrates containing the sulfuric acid. The hydration free energies for clusters containing protonated ammonia are ~2-8 kcal mole-1 more negative than those for (NH3)(H2SO4)(H2O)n. The comparison of hydration free energies for (NH4+)(H2SO4)(H2O)n and (H3O+)(H2SO4)(H2O)n shows that binary hydrate complexes formed over (H3O+) are bonded stronger than those formed over (NH4+). As may be seen from the comparison of hydration free energies for (NH4+)(H2SO4)(H2O)n containing the sulfuric acid with those for (NH4+)(H2O)n , the presence of the sulfuric acid does not enhance the hydration strength. The same pattern is observed in the case of (H3O+)(H2SO4)(H2O)n and (H3O+)(H2O)n. As a matter of fact, the hydration of (NH4+)(H2O)n and (H3O+)(H2O)n is 2-4 kcal mole-1 and 1.5-14.5 kcal mole-1 stronger than that of (NH4+)(H2SO4)(H2O)n and (H3O+)(H2SO4)(H2O)n. It is important to note that the difference in the hydration free energies for clusters formed over different core ions decreases quickly with the hydration number reaching ~-4.5 kcal mole-1 at n=5 that is ~6 times smaller than that at n=1.

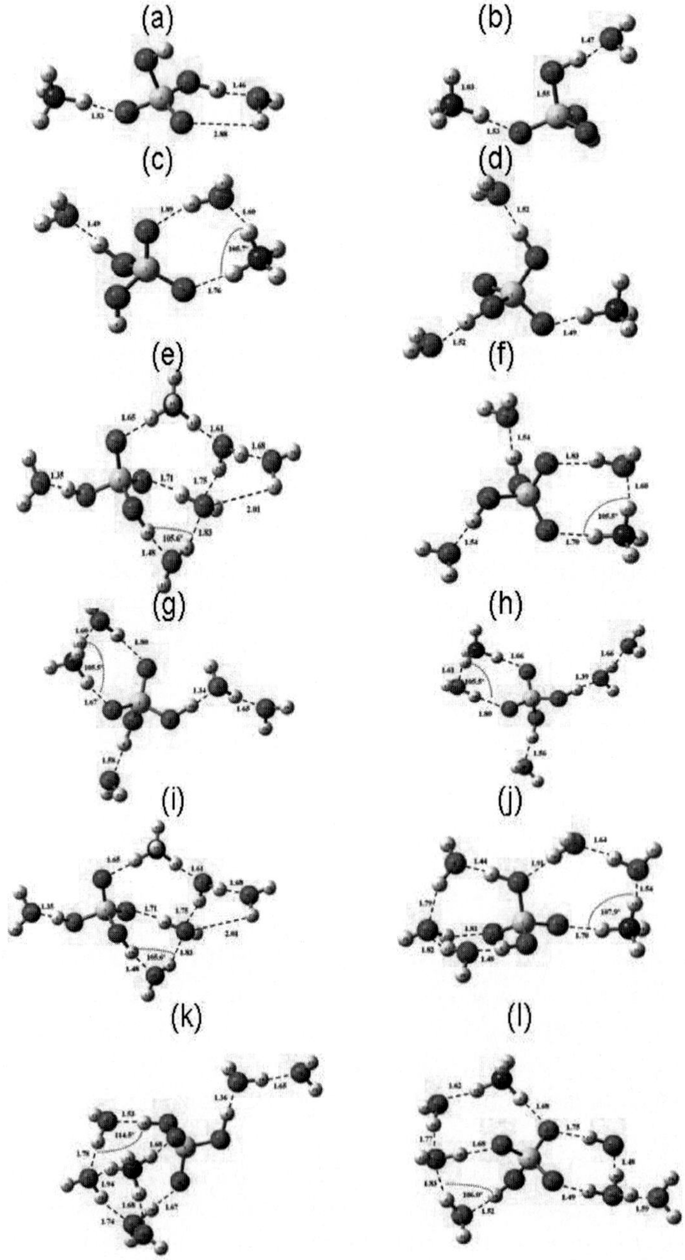

Figure 16. Most stable isomers (two in each class) of $(NH_4^+)(H_2SO_4)(H_2O)$ ((a); (b) 0.09 kcal mole^{-1}); $(NH_4^+)(H_2SO_4)(H_2O)_2$, ((c); (d) 1.80 kcal mole^{-1}); $(NH_4^+)(H_2SO_4)(H_2O)_3$, ((e); (f) 1.86 kcal mole^{-1}) ; $(NH_4^+)(H_2SO_4)(H_2O)_4$, ((g); (h) 0.07 kcal mole^{-1}); $(NH_4^+)(H_2SO_4)(H_2O)_5$, ((i); (j) 0.07 kcal mole^{-1}); $(NH_4^+)(H_2SO_4)(H_2O)_6$, ((k); (l) 0.07 kcal mole^{-1});

The thermochemistry of the sulfuric acid, the key atmospheric nucleation precursor, is critically important for the assessment of the efficiency of the ions as nucleation agents. Table 38 shows the comparison of the enthalpy, entropy and Gibbs free energy changes associated with the formation of (NH4+)(H2SO4)(H2O)n and (H3O+)(H2SO4)(H2O)n by the addition of the sulfuric acid. As seen from Table 38, the affinities of the sulfuric acid to (NH4+)(H2O)n and (H3O+) (H2O)n exhibit similar behavior. The difference in Gibbs free energies associated with the (NH4+) (H2SO4)(H2O)n and (H3O+)(H2SO4)(H2O)n formation is below ~ 4 kcal mole-1 for monohydrates and it drops to < 0.5 kcal mole-1 at n=5.

Table 37. Comparison of hydration enthalpies, entropies and Gibbs free energy associated with the formation of (NH4+)(H2SO4)(H2O)n , (NH4+)(H2O)n,, (NH3) (H2SO4)(H2O)n [107], (H3O+)(H2SO4)(H2O)n and (H3O+)(H2O)n [22]. Data by Piccard et al. [112] are given in parentheses. T=298.15K and P=101.3 KPa.

	ΔH	ΔG
$(NH_4^+)(H_2SO_4) + H_2O \Leftrightarrow (NH_4^+)(H_2SO_4)(H_2O)_1$	-17.91	-9.87
$(H_3O^+)(H_2SO_4) + H_2O \Leftrightarrow (H_3O^+)(H_2SO_4)(H_2O)_1$	-24.45	-14.27
$(NH_4^+) + H_2O \Leftrightarrow (NH_4^+)(H_2O)_1$	-21.84 (-20.28)	-13.3 (-14.11)
$(H_3O^+) + H_2O \Leftrightarrow (H_3O^+)(H_2O)_1$	-37.84	-28.89
$(NH_3)(H_2SO_4) + H_2O \Leftrightarrow (NH_3)(H_2SO_4)(H_2O)_1$	-10.96	-1.41
$(NH_4^+)(H_2SO_4)(H_2O)_1 + H_2O \Leftrightarrow (NH_4^+)(H_2SO_4)(H_2O)_2$	-17.69	-8.11
$(H_3O^+)(H_2SO_4)(H_2O)_1 + H_2O \Leftrightarrow (H_3O^+)(H_2SO_4)(H_2O)_2$	-16.84	-10.35
$(NH_4^+)(H_2O)_1 + H_2O \Leftrightarrow (NH_4^+)(H_2O)_2$	-15.83 (-17.04)	-10.62 (-8.63)
$(H_3O^+)(H_2O)_1 + H_2O \Leftrightarrow (H_3O^+)(H_2O)_2$	-22.81	-14.65
$(NH_3)(H_2SO_4)(H_2O)_1 + H_2O \Leftrightarrow (NH_3)(H_2SO_4)(H_2O)_2$	-14.07	-3.05
$(NH_4^+)(H_2SO_4)(H_2O)_2 + H_2O \Leftrightarrow (NH_4^+)(H_2SO_4)(H_2O)_3$	-14.90	-6.09
$(H_3O^+)(H_2SO_4)(H_2O)_2 + H_2O \Leftrightarrow (H_3O^+)(H_2SO_4)(H_2O)_3$	-18.63	-7.07
$(NH_4^+) H_2O)_2 + H_2O \Leftrightarrow (NH_4^+)(H_2O)_3$	-13.41 (-14.71)	-5.49 (-5.97)
$(H_3O^+)(H_2O)_2 + H_2O \Leftrightarrow (H_3O^+)(H_2O)_3$	-18.12	-10.36
$(NH_3)(H_2SO_4)(H_2O)_2 + H_2O \Leftrightarrow (NH_3)(H_2SO_4)(H_2O)_3$	-11.6	-2.96
$(NH_4^+)(H_2SO_4)(H_2O)_3 + H_2O \Leftrightarrow (NH_4^+)(H_2SO_4)(H_2O)_4$	-12.04	-4.25
$(H_3O^+)(H_2SO_4)(H_2O)_3 + H_2O \Leftrightarrow (H_3O^+)(H_2SO_4)(H_2O)_4$	-14.84	-6.65
$(NH_4^+)(H_2O)_3 + H_2O \Leftrightarrow (NH_4^+)(H_2O)_4$	-11.09 (-12.72)	-5.88 (-5.11)
$(H_3O^+)(H_2O)_3 + H_2O \Leftrightarrow (H_3O^+)(H_2O)_4$	-13.65	-5.54
$(NH_3)(H_2SO_4)(H_2O)_3 + H_2O \Leftrightarrow (NH_3)(H_2SO_4)(H_2O)_4$	-10.77	-2.28
$(NH_4^+)(H_2SO_4)(H_2O)_4 + H_2O \Leftrightarrow (NH_4^+)(H_2SO_4)(H_2O)_5$	-11.68	-1.63
$(H_3O^+)(H_2SO_4)(H_2O)_4 + H_2O \Leftrightarrow (H_3O^+)(H_2SO_4)(H_2O)_5$	-12.48	-4.47
$(NH_4^+) (H_2O)_4 + H_2O \Leftrightarrow (NH_4^+)(H_2O)_5$	-10.89	-0.31
$(H_3O^+) (H_2O)_4 + H_2O \Leftrightarrow (H_3O^+)(H_2O)_5$	-12.08	-4.17
$(NH_3)(H_2SO_4)(H_2O)_4 + H_2O \Leftrightarrow (NH_3)(H_2SO_4)(H_2O)_5$	-10.7	-0.01
$(NH_4^+)(H_2SO_4)(H_2O)_5 + H_2O \Leftrightarrow (NH_4^+)(H_2SO_4)(H_2O)_6$	-12.78	-2.44
$(H_3O^+)(H_2SO_4)(H_2O)_5 + H_2O \Leftrightarrow (H_3O^+)(H_2SO_4)(H_2O)_6$	-	-
$(NH_4^+)(H_2O)_5 + H_2O \Leftrightarrow (NH_4^+)(H_2O)_6$	-12.45	-3.52
$(H_3O^+)(H_2O)_5 + H_2O \Leftrightarrow (H_3O^+)(H_2O)_6$	-11.16	-3.50
$(NH_3)(H_2SO_4)(H_2O)_5 + H_2O \Leftrightarrow (NH_3)(H_2SO_4)(H_2O)_6$	-	-
$(NH_4^+)(H_2SO_4)(H_2O)_6 + H_2O \Leftrightarrow (NH_4^+)(H_2SO_4)(H_2O)_7$	-	-
$(H_3O^+)(H_2SO_4)(H_2O)_6 + H_2O \Leftrightarrow (H_3O^+)(H_2SO_4)(H_2O)_7$	-	-
$(NH_4^+)(H_2O)_6 + H_2O \Leftrightarrow (NH_4^+)(H_2O)_7$	-11.03	0.87
$(H_3O^+)(H_2O)_6 + H_2O \Leftrightarrow (H_3O^+)(H_2O)_7$	-9.79	-2.46
$(NH_3)(H_2SO_4)(H_2O)_6 + H_2O \Leftrightarrow (NH_3)(H_2SO_4)(H_2O)_7$	-	-

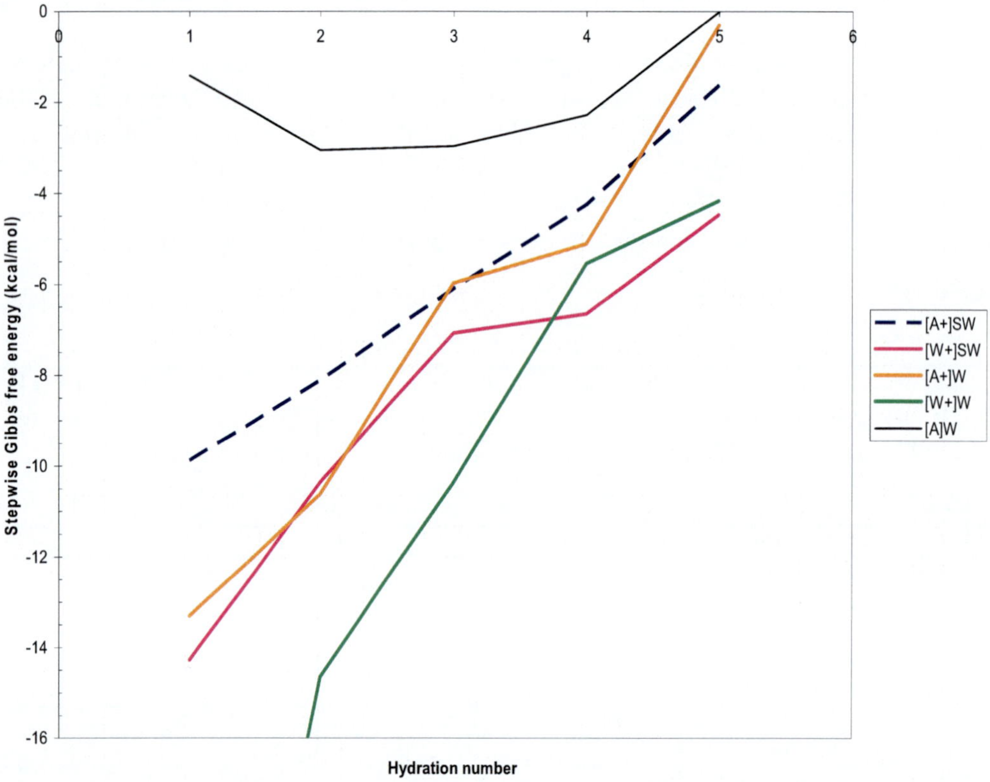

Figure 17. Hydration stepwise Gibbs free energy changes associated with formation of $(NH_4^+)(H_2SO_4)(H_2O)_n$ ([A+]SW), $(H_3O^+)(H_2SO_4)(H_2O)_n$ ([W+]SW), $(NH_4^+)(H_2O)_n$ ([A+]W), $(H_3O^+)(H_2O)_n$ ([W+]W), and $(NH_3)(H_2SO_4)(H_2O)_n$ ([A]SW).

Table 38 . Comparison of the enthalpy, entropy and Gibbs free energy changes associated with the formation of (NH4+)(H2SO4)(H2O)n and (H3O+)(H2SO4)(H2O)n by the addition of the sulfuric acid. T=298.15K and P=101.3 KPa.

	ΔH	ΔG
$(NH_4^+)(H_2O)_1 + (H_2SO_4) \Leftrightarrow (NH_4^+)(H_2SO_4)(H_2O)_1$	-17.55	-9.74
$(H_3O^+)(H_2O)_1 + (H_2SO_4) \Leftrightarrow (H_3O^+)(H_2SO_4)(H_2O)_1$	-24.02	-13.62
$(NH_4^+)(H_2O)_2 + (H_2SO_4) \Leftrightarrow (NH_4^+)(H_2SO_4)(H_2O)_2$	-19.41	-7.23
$(H_3O^+)(H_2O)_2 + (H_2SO_4) \Leftrightarrow (H_3O^+)(H_2SO_4)(H_2O)_2$	-18.05	-9.32
$(NH_4^+)(H_2O)_3 + (H_2SO_4) \Leftrightarrow (NH_4^+)(H_2SO_4)(H_2O)_3$	-20.90	-7.84
$(H_3O^+)(H_2O)_3 + (H_2SO_4) \Leftrightarrow (H_3O^+)(H_2SO_4)(H_2O)_3$	-14.77	-5.61
$(NH_4^+)(H_2O)_4 + (H_2SO_4) \Leftrightarrow (NH_4^+)(H_2SO_4)(H_2O)_4$	-21.85	-6.21
$(H_3O^+)(H_2O)_4 + (H_2SO_4) \Leftrightarrow (H_3O^+)(H_2SO_4)(H_2O)_4$	-19.75	-7.15
$(NH_4^+)(H_2O)_5 + (H_2SO_4) \Leftrightarrow (NH_4^+)(H_2SO_4)(H_2O)_5$	-25.09	-7.82
$(H_3O^+)(H_2O)_5 + (H_2SO_4) \Leftrightarrow (H_3O^+)(H_2SO_4)(H_2O)_5$	-20.15	-7.45

Table 39. Enthalpy and Gibbs free energy changes associated with (H3O)+(H2SO4)(H2O)n-1 + NH3 ↔ (NH4)+(H2SO4)(H2O)n, (H2SO4)(H2O)n + NH3 ↔ (NH3)(H2SO4)(H2O)n and (H3O)+(H2O)n-1+ NH3↔(NH4)+(H2O)n . T=298.15K and P=101.3 KPa.

	ΔH	ΔG
$(H_3O)^+(H_2SO_4)(H_2O)_0 + NH_3 \leftrightarrow (NH_4)^+(H_2SO_4)(H_2O)_1$	-40.59	-33.52
$(H_2SO_4)(H_2O)_1 + NH_3 \leftrightarrow (NH_3)(H_2SO_4)(H_2O)_1$	-16.71	-7.77
$(H_3O)^+(H_2O)_0 + NH_3 \leftrightarrow (NH_4)^+ (H_2O)_1$	-60.45	-52.01
$(H_3O)^+(H_2SO_4)(H_2O)_1 + NH_3 \leftrightarrow (NH_4)^+(H_2SO_4)(H_2O)_2$	-33.82	-27.35
$(H_2SO_4)(H_2O)_2 + NH_3 \leftrightarrow (NH_3)(H_2SO_4)(H_2O)_2$	-15.91	-6.9
$(H_3O)^+(H_2O)_1 + NH_3 \leftrightarrow (NH_4)^+ (H_2O)_2$	-38.44	-33.73
$(H_3O)^+(H_2SO_4)(H_2O)_2 + NH_3 \leftrightarrow (NH_4)^+(H_2SO_4)(H_2O)_3$	-31.88	-23.09
$(H_2SO_4)(H_2O)_3 + NH_3 \leftrightarrow (NH_3)(H_2SO_4)(H_2O)_3$	-17.68	-8.01
$(H_3O)^+(H_2O)_2 + NH_3 \leftrightarrow (NH_4)^+ (H_2O)_3$	-29.03	-24.57
$(H_3O)^+(H_2SO_4)(H_2O)_3 + NH_3 \leftrightarrow (NH_4)^+(H_2SO_4)(H_2O)_4$	-29.19	-20.72
$(H_2SO_4)(H_2O)_4 + NH_3 \leftrightarrow (NH_3)(H_2SO_4)(H_2O)_4$	-13.96	-6.88
$(H_3O)^+(H_2O)_3 + NH_3 \leftrightarrow (NH_4)^+ (H_2O)_4$	-19.23	-14.86
$(H_3O)^+(H_2SO_4)(H_2O)_4 + NH_3 \leftrightarrow (NH_4)^+(H_2SO_4)(H_2O)_5$	-22.24	-15.28
$(H_2SO_4)(H_2O)_5 + NH_3 \leftrightarrow (NH_3)(H_2SO_4)(H_2O)_5$	-14.66	-5.66
$(H_3O)^+(H_2O)_4 + NH_3 \leftrightarrow (NH_4)^+ (H_2O)_5$	-19.60	-14.22
$(H_3O)^+(H_2SO_4)(H_2O)_5 + NH_3 \leftrightarrow (NH_4)^+(H_2SO_4)(H_2O)_6$	-22.54	-13.25
$(H_2SO_4)(H_2O)_6 + NH_3 \leftrightarrow (NH_3)(H_2SO_4)(H_2O)_6$		
$(H_3O)^+(H_2O)_5 + NH_3 \leftrightarrow (NH_4)^+ (H_2O)_6$	-19.46	-9.85
$(H_3O)^+ (H_2SO_4)(H_2O)_6 + NH_3 \leftrightarrow (NH_4)^+(H_2SO_4)(H_2O)_7$	-13.65	-5.54

Table 39 presents enthalpies and Gibbs free energy associated with (H3O)+(H2SO4)(H2O)n-1 + NH3 ↔ (NH4)+(H2SO4)(H2O)n, (H2SO4)(H2O)n + NH3 ↔ (NH3)(H2SO4)(H2O)n and (H3O)+(H2O)n-1+ NH3↔(NH4)+(H2O)n . It is clear from Table 39 that the conversion of (H3O)+(H2O)n-1 into (NH4)+(H2O)n and (H3O)+(H2SO4)(H2O)n into (NH4)+(H2SO4)(H2O)n+1 are favorable thermodynamically in all the cases studied here. The high (~13-33 kcal mole-1) and (~10-52 kcal mole-1) affinity of ammonia to (H3O)+(H2SO4)(H2O)n-1 and (H3O)+(H2O)n-1 gives us a clear indication that in the real atmospheric conditions these reactions are kinetically controlled. The affinity of ammonia to positively charged clusters is much higher that that to neutral clusters of similar composition. This means that ammonia works as an efficient stabilizer of positively charged (H3O)+(H2SO4)(H2O)n clusters in the atmosphere.

The present thermochemical study of positively charged unary, binary and ternary clusters leads us to the following conclusions:

a) The protonation of ammonia leads to substantial enhancement in both the hydration strength and overall thermodynamic stability of unary and binary clusters.

b) Ammonia affects of positively charged and neutral aqueous clusters in different ways. While the presence of the sulfuric acid enhances the stability of neutral aqueous clusters, the stability of $(NH_4^+)(H_2O)_n$ cluster does not increase due to the presence of H_2SO_4 ;
c) The conversion of positively charged ions $(H_3O)^+(H_2O)_{n-1}$ and $(H_3O)^+(H_2SO_4)(H_2O)_n$ into $(NH_4)^+(H_2O)_n$ and $(NH_4)^+(H_2SO_4)(H_2O)_n$, respectively, by the addition of ammonia are favorable thermodynamically and are limited kinetically under typical atmospheric conditions. The efficiency of ammonia in stabilizing positively charged ions is much higher than that in stabilizing neutral clusters of similar composition.

Kinetic Nucleation Models Taking Advantage Of The Rigorous Clustering Thermochemistry And Detailed Simulations Of The Cluster Formation And Evolution In The Atmosphere

Most of theoretical nucleation studies in the past decades have been relying on the classical (equilibrium) approach based on the capillarity approximation. The classical approach does not explicitely treat clustering processes leading to the formation of critical embryos. The formation of new particles is actually a kinetic process involving forward and backward fluxes of clusters due to collisions of monomers with clusters and evaporation of monomers from clusters (see Fig. 2). Yu and colleagues pioneered the application of the kinetic approach to study the unary, binary, ternary, and ion-mediated nucleation processes [14-16, 19-21, 42, 52]. In the kinetic nucleation models, the formation and evolution of the clusters of various sizes (starting from monomers) are explicitly simulated.

The kinetic approach to nucleation has great advantages over classical theory because it allows for the incorporation of the thermodynamics of molecular clusters, and provides the detailed information about the size distributions of nucleating/growing clusters or particles. The incorporation of the thermochemistry of initial cluster formation, which can be derived using quantum-chemical methods (as shown in Section 5) and analysis of laboratory measurements, can significantly reduce the uncertainties in predicted nucleation rates. The updated kinetic unary, binary homogeneous and ion-mediation nucleation models [15, 21, 42, 52] have already taken into account some of clustering thermochemistry data that became available recently.

Due to the advancement in techniques of detection molecular clusters in the atmosphere, simultaneous measurements of pre-nucleation clusters, mobility spectrums of cluster ions, and the charged fraction of freshly nucleated nanoparticles have become available recently [38, 40]. A proper analysis of these new measurements could shed new light on the mechanisms governing atmospheric particle formation. The kinetic nucleation modeling is a nearly perfect choice for accomplishing this task [42]. Fig. 18 shows the evolution of the simulated particle size distributions for particles in several different categories on a nucleation event day (May 3, 2005) observed in boreal forests. The key parameters used to constrain the simulations are described in Yu and Turco [42]. The kinetic IMN model explicitly solves various microphysical processes controlling the formation and evolution of pre-nucleation clusters, including the evaporation of clusters, which depends strongly on the thermochemistry of initial steps of the cluster growth.

The present version of IMN model [42] assumes that only (H_2SO_4) and (H_2O) are involved in the nucleation process, although the subsequent condensation of organic species on nucleated particles is also considered. In the real atmosphere, ammonia and certain organic species may be involved in the formation of pre-nucleation clusters. To understand how these additional species may contribute to new particle formation in the atmosphere under different conditions, the nucleation models have to take into account the clustering of ammonia and organics with sulfuric acid clusters. The quantum-based thermochemistry data summarized in Section 5 are certainly useful for such purposes. The development of new and improved multicomponent kinetic nucleation models using the thermochemistry obtained from "first principles" are on-going research tasks.

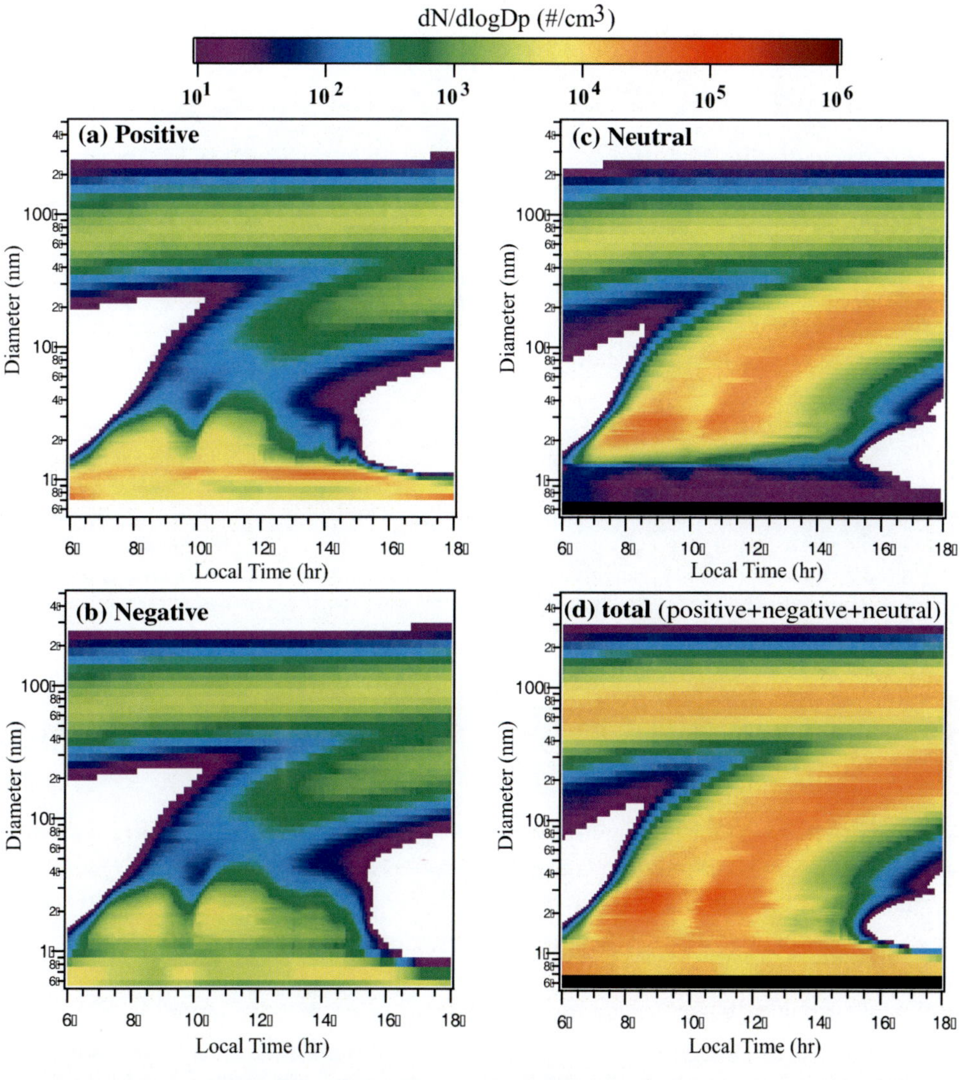

Figure 18. Simulated evolution of particle size distributions ($dN/d\log D_p$) for clusters/particles (starting from monomers, ~ 0.5 nm) of various types corresponding to the case study on May 3, 2005: (a) positively charged; (b) negatively charged; (c) neutral; and (d) total. (from Yu and Turco [42])

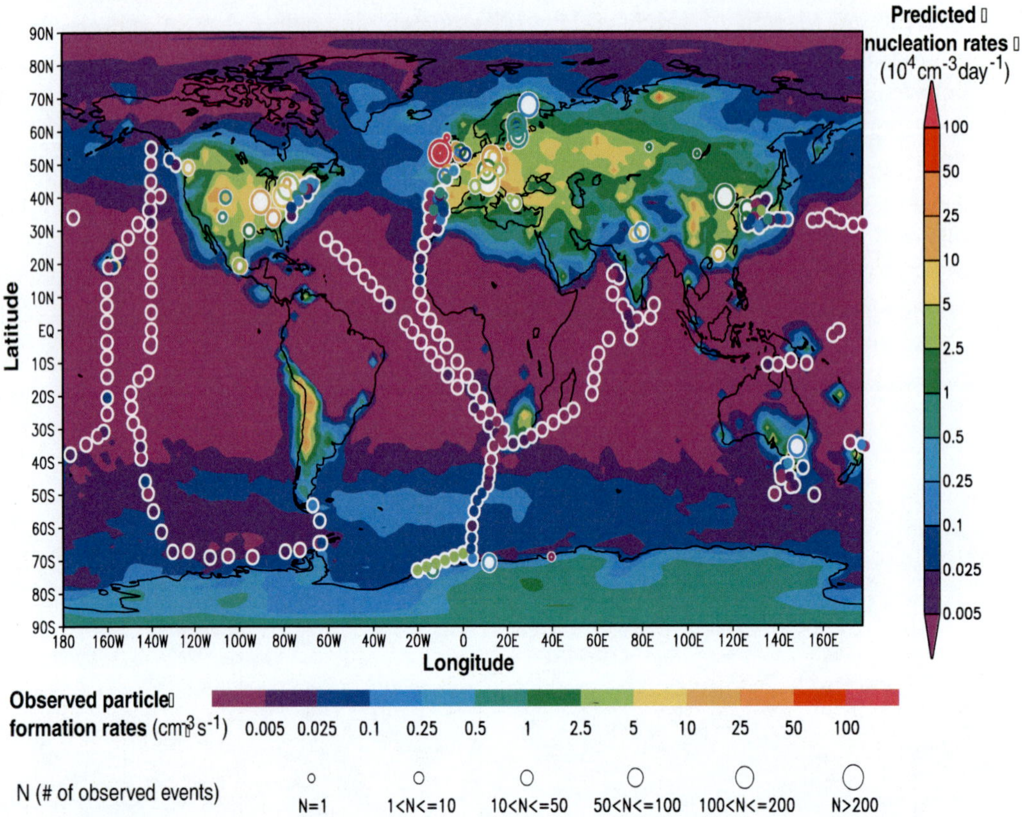

Figure 19. Simulated global distribution of annual mean ion-mediated nucleation rates averaged within the boundary layer (refer to the color bar on the right). Corresponding measured particle formation rates from ship observations are shown as color-filled circles (refer to the lower color bar, which gives the average local nucleation rate, typically over a window of several hours). Measurements over the land are indicated by unfilled circles, where circle size defines the number of nucleation events reported (refer to the scale at the bottom of the figure), while color gives the average nucleation rate over the event ensemble. (From Yu et al. [23])

Incorporation of nucleation module into global models and studies of new particle formation in the planetary atmosphere

New particle formation frequently observed throughout the troposphere is an important source of atmospheric particles and is one of the key processes that are need to be accurately represented in future generations of the climate models [23, 113]. Different nucleation theories/models predict quite different particle formation rates in the atmosphere. The number of controlling parameters and the dependence of nucleation rates on these parameters for different nucleation mechanisms are quite different. Therefore, it is important to identify the right nucleation mechanisms and to reduce the uncertainties in the existing nucleation models so that the aerosol radiative forcing can be confidently projected based on the future emission scenarios and climate change. As we pointed out earlier, the information on themochemistry of the initial steps of the formation of atmospheric molecular clusters is critically important for the development of new or improvement of the existing nucleation theories.

Nucleation mechanisms, which are constrained by reliable clustering themochemistry data, validated against available laboratory and/or field nucleation measurements and coupled with global models, can provide useful insights about the spatial and temporal variations of new particle formation in the planetary atmosphere. Fig. 19 gives one of such examples. The previously described IMN model, in which some of recently available thermochemistry data have been incorporated, has been coupled with a global chemical transport model to investigate the global nucleation [23]. It is clear that any further advancement in the understanding of molecular/quantum nature of clustering processes relevant to atmospheric nucleation can be readily used to study the atmospheric implications on the global scale.

SUMMARY

The thermochemistry of clusters formed in the Earth atmosphere during nucleation events is critically important for the determination of nucleation rates and resulting aerosol production related directly to global climate changes. Although the theory of atmospheric nucleation has been developing for several decades, a clear and insight understanding of the molecular nature of nucleation phenomena and mechanisms controlling the burst of new particles in the atmosphere is yet to be achieved. The thermochemistry is the cornerstone of the atmospheric nucleation theory, whose further progress is impossible without a quantum-mechanical theoretical treatment of the thermochemistry of nucleating and pre-nucleation clusters. Although the "first principles" atmospheric nucleation theory is unlikely to be finalized in the nearest future, the atmospheric science can already benefit from the application of the quantum approach to the thermochemistry of atmospheric clusters/complexes. The quantum theory can be used for the determination of the relative stability of atmospheric clusters, computations of the cluster distributions, reaction and nucleation pathways, the effects of the trace species on nucleation rates, and for the modification of existing nucleation models, which can be corrected using the data from quantum computations. Thermochemical data for the pre-nucleation clusters obtained using computational quantum methods can be conveniently incorporated in the framework of kinetic nucleation models providing explicit simulations of the observed nucleation events. The nucleation models constrained by quantum-derived thermochemical data for pre-nucleation clusters can be implemented, after the proper validation, in global models to study the new particle formation in the planetary atmosphere and various implications associated with the aerosol radiative forcing and climate change.

ACKNOWLEDGMENTS

The support of this study by NSF under grant by 0618124 and the NOAA/DOC under grant NA05OAR4310103 is gratefully acknowledged.

REFERENCES

. Health. Persp., 110, 440

[1] Saxon, D. and D. Diaz-Sanchez (2005), Nature Immunol. 6(3), 223.

[2] Vehkamäki H., et al. (2002), J. Geophys. Res., 107 (D22), 4622, doi:10.-1029/2002JD002184.

[3] Kulmala, M., Laaksonen, A., Pirjola, L. (1998), J. Geophys. Res., 103, 8301

[4] Yu, F. (2005), J. Chem. Phys., 122, 074501.

[5] Yu, F. (2007), J. Chem. Phys., 127, 054301.

[6] Yu, F. (2006), J. Geophys. Res., 111, D01204, doi:10.1029/2005JD005968.

[7] Napari, I., M. Noppel, H. Vehkamäki, and M. Kulmala (2002), J. Geophys. Res., 107(D19), 4381, doi:10.1029/2002JD002132.

[8] Ball, S. M., D. R. Hanson, F. L. Eisele, P. H. McMurry (1999), J. Geophys. Res., (D19), 23709-23718, 10.1029/1999JD900411.

[9] Yu, F., R. P. Turco (2000) , Geophys. Res. Lett., 27, 883

[10] Yu, F., R. P. Turco (2001) , J. Geophys. Res., 106, 4797

[11] Yu, F. (2006), Atmos. Chem. Phys. 6 (12) 5193

[12] Nadykto, A. B. et al. (2006), Phys. Rev. Lett. (2006) 125701.

[13] Yu, F., Z. Wang, G. Luo, R. P. Turco (2008), Atmos. Chem. Phys. 8, 2537-2554.

[14] Zhang, R., et al. (2004), Science, 304, 1487

[15] Nadykto, A.B., F. Yu (2007), Chem. Phys. Lett., 435, , 14

[16] Kim, T. O., T. Ishida, M. Adachi, K. Okuyama, J. H. Seinfeld (1998) Aerosol Sci,Techn., 29, 112

[17] Christensen, P. S., S. Wedel, and H. Livbjerg (1994), Chem. Eng. Sci., 49, 4605

[18] Diamond, G. L., J. V. Iribarne, D. J. Corr (1985) J. Aerosol Sc., 16, 43

[19] Nolan, P. M. (1987), Condensation nuclei formation by photooxidation of sulfur dioxide with OH radical in the presence of water vapor and ammonia, Thesis, Drexel University.

[20] Merikanto, J., I. Napari, H. Vehkamäki, T. Anttila, and M. Kulmala (2007), J. Geophys. Res., 112, D15207.

[21] Nadykto A.B. et al (2003) Chem. Phys. Lett., 382/1-2, 6

[22] Nadykto A.B, F. Yu (2003) J. Geophy. Res., 108(D23), 4717, doi:10.1029/ 2003JD-003664.

[23] Nadykto A.B. , F. Yu (2004) Atmos. Chem. Phys., 4, 385

[24] Yu, F. (2005) , J. Chem. Phys., 122, 084503.

[25] Froyd, K. D (2002), Ion induced nucleation in the atmosphere: Studies of NH3, H2SO4, and H2O cluster ions, Ph.D. thesis, Univ. of Colo., Boulder.

[26] Wilhelm, S., Eichkorn, S., Wiedner, D., Pirjola, L., and Arnold, F. (2004), Atmos. Environ., 38, 1735–1744. [37] Hirsikko, A. et al.(2007). Atmos. Chem. Phys. 7, 201

[27] Laakso, L., et al. (2007) Atmos. Chem. Phys. 7, 1333

[28] Yu, F., R. P. Turco (2007), in Nucleation and Atmospheric Aerosols, C. D. O'Dowd and P. E. Wagner (eds.), p392-396, Springer 2007..

[29] Kulmala, M. et al. (2007), Science, 318, 89-92.

[30] Yu, F., Z. Wang, G. Luo, and R. P. Turco (2007), Atmos. Chem. Phys. Discuss., 7, S6602–S6608 & S6754–S6766.

[31] Yu, F., and R. P. Turco (2008), Atmos. Chem. Phys. Discuss., 8, 5683-5723.
[32] Volmer M., A. Weber (1926) Z. Phys. Chem. 119, 277
[33] Becker, R., W. Doering (1935) Ann. Phys. 24, 719
[34] Schenter, G.K., Kathmann, S.M., Garrett, B.C. (1999) Phys. Rev. Lett, 82 (17), 3484
[35] Brodskaya, E.P., A.P. Lyubartsev, A. Laaksonen (2002) J. Phys. Chem. B 106, 6479
[36] Oh, K. J. , G. T. Gao X. C. Zeng (2001) Phys. Rev. Lett. 86, 5080
[37] Warshavsky, V.B., X.C. Zeng (2002) Phys. Rev. Lett. 89, 246104
[38] Kathmann, S.M., G. K. Schenter, B. C. Garrett (2007) Phys. Rev. Lett. 98, 109603
[39] Kathmann, S.M., G. K. Schenter, B. C. Garrett, (2007) J. Phys. Chem., C 111 (13) 4977
[40] Nadykto, A. B. et al. (2007) , Phys. Rev. Lett. 98(10), 109604.
[41] Du H., A.B. Nadykto, F.Yu, (2007) Phys. Rev.E, (2008), revised
[42] Helgaker, T.U., P. Jorgensen, J. Olsen (2000) Molecular Electronic-Structure Theory , Wiley
[43] Levine, I.N. (1999) Quantum Chemistry, Prentice Hall.
[44] Kerdcharoen T., U. Birkenheuer, S. Krüger , A. Woiterski and , N. Rösch (2003) Theor.Chem. Accounts, 109 (6), 285.
[45] Choe, Y.-K., Tsuchida, E., Ikeshoji, T. (2007) J. Chem. Phys. 126 (15), art. no. 154510
[46] Mikkelsen, K.V. (2006) Ann. Rev. Phys. Chem., 57, 365
[47] A.R Hoy, P.R. Bunker (1979) J. Molecular Structure 74, 1-7.
[48] G. Herzberg, (1966) Electronic spectra and electronic structure of polyatomic molecules, Van Nostrand, New York
[49] Hellwege, K.H. , Hellwege A.M. (Eds), (1976) Landolt-Bornstein: Group II: Atomic and Molecular Physics 7: Structure Data of Free Polyatomic Molecules, Springer-Verlag-Berlin
[50] Miller, Y. , Chaban, G.M. , Gerber R.B. (2005) J. Phys. Chem. A 109(29) 6565
[51] Re, S., Osamura, Y. , Morokuma, M. (1999) J. Phys. Chem. A 103, 3535
[52] Bandy, A.R. , Ianni, J.C. (1998) J. Phys. Chem. A 102(32), 6533
[53] Kuczkowski, R.L. , Suenram, R.D. , Lovas , F.J. (1981) J. Am. Chem. Soc. 103, 2561
[54] Taketsugu, T. ; Wales,D. (2002) Mol. Phys., 100(17), 2793-
[55] Kim, J., et al., (1995) J. Chem. Phys. 102(1), 310
[56] Tschumper, G.S. 1999, Ph. D. thesis, University of Georgia.
[57] Klopper, W., Rijdt, J.G., van Duijneveldt, F.B. (2000) Phys. Chem. Chem. Phys., 2, 2227
[58] Goldman, N. et al. (2002) J. Chem. Phys. 116(23), 10148
[59] Dyke, T.R., Mack, K.M., Muenter, J.S. (1977) J. Chem. Phys. 66(2), 498
[60] Odutola, J.A., Dyke,T.R. (1980) J. Chem. Phys. 72(9), 5062
[61] Kuchitsu, K., Morino,Y. (1965) Bulletin of the Chemical Society of Japan 38, 805
[62] Nielsen,I.M.B., Seidl,E.T, Janssen, C.L. (1999) J. Chem. Phys.110(19), 9435
[63] Liu K., Brown, M. G., Saykally, R. J. (1997) J. Phys. Chem. A 101, 8995
[64] Ding, C.-G., K. Laasonen, A. Laaksonen (2003), J. Phys. Chem. A., 107(41), 8648.
[65] Al Natsheh, A. et al. (2004) J. Phys. Chem. A, 108(41), 8914; J. Phys.Chem. A 110 (25) (2006) 7982
[66] Fiacco, D. L., S.W. Hunt, K. R. Leopold (2002) J. Am. Chem. Soc, 124(16), 4504.
[67] NIST Chemistry Webbook (2006)

[68] Shimanouchi, T. (1972) Tables of molecular vibrational frequencies, C.V. 1, NSRDS NBS-39
[69] Nadykto, A. B., H.Du , F.Yu (2007) Vibr. Spectr. 44, 2, 286
[70] Miller, Y.; Chaban, G.M.; Gerber, R.B. (2005) Chem. Phys., 313 (1-3), 213.
[71] Richter, F., Hochlaf, M. , Rosmus, P. , Gatti, F., Meyer, H.-D. (2004) J. Chem. Phys. 120 , 1306.
[72] Xantheas, S.S. Intern. Rev. Phys. Chem. 2006, 25 (4),.719.
[73] Diri,K, E.M. Myshakin, K.D.Jordan (2005) J. Phys. Chem. A, 109, 4005
[74] Dunn , M. et al. (2006) J. Phys. Chem. A, 110, 303
[75] Chaban,G.M., J.O.Jung , R.B.Gerber (2000) J. Phys. Chem. A, 104, 4952
[76] Dunn, M.E., Pokon, E.K., Shields, G.C. (2004) J. Am. Chem. Soc. 126(8), 2647
[77] Munoz-Caro, C. , Nino, A. (1997) J. Phys. Chem. A, 101(22): p. 4128-4135
[78] Curtiss, L.A., Frurip, D.J., Blander, M. (1979) J. Chem. Phys 71(6), 2703
[79] Kurdi L., E. Kochanski (1989) Chem. Phys. Lett. 158, 111
[80] Ianni, J. C., A. R. Bandy (1999) J. Phys. Chem. A, 103, 2801-2811.
[81] Ianni, J.C., A.R. Bandy (2000), J. Mol. Structure (Theochem), 497, 19
[82] Al Natsheh, A. et al. (2006) Chem. Phys., 324, 210;
[83] Al Natsheh, A. et al. (2006) Chem. Phys. Lett., 426, 20
[84] Kurten, T. et al. (2006) J. Phys. Chem. A 110, 7178.
[85] Kurten, T. et al. (2007) J. Geophys. Res. 112, D4, D04210, 10.1029/2006JD007391
[86] Kurten, T. et al. (2007) Boreal Env. Res. 12 (3) 431.
[87] Kurten, T. et al. (2007) Atmos. Chem. Phys. 7 (10), 2765
[88]]Arrouvel,C., V. Viossat , C. Minot (2005) J. Mol. Struct.: THEOCHEM 718, 71
[89]]Larson, L. J., A. Largent, F.-M. Tao (1999) J. Phys. Chem. A, 103, 6786.
[90]]Nadykto, A.B., F.Yu (2007), in Nucleation and Atmospheric Aerosols, K.O'Dowd and P.E.Wagner, Editors, Springer, 297.
[91] Nadykto A.B., F.Yu, J.Herb, (2008) Chem.Phys., revised,
[92] Nadykto, A.B., F.Yu ,A. Al. Natsheh, (2008) IJMS, revised
[93] Bianco, R., Wang, S., Hynes, J.T. (2005) J. Phys. Chem. B 109 (45) 21313
[94] Hanson, D. R., F. L. Eisele (2000) J. Phys. Chem. A, 104, 1715.
[95] Nadykto, A..B. et al. (2008) Adv. Quant. Chem. 55, 449
[96] Arnold, F., (1992) Ber. Bunsen-Ges. Phys. Chem. Chem. Phys. 96, 339.
[97] Ya. Kim , Yo. Kim (2002).Chem. Phys. Lett. 362, 419
[98] Froyd, K. D., Lovejoy, E. R. (2003) J. Phys. Chem. A, 107, 9800
[99] Froyd, K. D., Lovejoy E. R. (2003) J. Phys. Chem. A, 107, 9812
[100] Pickard IV, F.C., Dunn, M.E., Shields, G.C. (2005) J. Phys.Chem. A 109 (22), 4905.
[101] [113] Ghan, S. J. and S. E. Schwartz (2007), Bull. Amer. Meteorol. Soc. 88, 1059.

In: Thermochemistry and Advances in Chemistry Research ISBN: 978-1-60692-376-4
Ed: Anton G. Mirskiy

Chapter 3

TESTS OF MATERIALS. STUDY OF OPERATION MECHANISM AT HIGH-TEMPERATURE

G. E. Zaikov, A. A. Donskoi[1] and M. A. Shashkina[1]
N.M.Emanuel Institute of Biochemical Physics Russian
Academy of Sciences, 4 Kosygin str., Moscow 119991, Russia
[1] Institute of Aviation Materials, 17 Radio str., Moscow, 119105, Russia

Development of fire and heat shield materials with reduced combustibility consists of several stages.

Heat energy from an external source, delivered to the material surface, can increase its temperature so that degradation of material components or chemical reaction between them begins. Combustible degradation products formed are capable of igniting and combust independently of flame spreading. The problem of creating fire and heat shield materials is closely connected with the questions of material combustibility decrease.

Under normal conditions, heat transfer proceeds from higher heated (hot) bodies to less heated (cold) ones by means of heat conductivity or convection. Under fire conditions, irradiation is the main type of heat transfer from flame or incandescent material surface. The radiant flux falling on the body surface consists of three components [1 - 3].

The radiant flux permeates through the cover to the protected object surface and heats it up, which is unwanted for fire and heat shield materials. Heat flux absorbed by the material is accumulated in it or transforms into different types of energy [4].

The level and spectral characteristics of the radiant flux, absorption properties of the material, and reflecting power of the surface in relation to emission spectrum of falling flux affect the amount of energy absorbed by polymeric composite material.

Transmission and absorption coefficients depend on the chemical composition and structure of the substance. If the absorption coefficient is low, the radiant flux transmits throughout the material. If it is high, a thin layer of the material is rapidly heated up to a critical temperature of material decomposition. Therewith, the maximum temperature is observed on the surface protecting the layers below.

From positions of increasing fire resistance, it is desirable to increase reflecting power of materials, which is low for organic compounds, polymers in particular [4]. The reflecting power can be increased by both surface treatment and injection of fillers capable of reflecting the radiant energy in the infrared spectrum part into the compounding of composite material. Oxides and salt of various metals can be applied as such materials.

Fillers are often subdivided into active and passive ones. However, this division is rather relative, because one and the same filler may be a passive diluter of the condensed phase at low temperatures under different operation conditions, and under other conditions it may be transformed physically and chemically consuming heat energy for these processes.

At decreasing combustibility of polymeric materials in high-temperature flux, it is common practice to reduce the role of mineral fillers in the polymeric matrix to dilution of the condensed phase, change of thermophysical properties of the system and physical and chemical transformations. However, one may note a fortuitous coincidence that some compounds used as combustion decelerators are polymerization and polycondensation catalysts [5]. One may suggest these compounds to be capable of catalyzing reactions of cross-linking and coke formation under thermal degradation conditions.

1. Coke Formation as the Investigation Method of Fire and Heat Shield Materials and Covers

Covers from rubbers and rubber-like materials with the surface affected by high-temperature flux are capable of forming a coke-like solid residue with a porous structure. Such layer formed on the surface possesses low heat conductivity. This layer is gradually destroyed and substituted by a new layer of changed material formed on the surface. As underlying layers of the cover degrade thermally, volatile products are released transmitting though the "coke" and forming the border gas layer representing an additional heat resistance. It is noted that the use of rubbers has allowed a two-fold reduction in the weight of fire and heat shield covers.

There is an opinion that the ideal heat shield material is the one completely exhausted up to the end of high-temperature flux effect [6].

The importance of coke forming ability of the organic base of fire- and heat shielding materials is also outlined in ref. [7]: "Carbon properties relate directly to consideration of organic materials, because the main product of their thermal degradation under optimal conditions is carbon. It can be stated that application of plastics is just a special way of making carbon contact with incandescent gas flow". It is noted that the problem of application of protective materials at high temperatures has no unique solution: both chemical and physical aspects are of the same importance. Consideration of enthalpy of gases in the composition of solid propellant combustion products and melting temperatures of cover components are important, because if the latter is much lower than the temperature of cover heating, the phase transition heat is not used effectively enough due to rapidly carry off of the melt. Melting temperatures of components must be chosen so that melting proceeds not too rapidly, for example, refrasil melting at 2000 K is the most effective filler at solid propellant combusting at 3300 K. Fillers of the silicon dioxide type are suitable, because at this temperature their melting heat is effectively consumed in the neutral oxidative medium.

Ref. [8] shows calculation of heat transfer through the cover under the following assumptions: the cover surface obtains temperature of propellant combustion products immediately, the cover thickness does not change with time, the organic part of the material degrades at a definite temperature and known energy consumption, a stable coke-like layer is formed on the surface, the unchanged part of the cover possesses low heat conductivity, gaseous products possess no cooling capability, heat flux is perpendicular to the surface, and thermal constants are independent of temperature. The calculation formula is the following:

$$a = R\sqrt{\tau},$$

where a is the coordinate of the cover pyrolysis zone (the origin of coordinates is located on the surface contacting with the gas flow); τ is the duration of high temperature influence; R is a constant typical of the material.

Hence, the penetration depth of the thermal degradation zone into the cover is proportional to the square root of the duration of heating. Calculations show that degradation temperature increase causes an insignificant effect on the cover destruction rate, and the change of latent pyrolysis heat changes it significantly. Thermophysical constants of the unchanged part of the cover are of secondary importance, but it is useful if they provide for possibly higher temperature gradient.

Application of fillers or reinforcement complicates the mechanism of cover destruction, because several zones are formed in it; but the general character of cover operation remains practically unchanged. The coke-like layer should not increase heat conductivity of the cover, and fillers (oxides, nitrides, carbides, zirconates, and titanates) should change with absorption of maximum possible energy at high temperatures. Erosion behavior of such composites in oxygen-acetylene flame is better than of graphite, but this was not confirmed by tests in a high-rate flux.

Contrary to thermal degradation processes of natural products (mineral coal, crude oil and gas), which are used in industry for producing a great variety of materials, including carbon black production widely used in rubber industry, up to quite recent times, pyrolysis of high-polymeric materials was used only for studying their structure, kinetics and aging mechanism, stability at increased temperatures and under effect of oxidants, and identification in analyses of rubbers, in particular. Among great variety of works on the study of high polymers, investigations of their thermal degradation are few.

Hereinafter, by high temperatures we mean the ones at which “valence states, compounds formed and general properties of systems are significantly different from valence states, compounds and properties at room temperature” [9].

Initial works studying thermal degradation products of elastomers date back to 1929, when products of natural rubber refining in batches 7 kg each treated in iron vessels under atmospheric pressure and temperature of 973 K were studied in detail [10]. Among liquid products, 23 compounds possessing from 5 to 10 carbon atoms, including isoprene (10%) and dipentene (20%), were successfully identified. Later, thermal degradation of rubbers was studied mostly on smaller samples [11, 12].

Refs. [13, 14] present a detailed review of works on degradation of high polymers under the effect of various chemical and physical factors, and heat in particular. Depolymerization processes at moderate heating and thermal degradation of polybutadiene, polyisobutylene, butadiene-styrene rubber, natural rubber, polyethylene, polytetrafluoroethylene, etc. are discussed. Possible mechanisms of degradation and depolymerization are described.

Pyrolysis of polyisoprene and polyisobutylene is initiated at 300°C and terminated at 400°C; for polybutadiene, approximately at 350°C and 477°C; and for polyethylene, at 360°C and 475°C, respectively. Degradation of these polymers is almost complete: the residues after pyrolysis are 3.8% for polyisoprene, 0.3% for polybutadiene at 500°C, 0.4% for butadiene-styrene rubber at 455°C, and 1.4% for polyethylene at 475°C. These residues dissolve in benzene and cyclohexane, which is untypical of the carbon residue.

Ref. [15] also provides information about pyrolysis of butadiene-styrene rubber. It is noted that exothermal degradation is initiated at 380°C. At 430°C, it becomes so intense that ignition may occur.

If linear rubbers form no noticeable solid residue during pyrolysis, pyrolysis of phenol-formaldehyde resins gives up to 50% of it, and thermoreactive organosilicon resins give 85% [16].

In the case of pyrolysis of aliphatic compounds, the coke residue is absent or negligibly low. The presence of great amount of condensed ring structures produces a significant coke residue, because carbonaceous residues are formed at pyrolysis of cyclic compounds. Thus it becomes clear that preliminary cyclization during dehydration at high temperature is desirable for raising coking capacity of linear polymers.

Carbonization of samples of the materials studied can be held in both oxidation and reduction media. The medium composition is usually selected depending on expected operation conditions of the material. One of the investigation methods in reduction media is the following [16]. Samples of appropriate geometrical shape are placed into crucibles and covered by a coal charge. Temperature increase rate is 4°/min. As a temperature of 1,000°C is reached, the samples are exposed for 1 hour at this temperature. The amount of solid residue is calculated by the formula:

$$K = \frac{P_{fin}}{P_{ini}} \cdot 100\%,$$

where K is the amount of solid residue; P_{ini} is the initial sample weight; P_{fin} is the sample weight after test.

Proper geometrical shape allows a study of strength properties of the solid residue obtained.

2. The Role of Intumescence in the Problem of Fire Protection of Polymers

The fire shield mechanism of polymeric material components is quite complicated and manifold. To study the protective mechanism of the entire material and its separate components, the determining sign of the process and the main operation principle of the present material should be outlined. Intumescence can be this sign, which considers simultaneously processes of gasification and coke formation in polymeric material under high-temperature influence.

The technology of intumescence is quite new in the polymer science as the method providing for polymer protection from flame effects. Intumescent systems terminate polymer combustion in its early stage, i.e. at the stage of thermal degradation of the polymer accompanied by release of combustible gas products.

The intumescent process consists in combining coke formation and puffing up of combusting polymer surface. The foamed up cellular coke-like layer formed, the density of which reduces with temperature [17], protects the protective material from heat flux and flame effects.

2.1. Intumescence Chemistry

Intumescent additives usually include three components: an acidic component necessary for acidic catalytic effect, polyalcohols (as carbonizing compounds) and foaming agent. In the initial stage ($T > 280°C$), the acidic component interacts with carbonizing agent [18].

Carbonization proceeds at about 280°C. Therewith, Friedel-Craft reactions and free-radical processes take place [19]. Later, the foaming agent decays with release of gas products, which induces foaming up of the coke layer. Such intumescent material decomposes with temperature increase.

When protecting a polymer from the effect of high temperature, carbonized material participates in two chemical processes:

- Reactions between free-radical fragments of foamed material and radical products of the gas phase, which are the products of polymer composite degradation. These free-radical fragments of the coke layer can participate in termination reactions of radical chains formed during pyrolysis of polymeric composites in the condensed phase;
- Acid-catalytic reactions with oxidation products formed during thermooxidative degradation of the material.

In composition, the intumescent coke layer is a heterogeneous system. It represents the condensed phase, phosphorus-carbonized cells of which contain gaseous products. In turn, the condensed phase consists of solid and liquid phases (acid-catalyzed resins), which contain liquid and gaseous products of polymer degradation. Carbonized fractions of the condensed phase consist of polyaromatic fragments, formed into layers typical of graphite structures.

Further on, phosphorus-carbonized material forms definite areas in the material consisting of crystalline intermolecular polyaromatic layers linked by bridge bonds of polymeric chains and phosphate (poly-, di- or ortho-) groups, crystalline particles and the amorphous phase, disseminated into crystalline zones. This amorphous phase consists of small polyaromatic molecules, obtained by hydrolysis of phosphate fragments of alkyl chains – the decomposition products of material components and fragments of polymeric chains. Phosphorus-carbonized material possesses fire shield properties under the following conditions: it should coat completely the unchanged protected polymer and possess strength providing mechanical resistance of the cover formed.

2.2. Protection through Intumescence

The protection mechanism suggested is based on the coke layer effect as a physical barrier reducing heat and mass exchange between the gas and condensed phases. The presence of polymeric chain fragments in the intumescent layer tindicates that it absorbs combustible gaseous products of polymer pyrolysis. Moreover, the intumescent layer prevents diffusion of gaseous fuel into the flame zone, as well as restricts oxidant access from the surrounding air to the polymeric material.

Stability of the intumescent material limits formation of gaseous fuel and leads to spontaneous extinguish under standard conditions.

Recently, many intumescent additives have been studied [20 - 22], but they contained three main components: catalyst, coke forming and foaming up agents.

Coke formation catalysts are usually phosphorus-containing compounds. The the fact that they must be added in relatively large amounts (up to 15-20%) does not correlate with common ideas of catalysis. That is whyit can be suggested that these compounds may participate in formation of a new structure.

Ref. [23] shows analysis of the literature data on the study of fire shielding of some polymers (polypropylene, ethylene-butyl acrylate copolymer with maleic anhydride, poly(ethylene terephthalate), cotton, and polyacrylonitrile) containing ammonia polyphosphates and ethylene diammonia phosphates as fire shield additives combined with various synergic additives providing for the intumescent effect. Pentaerythritol, thimethylol melamine, hexabromocyclodecane, and triphenyl phosphate were studied as additives. The analysis indicates comparatively low efficiency of ammonium polyphosphates, which increases in the presence of synergists. The effect of these additives depends on the chemical structure.

Injection of compounds containing nitrogen, halogens, and antimony into the composites can increase efficiency of phosphorus-containing components. Studies of various systems showed synergism of the effect of these elements during intumescence.

For example, phosphorus-nitrogen bonds can participate in forming volumetric network structures, in which phosphorus will be fixed and thus diffusion will be complicated [24]. The element analysis and IR-spectroscopy carried out by Weil et al. confirmed existence of phosphorus-nitrogen bonds on the coke surface, formed during combustion of the composite containing ethylenevinyl acetate, pyrophosphate melamine, hexamethyltrisdioxyphosphorus, and melamine trioxide [25].

Antimony-halogens synergism for reducing combustibility of polymeric systems is commonly known. The halogen-antimony synergism is displayed in the condensed and the gas phases [26].

The main effect of the presence of antimony trioxide is observed in the gas phase. Antimony halides formed in the gas phase react with atomic oxygen, water and hydroxyl radicals, giving SbO and hydrogen halide. Dispersed solid SbO and Sb are formed in the flame zone. They catalyze recombination of hydrogen radicals. Moreover, there is an opinion that antimony halides decelerate halogen release from flame promoting the decrease of combustible components content.

Data on synergic interaction between bromine and chlorine exist in the literature [27 28]. In the majority of cases, the highest effect is observed when they are in equal concentrations and at a total content of 10 – 12%.

There are several notes about existence of bromine-phosphorus synergism in the literature. Studies of the oxygen index show that bromine compounds cause no noticeable effect in the gas phase, but act as foaming up agents during coke formation [23].

Comparison of phosphorus-bromine and antimony-bromine synergic efficiencies displays higher values for the second pair.

Ref. [29] suggests that phosphorus compounds are capable of causing the synergic effect with halogens. This is confirmed for the case of polymers containing oxygen that leads to a significant decrease of additive concentration used in the polymeric system.

2.3. The Role of Ammonium Polyphosphates in Protection of Polymers from Fire

Synergic efficiency of ammonium polyphosphates in polypropylene is higher than in other systems [23]. As such polymeric composite combusts in the presence of pentaerythritol, the following processes proceed: ammonia polyphosphate decomposition with ammonia and water release, pentaerythritol phosphorylation and polypropylene thermooxidation, dehydration, dephosphorylation, network cross-linking, carbonization and coke structure formation. Foaming up agent is delivered into the gas phase and decomposes to incombustible products.

Studying pyrolysis of the composition containing pentaerythritol and ammonia polyphosphate, Brauman [30] suggested that ammonium polyphosphate not only formed a protective layer, but also participated in chemical reactions in the condensed phase. These processes proceed at very high but different rates. Predominance of one or other reaction at any time depends on the relation of ratios that shows up on the coke properties during combustion.

Comparative analysis of systems containing different phosphates [31] shows that quite low values of the oxygen index are associated with formation of phosphorus oxides during combustion, which reduces phosphorus concentration in the coke. Addition of low amounts of zeolites restricts formation of condensed polyphosphate products and increases content of acid phosphate products in the intumescent material [32, 33].

However, higher thermostable metal-ammonium polyphosphates induce a destabilizing effect shown on the example of polyamide [34].

Good fire shield properties were observed for phosphamacyclomatrix inorganic polymer with high thermal stability [35]. Phospham structure is not detected: chemical analysis indicates a structure with formula of the type $(PN_2H)_x$.

Phospham is a combustion decelerator for polyamide, which is displayed by the oxygen index change, shown in Table 1.

The highest jump-like increase of the oxygen index is observed at injection of 10% of phospham. Further increase of its content causes no sharp rise of the index. The coke formed contains phospham; however, polyamide also participates in formation of solid residue, which confirms increased coke yield compared with the amount of injected modifier (Table 1).

Studies of polyalcohol-melamine coking system enabled the role of melamine in this process to be understood. Evaporation from the solid phase and degradation in flame are accompanied by an endothermal effect. Melamine also participates in phosphorylation at simultaneous foaming up during intumescence, because the following gaseous products are formed during melamine degradation: water, CO, CO_2, ammonia, and hydrocarbons promoting foaming up of the solid residue.

Table 1. Oxygen index and coke residue yield during combustion of polyamide containing phospham

Phospham, wt.%	Oxygen index, cond. units	Coke yield, wt.%
-	25.2	-
10	29.2	17.5
20	31.6	23.5
30	34.8	32.2

Studying intumescent additive consisting of ammonium polyphosphate and pentaerythritol in different polymeric matrices (polypropylene, polyethylene, polystyrene) using IR-spectroscopy, Sevostianov and Novikov [36] showed that chemical structure of the coke was hardly influenced by the type of polymeric matrix, and contents of carbon and phosphorus atoms in the solid residue corresponded to their amount in the intumescent additive.

Studying the composite based on propylene with the same additive, Delobel et al. [31] showed that polyphosphate chains were formed under thermal influence, and the solid residue contained increased amount of orthophosphate compounds. Ammonium polyphosphate substitution by diammonium pyrophosphate caused formation of pyrophosphate fragments instead of orthophosphate ones [37].

Study of the coke pore structure and the surface shows that the ratio of phosphorus to carbon increases with temperature (up to 500°C), whereas this ratio decreases in the main mass of the system. The oxygen-carbon ratio is of the same type, which suggested to the authors migration of phosphonates to the surface accompanied by their oxidation.

In ref. [38] it is shown that injection of zeolites into the polymeric system increases efficiency of intumescent additive reducing heat release and suppressing smoke formation. Authors of ref. [38] suggest that phosphorus-carbon structures are formed in the intumescent system, which become more stable in the presence of zeolite. Zeolite promotes formation of organic phosphates and/or aluminum phosphates in clusters of polymeric chains and thereby restricts depolymerization and, consequently, the amount of combustible gaseous products delivered into the flame zone. Moreover, it is shown that zeolite promotes formation of higher "coherent" structure in the polymeric material. Occurrence of a "coherent" macromolecular network and interaction with polymeric chains increase fire shield properties of the material. Actually, formation of polyaromatic structures in the intumescent protecting barrier makes the material stronger. The material surface becomes more flexible, which reduces probability of cracks occurring on the surface under high temperature effect. Therewith, diffusion of oxygen into polymeric matrix and combustible products of polymer degradation into the combustion zone is decelerated.

Polymers can also be used as carbonizing agents, which reduces the peak of heat release and delays sample ignition in time. Such resistance to heat flux effect can be explained by formation of a new fire shield layer, more resistant to cracking, formed in reactions of material components with injected polymer. Study of heat transfer in some systems shows that such protective layer is quite effective in limiting destruction of the material surface [23].

Ref. [23] discusses in detail the intumescent mechanism of material protection from flame and high temperature effects, composed of simpler simultaneously proceeding processes, including formation of degradation gas products and solid residue on the surface of sample or cover. These processes, the main ones in the intumescence mechanism, must be adjusted in time and by rate of proceeding for every polymeric linkage and conditions of thermal effect. During polymer combustion, a part of supplied heat energy is absorbed by the material and consumed for its degradation. Entering the gas phase, volatile products sustain combustion as fuel or oxidant. Obviously, the combustion rate should be determined by the amount of supplied heat and the amount that is used for heating up, phase transformations and polymer degradation.

Estimation of flame temperature effect on the rate of polymer degradation by the linear pyrolysis method consisted in studying processes of material degradation at one-side effect of heat energy, it showed that the rate of polymer degradation increases with the flux temperature. However, the degradation rate is essentially associated with the nature of the polymer [39].

The study of temperature profiles on linear pyrolysis of a series of polymers shows that surface temperatures of degrading polystyrene, polyethylene, foam polystyrene and epoxy resin increase with heat flux power, whereas the surface temperature of poly(methyl methacrylate) remains practically constant. In the case of the latter polymer, surface structure change was observed, which led to an increase in the sample specific surface.

The rate of linear pyrolysis increases and values of oxygen index decrease with surface temperature. Dependence of oxygen index on temperature is of nonlinear type and possesses curve bendings in the area of phase transitions, which is created by heat expended in these processes.

Measurements of temperature profiles in the condensed phase during combustion of poly(methyl methacrylate) and polystyrene show that surface temperatures and temperature fields near the surface, i.e. in the degradation zone, are identical. This indicates invariability of the combustion rate under critical conditions.

Hence, existence of the oxygen index is, in fact, stipulated by minimal rates of polymer degradation, at which the necessary amount of fuel is delivered to the gas phase. The role of fuel is played by volatile products of polymer degradation [39, 40]. Decrease of the oxygen index with temperature affecting the surface should mean reduction of the heat flux delivered from flame to the polymer surface. Obviously, the heat flux decrease is associated with the flame temperature decrease with oxygen concentration. One may suggest that gasification of polymers requires comparatively low amounts of heat energy, and if combustible volatile products are released during pyrolysis of polymers at a rate exceeding the minimum one, then the atmosphere possesses enough oxygen to sustain stable combustion [40].

Analysis of gaseous products selected from various flame zones of combusting poly(methyl methacrylate) in the nitrogen-oxygen mixture possessing various concentrations of components displays the presence of oxygen in the whole volume of the flame. The highest amount of oxygen entering reactions is observed in high-temperature zones and the ones above the surface. The latter indicates the possibility of proceeding of polymer thermooxidative degradation during combustion. Estimation of the linear pyrolysis rate of a series of polymers in heated gas flows with different oxygen concentrations shows that the presence of oxygen in flame can show up differently on the rate of polymer degradation. Oxygen concentration increase does not affect pyrolysis of polystyrene, induces an insignificant increase of poly(methyl methacrylate) degradation rate, but exerts a significant effect on polyethylene, which is associated with rapid oxidation reactions in the surface layer. Studies of epoxy resin at low oxygen concentrations show that the pyrolysis rate in the presence of oxygen is higher than in inert medium. However, starting from 3 vol.%, the degradation rate decreases abruptly, and therewith carbonized residue is formed on the polymeric sample surface [40, 41].

The notion of "carbonized" or "coke" residue is widely used in works on combustion of polymers. The term includes thermoresistant products of polymer pyrolysis, which, beside carbon, can contain nitrogen, oxygen, phosphorus and other elements [42, 43].

Formation of a carbonized surface layer promotes a decrease of polymeric material combustibility [44 - 47]. The main reasons for suppression of combustion due to carbonization are the following:

- The coke formed makes penetration of the heat energy into the condensed phase difficult;
- Carbonized layer prevents oxygen diffusion from the surroundings to degrading polymer;
- The presence of carbonized layer prevents exit of gaseous and liquid products to the surface.

Clear presentation of the role of separate process mechanisms, the degree of their participation and ratio in decreasing combustibility of polymeric materials has not yet been formulated.

At combustion of carbonizing polymers it is observed that, in some cases, flame combustion continues after the end of flame influence on the coke crust formation. One may suppose that, on the one hand, coke occurrence prevents heat permeation into the condensed phase decreasing the material combustibility and, on the other hand, the presence of porous structure promotes mass transfer of liquid degradation products to heated up surface of the material, which in its turn, promotes their combustion [44].

Behavior of a carbonizing polymeric system under the effect of radiant flux was studied on the model representing polymer samples covered by plates of porous thermoresistant material. It is found that the rate of poly(methyl methacrylate) degradation increases with heat flux intensity and reduces with the thickness of foam material. The rate of linear pyrolysis is also reduced with decrease in the condensed phase heating through.

The change of condensed phase heating rate has an essential effect on the mechanism of polymer pyrolysis, which is associated with competing processes of degradation and structuring. Study of degradation of polymers possessing reactive groups shows that degree of decomposition increases with the heating rate [48].

The change of combustion and pyrolysis conditions of polymers has an essential effect on toxicity of products formed. The gas chromatography method of studying gaseous degradation products shows the presence of definite ranges of oxygen concentration, in which the maximum amount of carbon oxide is released from different polymers [48 - 50].

Coke formed during combustion possesses porous structure, and gaseous degradation products permeate through the carbonized layer and enter the combustion zone. For liquid products, coke can be a "fuse", by which the liquid rises due to capillary forces and, entering the surface, sustains combustion.

Tests on linear pyrolysis of samples from poly(methyl methacrylate) covered by plates from porous thermoresistant material show that increase of protecting plate thickness at the same heat effect induces surface temperature increase and changes the level height of liquid degradation products in the foam material. The latter is associated with distribution of temperatures in the porous materials and its heat conductivity.

Liquid motion in foam materials adheres to the Darsy law [51]. Combustion rate of carbonizing polymers is determined by the rate of coke gasification and the transmission rate of gas and liquid products of polymeric material degradation through the carbonized layer. The amount of gas and liquid degradation products decreases with increase of carbonized residue yield. In the limit, the combustion rate may be determined by the rate of oxidation pyrolysis of the carbonized layer. Simultaneously, permeability of cokes formed during pyrolysis of the basic polymeric compound varies in a wide range. Calculations show [294] that at the present permeability, the carbonaceous layer becomes the obstacle for releasing volatile degradation products into the gas phase.

Permeability data [42] show that coke possesses about one-third of its volume as through pores with small diameter, by which liquids can rise by capillary forces. Viscosity of liquid products is the important factor affecting the rate of their motion by the carbonized layer [51, 52]. Calculations and experiments show that polymer melts can also transmit through the surface carbonized layer [52]. If polymer melting temperature is low, and the melt is of low viscosity, or liquid products are easily formed during pyrolysis, then coke formed on the surface cannot be the effective protection from fire [51].

Considering combustion of carbonizing polymers, it is desirable to separate two surfaces: the carbonized coke surface contacting the gas phase and the surface of degrading polymer contacting the coke. Analysis shows that the amount of heat absorbed by the polymer decreases with temperature rise of the coke surface, because the contribution of heat energy due to convection decreases and heat release by irradiation increases due to surface temperature rise. Heat losses for heating and gasification of the condensed phase also increase. Therewith, a great contribution is made by rise of enthalpy of gas products permeating through the coke, because heat capacity of gases is much higher than that of solids.

Existence of the minimal limit rates of combustion implies that creation of incombustible polymers requires striving for a decrease in the release rate of pyrolysis volatile products capable of igniting and possessing high heating capacity. Decrease of polymer combustibility is favored by processes proceeding in the condensed phase with high heat absorption. An effective measure for decreasing combustibility may be the carbonized layer formed on the surface [39].

Analyzing the above-said, the following ways of decreasing combustibility of carbonizing polymers can be outlined:

- Increase of carbonized residue yield, which decreases the amount of volatile products released into the combustion zone;

- Increase of coke thermoresistance increasing temperature of the material surface that promotes decrease of convective heat flux, and rise of heat energy scattering by the surface by irradiation and heat losses for material heating. The presence of coke on the surface forms a barrier for heat flux from flame and, consequently, it is desirable to increase the thickness of the carbonized layer and to decrease its heat conductivity;
- Decrease of coke permeability and increase of viscosity of polymer degradation liquid products to restrict their rise by carbonized product.

An empirical approach to decrease of polymer combustibility helped in detecting the main classes of combustion inhibitors. However, many problems connected with the mechanism of these compounds are not clear yet.

Chlorine- and bromine-containing compounds are widely applied to decelerate combustion of polymeric materials; fluorine and iodine derivatives are not used. Based on the interaction scheme of hydrogen halides with active radicals in the hydrocarbon flame, thermodynamic and kinetic calculations were executed using reference data. These calculations show that practically all the reactions with participation of hydrogen halides in the temperature range of 700 – 1,700°C are thermodynamically abandoned, and that hydrogen bromides and iodides are the most reactive compounds. This was confirmed experimentally by estimating oxygen indices of several polymers filled with zeolites, preliminarily saturated with hydrogen halides. Somewhat underestimated results were obtained for estimation of hydrogen iodide inhibiting ability, which is associated with its decomposition into components with temperature [53].

For reducing combustibility of polymers, phosphorus-containing compounds of various structures are the most widely used. Under combustion conditions, they promote formation of carbonized residue. Taking into account that many phosphorus-containing combustion inhibitors decompose to acids at heating and assuming the possibility of their oxidation, it is desirable to pay attention to phosphoric acids and their ammonium salts [48 - 58].

It is commonly assumed that phosphorus-containing compounds decrease combustion, because they promote carbonization of polymeric materials during pyrolysis and combustion. On the other hand, measurements of permeability of carbonized residues obtained during pyrolysis of phenol-formaldehyde resins to which ammonium monophosphate has been added show that the presence of a phosphorus-containing compound reduces the Darsy constant by several times [52, 59]. Consequently, one of the reasons for polymer combustibility decrease in the presence of phosphorus-containing compounds is the decrease of carbonized layer permeability.

Poly(vinyl alcohol) and epoxy resin combust in air, and at dynamic heating up to 500°C, i.e. to the level of combusting polymer surface temperature, they decomposed almost completely. Therewith, addition of phosphoric acids or their ammonium salts induces formation of a thermoresistant residue, combustibility of polymers being decreased simultaneously (the oxygen index of epoxy resin increases to 30) [60 - 64].

Combustion inhibition by phosphorus-containing compounds shows up not only in formation of carbonized layer, but also in decrease of coke permeability. This fact was confirmed by analyzing the coke residue obtained from phenol-formaldehyde resin and poly(phenylene dimaleimide), thermally treated by phosphoric acid, as well as in pyrolysis of polymeric materials modified by ammonium phosphate. Introduction of phosphorus compounds into composite material leads to a decrease of Darsy constants by almost 15 times, which shows up on combustibility decrease of polymeric material. For example, oxygen indices of materials based on epoxy resin, covered by coke plates treated by phosphorus-containing compounds, equal 76 and 52. If there was no treatment by phosphoric compounds performed, the limit of oxygen concentration is 45 and 35%, respectively. Permeability lowering in porous materials accompanying treatment by phosphoric compounds is probably associated with the fact that polyphosphates formed at high temperatures possess high viscosity and fill in the coke pores [65 - 67].

Some boric compounds also cause such effect on permeability of carbonized layer [50, 59]. Combustibility of the system based on polystyrene covered by a coke layer, obtained by pyrolysis of phenol-formaldehyde resin filled with boron oxide and cross-linked by hexamethylenetetramine, is estimated in the works mentioned. However, it is noted that permeability of cokes depends on temperature in the presence of boric compounds: it decreases first with temperature rise to 450°C, and then increases, which shows up on the material combustibility characterized by oxygen index. Oxygen index reaches 50 at 450°C, and thereafter reduces to 40 – 42. This is associated with the fact that at 450°C boric oxide becomes a limpid liquid, which probably covers coke pores and promotes permeability lowering. Further heating leads to lowering viscosity of films from boron-containing compounds. As a consequence, the surface of carbonized material is uncovered and burnt off.

Oxidation processes induce formation of carbonized residue during pyrolysis and combustion of some polymeric materials. The study of potassium hypochlorite (decomposing with oxygen release) influence on epoxy resin combustion shows that at low concentrations this substance increases coke residue yield and reduces material combustibility (oxygen index of epoxy resin equals 23 at 5 wt.% concentration). Increase of potassium hypochlorite concentration leads to an abrupt decrease of the oxygen index.

The oxidation process can be affected by multi-valent metal compounds. The studying of pyrolysis of materials based on epoxy resins in the presence of compounds with copper, tin, antimony, and cobalt shows that initiation of polymeric matrix degradation in the presence of tin compounds is shifted by 60 – 80°C towards high temperatures, and the solid residue yield is increased. Electron-spectroscopic studies of polymer oxidative degradation in the temperature range of 240 – 320°C show that addition of tin compounds significantly increases the induction period of paramagnetic particle occurrence, reduces the rate of their accumulation and total amount that shows up on thermooxidative degradation [68 - 72]. The presence of tin- and cobalt-containing compounds promotes formation of carbonized layer and, consequently, decrease of the combustion rate. Chromatographic studies also show quantitative change of the composition of pyrolysis volatile products.

Carbonization processes are also promoted by the presence of sulfogroups. However, the data on application of sulfocontaining compounds as inhibitors of polymeric material combustion, shown in the literature, are contradictory. Measurements of permeability of pyrolysis products of phenol-formaldehyde resin added by sulfponilamide show that the presence of sulfocontaining compound, similar to phosphorus-containing one, leads to a decrease of the Darsy constant almost by an order of magnitude. However, in this case, permeability lowering may be associated with foaming of the mixture of phenolic resin with sulfonilamide, which provides a way for liquid degradation products to erupt into the combustion zone [45]. Sulfur derivatives can participate in the coke composition and remain stable at high temperatures. Study of pyrolysis and foam formation of aryl sulfonilamide compounds, which are capable of forming thin foam, shows that they may have potential for creating fire shield materials and their operation mechanism should be studied.

Efficiency of carbonized layer increases with formation of foamed coke layer on the cover surface. A stable foamed coke can be formed in the case of coincidence of gas formation and polymeric system viscosity increase rates with future transition of polymer into solid. If these rates are different, the foam is unstable and subsides as it forms. In this case, protective properties of the foam are not displayed. Formation of a foam stable at heating requires system transition from melt to solid with sharp increase of the melt viscosity providing fixing of gas product bubbles in the product formed. Such viscosity increase at heating is possible by using components with several types of reactive groups, interaction of which at high temperatures leads to formation of non-melting compounds with spatial cross-linking of macromolecules. As an example, let us present the system containing para-aminobenzene sulfonilamide, described in ref. [73].

Thermogravimetric study shows that under conditions of isothermal heating of the given compound, two areas of intensive degradation are observed: in the range of 280 – 350°C with the mass loss of 20 – 50% and in the range of 450 – 550°C with the mass loss of 70 – 95%. Foaming up with gas products release, both volatile and nonvolatile substances, proceeds at 280 – 350°C, which corresponds to the first area of intensive degradation. In the second stage, nonvolatile foam material degrades.

The authors of ref. [73] have found that ammona release in the first stage probably proceeds under interaction of the electrophilic sulfur atom with a nucleophilic nitrogen atom of the amino group. Consequently, redistribution of electron densities at heating forms favorable conditions for polyamination and reamination to occur. Formation of such structures was confirmed by infrared spectroscopy data of volatile products obtained at 240°C, which display an absorption band at 3250 cm–1 typical of valence oscillations of secondary –NH- amine bond. Therewith, intensity of bands at 3200 and 3300 cm–1 typical of sulfonamide grouping is significantly reduced.

Ammonia release shifts the equilibrium to the right, and thereafter the second molecule of para-aminobenzene sulfonilamide can be attached forming an oligomer, which increases viscosity. Solid products formed at 250 – 260°C represent cross-linked insoluble spatial structures. Formation of spatially cross-linked insoluble products during pyrolysis becomes possible owing to interaction of the end amino group with secondary amino group [74].

As temperatures up to 360°C, sulfonilamide groups begin interacting with one another, as well as bonds in sulfonilamide compound backbone break with formation of foam cokes. It is found that SO3 is the main gaseous product of the reactions in the given temperature range. The amount of SO3 released increases with temperature.

According to the element analysis data of carbonized residue, a tendency to reduce sulfur content, but concentration of nitrogen and carbon with temperature is observed. IR- and mass-spectroscopy of solid degradation products shows bands occurring in the range of 1,400 cm–1 typical of R-SO2-OR sulfonates, and R-O-SO2-O-R esters of sulfonic acids. Asymmetric structure of R-SO2-S-R is confirmed by observing asymmetric oscillations of S-O bond accompanying symmetric ones in the range of 1,150 cm–1. Occurrence of an absorption band in the range of 1,040 – 1,020 cm–1 characterizes valence symmetric oscillations of RSO3H and RSO3H4 sulfoacids and their salts. Mass-spectrum data indicate that backbone and cross-linking breaks with formation of high-molecular compounds proceed simultaneously at para-aminobenzene sulfonilamide is heated. It is suggested that oligomer degradation proceeds as the result of S-N bond breaking, which are the weakest in the chain. This confirms the possibility of forming sulfonic acids, which as known are very unstable and transform into esters of thiosulfoacids. Being strong agents of dehydration, cyclization and cross-linking, sulfoacids formed can promote carbonization and lead to foam coke formation [74].

Based on the above-mentioned data and considering analysis of the literature, one may suggest that a significant role at thermal decomposition of para-aminobenzene sulfonilamide in the temperature range of 280 – 360°C is played by free radicals formation at the stage of the backbone rupture with their future recombination. One of the main reactions is the interaction of highly active phenyl radicals with aromatic rings, which leads to formation of a spatially cross-linked structure.

Hence, based on the mechanism suggested it can be concluded that condensation reactions with ammonia release proceed in the initial stage at thermal decomposition of para-aminobenzene sulfonilamide. Thereafter weak bonds rupture with SO3 release and formation of free radicals participating in the foam coke formation. The foam material formed is characterized by spatially cross-linked structures, which make an essential contribution to reduction of polymeric material combustibility.

Addition of dicarboxylic acids, for example, terephthalic acid, which interact with sulfonilamide or products of its decomposition under combustion conditions, increases thermal resistance of the foam. On evidence of thermogravimetric analysis under conditions of isothermal mode at 360°C, terephthalic acid is volatilized, sulfonilamide loses 55% of mass, and the sulfenamide + terephthalic acid mixture loses 25% of mass only, which indicates formation of higher thermoresistant products. As the result of chemical transformations proceeding at heating sulfonilamide mixture with terephthalic acid, the coke structure is changed, which is confirmed by X-ray patterns [74, 75], and a more regulated coke structure is observed. Traces of the crystalline phase typical of hexagonal graphite-like structure occur at 350°C that leads to a sharp increase of fire resistance of foam cokes.

Injection of fire shield additives into polymeric composites, shaped as mechanical admixtures, during the stage of component mixing is the most widespread method of increasing fire resistance of polymeric materials. However, an irregular distribution of components of the composite material is possible in this method. This leads to a significant instability of material properties.

A more effective method of conferring fire resistance or even full incombustibility on the polymeric material is polymer modification during the stage of its production by injecting combustion inhibitor into the polymeric chain structure.

As hydrogen halides inhibit combustion and bromine is one of the most active halogens, the supposition that vinyl bromide will decompose releasing hydrogen bromide is checked on a series of copolymers with vinyl bromide, synthesized specially for the purpose [319]. The study of the oxygen index of polymers obtained shows that compared with homopolymers, methyl methacrylate copolymers with butyl acrylate display a negligible decrease of combustibility, and styrene and acrylonitrile ones display a greater rise of the oxygen index. Thermogravimetric analysis shows that vinyl bromide injection reduces thermal resistance of polymers. In the case of copolymers based on styrene and acrylonitrile, a carbonized residue is formed during combustion and pyrolysis that reduces the degradation rate at one-side heating [76].

Of great interest are dehydrochlorination and aromatization of hydrocarbons in the presence of catalysts – titanium, cobalt, and aluminum oxides, and aluminum phosphates [5]. The mechanism of highly effective combustion decelerators of the intumescent type is associated with catalysis of coke formation proceeding with participation of polymer molecules or a carbon forming component, specially injected into the system [77].

Based on the thermal balance equation of polymer combustion, substances degrading or reacting with heat absorption should promote combustibility decrease. Decomposition enthalpies of a series of salt crystalline hydrates, hydroxides and other substances capable of transforming with heat energy absorption were measured on a differential scanning microcalorimeter. The study of the oxygen index of polymeric materials containing these compounds shows an increase in the presence of the latter [78 - 80].

Exothermal processes proceeding in the condensed phase are accompanied by combustion increase of polymeric materials.

The study of the influence of carbonaceous residue on combustion of polymeric material shows that a carbonized layer can serve as a powerful screen protecting from heat energy permeation into the lower layers of the material. The insulating effect increases with the layer thickness and decrease of its heat conductivity that can be achieved by forming a foamed up material under the fire effect. This principle is used in developing fire and heat shield materials. Foaming up covers increase in volume under the heat flow effect and form incombustible foam material [81 - 84].

Estimation of the temperature field in the expanding cover, temperature variations with the cover thickness, and dependencies on heat effects of the reactions proceeding in the cover shows the main contribution of the height of the foam material formed and its heat conductivity into the durability of fire and heat shield covers. Processes proceeding in covers with significant endoeffects are negligible in the absence of carbonized foamed layer [85].

In the initial stage of one-side heating of fire shield foaming covers, heat is absorbed and removed to underlying layers at a rate limited by thermophysical characteristics of the initial polymeric composite. The material degrades with heating, and the upper layers begin foaming up, when reaching a definite temperature. During the initial period, temperature on the protected surface rises rapidly. Thereafter the rate of heating decreases as the consequence of foam material formation and hardly changes for some time. Note that the protected surface temperature does not increase with heat flux intensification if such materials are used. Foaming up material realizes the effect of foaming self-control, when rise of the heat flux intensity induces increase of the heat insulating foam volume, which prevents an intensive heating through of the protected surface. Temperature increase of the surroundings promotes deeper heating through of the cover and involving a high volume of the material into foam formation.

The foam formation rate and height of foam coke layer formed increase with the initial cover thickness. However, a tendency to overestimate the height of the foamed material is observed. Measurements of the temperature field in foaming covers show that bottom layers are heated up at a low rate, and they do not reach the foam formation temperature for a long time, i.e. they play the role of thermal insulation [86].

Double-layer covers are used, in which the upper layer provides the function of foaming up heat shield material, and the bottom one represents a polymeric matrix filled with substances decomposing with heat absorption. Salt crystalline hydrates, boric acid, etc. are used as “cooling down” materials. Such combination allows the weight of the initial cover to be reduced by 25 – 30% preserving fire shield properties at the same level [80]. An efficient method of increasing fire shield properties of the covers is found in application of a foaming material with higher heat insulating properties as the bottom layer. In the initial stage of cover operation, this prevents heating of the protected object and increases the heating rate of the upper layer up to the foaming initiation. The latter also increases the rate of foam coke formation and its thickness.

The mechanism of foaming up fire shield covers differs from foam production from liquids or foam materials. The necessary condition for the material to foam up under fire effect is the initial composite material transition into the rubbery state with future irreversible transition into the solid state [81, 84, 87, 88].

Considering the scheme of foaming up, one should take into account kinetics of gas products release, diffusion of gas bubbles due to expulsion force, and removal of gas products with regard to permeability of the condensed phase. The surrounding pressure, resistance of polymeric matrix to gas bubbles formation and overcoming the surface tension forces may produce a significant effect on this process.

Stable foam cannot be formed under conditions of one-side heating, when composite surface temperature increases continuously, if the system does not transform irreversibly to the solid state at high temperatures. Such process is observed at carbonized residue formation. At the initial moment of flame influence, the polymer is in the solid state. That is why formation of a porous structure requires high pressure in gas bubbles during gas products release to stretch the polymer. Further on, elasticity modulus of the polymeric material is reduced and, consequently, gas bubbles are able to leave the condensed phase pressed out by the repulsion force. Temperature in the condensed phase increases under the flame influence, viscosity is reduced, and therewith, foam subsidence is observed. As the reaction proceeds with formation of rubbery compounds accompanied by increase of the composite rigidity, the foamed layer is fixed [86].

Hence, obtaining of a foamed material under the fire effect required synchronous gas release and increase of viscosity and rigidity of the composite. At high degrees of foaming transition of the system from the viscous flow state into the solid one proceeds at a low rate, and as the rate of changing viscoelastic properties increases, thin dispersed material is formed. The transition rate of the polymeric material into the solid state is determined by reactions of rigid ladder and three-dimensional structures, the occurrence of which requires the presence of two or more reactive groups in the composite material [65, 66, 89]. Mixtures of polyatomic alcohols with phosphoric acid or compounds forming phosphoric acids during decomposition at increased temperatures may form components of such systems. Injection of compounds containing amino groups into such systems promotes increase in coke yield during pyrolysis. Addition of amines mixed with pentaerythritol and phosphoric acid shifts the initiation of coke formation towards lower temperatures. Interaction between alcohols and phosphoric acid increases system viscosity, which shows up favorably on the foam formation. Injection of nitrogen-containing compounds increases the melt viscosity and reduces the foaming up temperature, which shows up on the foam coke obtained. Further heating induces formation of double bonds and cross-linked structures [88, 90].

REFERENCES

[1] Deribere M., *Practical Application of Infrared Rays*, Gosenergoizdat, Moscow-Leningrad, 1959, 440 p. (Rus)

[2] Polezhaev Yu.V. and Yurevich F.B., *Heat Protection*, Ed. A.V. Lyikov, Energiya, Moscow, 1976, 380 p. (Rus)

[3] Margolin I.V. and Rumyantsev N.P., *Grounds of Infrared Engineering*, Voenizdat Min. Oborony, 1957, 308 p. (Rus)

[4] Lecont J., *Infrared Radiation*, Izdat. Fiz.-Mat. Literatury, Moscow, 1958, 584 p. (Rus)

[5] Dolgov B.N., *Catalysis in Organic Chemistry*, Moscow-Leningrad, Goskhimizdat, 1949, 560 p. (Rus)

[6] Ridder R.C., *Plast. Eng.*, 1977, vol. 33(1), pp. 38 – 42.

[7] Grunfest I.J., *British Plastics*, 1958, vol. 31(2), pp. 530, 531, 539.

[8] Hourt W.C., *Ind. Eng. Chemistry*, 1960, vol. 52(9), pp. 761 – 763.

[9] Bruer L. and Sirsi A.V., *Khimia Vysokikh Temperatur*, 1958, vol. 27(8), pp. 966 – 989. (Rus)

[10] Midgley T. and Henne A.L., *J. Chem. Soc.*, 1929, vol. 51(4), pp. 1215 – 1226.

[11] Burchfield H.P., *Ind. Eng. Chem. Anal. End.*, 1944, vol. 16(7), pp. 424 – 426.

[12] Wall L.A., *J. Res.Nat. Bur Stand.*, 1948, vol. 41(4), pp. 315 – 322.

[13] Grassi N., *Chemistry of Polymer Degradation Processes*, Moscow, Inostrannaya Literatura, 1959, 252 p. (Rus)

[14] Madorsky S.L., Strams S., Thompson D., and Williamson L., *J. Polym. Sci.*, 1968, vol. 4(5), p. 639.

[15] Simon V.L., Nitrile Rubbers, In Coll.: *Synthetic Rubber*, Ed. G.S. Witby, Leningrad, Goskhimizdat, 1957, pp. 777 – 822. (Rus)

[16] Madorsky S.L. and Strans S., *WADS Technical Report*, 1959, pp. 59 – 64.

[17] Bourbigot S., Morice L., and Leroy J., *Fire Retardancy of Polymers*, Ed. M. Le Bras, G. Camini, S. Bourbigot, and R. Delobel, 1988, No.224, Cambridge, pp. 129 – 139.

[18] Le Bras M., Bourbigot S., Le Tallec Y., and Laurens J., *Polym. and Stab.*, 1997, vol. 56, pp. 11 – 21.

[19] Le Bras M., Bourbigot S., Delporte C., Siat C., and Le Tallec J., *Fire and Materials*, 1996, vol. 20, pp. 191 – 203.

[20] Sharf D., In: *Rec. Adv. in Fr. Of Polym. Mat.*, vol. 2, Ed. M. Lewin, BCC, 1991, p. 55.

[21] Sharf D., Nalpta R., Heflin R., and Wus T., *Fire Safety J.*, 1992, vol. 8(19), p. 103.

[22] Camio G., Costa L., and Troassarelli L., *Polym. Degr. Stab.*, 1984, vol. 8, p. 243; *Polym. Degr. Stab.*, 1985, vol. 12, p. 203.

[23] Ruban L.V. and Zaikov G.E., The role of Intumescence in Fire Protection of Polymers, *Plasticheskie Massy*, 2000, No. 1, pp. 39 – 43. (Rus)

[24] Langley J., Drews M.J., and Barker R.T., *J. Appl. Polym. Sci.*, 1974, vol. 25, p. 243.

[25] Zhu W. *et al.*, *J. Appl. Polym. Sci.*, 1996, vol. 62, p. 2267.

[26] Costa L., Goberti L., Paganetto P., Camio G., and Squarzi P., *Proc. 3rd Meeting on FR Polymers*, Torino, 1989, p. 19.

[27] Cleave R.F., *Plastics and Polymer*, 1979, vol. 20, p. 783.

[28] Markezich R.L. and Mundhenke R.F., In: *Rec. Adv. in Fire Retardance of Polym. Mat.*, Ed. M. Lewin, BCC, 1995, vol. 66, p. 177.

[29] Dombrowski R. and Huggard M., In: *4th Intern. Sumposium, Additives-95*, Clearwater Beach, FL, 1996, p. 1.
[30] Brauman S., *J. Fire Retardant Chem.*, 1980, vol. 7(2), p. 61.
[31] Delobel R., Le Bras M., Quassou N., and Alistiqsa F., *J. Fire Sci.*, 1990, vol. 8(3-4), p. 85.
[32] Bourbigot S., Le Bras M., Delobel R., and Breant P., *Polym. Degr. Stab.*, 1996, vol. 54, p. 275.
[33] Bourbigot S., Le Bras M., Breant P., Tremillon J.-M., and Delobel R., *Fire and Materials*, 1996, vol. 20, p. 145.
[34] Selevich A.F., Levchik G.F., Lesnikovich A.I., and Levchik C.V., *Belarus Patent* No. 00260-016 (Belorussian University), 1993. (Rus)
[35] Levchik S.V., Levchik G.F., Camino G., Costa L., and Lesnikovich A.I., *Fire Mater.*, 1996, vol. 20, p. 183.
[36] Sevostyanov M. and Novikov S., *Vysokomol. Soedin.*, 1991, vol. 33, p. 1568. (Rus)
[37] Bourbigot S., Le Bras M., and Delobel R., *Appl. Surf. Sci.*, 1994, vol. 81, p. 299.
[38] Bourbigot S., Le Bras M., and Le Laureyns Y., *Polym. Degr. and Stab.*, 1997, vol. 56, p. 11.
[39] Zhubanov T.B. and Gibov K.M., Minimal Border Rates of Polymer Combustion, *Collection of Reports on International Conference "Flammability of Polymers"*, Smolensk, 1985, pp. 42 – 43. (Rus)
[40] Zhubanov T.B., Gibov K.M., and Zhubanov B.A., On Connection between Oxygen Indices and Border Combustion Rates of Polymers, *Vysokomol. Soedin.*, 1986, vol. 28B, p. 250. (Rus)
[41] Zhubanov T.B. and Gibov K.M., Polymer Combustion Rate as the Estimation Measure of its Combustibility, *Thes. XXII All-Union Conference on High-molecular Compounds*, Alma-Ata, Oct., 1985, p. 42. (Rus)
[42] Zhubanov B.A., Abdicarimov M.N., and Gibov K.M., Pyrolysis Processes under Combustion Conditions, *Coll. Rep. Intern. Conference "Nehorl avost Polymernykh Materialov*", Bratislava, 1976, pp. 30 – 31. (Rus)
[43] Zhubanov T.B. and Gibov K.M., Linear Pyrolysis of Carbonizing Polymers. Model Systems, *Materials of VIII All-Union Symposium on Combustion and Explosion*, Oct., 1986, Tashkent, Coll.: *Combustion of Heterogeneous and Phase Systems*, Chernogolovka, 1986, pp. 101 – 103. (Rus)
[44] Zhubanov T.B., Gibov K.M., and Shapovalova L.N., Capillary Events During Combustion of Carbonized Polymers, In: *Fire Shield Polymeric Materials, Problems of Estimating Properties, Meeting Thes.*, Tallinn, 1981, p. 93. (Rus)
[45] Shapovalova L.N., Study of Foam Formation in Polymeric Composites. *Proc. Scientific Conference of Young Scientists of the Institute of Chemical Sciences*, Alma-Ata, 1983, pp. 135 – 136. (Rus)
[46] Shapovalova L.N. and Karzhubaeva R.G., Pyrolysis and Combustion of Sulfur-containing Polymeric Materials, In: *Polymer Combustion and Creation of Limitedly Combusting Materials, Thes. Rep. 5th All-Union Conference*, Volgograd, 1983, p. 40. (Rus)
[47] *Patent* No. 1,215,343. Acrylonitrile (Styrene) Compolymers with Sulfonilamide Methacrylate Possessing Reduced Combustibility and Method of Their Production. (Rus)

[48] Druz N.N. and Gibov K.M., Toxicity of Pyrolysis and Phenoplast Combustion Gas Products, *Thes. Rep. Coord. Meeting on Phenoplasts*, Kemerovo, 1983, p. 158. (Rus)

[49] Druz N.N. and Gibov K.M., Influence of Polymer Pyrolysis Conditions on Carbon Oxide Formation, *Thes. Rep. All-Union Conference on Polymer Combustion and Creation of Limitedly Combustible Materials*, Volgograd, 1983, p. 91. (Rus)

[50] Druz N.N. and Gibov K.M., Toxicity of Polymeric Material Combustion Products, *Proc. IKhN AN KazSSR "Chemistry and Physics of Polymers"*, Alma-Ata, 1984, p. 145. (Rus)

[51] Gibov K.M., Zhubanov B.A., and Shapovalova L.N., Influence of Carbonized Surface Layer Porosity on Combustion of Polymers, *Vysokomol. Soedin.*, 1984, vol. 26B(2), pp. 108 – 110. (Rus)

[52] Gibov K.M., Zhubanov B.A., and Shapovalova L.N., Mass Transfer of Pyrolysis Products through Carbonized Layer during Combustion of Polymers, *Thes. Rep. First Symposium on Macroscopic Kinetics and Chemical Gas Dynamics*, Chernogolovka, 1984, vol. 1, Part 2, p. 4. (Rus)

[53] Gibov K.M., Nazarova S.A., and Zhubanov B.A., Deceleration of Polymer Combustion by Hydrogen Halides, *Izv. AN KazSSR, Ser. Khim.*, 1978, No. 1, p. 60. (Rus)

[54] Druz N.N. and Gibov K.M., Toxicity of Polymer Material Combustion Products. *Proc. IKhN AN KazSSR. Chemistry and Physics of Polymers*, Alma-Ata, 1984, vol. 62, p. 145. (Rus)

[55] Zhubanov Yu.A., Gibov K.M., Nikitina I.I., and Sarsembinova B.T., On Some Features of Thermal Transformations of Ammonium Phosphates in Polymeric Materials, *Thes. Rep. VI All-Union Conference "Phosphates-84"*, Alma-Ata, 1984, p. 45.

[56] Nikitina I.I., Gibov K.M., Kan A.A., and Zhubanov B.A., Study of High-temperature Reactions Between Phosphoric Compounds and Hydroxyl-containing Substances, Scientific Collection: *Theoretical and Applied Aspects of Fire Protection of Wooden Materials, Institute of Wood Chemistry AS LatSSR*, Riga, Zinatne, 1985, p. 93. (Rus)

[57] Nikitina I.I., Gibov K.M., and Kan A.A., On Phosphoric Compounds Reactions During Combustion of Polymeric Materials. *Thes. Rep. XXII All-Union Conference on High-molecular Compounds*, Alma-Ata, Oct., 1985, p. 53. (Rus)

[58] Sarsembinova B.T., Nikitina I.I., Gibov K.M., and Zhubanov B.A., On the Mechanism of Phosphorus-containing Decelerators of Polymer Combustion, Scientific Collection: *Studies of Monomers and Polymers ICS AS KazSSR*, Alma-Ata, 1986, vol. 66, p. 158. (Rus)

[59] Shapovalova L.N. and Zhubanov T.B., On the Role of Phosphorus- and Boron-Containing Compounds in Fire Shield Covers, *Thes. Rep. Republ. Scientific and Practical Conference on "Introduction of Scientific-Research and Production-Technical Works on Chemical Technology"*, Karaganda, 1985, p. 141. (Rus)

[60] Ksandopulo G.I., Chuvasheva S.P., Kononenko K.M., and Gibov K.M., Deceleration of Epoxy Resin Combustion by Compounds Containing Halogens and Phosphorus, *Collection of Reports of All-Union Meeting on the Inhibition Mechanism of Chain Gas Reactions*, Kaz. S.M. Kirov State University, il Alma-Ata, 1971, pp. 223 – 229. (Rus)

[61] Akulova D.V., Ivanov B.A., Ksandopulo G.I., Chuvasheva S.P., and Gibov K.M., Some Problems of Polystyrene Combustion Deceleration, *Thes. Rep. III All-Union Scientific-Technical Conference "Combustion Processes and Problems of Fire Extinguishing"*, VNIIPO, Moscow, 1972, p. 58. (Rus)

[62] Akulova D.V., Chuvasheva S.P., and Gibov K.M., On Pyrolysis Influence on Polystyrene Combustion, *Collection of Works on Applied and Theoretical Chemistry*, Kaz. S.M. Kirov State University, Iss. 5, Alma-Ata, 1974, pp. 133 – 139. (Rus)

[63] Gibov K.M., Kapyirina V.Ya., and Dovchilin T.Kh.. Fire Shield Composites Based on Epoxy Resin, *Plastmassy*, 1977, No. 12, p. 46. (Rus)

[64] Abdikarimov M.N., Gibov K.M., and Zhubanov B.A., Pyrolysis and Combustion of Epoxy Resin in the Presence of Amines Hydrochlorides and Sulfates, *Izvestiya AN KazSSR, Ser. Khim.*, 1980, No. 2, p. 42. (Rus)

[65] Gibov K.M., Nikitina I.I., and Gadranova Kh.S., On Condensation Reactions During Combustion of Polymers, *Proc. VIII Intern. Microsymposium on Polycondensation*, Alma-Ata, 1980, p. 44. (Rus)

[66] Gibov K.M., Zhubanov B.A., Nikitina I.I., and Gadranov Kh.S., Some Features of Polymer Carbonization, *Works of ICS AS KazSSR "Synthesis of Monomers and Polymers"*, Alma-Ata, Nauka, 1982, vol. 57, pp. 54 – 77. (Rus)

[67] Nikitina I.I., Gibov K.M., and Galyizhanov, On the Influence of Some Carbonizing Additives on Polymer Combustibility Reduction, *Izvestiya AN KazSSR, Ser. Khim.*, 1986, No. 2, p. 73. (Rus)

[68] Ksandopulo G.I., Chuvasheva S.P., and Gibov K.M., On the Influence of Some Additives on Foam Polystyrene Combustion, *Collection of Scientific Works on Chemistry*, Kaz. S.M. Kirov State University, Alma-Ata, 1971, Iss. 1, Part 2, pp. 40 – 55. (Rus)

[69] Ksandopulo G.I., Sumarokova G.I., Kononenko K.M., Gibov K.M., Chuvasheva S.P., and Surpina D., Influence of Complex Tin Compounds with Amines on Combustion of Condensed Systems, *Coll. Rep. II All-Union Scientific-Technical Conference "Combustion Processes and Problems of Fire Extinguish"*, VNIIPO, Moscow, 1972, pp. 43 – 48. (Rus)

[70] Chuvasheva S.P. and Gibov K.M., Influence of Cobalt Salts on Epoxy Resin Combustion, *Collection of Scientific Works on Applied and Theoretical Chemistry*, Kaz. S.M. Kirov State University, Alma-Ata, 1974, Iss. 5, pp. 139 – 145. (Rus)

[71] Ksandopulo G.I., Pivovarov A.P., and Gibov K.M., Study of Variable Valence Metal Salt Additives Influence on Epoxy Resin Pyrolysis by ESR Method, *Collection of Scientific Works on Applied and Theoretical Chemistry*, Kaz. S.M. Kirov State University, Alma-Ata, 1974, Iss. 5, pp. 145 – 151. (Rus)

[72] Chuvasheva S.P., Gibov K.M., and Ksandopulo G.I., Influence of Tin and Cobalt Salts on Epoxy Resin Pyrolysis Under Combustion Conditions, *Coll. Rep. IV All-Union Scientific-Technical Conference on "Problems of Combustion and Fire Extinguish"*, VNIIPO, Moscow, 1975, p. 63. (Rus)

[73] Shapovalova L.N. and Smirnova T.Ya., Features of Polycondensation and Pyrolysis Mechanism of Para-aminobenzene Sulfonilamide, *Thes. Rep. 23rd Conference on High-molecular Compounds*, Alma-Ata, 1985, p. 83. (Rus)

[74] Shapovalova L.N. and Cherdabaev A.Sh., On Foam Material Structure from Para-aminobenzene Sulfanilamine, *Thes. Rep. of Young Scientists, AS KazSSR "Youth and Scientific-Technical Progress"*, Alma-Ata, 1986, p. 35. (Rus)

[75] Karachubaeva R.G., Shapovalova L.N., and Gibov K.M., Radical Copolymerization of Methacrylic Acid Sulfonilamide With Several Vinylic Monomers, *Intern. Symp. on the*

Radical Polymerization, Kinetics and Mechanisms, Preprints, Italy, 1987, pp. 159 – 162. (Rus)

[76] Gibov K.M., Zhubanov B.A., Abdicarimov M.N., and Asodchi E.F., Linear Pyrolysis and Combustion Vinylic Polymers with Vinyl Bromide, Proc. VI All-Union Symposium on Combustion and Explosion, Collection: *Combustion of Condensed and Heterogeneous Systems*, Alma-Ata, Chernogolovka, 1980, pp. 55 – 58. (Rus)

[77] Zhubanov B.A., Dovlichin T.Kh., and Gibov K.M., Influence of Oxygen Concentration on Diffusional Combustion of Poly(methyl methacrylate), *Vysokomol. Soedin.*, 1975, vol. 17B, p. 746. (Rus)

[78] Chuvasheva S.P., Ksandopulo G.I., Kononenko K.M., and Gibov K.M., Deceleration of Polymeric Materials Combustion, *Coll. Rep. II All-Union Scientific-Technical Conference on "Combustion Processes and Problems of Fire Extinguish"*, VNIIPO, Moscow, 1972, pp. 56 – 62. (Rus)

[79] Nikitina I.I., Gibov K.M., Galyimzhanova S.A., and Paltseva N.G., On the Influence of Some Additives on Polymer Combustion Decrease, *Thes. Rep. Scientific-Technical Conference on "Polymeric Materials in Mechanical Engineering"*, Izhevsk, 1983, p. 37. (Rus)

[80] Gibov K.M. and Paltseva N.G., On the Role of Endothermal Processes in Reducing Polymer Combustibility, *Thes. Rep. 1st All-Union Symposium on Macroscopic Kinetics and Chemical Gas Dynamics*, Alma-Ata, Chernogolovka, 1984, vol. 1, Part 2, p. 5. (Rus)

[81] Gibov K.M., Zhubanov B.A., Dovlichin T.Kh., and Mamleev V.Sh., On the Mecanism of Fire Shield Effect of Foaming Polymeric Covers, *Coll. Rep. Intern. Conference "Nehorl avost Polymernych Materialov"*, Bratislava, 1976, pp. 69 – 71. (Rus)

[82] Gibov K.M., Zhubanov B.A., and Dovlichin T.Kh., Fire Shield Polymeric Covers, "Chemistry and Physical Chemistry of Polymers", *Trudy IkhN AN KazSSR*, Alma-Ata, Nauka, 1979, vol. 49, pp. 43 – 56. (Rus)

[83] Smirnova T.Ya., Ergazieva K.I., Nikitina I.I., Jadranova Zh.S., and Gibov K.M., Application of Oligomeric Systems to Creation of Foaming Covers, *Thes. Rep. II All-Union Conference on Chemistry and Physical Chemistry of Oligomers*, Alma-Ata, Chernogolovka, 1979, p. 59. (Rus)

[84] Nikitina I.I., Gibov K.M., Kan A.A., and Zhubanov B.A., Fire Shield Foaming Covers, *Coll. Rep. Intern. Conference "Flammability of Polymers"*, Smolenict, 1985, pp. 20 – 21. (Rus)

[85] Kan A.A., K.M. Gibov, and Sarsembinova B.T., Study of Fire Shield Foaming Covers Behavior at One-side Heating, *Izvestiya AN KazSSR, Ser. Khim.*, 1982, No. 3, p. 80. (Rus)

[86] Gibov K.M., Zhubanov, Dovlichin T.Kh., Mamleev V.Sh., and Nikitina I.I., Heat Mode of Foaming Fire Shield Covers, *Izvestiya AN KazSSR, Ser. Khim.*, 1972, No. 5, p. 36. (Rus)

[87] Gibov K.M., Zhubanov B.A., Nikitina I.I., Kan A.A., and Krasnikov A.V., *Thes. Rep. 1st All-Union Conference on Composite Materials and Their Application in National Economy*, Tashkent, 1980, p. 87. (Rus)

[88] Kan A.A. and Gibov K.M., Foam Coke Formation at High-temperature Pyrolysis of Polymers, *Thes. Proc. All-Union Conference "Combustion of Polymers and Creation of Limitedly Combustible Materials"*, Volgograd, 1983, p. 28. (Rus)

[89] Karxhaubaeva R.G., Gibov K.M., Shapovalova L.N., Ergozhin E.E., and Chigir L.V., Synthesis and Study of Sulfur-containing Copolymers with Reduced Combustibility, *Collection of Works on Chemistry*, Kaz. S.M. Kirov State University, Alma-Ata, 1984, Iss. 8, pp. 49 – 56. (Rus)

[90] Gibov K.M., Zhubanov B.A., Nikitina I.I., and Jadranova Zh.S., Carbonization of Fire Shield Foaming Systems, *High School Collection of Scientific Works "Chemistry and Technology of Elementorganic Semi-products and Polymers*, Volgograd, Polytechnical Intitute, 1981, pp. 30 – 35. (Rus)

In: Thermochemistry and Advances in Chemistry Research ISBN: 978-1-60692-376-4
Ed: Anton G. Mirskiy

Chapter 4

PLASMA ASSISTED THERMOCHEMICAL TREATMENTS

Ayhan Çelik, Akgün Alsaran, and Mehmet Karakan
Ataturk University, Engineering Faculty, Mechanical Engineering Department, 25240, Erzurum-TURKEY

ABSTRACT

Plasma surface treatments are widely used in industrial applications. Plasma assisted surface treatments improve the wear, corrosion resistance and fatigue strength by the change at the properties of surface and near-surface region of material. The variety of surface treatment is too large for this chapter to cover all of them in detail. The scope of this chapter is limited on thermochemical processes assisted by plasma which have been widely applied by manufacturing sector in Europa, North America and Japan during the last two decades. These processes involve heating into the ferrite or austenite regions of a materials part in an environment that allows elements such as C, N and B to be diffused through surface of materials. Firstly, plasma state and glow discharge are given in this chapter and, then thermochemical processes assisted by plasma are explained in detail. This chapter is especially focused on compound layer formed during thermochemical process.

1. INTRODUCTION

The modern use of metals in technology has increased the demand to high performance materials. Though the special properties of ceramics or other high tech materials have displaced ferrous alloys in some fields, steels are still used in most applications. Their properties have been improved by means of structural modification techniques.

The surface of steels is treated by a varity of methods for many reasons, such as painting and coating to prevent corrosion, shot-peening to develop a harder surface and comparative residual stress, quench to form a layer of martensite, processes changing the chemistry of the surface to improve mechanical and tribological properties and thin film coating in order to improve tribological and corrosion properties. The classification of surface treatmens is given in Table 1.

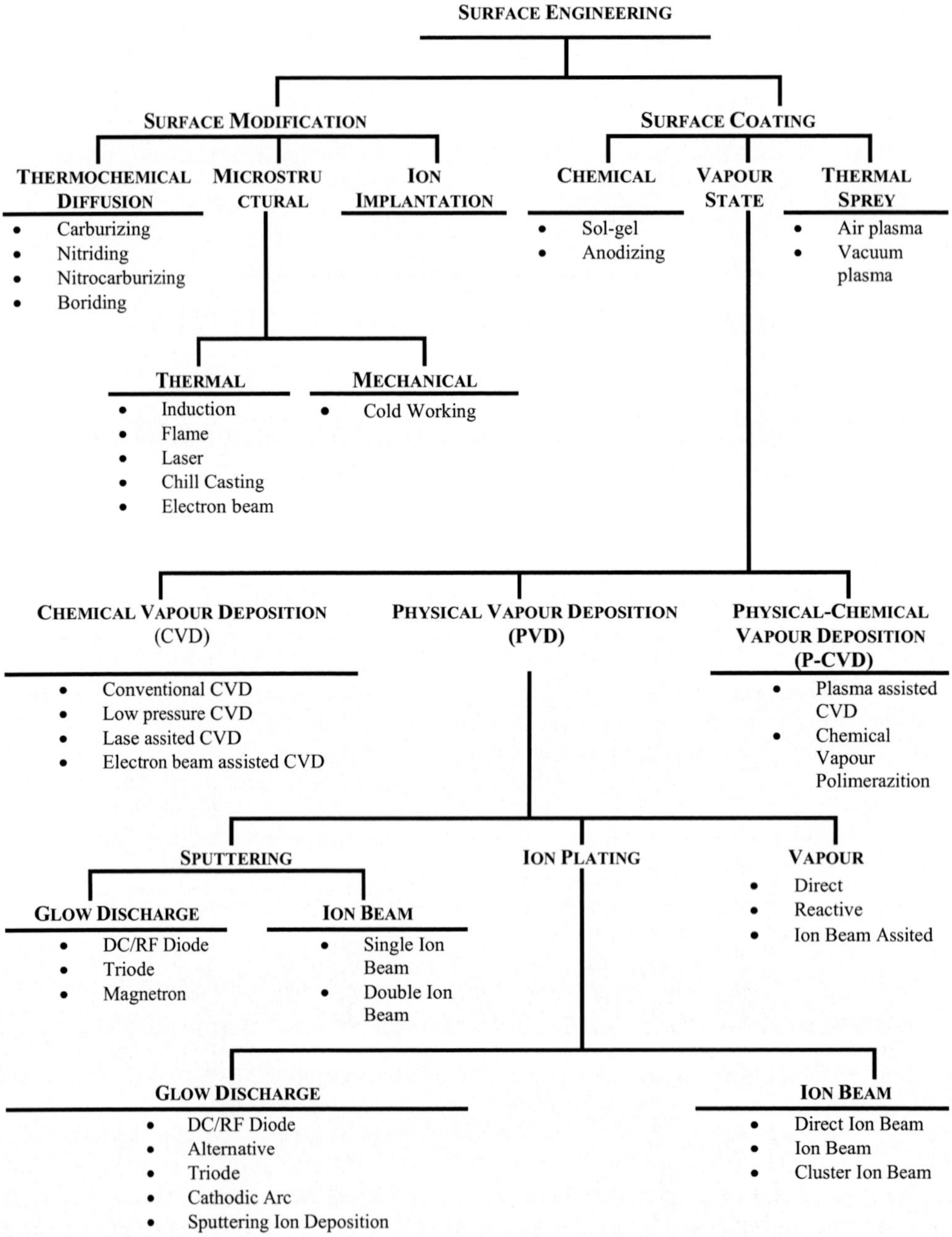

Figure 1. Classification of surface treatments used in technology.

Table 1. Characteristics of plasma assisted thermochemical treatments.

Process	Principle	Temperature (°C)	Processing Pressure (mbar)	Processing Characteristics	Suitable Steels	Applications
Plasma Nitriding	Interstitial solution of nitrogen	400-600	1-10	Very low distorsion, short treatment time	Alloy and tool steel which contain sufficient nitride forming elements such as Al, Cr, V, Mo	Severe surface stress conditions, good corrosion resistance, load bearing capacity, maximum hardness, good wear resistance
Plasma Carburizing	Solution of carbon, followed by quenching to form martensitic structure	800-1000	5-20	Care needed as high temperature may cause distorsion	Mild, low carbon and low alloy steel	High surface stress conditions
Plasma Nitrocarburizing	Formation of ε layer through intersitial solution of carbon and nitrogen	570-640	1-10	Very low distorsion, iImposible to machine after processing	Many steel from low carbon to tool steel	Low to medium surface stress condition, good wear resistance, post oxidation and impregnation gives good corrosion resistance
Plasma Boriding	Formation of FeB layer on surface of materials	800-1050	1-10	All post coating heat treatment must be in an oxygen free, toxic	Most steel from mild to tool steels expect austenitic stainless grades	Low to high surface stress conditions depending on substrate steel, excellent wear resistance

Themochemical treatment is a diffusion process. Although thermochemical diffusion treatments can be carried out using solid, liquid, or gaseous media, technical, economic and environmental constraints force the use of the plasma technology. Plasma thermochemical processes are more economical, result in much faster diffusion, and have lower process temperature, shorter treatment time and lower gas consumption when compared to other conventional processes. Characteristics of plasma assisted thermochemical treatments are generally given in Table 1 [1].

1.1. Plasma State

Plasma is sometimes called "the fourth state of matter" beyond the familiar three solid, liquid and gas (Fig. 2). A plasma is a collection of particles consisting of neutral gas atoms or molecules, electrons, ions, exited molecules, free radicals and photons. Plasma state is usually obtained when sufficient energy, higher than the ionization energy, is added to atoms of a gas, causing ionization and production of ion and free electron (Fig. 3). There are various ways to supply the necessary energy for plasma generation such as thermal energy, compensation, energetic beams and glow discharge. The most commonly used method is glow discharge.

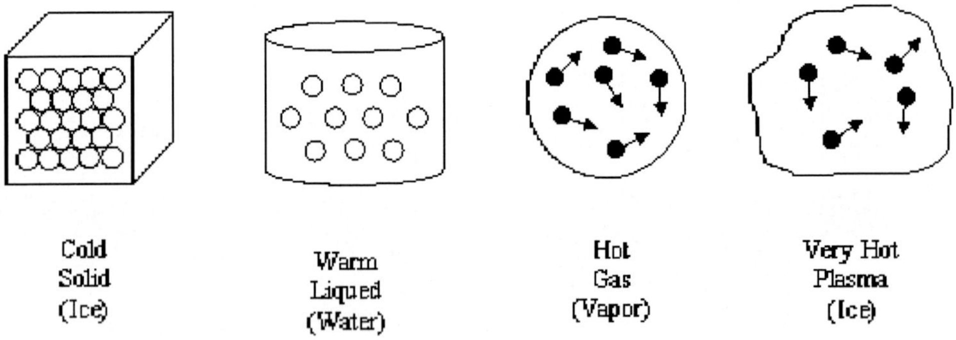

Figure 2. Atom order in four states of matter.

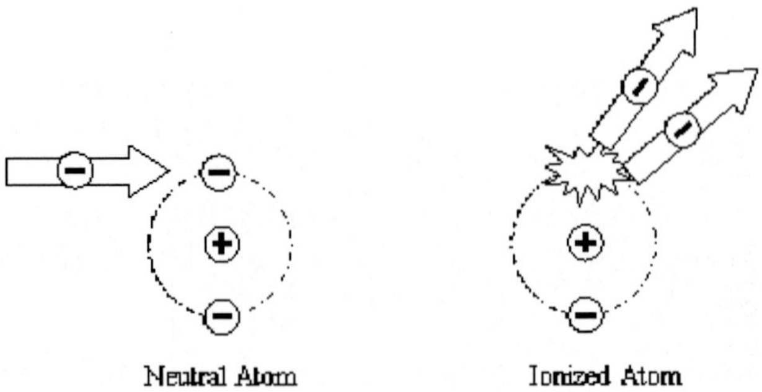

Figure 3. Ionization mechanism (Ionizations occur after knocking of neutral atoms or molecules by energetic electrons).

It was so far implicitly assumed that the plasma is homogeneous. However, plasma has a density gradient and has to be considered as having a nonuniform distrubution of ions and electrons in a dense background of neutrals. Because of the concentration gradients, the ions will tend to move by diffusion towards regions of lower density. As the plasma spreads out as a results of concentration gradient and electric field forces, the individual charged particles diffuse by undergoing a random walk and collide frequently with neutral atoms. The electrical conductivity of plasma caused by the movement of charged particles is therefore controlled by the diffusion of the charged particles through it [2].

1.2. Glow Discharge

In the surface engineering, the glow discharge is generated by either a direct current or a radio frequency source. However, DC glow discharge has been extensively used in thermochemical processing. DC glow discharge is produced by appliying a DC voltage between two conductive electrodes inserted into a gas at low pressure. As the electrical field is increased with applied voltage in discharge tube, available free electrons are accelerated, thereby gaining kinetic energy. Concomitantly, the electrons lose energy in collisions with the atoms or molecules of the gas. This phenomenon is called glow discharge. In order to appreciate this development, it is necesary to understand the voltage-current characteristics of plasma (Fig. 4) [3].

When the applied voltage is low, the current through chamber is produced by the collections the available free charges and is negligibly small. When the voltage increases, and more charged particles are created by the ionization of the gas, the current increasing steadily, while the voltage reaches a limit determined by the output impedance of the power supply. This region of the I-V characteristics is known as the Townsend discharge [2].

When the applied voltage reaches a certain threshold value, marked V_s in Fig. 4, an avalanche process occurs as a results of three simultaneous processes; the emission of secondary electron from cathote, newly formed ions, and recombinanation. When the number of electrons is sufficient to produce just enough ions to regenerate the number of the lost electrons, a steady state is reached in which an equilibrium is establish between the rate of the formation of ions and the rate of their combination with electrons. At this stage the discharge is thus established. The gas begins to glow, the voltage drops, and current rises abruptly. The mode of the discharge at this point is called normal glow.

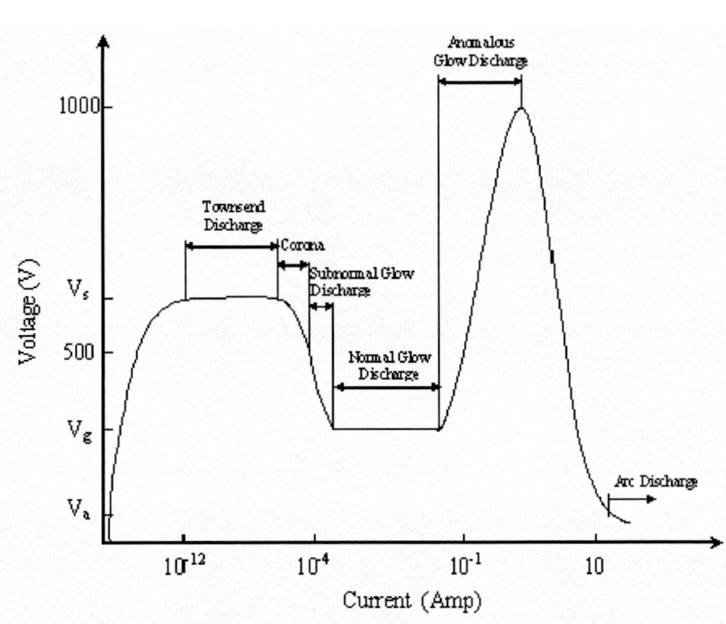

Figure 4. Current-Voltage characteristic for a DC discharge.

Threshold voltage required to produce the glow discharge, V_s, is called the breakdown voltage. After the voltage reaches the breakdown value V_s and the glow discharge is established, initially at low power, a discharge covers only the area near the rim of cathode. When the power raised, the current increases and the discharge spreads over whole surface of the cathode. When the current of discharge increases and passes the point of complete cathode coverage, the voltage begins to rise. This corresponds to the abnormal discharge, that is the mode used in glow discharge processing. At this stage, the discharge currrents is limited by the surface area of the electrodes and by the resistance of the power supply. As the current and voltage increase, the cathode current density also increases, ultimately heating the cathode and causing incandescence and thermionic emission. Once the cathode is hot enough to emit electrons thermionically, the discharge will change to arc regime [2]. Indeed, it was technical difficulties associated with the maintenance of the unstable, abnormal glow discharge that had deterred its initially industrial applications. Today, however, these problems are solved largely after development of power electronics and microprocessor based control [1].

2. Plasma Assisted Thermochemical Processes

2.1. Plasma Nitriding

Plasma nitriding is a surface hardening process which provides several advantages compared with conventional nitriding technologies. This technique was developed in 1932 by Wehneld and Berghaus as an alternative process to other methods. Surface treatment with plasma nitriding (also known as glow discharge or ion nitriding) is used to increase fatigue strength, wear and corrosion resistance of ferrous alloy steel. The advantages of plasma nitriding are given as follows;

- High surface hardness
- Shorter processing time
- Minimum distortion
- Process temperature can be as low as 350 ^{o}C
- Eliminated environmental problems
- The control of formation of phase composition and structure of nitride layers
- Reducing operating costs, low energy and gas consumption
- Improve corrosion resistance
- The ability to provide uniform case on complex geometries
- Partial hardness can be done by using sheet masks
- The opportunity to reduce thickness of compound layer (compound layer thickness)
- Ion bombardment provides the heat for the process

2.1.1. Process Equipment Description

Plasma nitriding system consists of a cylindrical vacuum chamber, vacuum pump, gas distribution system and electrical power supply. Experimental schematic diagram of the plasma nitriding system is shown in Fig. 5.

Ion nitriding is generally performed at absolute pressure levels of 1 to 10 torr. Initially the system pumped down where base pressure reach to 3.10^{-3} mbar and then pumping speed from plasma chamber being regulated by a valve.

The process carried out using a mixture of nitrogen, hydrogen and, at times, small amounts of methane. Ion nitriding is the result of a reaction caused by a high voltage glow discharge between the metal work being processed (the cathode) and the vacuum chamber wall (the anode) [4].

In vacuum, high-voltage is used to form a plasma, through which nitrogen ions are accelerated to impinge upon the workpiece. This ion bombardment heats the workpiece, cleans the surface and, provides active nitrogen [6].

Applied voltage between the cathode and anode value change 500-1500 V. Under the influence of this voltage, the nitrogen gas is dissociated, ionized and accelerated toward the workpiece (the cathode). Within a short distance of the workpiece, the positively charged nitrogen ion then acquires an electron from the cathode and thus emits a photon. This photon emission during the return of nitrogen ions to their atomic state results in the visible glow discharge that is characteristic of plasma techniques [6].

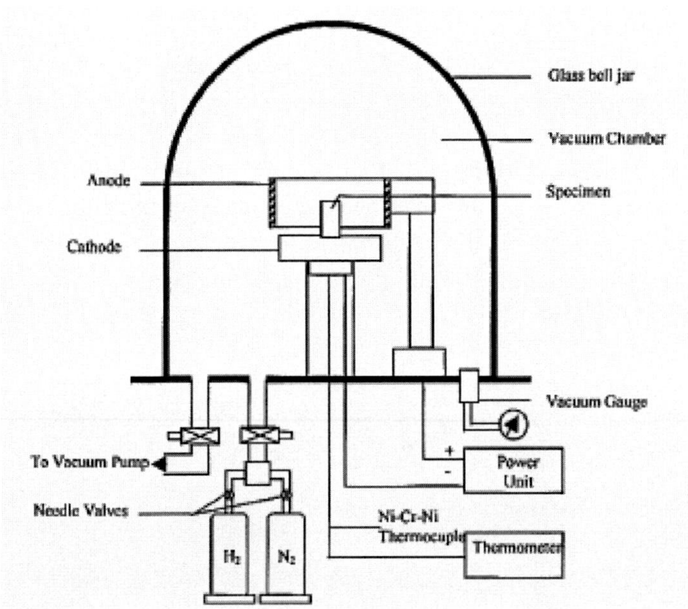

Figure 5. Schematic diagram of the ion nitriding system [5].

The temperature is measured at the bottom of the specimens using thermocouple. The treatment temperature of plasma nitriding is between 350 and 600 °C, depending on the composition of material, gas mixture and voltage.Treatment times vary from 5 minutes [7] to 20 hours, depending upon on the grade of the steel and desired depth of hardening [8].

2.1.2. The Mechanism of Plasma Nitriding

This process has been explained by Edenhofer, plasma consists of electrical charged particles, ions and electrons. A few of charged carriers which are present in any gas are accelerated in electrical field formed between cathode and anode.

New charged carriers are thus continiously produced, the electrons being accelerate towards anodically connected wall of the chamber and positive ions towards the cathodically connected workpiece [8]. Plasma nitriding is performed as a cathodic process, accompanied by the bombartment of positive ions. The target typically serves as the cathode of a DC glow discharge, which is maintaned at a pressure of several hundred Pa, and a voltage between 1000-1500 V [9].

When the cathode is bombarded by positive ions, metallic and nonmetallic atoms (Fe, Cr, Mo,…, C, O, N) are removed because of sputtering. The latter reacts with the nitrogen atoms in the plasma, to form nitride (FeN). These nitrides are deposited at the workpiece, however they are unstable and react further with the target (Fe) atoms [9]. Finally, they decomposes into the lower nitrides Fe_2N, Fe_3N and Fe_4N (Fig. 6).

2.1.3. Case Structures and Formation

Ion nitrided steel surfaces consist of two parts. The outermost layer, which is very thin and consists mainly of iron nitrides, is not attacked by an alcoholic nitric acid etch, so that it is white on the micrograph. For this reason the layer is usually referred to as white layer. Because of the structure and special character of this layer, the more recent term used is compound layer. Beneath the compound layer is the so-called diffusion zone, where the nitrogen has mainly been incorporated into the existing iron lattice as interstitial atom or as a finaly dispersed alloy nitride precipitate (Fig. 7)[10].

Compound and diffusion layer thickness depend on temperature, nitriding time, material and gas mixture.

Compound layer thickness depends on temperature and nitriding time. Çelik and Karadeniz [12] investigated the compound layer thickness in the temperature range of 500-600 °C, time (1, 4, 10 h) and 50% N_2-50% H_2 gas mixtures for AISI 4140 steel. They observed that the thickness of the compound layer increases with increasing time at 500 °C and 550 °C. However, it begins to decrease with increasing time at 600 °C. Fig. 8 shows X-Ray analysis of specimens ion nitrided at 600 °C for 1, 4 and 10 h. The changes in the compound layer thickness with increasing time at various temperatures are shown in Fig. 9. It can be seen that the compound layer thickness of the 600 °C specimen increases up to 4 h and then decreases. Increasing time and temperature cause a decrease in thickness. The occurance of sputtering requires a sufficently high temperature and long enough treatment time. Sputtering causes a decrease in the compound layer thickness.

Diffusion and compound layer thickness increase with increasing temperature. At 400 °C the white layer is very thin and is not continuous over the entire surface. At the highest temperature investigated, 600 °C, the white layer is very thick and severely cracked[13].

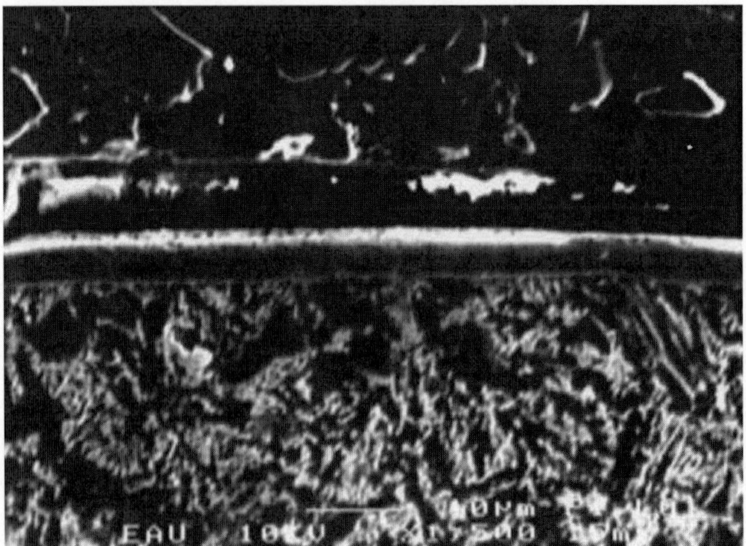

Figure 6. Mechanism of compound layer formed during plasma nitriding.

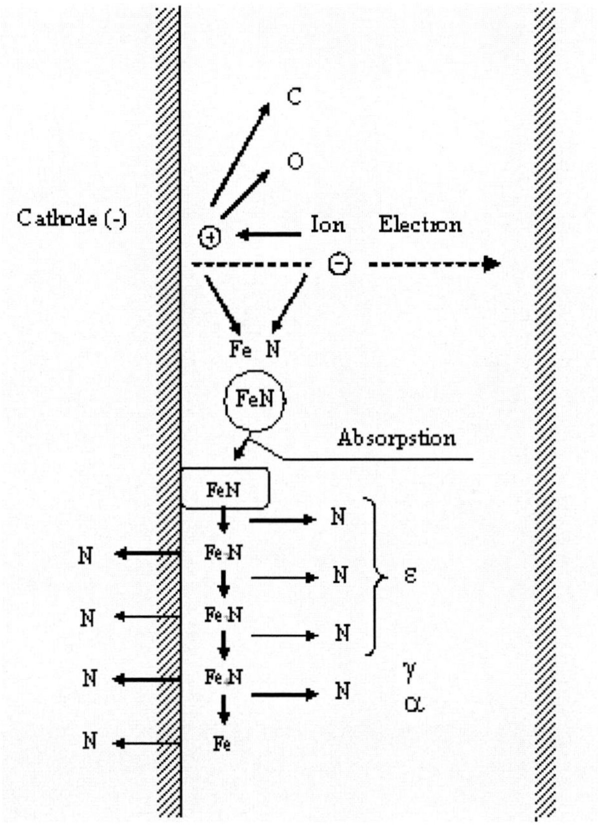

Figure 7. SEM micrograph of plasma nitrided specimen for process time of 0.5 h, a temperature of 500°C and a gas mixture of 30% N_2+70% H_2 [11].

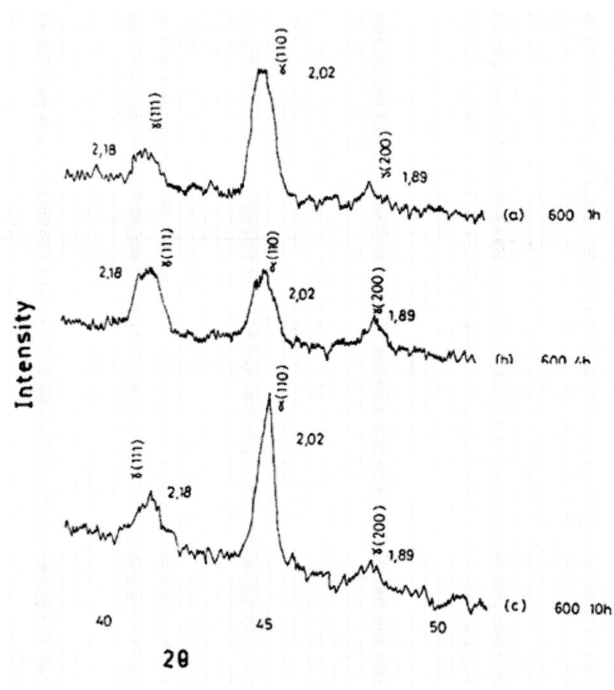

Figure 8. X-Ray analysis of AISI 4140 steel iron nitrided at 600 °C for (a) 1, (b) 4 and (c) 10 h [12].

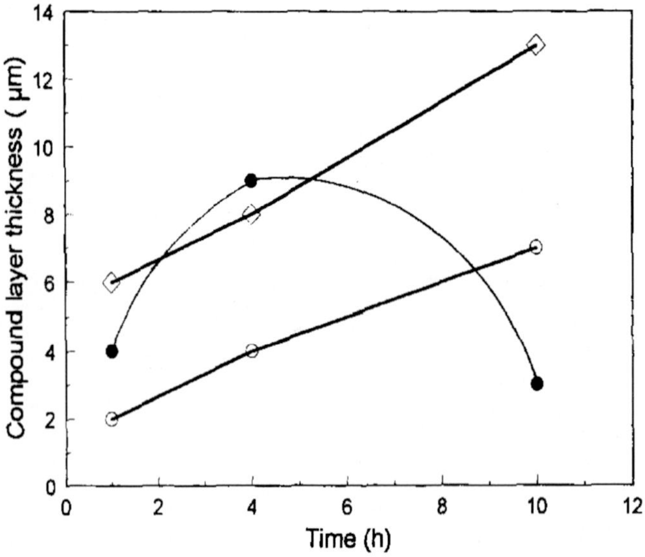

Figure 9. Variation in thickness of compound layer with treatment time at various temperatures: ● 600 °C ; O 500 °C; ◊ 550 °C [12].

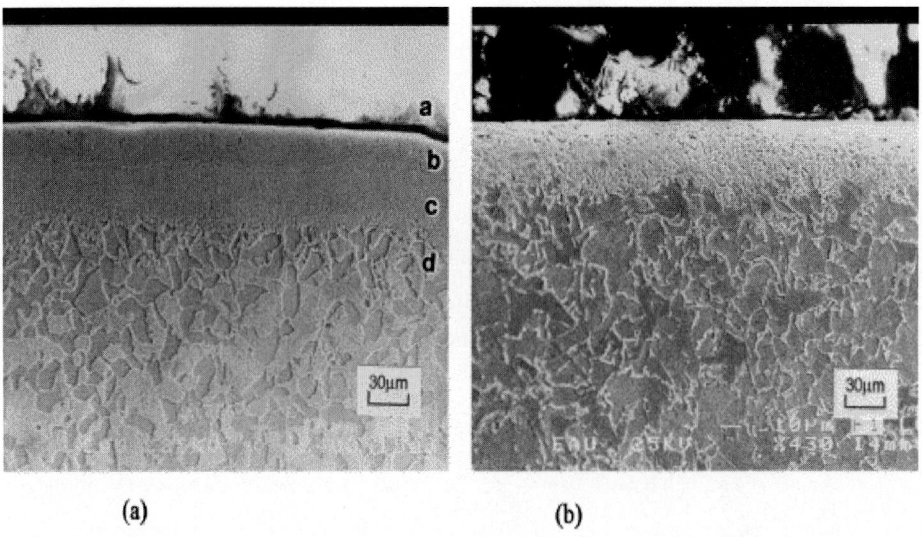

Figure 10. SEM micrographs of AISI 8620 steel ion-nitrided at 600°C for (a) 1 h and (b) 8 h; (a, mounting compound; b, compound layer; c, braunite; and d, diffusion layer) [15].

Braunite is a eutectoid mixture of Fe_4N and ferrite [14] which formed during treatment at 600 °C. Çelik et all. investigated this layer [15]. SEM micrographs of ion nitrided specimens at 600 °C for 1 and 8 h are shown in Fig. 10a and b, respectively. When Fig. 10a was examined, it can be seen that two dense zones are formed. There is a very thin line separating the outermost compound layer (region b) and the underlying braunite (region c). The thickness of this layer decreases with a long treatment time.

2.1.4. Effect on the Mechanical Properties

The mechanical properties of ion nitrided material is changes, such as hardness, fatigue strenght, wear resistance, corrosion behavior. Fig. 11 shows the surface hardness as a function of temperature for the treatment time. Surface hardness decreases with increasing temperature and ion nitriding time.

At lower temperatures such as 400 °C, surface hardness is low. The reason of that, at the lowest temperature the nitrogen uptake and precipitate density are low and therefore the amount of hardening is also low. At higher temperature and/or longer treatment times, both precipitate growth and tempering of the matrix become important [13].

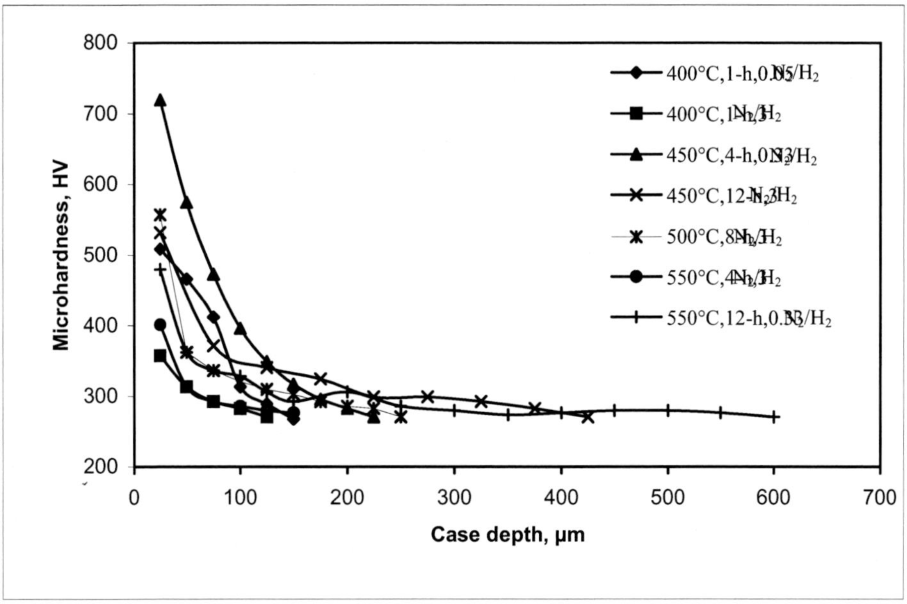

Figure 11. The hardness profiles of AISI 5140 low-alloy steel ion nitrided at different process parameters [16].

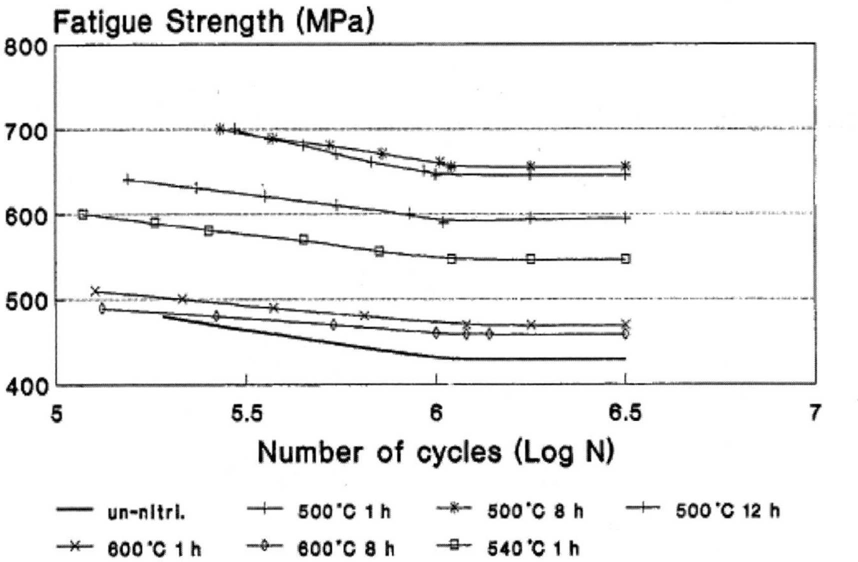

Figure 12. Fatigue life curves AISI 4140 steel ion nitrided at different temperature [17].

Fatigue strength of materials are improved with the ion nitriding process. Çelik and Karadeniz [17] studied on this subject. This study has demonstrated that the fatigue strenght of material depends on the surface hardness of diffusion depth, because the origin of fatigue cracks moves into the core owing to the high surface hardness of the specimens. The fatigue strenght of an ion nitrided material increases with increasing surface hardness. (Fig .12)

Plasma nitriding can improve the wear resistance of 316 austenitic stainless steel [18]. The relation of the nitriding parameters with compound layer thickness and wear rate are illustrated in Fig. 13. Increasing compound layer thickness increases the wear of steels because the presence of a hard and brittle compound layer on the surface causes an increase in wear of specimen surface due to the fracture of the compound layer with high stress and hard abrasive particles formation in the initial stage of sliding [19].

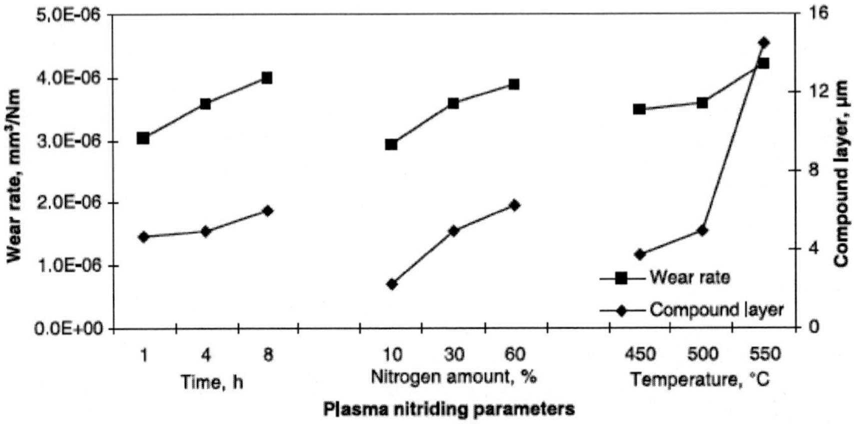

Figure 13 The relation of the ion-nitriding parameters with compound layer thickness and wear rate.

3. Plasma Carburizing

Carburizing is the addition of carbon to the surface of low-carbon steels at temperatures generally between 850 and 950°C, at which austenite, with its high solubility of carbon, has a stable crystal structure. Hardening is accomplished when the high-carbon surface layer is quenched to form martensite so that a high-carbon martensitic case with good wear and fatigue resistance is superimposed on a tough, low-carbon steel core. Carburizing steels for case hardening usually have base-carbon contents of about 0.2%, with the carbon content of the carburized layer generally being controlled at between 0.8 and 1% C. However, surface carbon is often limited to 0.9% because too high carbon content can result in retained austenite and brittle martensite [20].

Carburizing methods include gas, vacuum, plasma (ion), salt bath, and pack carburizing [3, 21,24]. These methods introduce carbon by use of an atmosphere (atmospheric gas, plasma, and vacuum), liquids (salt bath), or solid compound (pack) [21]. Pack carburizing, where parts are packed in a blend of coke and chare coal with activators, and then heated in a closed container. This is an old, labor-intensive process, but it is still practiced in tool room because facility requirements are minimal. Vacuum and plasma carburizing, both of which utilize a vacuum chamber with a partial pressure of hydrocarbon gas as the source of carbon. The prime advantage to these processes is the absence of oxygen in the furnace atmosphere. Salt bath carburizing, in which the baths contain cyanides, cyanates, or carbon-carbonate blends. These are particularly useful to produce thin carburized cases because the carburizing time can be precisely controlled [25].

Historically, the first-reference to plasma carburizing appears in a US patent written by Egan in 1934. This early patent claimed that dramatic improvements in the rate of carburizing could be achieved by this method. Little appears in the literature up to 1960 when Vanin and his colleagues concluded that this types of process was of interest but did not appear to offer any significant advantages over the methods available at the time. More recently Edenhofer, Collignon et al. and Grube have demonstrated that plasma carburizing is a viable process which has a number of advantages over the conventional techniques [3]. In the last decade much effort has been devoted to the development of a plasma carburizing treatment.

3.1. The Mechanism and Process Equipment of Plasma Carburizing

In principle, plasma carburizing or ion carburizing is analogous to ion nitriding, except that ion nitriding can be conducted at much lower temperatures in steels. The specimen is made cathode with respect to the chamber in a DC electrical circuit. At a suitable pressure in the range of 1 to 20 torr a glow discharge plasma is established between the cathode and anode by the application of high voltage. To initiate the plasma a voltage between 500 and 1000 V may be required, depending on the total pressure in the vacuum chamber, mixture of gases employed, and the temperature. In ferrous alloys, ion carburizing is carried out in the austenitic phase region and produces a martensitic structure after quenching from carburizing temperature [26]. Today, plasma carburizing is best performed using low-pressure techniques under 20 torr and typically at temperature between 790 and 1040°C. As illustrated in Fig.14, hydrocarbon gases currently being used for vacuum carburizing are acetylene (C_2H_2), propane (C_3H_8), and, to a lesser degree, ethylene (C_2H_4). Methane (CH_4) is not used, because it is nearly nonreactive at this low pressure, unless the temperature is at or above 1040°C [24]. Schematic diagram of the plasma carburizing system is given in Fig. 15 and schematic diagram of the plasma carburazing processes are shown in Fig. 16.

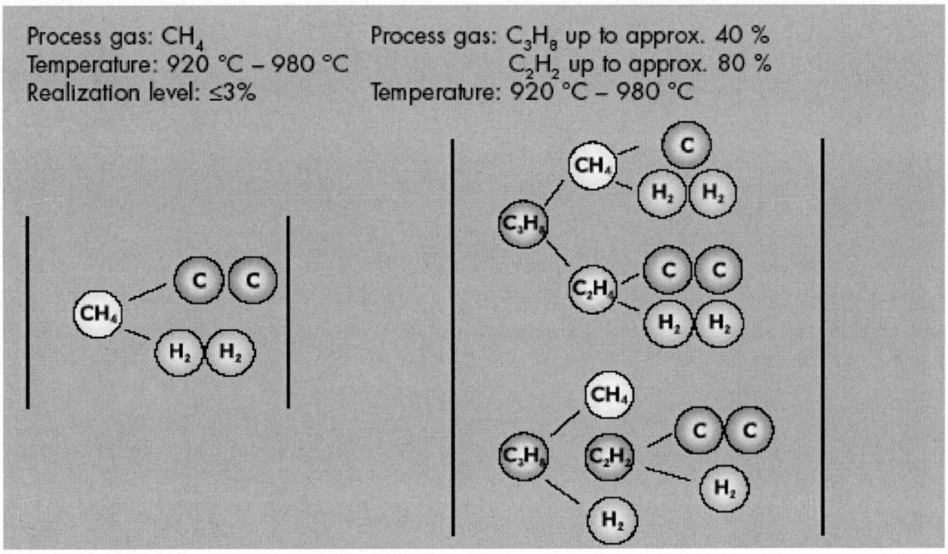

Figure 14. Reaction during plasma carburizing [27].

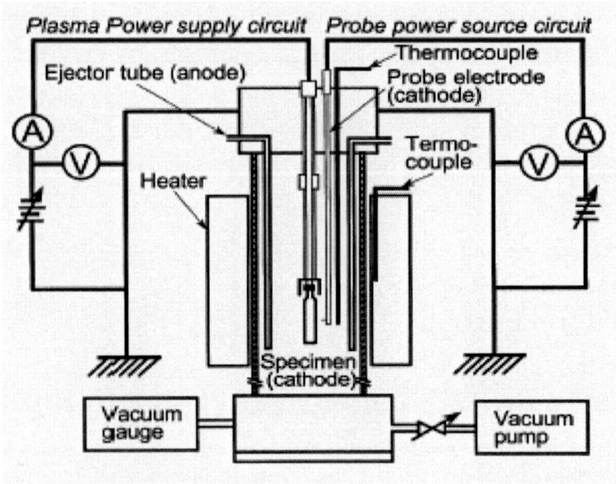

Figure 15. Schematic diagram of plasma carburizing system [28].

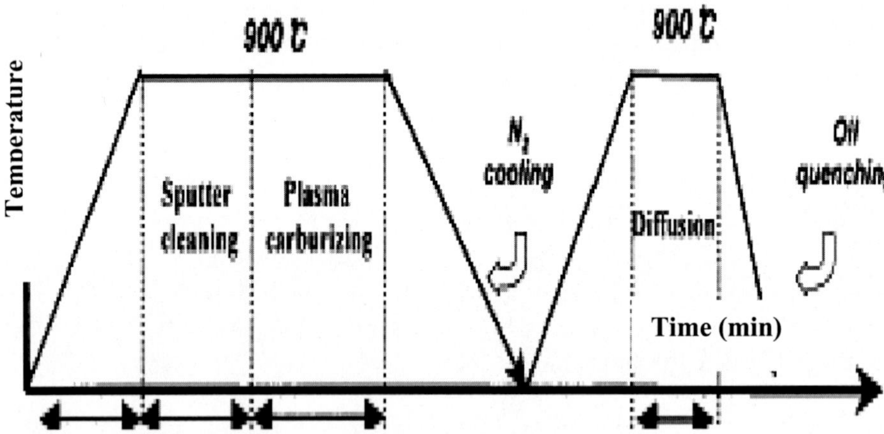

Figure 16. Schematic diagram of the plasma carburizing process [29].

3.2. The Advantages of Plasma Carburizing

Plasma carburizing is a proven method of pure carburizing and pure diffusion in which carbon penetrates into the surface of the steel being processed without interference from external influences, such as gas chemistry or surface contaminants. Advantages of plasma carburizing include several technical factors (such as carburizing and case uniformity) as well as economic and environmental factors.

These advantages of vacuum carburizing include [23, 24, 27]:

- Easy integration into manufacturing; Process is clean, safe, simple to operate, and easy to maintain. Also, working conditions are excellent (that is, there are no open flames, heat, and pollution).
- Full automation capability using recipe or part-number control of heat treating cycles
- Capability of higher temperatures and flexible cycles, due to the types of equipment and the nature of the process.
- Precise process control achieved using computer simulations, which allow adjustments to established cycles.

- Consumption of energy by the equipment and process only when needed, due to the nature of the vacuum operation.
- High carburizing rate, improved case uniformity, and environmental improvements

3.3. Microstructures and Properties of Plasma Carburized Components

The object of carburizing is to produce a high-hardness, high-carbon case on a low-carbon steel core with low-to-medium hardness. This objective is accomplished by the introduction of carbon into austenite by various gas-metal surface reaction and the diffusion of carbon through the austenite from the surface into the core. The austenite is then quenched and converted to microstructural gradients by solid-state phase transformation during cooling. The microstructure gradient in turn produce hardness gradient [30]. In the case of plasma carburizing the saturation of surface is a function of time (as illustrated Fig. 17), temperature, and the chemical composition of the materials. The case depth and carbon gradient increase as the process time and temperature increase [26, 30].

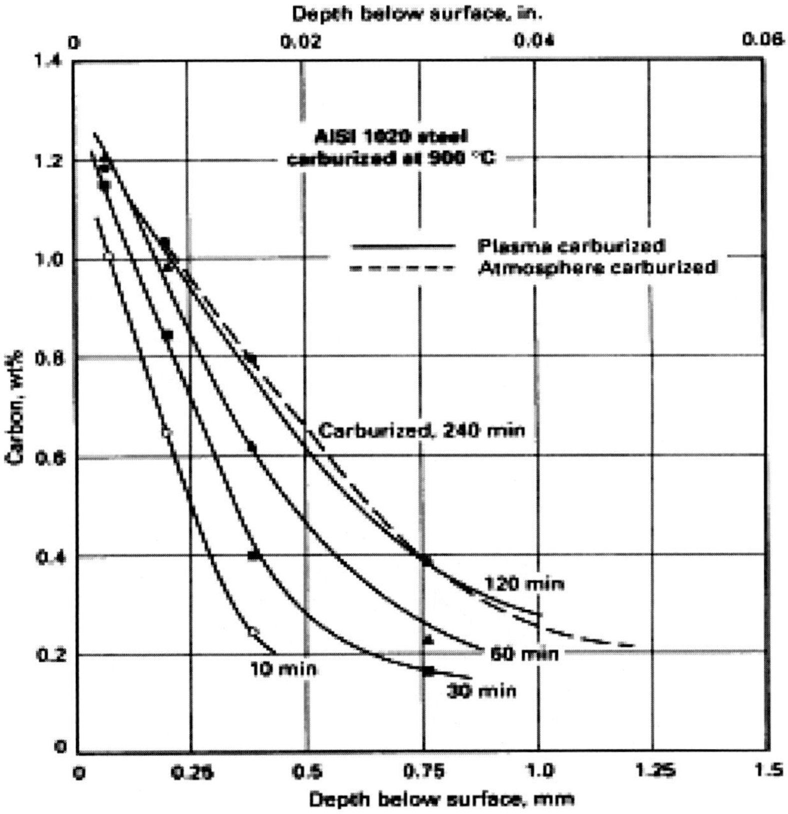

Figure 17. Carbon concentration profiles in AISI 1020 steel after ion carburizing for 10, 20, 30, 60, and 120 min at 900 °C. Carbon profile after atmosphere carburizing for 240 min at 900 °C (1650 °F) shown for comparison [31].

Recent sliding tests demonstrated that the precipitation-free carburized layers possess excellent sliding wear resistance under dry conditions and have the ability to prevent surface plastic deformation and severe adhesive wear. Sufficient case depth and case hardness must be provided to prevent indentation or case crushing under the anticipated contact loads. Coarse primary carbides can be helpful in resisting abrasive wear [32].

There are many factors that affect the fatigue resistance of carburized steel. These factors include effective case depth, hardness, residual stress, surface finish, grain size, morphology and distribution of carbides, intergranular oxidation, microcracking and the presence of retained austenite [29, 30]. The fatigue strength of the carburized materials increases considerably compared with untreated materials. The case depth dependence of fatigue strength is seen in the finite life region where the fatigue lives of the material with thick case are slightly longer than those of the material with thin case, while there is no discernible difference in the fatigue limit that is 30% increase with respect to the fatigue limit of the untreated material. Fatigue cracks initiates at a very early stage of fatigue life near the boundary between the carburized case and the core material due to cyclic slip deformation and thus fatigue life is dominated by small crack growth. The crack growth at the specimen surface is discontinuous. This is due to intermittent brittle fracture of the carburized case because of preceded crack growth in the subsurface [33]. As illustrated in Fig. 18, fatigue limit of the plasma carburized steel is higher than those of the gas carburized steel in the same effective case depth [29].

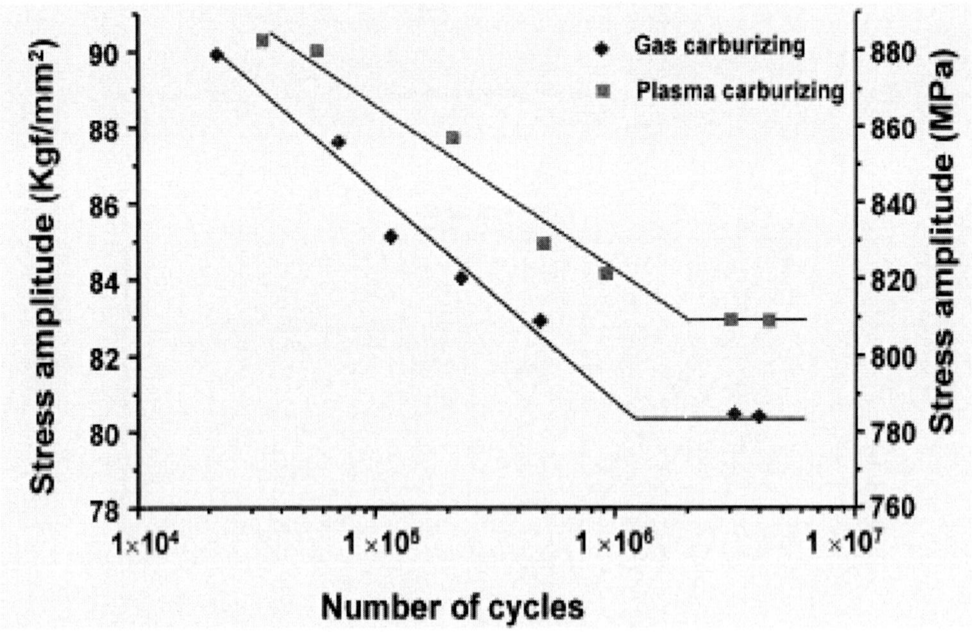

Figure 18. Results of high cycle test on plasma and gas carburized steel [29].

3.4. Plasma Nitrocarburizing

Nitrocarburizing is a thermochemical process in which nitrogen and carbon diffuse into the surface of ferrous materials at certain elevated temperatures. As a result of nitrocarburizing, two different structures occur on the surface of steel, known as the compound layer and diffusion region from surface to core respectively. Provided that the nitrogen and carbon activities imposed by the nitrocarburizing agent on the ferrous material surfaces are sufficiently high, a compound layer consisting predominantly of ϵ-$Fe_{2\text{-}3}$(N, C) and/or γ'-Fe_4(N, C) phases is formed at the surface. The compound layer provides the materials with good physical and chemical properties against wear and atmospheric corrosion. The diffusion region brings about an improvement of fatigue strength when compared to an untreated material. In this region, N and C atoms are dissolved interstitially in the ferritic lattice, and form the nitride precipitates.

Nitrocarburizing can be applied in solid, liquid and plasma atmospheres. Nitrocarburizing treatments were first introduced in the early 1950. Until the late 1960s, the only nitrocarburizing processes available used a medium of molten cyanide-based salts to confer the property improvements required. However, since then concern about the overall environmental aspects of heat-treatment processing with cyanide-based salts has been increasing. Consequently, there has been intense interest in the development of cyanide-free nitrocarburizing treatment as technically and economically viable alternative to the cyanide-based processes [34]. Whilst the widely used gaseous nitrocarburizing technique, although less harmful to the environment, also produces a significant amount of exhaust gases and fumes and suffers from the risk of explosion if safety precautions are not rigorously followed, due to the use of a combustible atmosphere for the treatment. In the past decade, it has been recognized that plasma-based thermochemical treatments exhibit significantly fewer environmental problems than the related gaseous processes. Therefore, efforts have been made to develop plasma nitrocarburizing techniques for industrial applications [35, 36].

Plasma nitrocarburizing involves a plasma diffusion treatment in an atmosphere of nitrogen, hydrogen, and a carbon-containing gases, usually methane or carbon dioxide. The plasma nitrocarburizing is more economical, environment friendly and results faster nitrogen and carbon diffusion, lower gas consumption compared to other nitrocarburizing techniques. It is applied to various materials such as carbon, alloy, tool steels, stainless steels, cast irons and sintered materials [34, 37, 39].

Plasma nitrocarburizing treatment is classified into two groups as ferritic and austenitic according to process temperature. While treatments realized under eutectic point (592°C) at Fe-N diagram is named as ferritic nitrocarburizing, the treatments over this temperature is austenitic nitrocarburizing [40, 41]. Ferritic nitrocarburizing offers especially good results for the applications required high hardness and excellent tribological properties in low alloy steels including nitride formed elements. Austenitic nitrocarburizing is applied to produce much thicker compound layer and to improve load capacity by increasing the hardenabletibility of nitrogen diffusion zone for plain carbon steels [36].

3.4.1. The Mechanism of Plasma Nitrocarburizing

During nitrocarburizing, the explanation interested in the growing of compound layer is not completely clear as that of expressed for nitriding. Since, the phase development of compound layer changes continuously depending on growing. It has been generally studied the models suggested on ion nitriding are valid or invalid and stated the same mechanisms is usable for nitrocarburizing because the process mechanism for plasma nitrocarburizing can not completely clarified yet. Although there is no entirely a certain model on this subject, the most of the models clarifying ion nitriding phenomena is given by Edenhofer [37, 39].

Other approach relating to the process mechanism of nitrocarburizing is suggested by Somers [42]. According to him, the phase observed initially on the surface is cementite for all cases. During the initial stage of cementite formation, cementite amount reduces and thus it decomposes entirely, and finally a dominant ε phase structure in compound layer retains after long treatment. Also, γ' phase content especially increases in compound layer region near substrate. While the development of composition/depth profiles and C content on surface decreases with time, N content on surface increases with time. It was observed that cementite vanished with continuing of nitrocarburizing process and replaced with ε-carbonitride near substrate and γ' (carbo) nitride [43, 44].

3.4.2. Structures and Properties Resulting from Plasma Nitrocarburizing

a) Compound layer

On the surface of plasma nitrocarburized components, a compound layer including ε-Fe_{2-3}(N, C) and/or γ'-Fe_4(N, C) phases exist. It is generally believed that ε-Fe_{2-3}(CN) carbonitride shows very good wear and corrosion resistance. However, it is difficult to form this structure. Investigations are dominated on compound layer with monophase (ε-carbonitride). Previous investigations have been carried out in methane atmosphere. However, it was observed that the compound layer with monophase ε was difficult. For this reason, CO_2 have been mainly used for nitrocarburizing process for last five years. The thickness of compound layer consisting polyphase carbonitrides is important for wear resistance, since compound layer was spalled and debonded during wear. Fig. 19 shows XRD results of plasma nitrocarburized AISI 4140 steel as a function of the process time. The compound layer formed during nitrocarburizing consists of polyphase iron nitrides or iron carbonitrides such as γ'-Fe_4N or Fe_4(C, N) and ε-$Fe_{2-3}N$ or Fe_{2-3}(C, N). It is difficult to determine whether the phases on the compound layer are nitrides or carbonitrides as the lattice spacing in iron nitrides or carbonitrides are nearly the same. While the intensity of ε (101) carbonitride reached the maximum values for a time of 4 h, and then decreased, the intensities of ε(002)/γ'(111) and γ'(200) increased with increasing process time [45].

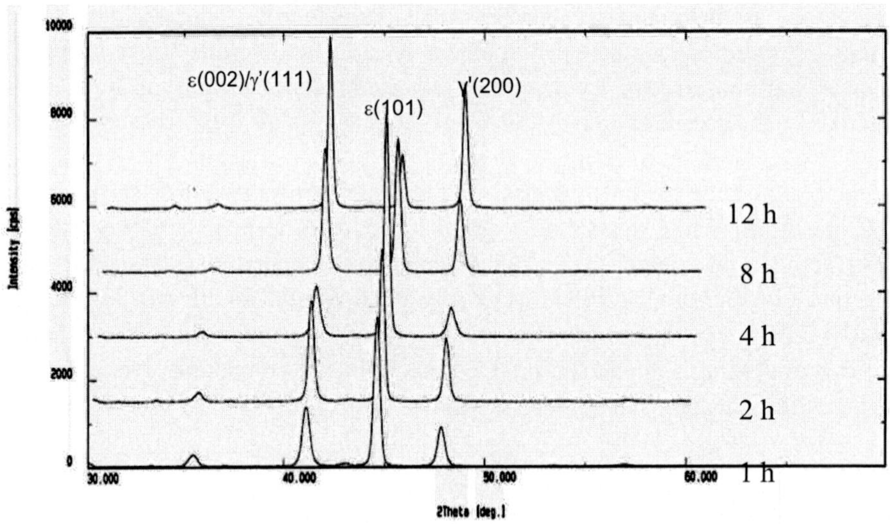

Figure 19. XRD results of plasma nitrocarburized AISI 4140 steel at 570°C and a gas mixture of 49%N_2 + 49%H_2+2%CO_2 for different time [45].

Fig. 20 shows the cross section of nitrocarburized specimen. The compound layer has a porous structure, which is of a columnar morphology. The pores at the outer surface are relatively large and open to the surface. The reason of the porous structure is the formation of molecular nitrogen. During plasma nitrocarburizing, the diffusing carbon atoms occupy interstitial positions in the ε iron lattice and repel atomic nitrogen. The expelled nitrogen atoms tend to move toward the lower energy position such as grain boundaries, where vacancies are abundant. Then, the N atoms accumulated at grain boundaries form molecular nitrogen which further develops micropores [35, 37, 46].

The compound layer thickness is a function of process time, gas mixture and temperature. The compound layer thickness increases up to a certain time and then decreases because of sputtering and denitriding [47, 48]. The compound layer thickness decreases with increasing of H_2 amount in gas mixture, since a large amount of H_2 in gas mixture increases denitriding rate. The thickness of compound layer increases with increasing temperature. The development of compound layer thickness is fast up to 550°C, then the growth rate becomes slow after 550°C [48, 49].

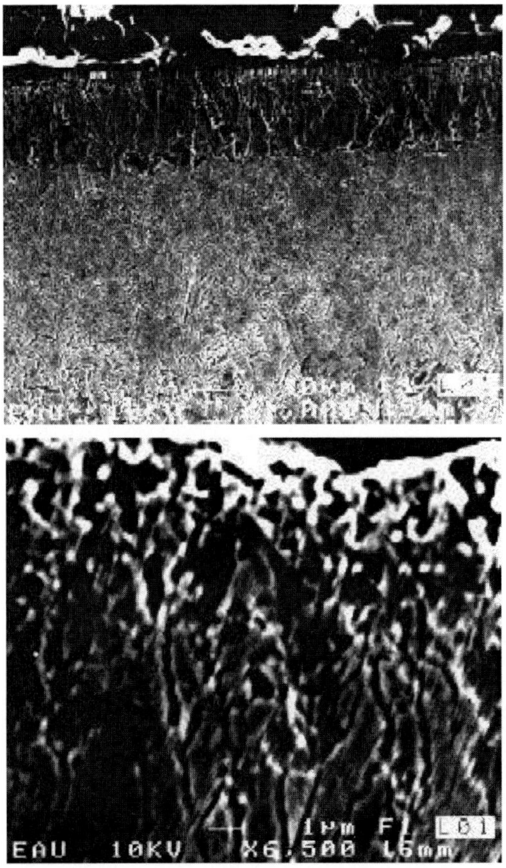

A

B

Figure 20. SEM micrographs of plasma nitrocarburized AISI 4140 steel at 570°C and a gas mixture of 49% N2+49%H2+2%CO2 for 4 h (a) the cross-section of the nitrocarburized specimen (A: Compound layer, B: Diffusion layer) (b) showing presence of porosity in compound layer.

The wear rate of nitrocarburized component improves due to the compound layer (Fig. 21). Compound layer resulting from nitrocarburizing processes are known to be beneficial to wear. However, the wear performance of compound layer depends on the microstructure and composition of layer. It has been also reported that thicker compound layers do not always result in better wear performance, because thicker compound layers include some pores and are more brittle, and can therefore be easily broken off [50, 51]. Investigations have shown that the wear resistance of compound layer is strongly affected by the volume of ϵ-$Fe_{2\text{-}3}$(N, C). In most cases, resistance increases with increasing ϵ phase content [52, 53]. There are two major categories of wear mechanisms, adhesive and abrasive wear. Adhesive wear occurs when the contact pressure between rolling surfaces is high enough to cause local plastic deformation and welding between the contacting asperities. Abrasive wear occurs when hard debris particles are indented and make a groove in the rolling surface of the materials. Therefore, abrasive wear is convincingly related to the mechanical strength of the compound layer and to the adhesion between the compound layer and the substrate [54].

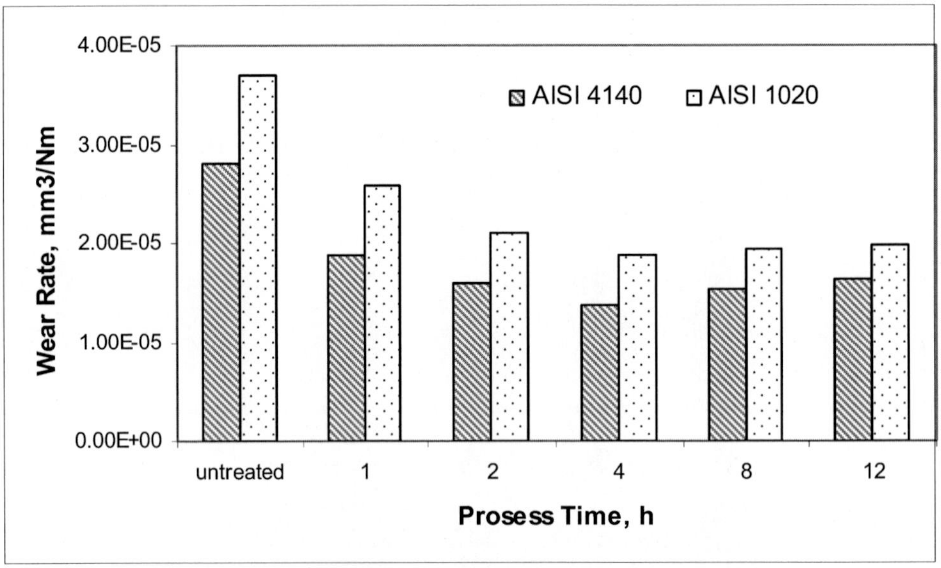

Figure 21. Effect of process time on wear rate and surface hardness of plasma nitrocarburized AISI 4140 and 1020 steel for 49%N_2+49%H_2+2%CO_2 gas mixtures and 570°C [49].

Compound layer enhances surface resistance to corrosion. The post oxidation of plasma nitrocarburized component was successfully applied to improve both tribological and corrosion properties [55, 58]. It has been established that the quality and particularly the corrosion properties of nitrocarburized steel components can be further improved by oxidation treatments at the later stage of plasma nitrocarburizing, to form a thin oxide film on top of the compound layer. The corrosion characteristics of the compound layer can be further improved by the applications of such a thin oxide film [36]. The process parameters of the oxidation such as time, temperature and medium define the type of oxidation modification (Fe_3O_4, Fe_2O_3, FeO). Due to its low friction coefficient and chemical stability, only magnetite Fe_3O_4 is used to improve the wear and corrosion properties of surface structure [55, 57].

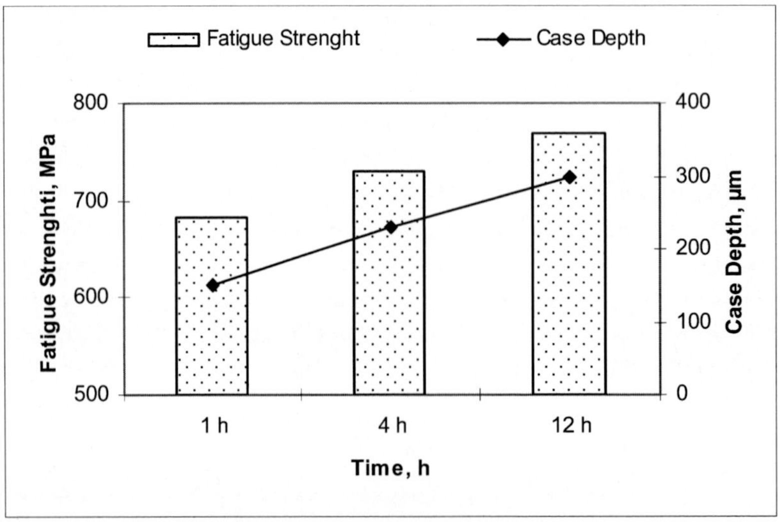

Figure 22. The relations of fatigue strength and case depth of AISI 4140 steel ferritic plasma nitrocarburized at different process times for 49%N_2+49%H_2+2%CO_2 gas mixtures and 570°C [49].

Fatigue strength can be also significantly improved by plasma nitrocarburizing (Fig. 22). The fatigue behaviour of plasma nitrocarburizing is the same as plasma nitriding. The formation of precipitates in the diffusion layer tends to increase the hardness and create compressive residual stresses. These beneficial stresses lower the magnitude of the applied tensile stresses and hence increase the fatigue life of the component [59, 60]. In the studies that the effect of nitriding and nitrocarburizing to fatigue behaviour is searched, it was seen that increasing diffusion layer thickness has an important for increasing fatigue strength [16, 49, 60, 61]. Furthermore, compound layer has not a significant effect on fatigue [16, 62, 63].

b) Diffusion layer

In the region underneath the compound layer, so-called a diffusion zone develops. The diffusion region brings about an improvement of fatigue strength when compared to an untreated material. In this region, N and C atoms are dissolved interstitially in the ferritic lattice and form nitride precipitates. The case depth increased with increasing process time as expected for a diffusion controlled growth. The maximum surface hardness was also obtained for 4 h. Since the precipitate growth and the materials were tempered for long process time, the surface hardness decreased after 4 h (Fig 23)[49].

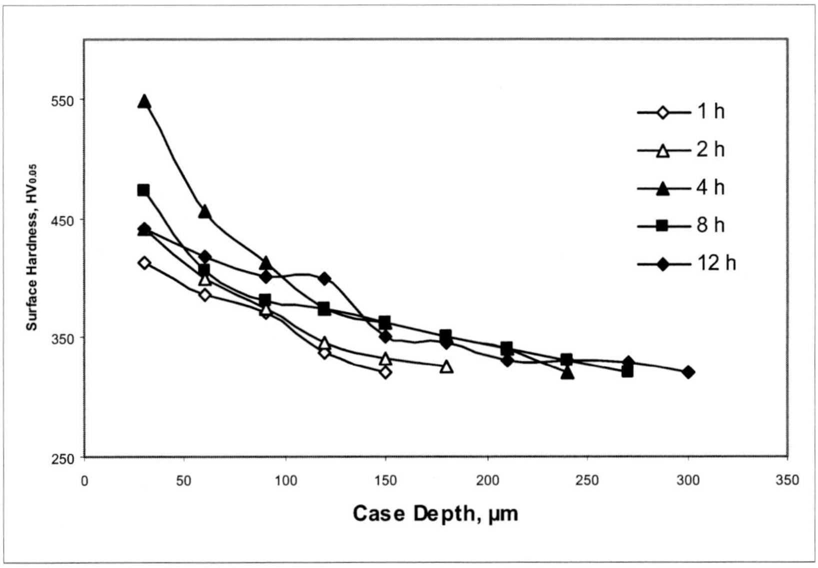

Figure 23. The hardness profiles of AISI 4140 steel ferritic plasma nitrocarburized at different process times for $49\%N_2+49\%H_2+2\%CO_2$ gas mixtures and 570°C.

c) Austenite layer

The surface structure of the austenitic plasma nitrocarburized components is normally a top surface compound layer including ϵ-$Fe_{2-3}(N, C)$ and/or γ'-$Fe_4(N, C)$ phases beneath which is the nitrogen austenite followed by a nitride-containing diffusion zone (Fig. 24).

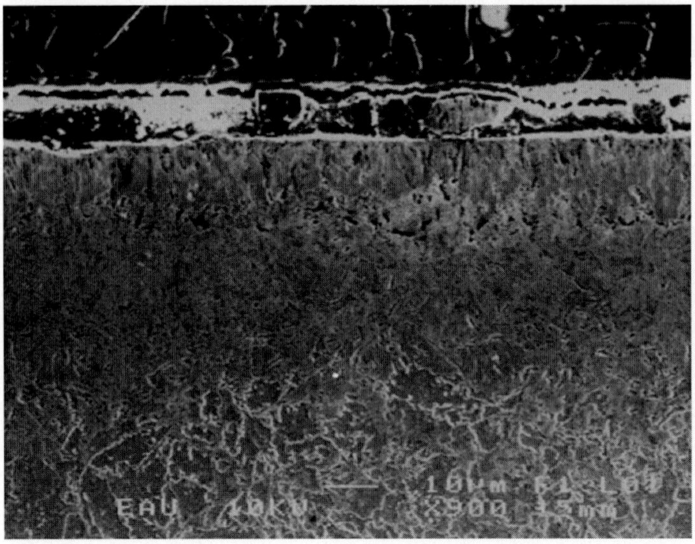

Figure 24. SEM micrographs of plasma nitrocarburized AISI 4140 steel at 640°C and a gas mixture of 49% N2+49%H2+2%CO2 for 4 h (A: Compound layer, B: Austenite layer C: Diffusion layer) [49].

The cross-sectional microstructure indicates that the nitrogen diffusion rate in the steel is much faster than that of carbon. The formation of the nitrogen austenite results solely from the nitrogen diffusion from the carbonitride compound layer to the ferritic matrix [39]. The hardening response of the underneath nitrogen diffusion zone is relatively poor in plain carbon steels. Under high-intensity loading conditions, such a surface and subsurface system will fail prematurely due to its relatively low load bearing capacity. In order to solve this problem, attempts have been made to raise the nitrocarburizing temperature above the Fe-N eutectoid point. This process, termed austenitic nitrocaburizing, produces not only a hard surface layer and diffusion zone, but also an iron-carbon – nitrogen austenite in the subsurface which is then transformed to martensite or bainite, by quenching and/or sub-zero treatment, resulting in a higher subsurface hardness compared with ferritic treatments [36].

3.5. Plasma Boriding (Boronizing)

Boriding is a thermochemical diffusion process, where the surface of the metal is enriched with boron. Boron atoms diffuse into the surface of metal to produce hard boride layers. It is a diffusion process whereby a single phase Fe_2B or a double inter metallic phase (Fe_2B, FeB) is obtained by diffusing boron atoms into the surface of metallic materials (Fig. 25). Boronized steels are characterized by their increased surface hardness, increased wearing and excellent corrosion resistance. Normally boriding is applied to all ferrous materials, nickel alloys, titanium alloys, and sintered carbides [64,66].

When ferrous based materials are boronized at temperatures of 800-1000°C for periods varying between 1 to 8 h, (Fe_2B+FeB) or Fe_2B iron-boride phases are formed at the materials surface, and a boride layer of up to 2000 HV hardness is produced [68]. The boriding temperature depends on the alloy content of the steel substrate [69]. The boriding time is adjusted according to the desired layer thickness. It is seen that hardness of boride layers have three different zones (Fig 26). The diffusion depth of technically useful layer is, depending on the working part, between 10 and 200 μm.

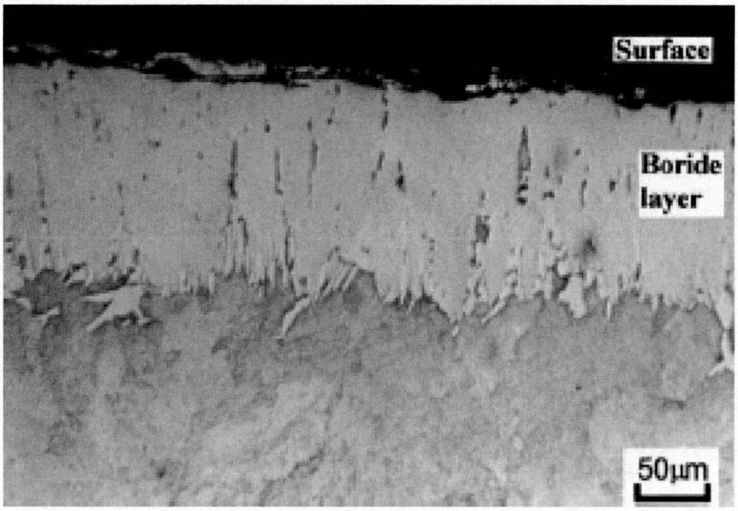

Figure 25. An optical micrograph (cross-sectional view) of borided AISI W1 steel at 1223 K for 6 h [67].

Process methods for boriding include pack, salt bath, gas and plasma. Pack boriding process has the disadvantages such as the control difficulty of its parameters, impossibility of automation, manual working obligation and the detrimental effect of residual products to environment [69, 71]. In the same way, the disadvantages such as the toxicity, fulmination, environmental pollution of salt bath and gas boriding limit its usage [72].

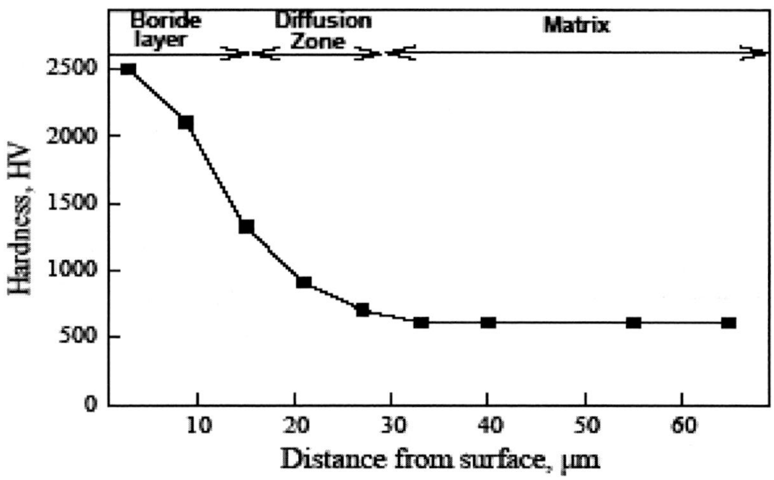

Figure 26. The hardness variation of borided AISI 440C stainlesssteel as a function of distance [73].

Plasma boriding has many advantages such as no environmental pollution, the possibility of controlling the phase composition and the thickness of the boride layers sensitively and saving in energy and gas consumption unlike conventional processes [72, 74, 75]. However, the plasma boriding process has its own limitations. B_2H_6 and BCl_3 gases have been used as a boron source gas, but these gases are relatively expensive, toxic, and explosive. Corrosion in the vacuum chamber by BCl_3 is another serious problem in plasma boriding [65, 72, 74]. Plasma boriding remains in the development stage [74].

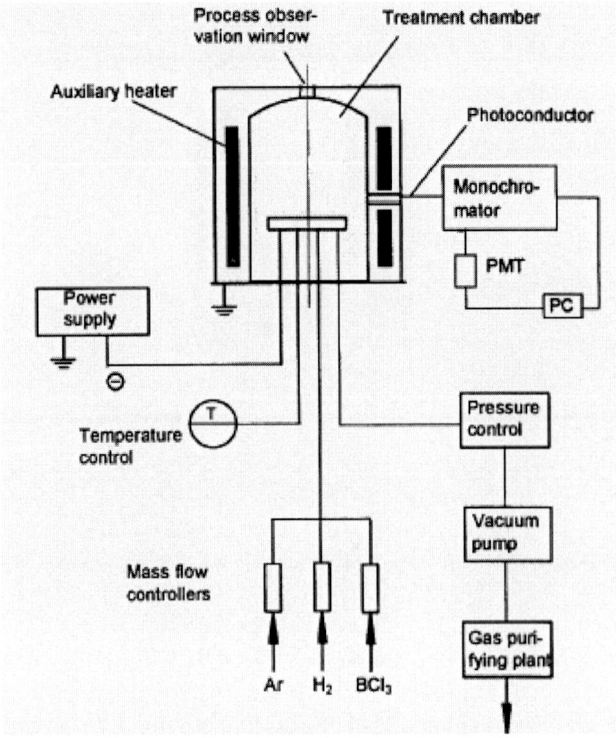

Figure 27. Schematic diagram of the plasma boriding plant [71].

3.5.1. The Mechanism and Process Equipment of Plasma Boriding

Plasma boriding is a boriding process applied on ferrous and nonferrous metallic materials in the plasma formed in a low pressure such as about 10^{-2} Pa and at temperature between 800-1000°C by using BCl_3, B_2H_6, BF_3, $B(OCH_3)_3$ (trimetylborate) as boride source together Ar, H_2 gases [65, 70, 72, 74]. It is suggested that BCl_3 as boride source is the most convenient gas for plasma boriding process [69, 75, 77]. The microstructure and the growing of ironboride layers, process temperature, gas mixture ratios, material compositions, the changing ratios of process pressure and applied current intensity can be controlled. A schematic diagram of the plasma boriding system is shown in Fig. 27.

3.5.2. Microstructures and Properties of Plasma Boronized Components

The microstructure and the growth rate of boride layer are influenced by the treatment temperature, the gas mixture and the materials composition. With longer process time the boride layer depth increases according to the so-called square root of time law, which is well known for diffusion-controlled processes. The hardness of boride layer increases because of the formation of FeB phases at prolonged boriding time at high temperatures [78].

The polyphase boride coatings thermochemically grown on iron and constructional steel are constituted by an outer layer of FeB and an inner layer of Fe_2B, with an extent of crystallographic order that is considerably different for regions located at different depths from the external surface [73, 79]. The single-phase boride layer consists of Fe_2B, while the double-phase layer consists of an outer dark etching phase of FeB and an inner bright etching phase of Fe_2B. The boron-rice FeB phase is considered undesirable, in part, because FeB is more brittle than the iron subboride Fe_2B. Also, because FeB and Fe_2B are formed under tensile and compressive residual stresses, respectively, crack formation is often observed at or in the neighborhood of the FeB/ Fe_2B interface of a double phase layer (Fig 28). Therefore, the boron-rice FeB phase should be avoided or minimized in the boride layer [75].

The wear rate of the coatings was initially high, both under sliding and abrasive conditions, because of the presence of an outer, thin and friable layer of disordered crystals. Then the wear rate decreases because of the resistance offered by ordered crystals first of FeB and then of Fe_2B, until it reaches a minimum value in the inner, very compact regions of the coatings constituted by highly ordered crystals of Fe_2B [79]. It is noted that the fracture toughness value of Fe_2B phase is almost four-fold greater than that of the FeB phase [80].

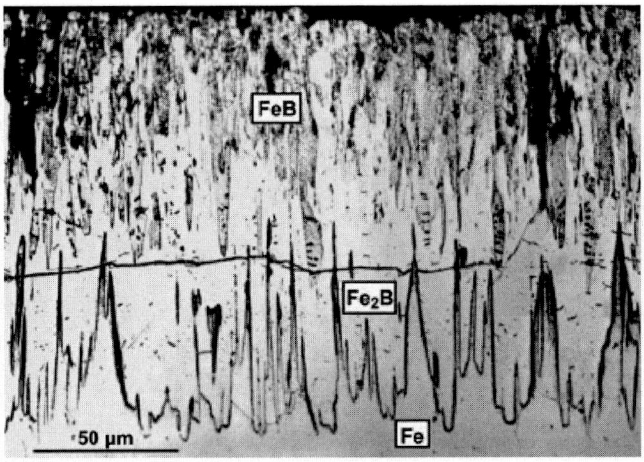

Figure 28. Cross-section of a polyphase boride coating grown on iron, showing the columnar morphology of both FeB and Fe_2B layers, and a crack preferentially propagating along the FeB–Fe_2B interface [79].

3.6. Duplex Treatment

Duplex treatment are carried out two types as before and after thin ceramic film coating. This two different processes are performed in the same or different units as conventional plasma diffusion unit and PVD treatment. There are few a papers including thermochemical process applied on thin film deposited on substrate [81]. The handicap of the method is the fact that the ceramic film deposited on surface is obstracts the diffusion. In additon, before boriding the nitring are carried out to hinder suddenly increase of hardness between compound layer and substrate.

Despite of the rapid growth in the application of hard ceramic coating in industry, very few applications have been found in engineering, because of commonly used low alloy steels with relatively low strength. What limits the common use of hard thin ceramic coating especially in the case of machine element is the load carrying capacity of the coating-substrate system.

- Compound layer (may cause black layer between compound layer and thin film) (Fig 29)
- The surface roughness
- The coating parameters such as bias voltage, coating pressure, current
- The temperature of thermochemical process
- Adhesion
- The orientation of thin ceramic film (may affect the compound layer)

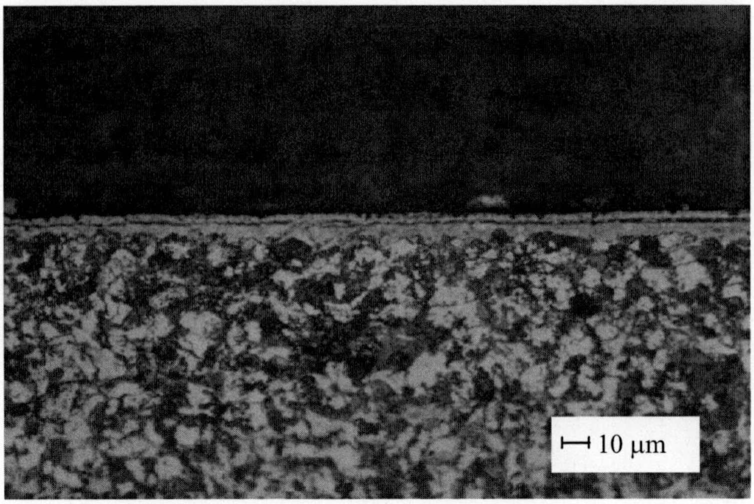

Figure 29. Optical micrographs of duplex treated specimens, 6 A, 50 V, 0.4 Pa/nitrided AISI 5140 [85].

The substrate must have sufficient hardness and flow strength to support the coating without plastic deformation when subjected to a high intensity loading. If the substrate does not have the strength to carry the contact load, plastic deformation will lead to premature failure of the coating [82, 83]. A duplex surface treatment combined a diffusion process and a thin ceramic film coating has been recently developed in an attempt to further improve the tribological performance and corrosion resistance [84, 89]. The restrictions during duplex treatment mentioned are as follws;

The post-oxidizing process can be applied after thermochemical processes. During thermochemical process, the formation of a brittle and porous compound layer affects the wear and corrosion resistance, and restricts the applications of this technique. Relatively little work has been done concerning oxidation treatment after plasma nitriding to remove this disadvantages.

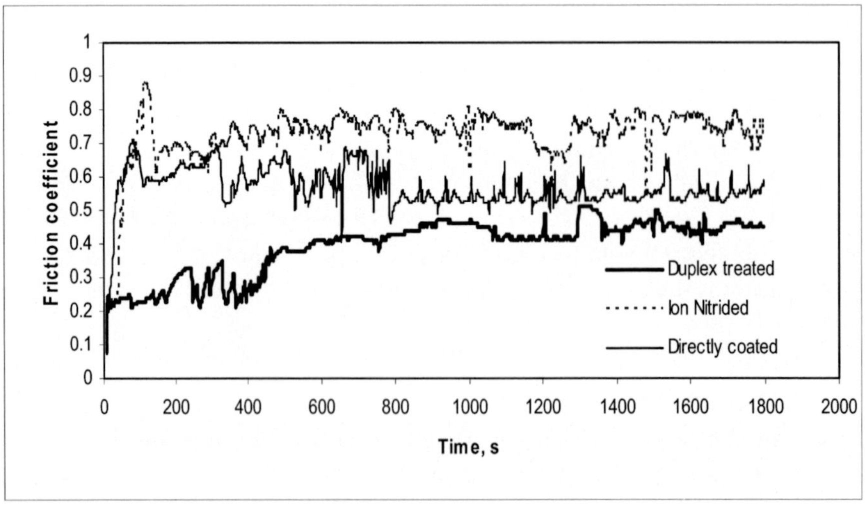

Figure 30. The friction coefficient with time of ion nitrided, directly TiN coated (8 A, 30 V, 0.4 Pa) and duplex treated specimens (8 A, 30 V, 0.4 Pa/nitrided AISI 5140) [90].

Post-oxidation treatment is commonly applied after nitrocarburizing process [36, 57, 58]. Recent reports have demonstrated that using the plasma nitriding process is also possible to form an oxide film on the surface of materials [56]. Plasma oxidation was performed at temperatures of approximately 500°C using gas mixtures of oxygen and air to produce an oxide layer [56, 91].

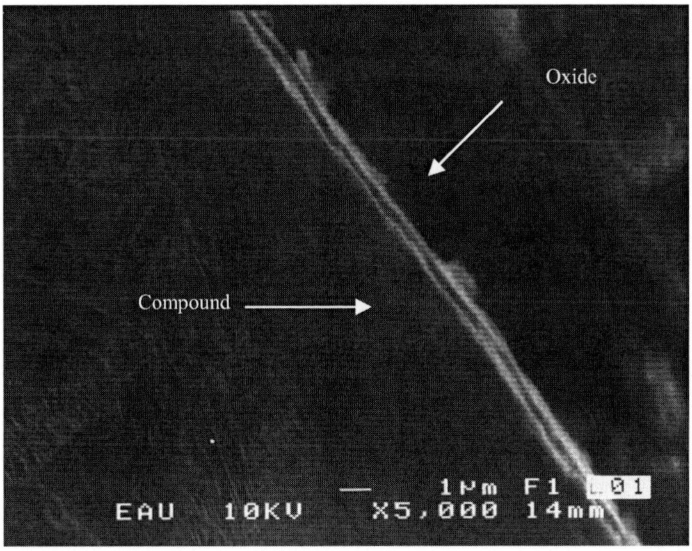

Figure 31. SEM micrograph of post-oxidized AISI 5140 steel for 30 min after plasma nitriding [92].

Fig. 31 shows SEM micrographs of plasma nitrided and post-oxidized AISI 5140 steel. The oxide layer can be identified easily under metallographic examination since it is dark gray. With post-oxidizing, a continuous oxide layer of 0.4 μm was formed on the compound layer of 5 μm. Post-oxidation treatment after plasma nitriding and nitrocarburizing provide a beneficial effect in reducing the friction coefficient, wear rate and corrosion rate (Fig .32 and 33). However, the failure of the oxides may occur by spallation with increasing ofoxidation time.

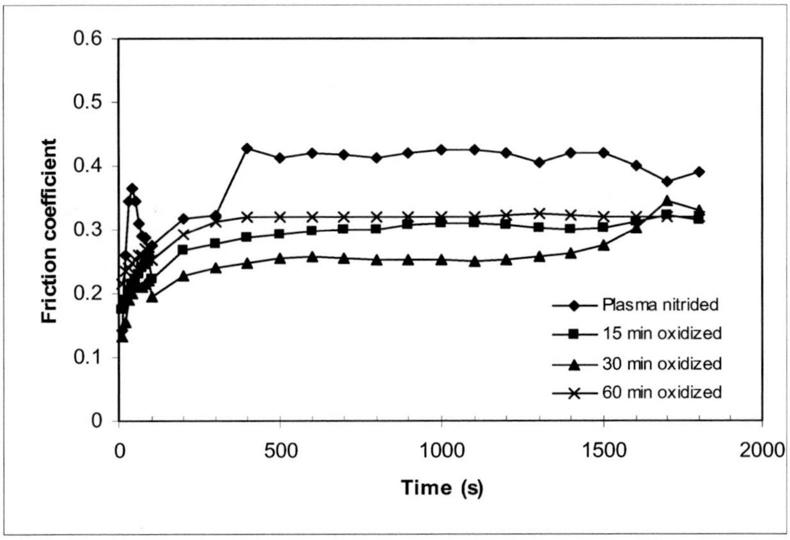

Figure 32. The effect of the plasma nitriding and post-oxidizing for different times on the friction coefficient [92].

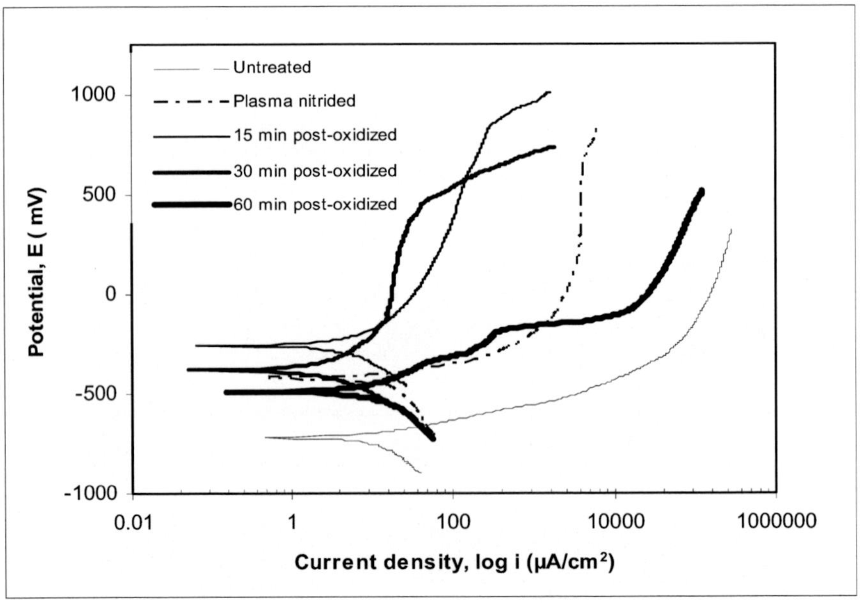

Figure 33. Potentiodynamic polarization curves of untreated, plasma nitrided and nitrided + post-oxidized for 15, 30, 60 min. AISI 5140 steel [92].

References

[1] Strafford, K.N.; Smart, R.St.C.; Sare, I.; Subramanian, C. *Surface Engineering Processes and Applications;*, 1-56676-154-9; Technomic Publishing Company, USA, 1995; pp. 24-25.

[2] Grill, A.; *Cold Plasma in Materials Fabrications;* 0-7803-1055-1; IEEE Press, USA, 1994; pp.17-27.

[3] Stainess, A.M. 1985, Thermochemical treatments in a glow discharge enviroments. *Metalls and Materials.*, vol, 1, 739-744.

[4] Solar Atmoshperes,www.solaratm.com

[5] Celik, A.; Alsaran, A., 2001, Structural characterization of ion-nitrided AISI 5140 low-alloy steel. *Materials Characterization,* vol. 47-3/4, 207-213.

[6] James M.O.; Goodman, D. Plasma (Ion) Nitriding, Surface Combution , Inc.

[7] Metin, E.; Inal, O.T. 1987, Formation and growth of iron nitrides during ion nitriding, *Journal of Materials Science,* vol, 22, 2783-2788.

[8] Edenhofer, B., 1974, Physical and Metallurgical Aspects of Ionnitriding, *Heat Treatment of Metals,* vol,1, 23-28.

[9] Bogaerts, A.; Neyts, E.; Gijbels, R.; Van der Mullen, J. 2002, Gas discharge plasmas and their applications, *Spectrochimica Acta Part B,* vol, 57, 609-658.

[10] Edenhofer, B. 1974, Physical and Metallurgical Aspects of Ionnitriding, *Heat Treatment of Metals,* vol, 2, 59-67.

[11] Alsaran, A.; Kaymaz, I.; Celik, A.; Yetim, A:F.; Karakan, M., 2004, A repair process for fatigue damage using plasma nitriding, *Surface and Coating Technology,* vol, 186/3, 333-338.

[12] Çelik, A.; Karadeniz, S., 1996, Investigation of compound layer formed during ion nitriding of AISI 4140 steel, *Surface and Coating Technology,* vol. 80, 283-286.

[13] Robino C.V.; Inal, O.T., 1983, Ion nitriding behavior of several low alloy steels, *Materials Science and Engineering,* vol, 59, 79-90.

[14] Edenhofer, B.; Bewley, T.J. in P. M. Unterweiser (ed.) , Heat Treatment '76, American Society for Metals , Metals Park, OH, 1977, p.7.
[15] Çelik, A.; Efeoglu, I.; Sakar, G., 2001, Microstructure and structural behavior of ion-nitrided AISI 8620 steel, *Materials Characterization,* vol, 46, 39-44.
[16] Alsaran, A.; Karakan, M.; Celik, A., 2001, The investigation of mechanical properties of ion-nitrided AISI 5140 low-alloy steel, *Materials Characterization,* vol, 48/4, 323-327.
[17] Çelik, A.; Karadeniz, S., 1995, Improvement of the fatigue strength of AISI 4140 steel by an ion nitriding process, *Surface and Coating Technology,* vol, 72, 169-173.
[18] Li, C.X.; Bell, T., 2004, Sliding wear properties of active screen plasma nitrided 316 austenitic stainless steel, *Wear,* vol, 256/11-12, 1144-1152.
[19] Alsaran, A., 2002, Determination of tribological properties of ion-nitrided AISI 5140 steel, *Materials Characterization*, vol, 49/2, 171-176.
[20] Carburizing, www. Key-to-Steel.com
[21] Surface Hardening of Steels, Chapter 1, Process Selection Guide, 1-15, www.asm-intl.org
[22] Lampman, S., 1991, Introduction to Surface Hardening of Steels. *ASM Handbook*, vol, 4, 259-267.
[23] Grube W.L.; Verhoff, S., 1991, Plasma (Ion) Carburizing. *ASM Handbook*, vol, 4, 352-362.
[24] Surface Hardening of Steels, Chapter 3, *Vacum and Plasma Carburizing*, 91-114, www.asm-intl.org
[25] Stickels, C.A., 1991, Carburizing, *ASM Handbook,* vol, 4, 873-877.
[26] Pehlivantürk, N.Y.; İnal, O.T., 1988, Ion/Plasma Carburizing; Theory and Practice, *Adv. Materials and Manufacturing Processes*, vol, 3, 551-573.
[27] http://www.ald-vt.com/home/pdf/bibel/Productgroups/ht_casehardening.pdf
[28] Okumiya M.; Tsunekawa Y.; Matsumoto T.; Tanaka K.; Nagai T., 2003, *Carbon content control in plasma carburizing using process parameter and probe methods*, *Surface and Coatings Technology,* vol, 174, 1171-1174.
[29] Shin D.M.; Lee C.Y.; Lee K.S., 2002, Fatigue properties of plasma carburized low carbon Cr-Mo steel, *ISIJ International,* vol, 42, 1042-1048.
[30] Kraussss, G., 1991, Microstructures and properties of carburized steels, *ASM Handbook,* vol, 4, 363-373.
[31] Grube, W.L., 1980, *J. Heat. Treat.* Vol, 3, 40-49.
[32] Sun Y.; Bell T., 2002, Effect of layer thickness on the rolling-sliding wear behaviour of low-temperature plasma-carburized austenitic stainless steel, *Tribology Letters,* vol, 13, 29-34.
[33] Tokaji K.; Kohyama K.; Akita M., 2004, Fatigue behaviour and fracture mechanism of a 316 stainless steel hardened by carburizing*, Int. J. Fatigue,* vol, 26, 543-551.
[34] Bell, T., 1997, Gaseous and Plasma Nitrocarburizing. *ASM Handbook,* vol, 4, ASM International, Materials Park, Ohio, 425-436.
[35] Haruman, E.; Bell, T.; Sun, Y., 1992, Compound layer characteristics resulting from plasma nitrocarburasing in atmospheres containing carbon dioxide gas additions, *Surface Engeneering,* vol, 8, 275-282.
[36] Bell, T.; Sun, Y.; Suhadi, A., 2000, Environmental and technical aspects of plasma nitrocarburising. *Vacuum,* vol, 59, 14-23.
[37] Li, S.; Manory, R.R., 1996, Surface morphology and compound layer pores of plasma nitrocarburized low carbon steel, *Metallurgical and Materials Transactions A*, vol, 27A, 135-143.
[38] Rembges W.; Oppel, W., 1993, Process control of plasma nitriding and plasma nitrocarburizing in industry, *Surf Coat Tech*, vol, 59, 129-134.

[39] Li, S.; Manory, R.R., 1999, Comparison of compound layer nucleation mechanisms in plasma nitriding and nitrocarburizing: the effect of CH_n species, *Journal of Materials Science*, vol, 34, 1045-1049.

[40] Bell, T.; Kınalı, M.; Munstermann, G., 1987, Physical Metallurgy Aspects of the Austenitic Nitrocarburising Process, *Heat Treatment of Metals,* vol, 2, 47-51.

[41] Li, S. ; Manory, R. R.; Hensler, J. H., 1995, Compound layer growth and compound layer porosity of austenite plasma nitrocarburised non-alloyed steel, *Surface and Coatings Technology,* vol, 71, 112-120.

[42] Somers, M.A.J., 2000, Thermodynamics, kinetics and microstructural evolution of the compound layer; a comparison of the states of knowledge of nitriding and nitrocarburising, *Heat Treatment of Metals,* vol, 4, 92-102.

[43] Du, H.; Agren, J., 1996, Theoretical treatment of nitriding and nitrocarburizing of Iron, *Metallurgical and Materials Transactions A,* vol, 27A, 1073-1080.

[44] Du, H.; Somers, M. A. J. ; Agren, J., 2000, Microstructural and Compositional Evolution of Compound Layers during Gaseous Nitrocarburizing, *Metallurgical and Materials Transactions A,* vol, 31A, 195-211.

[45] Karakan, M.; Alsaran, A.; Çelik, A., 2004, Effect of process time on structural and tribological properties of ferritic plasma nitrocarburised AISI 4140 steel, *Materials &Design,* vol, 25, 349-353.

[46] Malinova, T.; Malinov, S.; Pantev, N., 2001, Simulation of microhardness profiles for nitrocarburized surface layers by artificial neural network, *Surface and Coatings Technology,* vol, 135, 258-267.

[47] Cho, H.S.; Bell, T., 2002, The influence of treatment time on plasma nitrocarburised compound layers for pure iron, *Metals and Materials International,* vol, 8, 99-102.

[48] Alsaran, A.; Karakan, M.; Bulbul, F.; Çelik, A.; Efeoğlu, I., 2003, Study on compound layer formed during plasma nitrocarburizing of AISI 5140 steel, *J. Materials Science Letters,* vol, 22-24, 1759-1761.

[49] Karakan, M., 2004., *Ph.D. Thesis,* Ataturk University, Erzurum/Turkey (in Turkish).

[50] Kim, H.J.; Kweon, Y.G., 1994, The improvement of three-body abrasive wear of plain carbon steel by thermochemical treatments, *Wear,* vol, 174, 201-208.

[51] Karamis, M.B., 1991, An investigation of the properties and wear behaviour of plasma-nitrided hot-working steel (H13), *Wear,* vol, 150, 331-342.

[52] Hoffmann, F.T.; Mayr, P., 1997. Nitriding and Nitrocarburizing, *ASM Handbook,* vol, 6, ASM International, Materials Park, Ohio, pp. 878-883.

[53] Qiang,Y.H.; Ge, S.R.; Xue, Q.J., 2000, Microstructure and tribological properties of complex nitrocarburized steel, *J. of Materials Processing Tech.,* vol, 101, 180-185.

[54] Chiu, L.H.; Wu, C.H.; Chang, H., 2002, Wear behavior of nitrocarburized JIS SKD61 tool steel, *Wear,* vol, 253, 778-786.

[55] Zlatanovic, M.; Popovic, N.; Bogdanov, Z.; Zlatanovic, S., 2003, Plasma post oxidation of nitrocarburized hot work steel samples, *Surface and Coatings Technology,* vol, 177-178, 277-283.

[56] Lee, K.H.; Nam, K.S.; Shin, P.W.; Lee, D.Y.; Song, Y.S., 2003, Effect of post-oxidizing time on corrosion properties of plasma nitrocarburized AISI 1020 steel. *Materials Letters,* vol, 57, 2060-2065.

[57] Hoppe, S., 1998, Fundamentals and applications of the combination of plasma nitrocarburizing and oxidizing, *Surface and Coatings Technology,* vol, 8, 1199-1204.

[58] Sun, Y., 2002, Structures and properties of oxide films formed on plasma nitrocarburised steels by post-oxidation treatments, *Heat Treatment of Metals,* vol, 1, 15-19.

[59] Costa, J.D., Ferreira, J.M.; Ramalho, A.L., 2001, Fatigue and fretting fatigue of ion-nitrided 34CrNiMo6 steel, *Theoretical and Applied Fracture Mechanics,* vol, 35, 69-76.

[60] Genel, K.; Demirkol, M.; Capa, M., 2000, *Effect of ion nitriding on fatigue behaviour of AISI 4140 steel, Materials Science and Engineering A*, vol, 279, 207-216.

[61] Pellizzari, M.; Molinari, A.; Straffelini, A., 2003, Thermal fatigue resistance of gas and plasma nitrided 41CrAlMo7 steel, *Materials Science and Engineering A,* vol, 352, 186-194.

[62] Ashrafizadeh, F., 2003, Influence of plasma and gas nitriding on fatigue resistance of plain carbon (Ck45) steel, *Surface and Coatings Technology,* vol, 173-174, 1196-1200.

[63] Li, C.X.; Sun, Y.; Bell, T., 2000, Factors influencing fretting fatigue properties of plasma-nitrided low alloy steel, *Materials Science and Engineering A,* vol, 292, 18-25.

[64] Yan, P.X.; Zhang, X.M.; Xu, J.W.; Wu, Z.G.; Song, Q.M., 2001, High-temperature behaviour of the boride layer of 45# carbon steel, *Materials Chemistry and Physics,* vol, 71,107-110.

[65] Hunger, H.J.; Löbig, G.. 1997, Generation of Boride Layers on Steel and Nickel Alloys by Plasma Activation of Boron Trifloride, *Thin Solid Films,* vol, 310, 244-250.

[66] Matuschka, A.G., 1980, Boronizing, C. Hanser, München.

[67] Genel, K.; Ozbek I.; Bindal, C., 2003, Kinetics of boriding of AISI W1 steel, *Materials Science and Engineering A,* vol, 347, 311-314.

[68] Meriç, C.; Sahin, S.; Yilmaz, S.S., 2000, Investigation of the effect on boride layer of powder particle size used in boronizing with solid boron-yielding substance, *Materials Research Bulletin,* vol, 35, 2165-2172.

[69] Cabeo, E.R.; Laudien, G.; Biemer, S.; Rie, K.T.; Hoppe, S., 1999, Plasma-assisted Boriding of Industrial Components in a Pulsed D.C. Glow Discharge, *Surface and Coating Technology,* vol, 116/119, 229-233

[70] Küper, A.; Qiao, X.; Stock, H.R.; Mayr, P., 2000, A novel approach to gas boronizing, *Surface and Coating Technology,* vol, 130, 87-94.

[71] Kaestner, P.; Olfe, J.; Rie, K.-T., 2001, Plasma-assisted Boriding of Pure Titanium and TiAl6V4, *Surface and Coating Technology,* vol, 142/144, 248-252.

[72] Yoon, J.H.; Jee, Y.K.; Lee, S.Y., 1999, Plasma Paste Boronizing Treatment of The Stainless Steel AISI 304, *Surface and Coating Technology,* vol, 112, 71-75.

[73] Ozbek, I.; Sen, S.; Ipek, M.; Bindal, C.; Zeytin, S.; Ucisik, A. H., 2004, A mechanical aspect of borides formed on the AISI 440C stainless-steel, *Vacuum*, vol, 73, 643–648.

[74] Nam, K.S.; Lee, K.H.; Lee, S.R.; Kwon, S.C., 1998, A Study on Plasma-assisted Boriding of Steels, *Surface and Coating Technology,* vol, 98, 886-890.

[75] Sinha, A.K., 1991, Boriding (Boronizing), *ASM Handbook,* vol, 4, 437-447.

[76] Dearnley, P.A.; Farrell, T.; Bell, T., 1986, Developments in plasma boronizing, *J. Materials for Energy Systems,* vol, 8/2, 128-131.

[77] Küper A., 2003, Plasma-assisted boronizing, *Advanced Materials & Processes,* 61/3, 21-23.

[78] Sen, U.; Sen, S.; Yilmaz, F., 2004, Structural characterization of boride layer on boronized ductile irons, *Surface and Coatings Technology,* vol, 176, 222–228.

[79] Martini, C.; Palombarini, G.; Poli, G.; Prandstraller, D.; 2004, Sliding and abrasive wear behaviour of boride coatings, *Wear,* vol, 256, 608–613.

[80] Ozbek, I.; Bindal, C., 2002, Mechanical properties of boronized AISI W4 steel, *Surface and Coatings Technology*, vol, 154, 14–20.

[81] Efeoglu, I.; Celik, A., 2001, Mechanical and structural properties of AISI 8620 steel TiN coated, nitrided and TiN coated+nitrided, *Materials Characterization,* vol, 46/4, 311-316.

[82] Sun, Y.; Bell, T., 1991, Plasma surface engineering of low alloy steel, *Mater. Sci. and Eng.,* vol, A140, 419-434.

[83] Rie, K.T.; Broszeit, E., 1995, Plasma diffusion treatment and duplex treatment-recent development and new application, *Surf. and Coat. Tech.,* vol, 76-77, 425-436.

[84] Yılbaş, B.S.; Nizam, S. M., 2000, Wear behavior of TiN coated AISI H11 and AISI M7 twist drills prior to plasma nitriding, *J. Mater. Process. Tech.,* vol, 105, 352-358.

[85] Alsaran, A.; Celik, A.; Celik, C.; Efeoglu, I, 2004, Optimization of coating parameters for duplex treated AISI 5140 steel, *Mater. Sci. and Eng. A.,* vol, 371/1-2, 145-152.

[86] Sun, Y.; Bell, T., 1992, Combined plasma nitriding and PVD treatment, *Trans. Inst. Metal Finish,* vol, 70/1, 38-44.

[87] Gredic, T.; Zlatanovic, M.; Popovic, N.; Bogdanow, Z., 1993, Effect of plasma nitriding on the properties of (Ti, Al)N coatings deposited onto hot working steels substrates, *Thin Solid Films,* vol, 228, 261-266.

[88] Van Stappen, M.; Malliet, B.; Stals, L; De Schepper, L.; Roas, J.R.; Celis, J.P., 1991, Characterization of TiN coatings deposited on plasma nitrided tool steel surface, *Mater. Sci. Eng.,,* vol, A140, 554-562.

[89] Alsaran, A.; Çelik, A.; Çelik, C., 2002, Determination of optimum condition for the plasma nitriding process, *Surf. and Coat. Tech.,* vol, 160/2-3, 219-226.

[90] Alsaran, A., 2001, *Ph D Thesis*, Ataturk University, Erzurum/Turkey (in Turkish).

[91] Borgioli, F. ; Galvanetto, E. ; Fossati, A. ; Bacci, T., 2003, Glow-discharge nitriding and post-oxidising treatments of AISI H11 steel, *Surf. Coat. Tech.,* vol, 162, 61-66.

[92] Alsaran, A., Altun H., Karakan, M., Çelik, A., 2004, Effect of post-oxidizing on tribological and corrosion behaviour of plasma nitrided AISI 5140 steel, *Surface and Coating Technology*, vol, 176/3, 344-348

In: Thermochemistry and Advances in Chemistry Research ISBN: 978-1-60692-376-4
Ed: Anton G. Mirskiy

Chapter 5

THERMOCHEMICAL PROPERTIES, NON-ISOTHERMAL DECOMPOSITION REACTION KINETICS AND QUANTUM CHEMICAL INVESTIGATION OF 2,6-DIAMINO-3,5-DINITROPYRAZINE-1-OXIDE (LLM-105)

Zhao Feng-Qi,[a] Hu Rong-Zu,[a] Gao Hong-Xu[a] and Ma Hai-Xia[b]
[a] Xi'an Modern Chemistry Research Institute Xi'an, Shaanxi, China
[b] Department of Chemical Enineering, Northwest University, Xi'an, Shaanxi, China

ABSTRACT

The constant-volume combustion energy, $\Delta_c U$ (LLM-105, s, 298.15 K), and kinetic behavior of the decomposition reaction of the title compound (LLM-105) are determined by a precise rotating bomb calorimeter, TG/DTG and DSC, respectively. Its standard enthalpy of combustion, $\Delta_c H_m^\theta$(LLM-105, s, 298.15 K), standard enthalpy of formation, $\Delta_f H_m^\theta$(LLM-105, s, 298.15 K) and kinetic parameters of the exothermic main decomposition reaction in a temperature-programmed mode [the apparent activation energy (E_a) and pre-exponential factor (A)] are calculated. The values of $\Delta_c U$ (LLM-105, s, 298.15 K), $\Delta_c H_m^\theta$ (LLM-105, s, 298.15 K), and $\Delta_f H_m^\theta$(LLM-105, s, 298.15K) of LLM-105 are (-9729.89±3.88)$J\cdot g^{-1}$, (-2091.48±0.84)$kJ\cdot mol^{-1}$ and (-54.22±0.99) $kJ\cdot mol^{-1}$, respectively. The kinetic model function in differential form and the value of E_a and A of the decomposition reaction of LLM-105 are $(1-\alpha)^{0.71}$, 154.41$kJ\cdot mol^{-1}$ and $10^{11.22}$ s^{-1}, respectively. The critical temperature of thermal explosion of LLM-105 is 328.19°C. The values of $\Delta S^{\neq}$, $\Delta H^{\neq}$ and $\Delta G^{\neq}$ of this reaction are -35.65$J\cdot(mol\cdot K)^{-1}$,149.58$kJ\cdot mol^{-1}$ and 170.28$kJ\cdot mol^{-1}$, respectively. The compatibility of LLM-105 with components in propellant is investigated by vacuum stability test. The experimental data indicate that the binary systems LLM-105/RDX (cyclotrimethylenetrinitramine), LLM-105/DINA (N-nitro-dihydroxyethylamine-dinitrate), LLM-105/AP (ammonium perchlorate), LLM-105/β-Pb salt (lead 2,4-dihydrxybenzoate), LLM-105/(NC/NG mixture), LLM-105/DEP (diethylphthalate) have good compatibility. The theoretical investigation on LLM-105 as structure unit is carried out by DFT-B3LYP/6-311++G method, and the atomic net charges and the population analysis are discussed.

Keywords: 2,6-Diamino-3,5-dinitropyrazine-1-oxide(LLM-105), combustion energy, standard combustion enthalpy, standard enthalpy of formation, decomposition, kinetics

1. INTRODUCTION

2,6-Diamino-3,5-dinitropyrazine-1-oxide (LLM-105) is a new nitrogen-rich energetic material with higher energy and lower sensitivity. Its crystal density is 1.913g·cm^{-3}. The detonation velocity and specific impulse are about 8560m·s^{-1} and 2122.68 N·s/kg (7MPa), respectively. It has the potential for possible use as an energy ingredient of propellants and explosives from the point of view of the above-mentioned high performance. Its preparation [1], properties [2] have been reported. Herein, we report its combustion energy, standard combustion enthalpy, standard enthalpy of formation obtained by a precise rotating bomb calorimeter, kinetic model function and kinetic parameters of the decomposition reaction studied with TG-DTG and DSC, the compatibility of LLM-105 with components in propellant obtained by vacuum stability test (VST) and the theoretical results obtained by the quantum chemical investigation method. This is quite useful in the evaluation of its thermochemical properties and thermal stability under non-isothermal condition and in the study of its thermal changes at high temperature and in the analysis of compatibility of energetic materials.

2. EXPERIMENTAL

2.1 Material

According to the method reported in the literature [1], LLM-105 was prepared by methoxylation, nitration, amination and oxidation using 2, 6-dichloropyrazine as start material in Scheme 1 and purified by crystallization from 50/50 dimethylsulfoxide /H_2O soluvent for analysis. Sample was kept in vacuum desiccator before use.

2.2 Apparatus and Experimental Conditions

Elemental analyses of carbon, hydrogen and nitrogen were carried out with an Elementar vario EL-III microanalyzer. The infrared spectrum was recorded on a 60 SXR-FTIR spectrophotometer (Nicolet Co., USA) over the frequency range 4000-400cm^{-1} using the KBr pellet technique. The ^{1}H NMR, ^{15}N NMR, and ^{13}C NMR spectra were performed on a Bruker AV 500MHz superconducting NMR spectrometer (Bruker Biospin International AG., Switzerland). The mass spectrum studies were conducted on a mass spectrometer (Hewlett Packard co., USA). The compatibility test of LLM-105 with component used in formulations was carried out according to the method reported [3].

The constant-volume combustion energy of LLM-105 was determined by a precise rotating bomb calorimeter (RBC-type II) [4]. The main experimental procedures were described previously [4,5]. The initial temperature was regulated to 25.0000±0.0005°C, and the initial oxygen pressure was 2.5 MPa.

The correct value of the heat exchange was calculated according to Linio-Pyfengdelel- Wsava formula [6].

The calorimeter was calibrated with benzoic acid of 99.999% purity. It had an isothermal heat of combustion at 25°C of -26434±3 J·g^{-1}. The energy equivalent of calorimeter was determined to be 17936.01±9.08 kJ·K^{-1}. The precision of the measurements was in 4.68×10^{-4}.

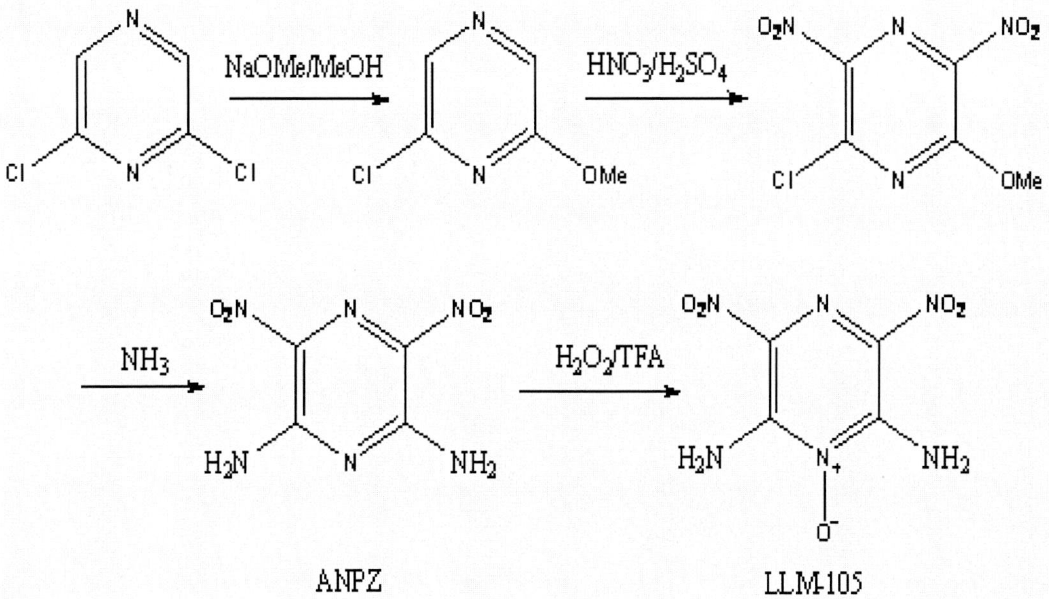

Scheme 1.

The analytical methods of final products (gas, liquid and solid) were the same as these in Ref. [4]. The analytical results of the final products showed that the combustion reactions were complete. As a result, either carbon deposits or carbon monoxide formed during the combustion reactions and the amount of NO_x in the final gas phase were negligible.

The DSC and TG-DTG experiments of LLM-105 were performed using a model Q600SDT (TA, USA) under a nitrogen atmosphere at a flow rate of 100 mL·min^{-1} and the amount of used sample was about 0.7 mg. The heating rates used were 2.5, 5, 10, 15 and 20°C·min^{-1} from ambient temperature to 420 °C.

3. Results and Discussion

3.1 Characterization

Composition Analysis of LLM-105

The elementary analyses results of LLM-105 are as follows: w (calcd): C22.23, H1.87, N38.89; w (found): C22.46, H1.49, N38.76, which is identified as the formula of $C_4H_4O_5N_6$.

IR-Spectroscopy

The IR characteristic absorption peaks in Figure 1 are assigned as(KBr, cm^{-1}): 3422, 3403(m, υ_{N-H}^{as}), 3282, 3226(m, br, υ_{N-H}^{s}), 1565(s, $\upsilon_{NO_2}^{as}$), 1337(s, $\upsilon_{NO_2}^{s}$), 1266, 1246 (s, br, $\upsilon_{N-O}+\upsilon_{C-N}$), 1616, 1490, 1454, 1379(m, s, m, m, $\upsilon_{C=C}$, $\upsilon_{C=N}$, υ_{ring}), 1647(s, δ_{NH_2}), 890(m, δ_{N-O}), 815(s, δ_{C-N}), and 597-532(m, br, δ_{NO_2}).

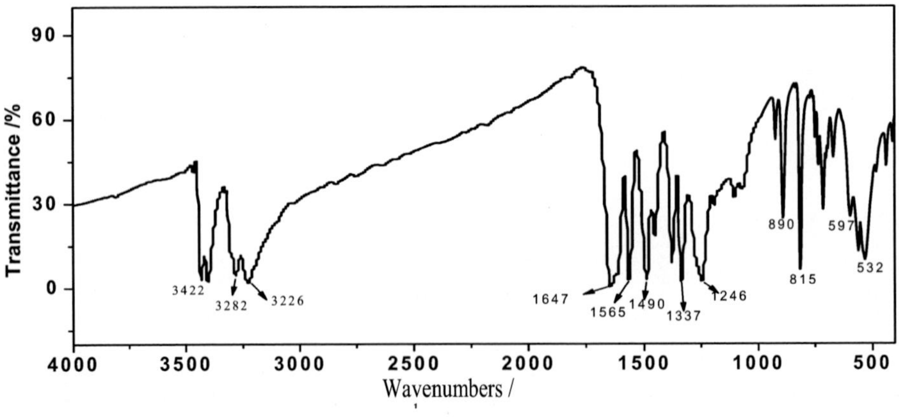

Figure 1. IR spectrum of LLM-105 at room temperature

NMR-Spectroscopy

The ^{1}H NMR spectrum of LLM-105 in Figure 2 shows two peaks at 8.784 and 9.064 ppm for both amino groups which demonstrates their hindered rotation around the C-N bond on the NMR time scale, revealing an effect of hydrogen-bonding to the neighbouring nitro groups [1]. Two small peaks at 8.220 and 8.610 ppm are attributable to 2,6-diamino-3,5-dinitropyragine(ANPZ) which shows the same effect. The purity of LLM-105 is more than 99.5% ($1-\frac{0.04}{9.99}=99.6\%$) according to ^{1}H NMR.

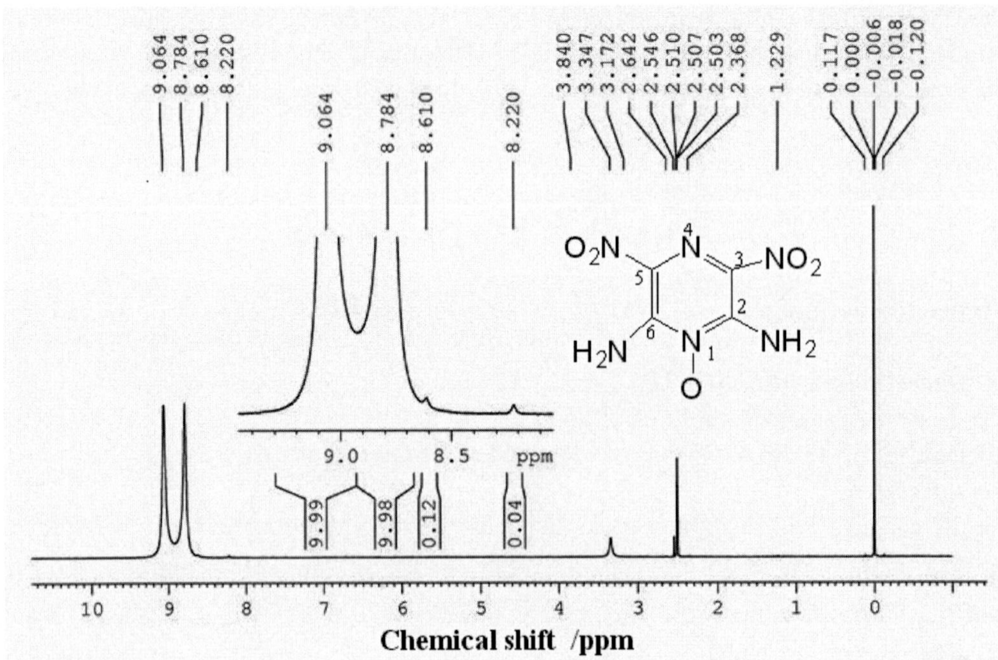

Figure 2. ^{1}H NMR spectrum of LLM-105 (DMSO-d6 as solvent, DMSO as reference standard, 500MHz)

Figure 3 shows the ^{15}N NMR spectrum composed of four different signals. The signals at -292.82, -138.94, -107.16 and -16.70 ppm are attributed to the amino groups, N-oxide group, second ring-nitrogen and nitro groups, respectively.

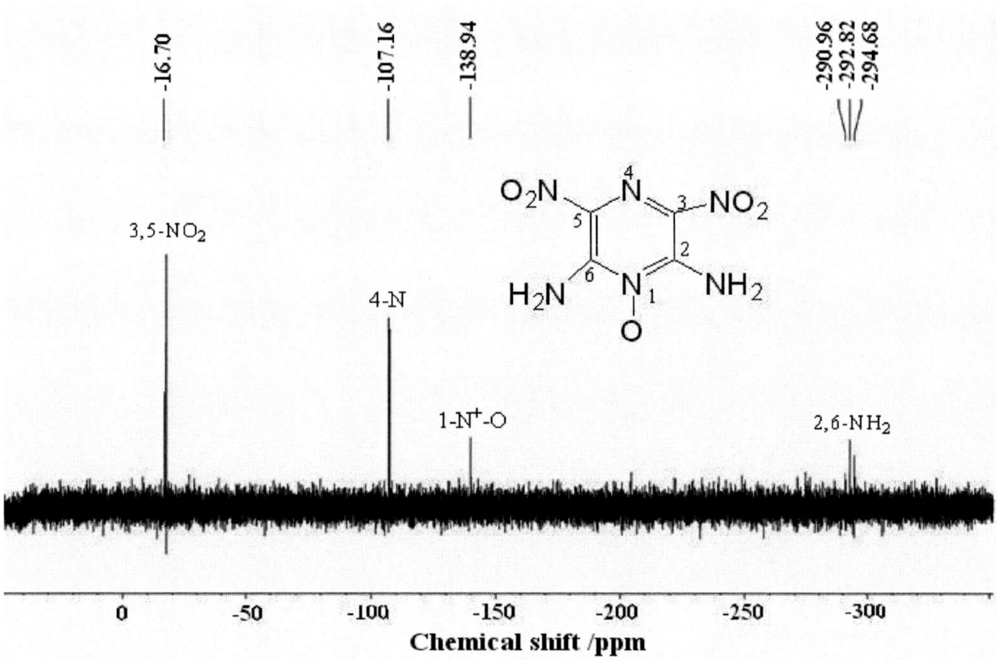

Figure 3. ^{15}N NMR spectrum of LLM-105(DMSO-d6 as solvent, DMSO as reference standard, 500MHz)

The ^{13}C NMR spectrum in Figure 4 indicates an up-field chemical shift to 124.82 ppm of the carbon-atoms attached to the nitro groups. The C-NH_2 absorbs at 144.56 ppm.

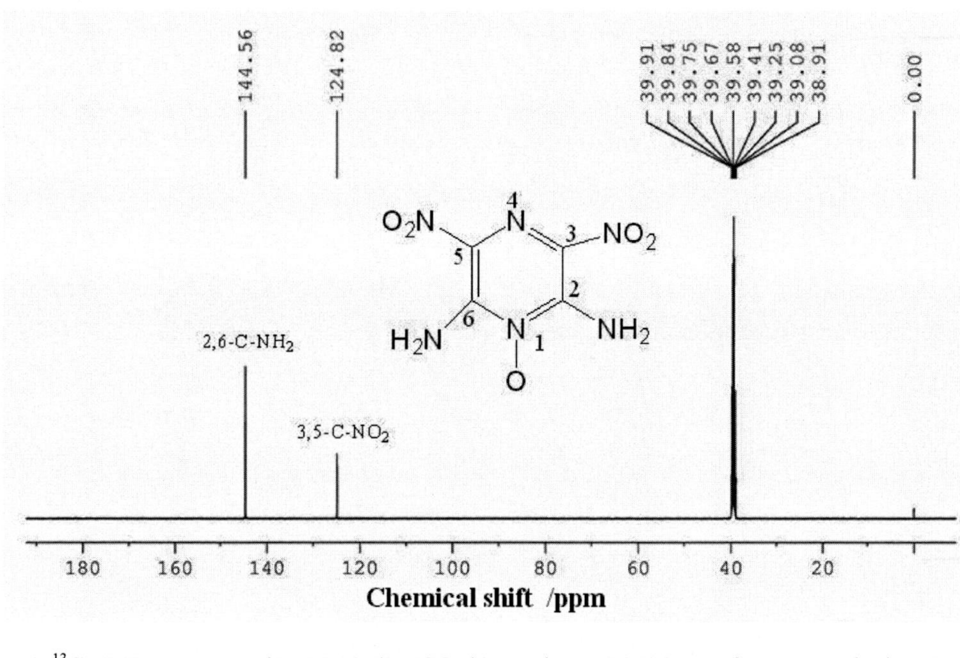

Figure 4. ^{13}C NMR spectrum of LLM-105(DMSO-d6 as solvent, DMSO as reference standard, 500MHz)

Mass Spectroscopy

The molecular ion (M) peak of LLM-105 can be found in the spectrum in Figure 5 with high abundance.

3.2 Thermochemical Properties

Combustion Energy

The determination method of combustion energy for LLM-105 was the same as that for the calibration of the calorimeter with benzoic acid. The combustion energies of LLM-105 samples were calculated by the formula

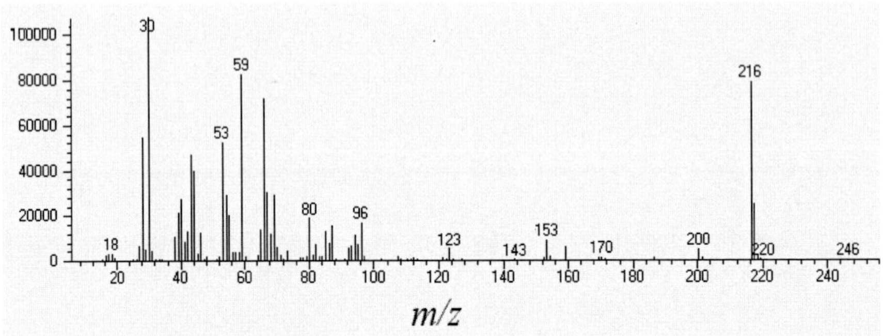

Figure 5. Mass spectrum of LLM-105

$$\Delta_c U(\text{LLM-105,s}) = \frac{W\Delta T - aG - 5.983b}{m} \tag{1}$$

where $\Delta_c U$ (LLM-105, s) denotes the constant-volume combustion energy of the samples, W is the energy equivalent of the RBC-type □ calorimeter (in $J \cdot K^{-1}$), ΔT the correct value of the temperature rising, a the length of actual Ni-Cr wire consumed (in cm), G the combustion enthalpy of Ni-Cr wire for ignition (0.9 $J \cdot cm^{-1}$), 5.983 the formation enthalpy and solution enthalpy of nitric acid corresponding to 1 mL of 0.1000 $mol \cdot L^{-1}$ solution of NaOH(in $J \cdot ml^{-1}$), b the volume in mL of consumed 0.1000 $mol \cdot L^{-1}$ solution of NaOH and m the mass in g of the sample. The results of the calculations were given in Table 1.

Standard Combustion Enthalpy

The standard combustion enthalpy of LLM-105, $\Delta_c H_m^\theta$ (LLM-105, s, 298.15 K), was referred to the combustion enthalpy change of the following ideal combustion reaction at 298.15 K and 100 kPa.

$$C_4H_4N_6O_5(s) + 5/2O_2(g) \rightarrow 4CO_2(g) + 2H_2O + 3N_2(g) \tag{2}$$

The standard combustion enthalpy of LLM-105 was calculated by the following equations:

$$\Delta_c H_m^\theta \text{ (LLM-105, s, 298.15 K)} = \Delta_c U \text{ (LLM-105, s, 298.15 K)} + \Delta n RT \tag{3}$$

$$\Delta n = n_g \text{ (products)} \rightarrow n_g \text{ (reactants)} \tag{4}$$

where n_g is the total amount in mole of gases present as products or as reactants, $R = 8.314\ J \cdot K^{-1} \cdot mol^{-1}$, $T = 298.15$ K. The result is (–2091.48±0.84) $kJ \cdot mol^{-1}$.

Table 1. Experimental results for the combustion energies of samples

No.	Mass of sample m /g	Calibrated heat of combustion wire Q_c /J	Calibrated heat of acid Q_N /J	Calibrated ΔT /K	Combustion energy of sample $-\Delta_c U$ /J·g^{-1}
1	1.15460	12.60	38.49	0.6354	9737.75
2	1.10536	12.60	36.99	0.6076	9725.84
3	1.15962	12.60	39.00	0.6375	9727.34
4	1.14587	12.60	38.67	0.6311	9745.08
5	1.16475	12.60	39.06	0.6400	9722.60
6	1.12361	12.60	37.66	0.6173	9720.72
mean					9729.89±3.88

Standard Enthalpy of Formation

The Standard enthalpies of formation of LLM-105, $\Delta_f H_m^\theta$(LLM-105, s, and 298.15 K), were calculated by Hess's law according to the above thermochemical equations (2):

$\Delta_f H_m^\theta$(LLM-105, s, 298.15 K) = 4$\Delta_f H_m^\theta$(CO_2, g, 298.15 K) + 2$\Delta_f H_m^\theta$(H_2O, l, 298.15 K) - $\Delta_c H_m^\theta$(LLM-105, s, 298.15 K)

when $\Delta_f H_m^\theta$(CO_2, g, 298.15 K) = (-393.51 ±0.13) kJ·mol^{-1}[7], $\Delta_f H_m^\theta$(H_2O, l, 298.15 K) = (-285.83±0.042) kJ·mol^{-1}[7], the results obtained is (-54.22±0.99) kJ·mol^{-1}.

3.3 Thermal Decomposition Kinetics

Typical DSC, TG-DTG and T vs. α curves for LLM-105 at a heating rate of 10°C·min^{-1} are shown in Figures 1, 2 and 3. DSC curve shows that two exothermic peaks before 420°C. A major exothermic peak at 359.84°C is due to decomposition of LLM-105. An exothermic process at temperatures higher than 369.95°C is caused by the further decomposition of the decomposition products in coacervate phase. The TG curve shows only one stage mass loss process. The stage began at about 199.54°C and completed at 347.02°C accompanied with 89.03% mass loss. It indicates that the decomposition of LLM-105 is much more complete.

In order to obtain the kinetic parameters [apparent activation energy (E_a) and pre-exponential factor (A)] of the decomposition reaction for LLM-105, the five model-free isoconversional methods [Eqs. (5)-(9)] were employed. These methods are as follows:

Differential methods

Kissinger equation [8]

$$\ln\frac{\beta}{T_p^2} = \ln\frac{AR}{E} - \frac{E}{RT_p} \tag{5}$$

differential isoconversional non-linear[NL-DIF] equation [9]

$$\left| \sum_{i}^{n} \sum_{j \neq i}^{n} \frac{\beta_i \left(\frac{\mathrm{d}\alpha}{\mathrm{d}T} \right)_i \exp\left(E_a / RT_{\alpha,i}\right)}{\beta_j \left(\frac{\mathrm{d}\alpha}{\mathrm{d}T} \right)_j \exp\left(E_a / RT_{\alpha,j}\right)} - n(n-1) \right| = \min \tag{6}$$

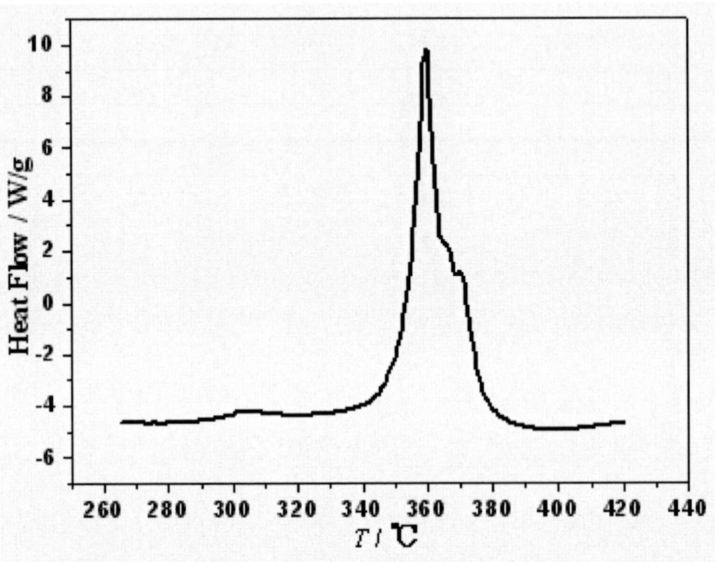

Figure 6. DSC curve for the decomposition of LLM-105 at a heating rate of 10 ° C·min^{-1}

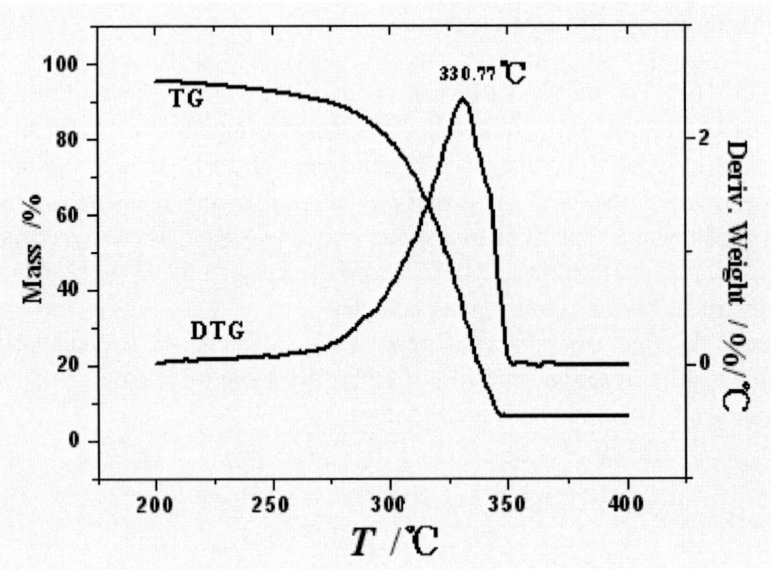

Figure 7. TG-DTG curve for the decomposition of LLM-105 at a heating rate of 10 ° C·min^{-1}

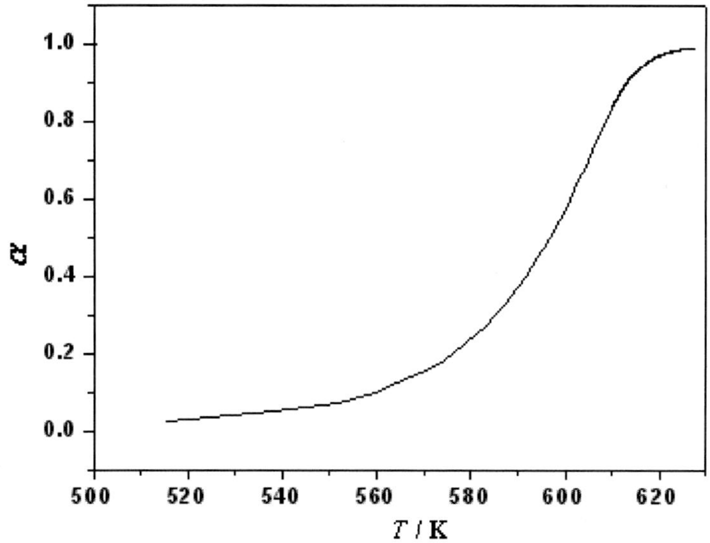

Figure 8. T vs α curve for the decomposition of LLM-105 at a heating rate of 10 °C·min^{-1}

Integral methods

Flynn-Wall-Ozawa (F-W-O) equation [10]

$$\lg \beta = \lg\left(\frac{AE}{RG(\alpha)}\right) - 2.315 - 0.4567\frac{E}{RT} \tag{7}$$

integral isoconversional non-linear[NL-INT] equation [11]

$$\left|\sum_{i}^{n}\sum_{j\neq i}^{n}\frac{\beta_j I(E_\alpha, T_{\alpha,i})}{\beta_i I(E_\alpha, T_{\alpha,j})} - n(n-1)\right| = \min \tag{8}$$

integral isoconversional non-linear[NL-INT$_0$] equation [12]

$$\left|\sum_{i}^{n}\sum_{j\neq i}^{n}\frac{\beta_j (T_i - T_{0,i})\exp(-E_a/RT_i)}{\beta_i (T_j - T_{0,j})\exp(-E_a/RT_j)} - n(1-n)\right| = \min \tag{9}$$

where α is the conversion degree, T the absolute temperature, T_0 the initial point at which DTG curve deviates from the baseline, E_α the apparent activation energy, β the heating rate, R the gas constant, T_p the peak temperature of DTG curve, E the apparent activation energy, A the pre-exponential factor, $G(\alpha)$ the integral mechanism function.

From the original data in Table 2, the value of E_k obtained by Kissinger's method [Eq.(5)] was determined to be 142.60 kJ·mol^{-1} and A $10^{10.14}$ s^{-1}. The linear correlation coefficient (r_k) is 0.9805. The values of E_o and r_o obtained by Ozawa's method [Eq.(7)] are 141.10 kJ·mol^{-1} and 0.9829, respectively.

Table 2. Maximum peak temperature (T_p) of the exothermic decomposition reaction for LLM-105 determined by the DTG curves at various heating rates (β)

β/(°C•min^{-1})	2.5	5	10	15	20
T_p /°C	310.46	319.33	330.77	346.55	347.82

By substituting the original data, β_i, T_{0i}, T_i, α_i and $(d\alpha/dT)_i$, i=1,2, ⋯n, tabulated in Tables 3-7 from TG-DTG curves into Eqs. (6)-(9), the values of E_α for any given value of α in Table 8 are obtained. The average value of E_α in the α range of 0.125 to 0.825 in Fig. 9 is in good agreement with the calculated values obtained by Kissinger's method and Ozawa's method. The E values calculated using Eqs.(5)-(9) were used to check the validity of activation energy by the other methods.

Table 3. Data of LLM-105 determined by TG[a]

Data point	T_i/K	α_i	$(d\alpha/dT)_i \cdot 10^{-3}$ / K^{-1}
1	547.21	0.125	9.09
2	550.27	0.150	9.71
3	552.89	0.175	10.38
4	555.17	0.200	11.52
5	557.23	0.225	12.71
6	559.01	0.250	13.96
7	560.84	0.275	15.19
8	562.41	0.300	16.46
9	563.87	0.325	17.73
10	565.24	0.350	18.95
11	566.45	0.375	20.15
12	567.72	0.400	21.38
13	568.85	0.425	22.52
14	569.94	0.450	23.66
15	570.95	0.475	24.89
16	571.93	0.500	26.04
17	572.88	0.525	27.19
18	573.78	0.550	28.23
19	574.65	0.575	29.26
20	575.47	0.600	30.53
21	576.30	0.625	31.60
22	577.08	0.650	32.39
23	577.80	0.675	33.38
24	578.57	0.700	34.35
25	579.29	0.725	35.09
26	579.99	0.750	35.94
27	580.68	0.775	36.80
28	581.34	0.800	37.20
29	582.00	0.825	37.60

[a]T_o = 481.49 K; β = 0.04167 °C·s^{-1}

Table 4. Data of LLM-105 determined by TG[a]

Data point	T_i/K	α_i	$(\mathrm{d}\alpha/\mathrm{d}T)_i \cdot 10^{-3}/\mathrm{K}^{-1}$
1	555.11	0.125	8.36
2	558.44	0.150	8.87
3	561.24	0.175	9.46
4	563.80	0.200	10.59
5	566.07	0.225	11.79
6	568.05	0.250	12.99
7	569.87	0.275	14.24
8	571.57	0.300	15.45
9	573.14	0.325	16.59
10	574.58	0.350	17.76
11	575.94	0.375	18.96
12	577.23	0.400	20.12
13	578.43	0.425	21.26
14	579.58	0.450	22.37
15	580.66	0.475	23.42
16	581.70	0.500	24.46
17	582.71	0.525	25.50
18	583.67	0.550	26.53
19	584.60	0.575	27.57
20	585.48	0.600	28.55
21	586.34	0.625	29.47
22	587.17	0.650	30.31
23	587.99	0.675	31.10
24	588.78	0.700	31.85
25	589.56	0.725	32.55
26	590.31	0.750	33.14
27	591.06	0.775	33.52
28	591.79	0.800	33.73
29	592.53	0.825	33.86

[a] T_o = 496.70 K; β = 0.08333 °C·s^{-1}

Table 5. Data of LLM-105 determined by TG[a]

Data point	T_i/K	α_i	$(d\alpha/dT)_i \cdot 10^{-3}$ / K^{-1}
1	564.07	0.125	6.06
2	569.36	0.150	6.74
3	572.92	0.175	7.39
4	575.99	0.200	8.66
5	578.64	0.225	9.91
6	581.02	0.250	11.05
7	583.17	0.275	12.17
8	585.14	0.300	13.27
9	586.94	0.325	14.35
10	588.62	0.350	15.54
11	590.18	0.375	16.64
12	591.66	0.400	17.55
13	592.95	0.425	18.55
14	594.33	0.450	19.59
15	595.59	0.475	20.49
16	596.78	0.500	21.41
17	597.90	0.525	22.50
18	598.99	0.550	23.49
19	600.04	0.575	24.37
20	601.04	0.600	25.19
21	602.01	0.625	25.95
22	602.96	0.650	26.57
23	603.88	0.675	27.01
24	604.80	0.700	27.27
25	605.70	0.725	27.09
26	606.61	0.750	26.30
27	607.53	0.775	25.19
28	608.55	0.800	24.58
29	609.69	0.825	23.66

[a] T_o = 472.69 K; β = 0.16667 °C·s^{-1}

Table 6. Data of LLM-105 determined by TG[a]

Data point	T_i/K	α_i	$(d\alpha/dT)_i \cdot 10^{-3}/K^{-1}$
1	568.54	0.125	6.75
2	572.67	0.150	7.18
3	576.08	0.175	7.71
4	579.33	0.200	8.69
5	582.04	0.225	9.77
6	584.44	0.250	10.92
7	586.61	0.275	12.10
8	588.59	0.300	13.20
9	590.39	0.325	14.30
10	592.07	0.350	15.40
11	593.66	0.375	16.38
12	595.11	0.400	17.31
13	596.51	0.425	18.22
14	597.86	0.450	19.09
15	599.14	0.475	19.97
16	600.38	0.500	20.92
17	601.54	0.525	21.93
18	602.66	0.550	22.99
19	603.72	0.575	24.05
20	604.74	0.600	25.10
21	605.71	0.625	26.16
22	606.65	0.650	27.17
23	607.55	0.675	28.17
24	608.42	0.700	29.14
25	609.27	0.725	30.04
26	610.08	0.750	30.88
27	610.88	0.775	31.64
28	611.66	0.800	32.02
29	612.42	0.825	32.34

[a] T_o = 495.77 K; β = 0.2500 °C·s^{-1}

Table 7. Data of LLM-105 determined by TG[a]

Data point	T_i/K	α_i	$(d\alpha/dT)_i \cdot 10^{-3}/K^{-1}$
1	571.60	0.125	5.63
2	576.70	0.150	6.08
3	580.83	0.175	6.57
4	584.43	0.200	7.51
5	587.57	0.225	8.51
6	590.31	0.250	9.52
7	592.80	0.275	10.56
8	595.09	0.300	11.61
9	597.15	0.325	12.67
10	599.04	0.350	13.78
11	600.79	0.375	14.87
12	602.40	0.400	15.93
13	603.91	0.425	17.00
14	605.35	0.450	18.02
15	606.69	0.475	18.98
16	607.97	0.500	19.92
17	609.19	0.525	20.85
18	610.37	0.550	21.72
19	611.50	0.575	22.55
20	612.58	0.600	23.36
21	613.63	0.625	24.16
22	614.65	0.650	24.90
23	615.64	0.675	25.58
24	616.60	0.700	26.27
25	617.54	0.725	26.95
26	618.46	0.750	27.53
27	619.35	0.775	28.01
28	620.23	0.800	28.30
29	621.10	0.825	28.52

[a] T_0 = 456.89 K; β = 0.33333 °C·s^{-1}

Table 8. Apparent activation energies of thermal decomposition of LLM-105 obtained using isoconversional methods and the data taken from Figure 1

No	α	E_α /(kJ·mol^{-1})			
		NL-DIF method [Eq.(6)]	NL-INT method [Eq.(8)]	NL-INT$_0$ method [Eq.(9)]	Flynn-Wall-Ozawa [Eq.(7)]
1	0.050	226.08	271.69	280.50	–
2	0.075	200.92	274.91	273.04	269.80
3	0.100	173.64	231.31	228.30	228.57
4	0.125	169.12	209.60	206.68	208.00
5	0.150	158.54	194.97	192.08	194.13
6	0.175	153.53	186.83	184.04	186.44
6	0.175	153.53	186.83	184.04	186.44
7	0.200	151.38	180.13	177.51	180.10
8	0.225	150.27	175.52	173.01	175.74
9	0.250	148.77	171.45	169.04	171.91
10	0.275	148.44	169.08	166.71	169.68
11	0.300	147.42	166.54	164.24	167.29
12	0.325	146.86	164.58	162.33	165.45
13	0.350	146.82	162.89	160.69	163.86
14	0.375	146.14	161.21	158.98	162.20
15	0.400	145.69	160.25	158.14	161.39
16	0.425	145.49	159.25	157.19	160.46
17	0.450	145.00	158.16	156.13	159.44
18	0.475	144.12	157.17	155.19	158.51
19	0.500	143.64	156.31	154.38	157.71
20	0.525	143.50	155.66	153.76	157.11
21	0.550	143.36	154.93	153.06	156.43
22	0.575	143.11	154.30	152.46	155.84
23	0.600	142.26	153.64	151.84	155.23
24	0.625	142.03	153.17	151.39	154.80
25	0.650	142.08	152.60	150.85	154.27
26	0.675	141.44	151.97	150.24	153.68
27	0.700	140.97	151.59	149.87	153.33
28	0.725	140.52	151.11	149.41	152.88
29	0.750	139.26	150.59	148.90	152.40
30	0.775	137.69	150.11	148.45	151.96
31	0.800	135.14	149.41	147.76	151.31
32	0.825	132.40	148.62	146.97	150.56
33	0.850	129.56	147.66	146.02	149.66
34	0.875	126.86	146.45	144.83	148.52
35	0.900	122.77	145.46	143.85	147.60
36	0.925	118.34	143.89	142.29	146.11
37	0.950	112.15	142.32	140.73	144.64
38	0.975	107.24	140.49	138.91	142.92
39	1.000	99.56	136.80	135.24	139.46
		mean in the range of 0.125 to 0.825 :			
		145.34 kJ·mol^{-1}	162.47kJ·mol^{-1}	160.39kJ·mol^{-1}	163.52kJ·mol^{-1}

Table 9. Common mechanism functions $f(\alpha)$ and $G(\alpha)$ in non-isothermal reaction kinetics

No.	Name of function	$f(\alpha)$	$G(\alpha)$	Mechanism
1	Parabola law	$\alpha^{-1/2}$	α^2	one-dimensional diffusion, ID
2	Valensi equation	$[-\ln(1-\alpha)]^{-1}$	$\alpha+(1-\alpha)\ln(1-\alpha)$	two-dimensional diffusion, 2D
3	Jander equation	$4(1-\alpha)^{\frac{1}{2}}\left[1-(1-\alpha)^{\frac{1}{2}}\right]^{\frac{1}{2}}$	$\left[1-(1-\alpha)^{\frac{1}{2}}\right]^{\frac{1}{2}}$	two-dimensional diffusion, 2D, $n=1/2$
4	Jander equation	$(1-\alpha)^{\frac{1}{2}}\left[1-(1-\alpha)^{\frac{1}{2}}\right]^{-1}$	$\left[1-(1-\alpha)^{\frac{1}{2}}\right]^{2}$	two-dimensional diffusion, 2D, $n=2$
5	Jander equation	$6(1-\alpha)^{\frac{2}{3}}\left[1-(1-\alpha)^{\frac{1}{3}}\right]^{\frac{1}{2}}$	$\left[1-(1-\alpha)^{\frac{1}{3}}\right]^{\frac{1}{2}}$	three-dimensional diffusion, 3D, $n=1/2$
6	Jander equation	$\frac{3}{2}(1-\alpha)^{\frac{2}{3}}\left[1-(1-\alpha)^{\frac{1}{3}}\right]^{-1}$	$\left[1-(1-\alpha)^{\frac{1}{3}}\right]^{2}$	three-dimensional diffusion, spheres symmetry, 3D, $n=2$
7	G.-B. equation[a]	$\frac{3}{2}[(1-\alpha)^{-\frac{1}{3}}-1]^{-1}$	$1-\frac{2}{3}\alpha-(1-\alpha)^{\frac{2}{3}}$	three-dimensional diffusion, 3D
8	Anti-Jander equation	$\frac{3}{2}(1+\alpha)^{\frac{2}{3}}\left[(1+\alpha)^{\frac{1}{3}}-1\right]^{-1}$	$\left[(1+\alpha)^{\frac{1}{3}}-1\right]^{2}$	three-dimensional diffusion, 3D
9	Z.-L.-T. equation[b]	$\frac{3}{2}(1-\alpha)^{\frac{4}{3}}\left[(1-\alpha)^{-\frac{1}{3}}-1\right]^{-1}$	$\left[(1-\alpha)^{-\frac{1}{3}}-1\right]^{2}$	three-dimensional diffusion, 3D
10	Avrami-Erofeev equation	$4(1-\alpha)[-\ln(1-\alpha)]^{\frac{3}{4}}$	$[-\ln(1-\alpha)]^{\frac{1}{4}}$	random nucleation and subsequent growth, $n=1/4$, $m=4$
11	Avrami-Erofeev equation	$3(1-\alpha)[-\ln(1-\alpha)]^{\frac{2}{3}}$	$[-\ln(1-\alpha)]^{\frac{1}{3}}$	(same as above) $n=173$, $m=3$
12	Avrami-Erofeev equation	$\frac{5}{2}(1-\alpha)[-\ln(1-\alpha)]^{\frac{3}{5}}$	$[-\ln(1-\alpha)]^{\frac{2}{5}}$	(same as above) $n=2/5$

Table 9. Continued

No.	Name of function	$f(\alpha)$	$G(\alpha)$	Mechanism
13	Avrami-Erofeev equation	$[-\ln(1-\alpha)]^{\frac{1}{2}}$	$2(1-\alpha)[-\ln(1-\alpha)]^{\frac{1}{2}}$	(same as above) $n = 1/2$, $m = 2$
14	Avrami-Erofeev equation	$\frac{3}{2}(1-\alpha)[-\ln(1-\alpha)]^{\frac{1}{3}}$	$[-\ln(1-\alpha)]^{\frac{2}{3}}$	(same as above) $n = 2/3$
15	Avrami-Erofeev equation	$\frac{4}{3}(1-\alpha)[-\ln(1-\alpha)]^{\frac{1}{4}}$	$[-\ln(1-\alpha)]^{\frac{3}{4}}$	(same as above) $n = 3/4$
16	Avrami-Erofeev equation	$1-\alpha$	$-\ln(1-\alpha)$	(same as above) $n = 1$, $m = 1$
17	Avrami-Erofeev equation	$\frac{2}{3}(1-\alpha)[-\ln(1-\alpha)]^{-\frac{1}{2}}$	$[-\ln(1-\alpha)]^{\frac{3}{2}}$	(same as above) n=3/2
18	Avrami-Erofeev equation	$\frac{1}{2}(1-\alpha)[-\ln(1-\alpha)]^{-1}$	$[-\ln(1-\alpha)]^{2}$	(same as above) $n = 2$
19	Avrami-Erofeev equation	$\frac{1}{3}(1-\alpha)[-\ln(1-\alpha)]^{-2}$	$[-\ln(1-\alpha)]^{3}$	(same as above) $n = 3$
20	Avrami-Erofeev equation	$\frac{1}{4}(1-\alpha)[-\ln(1-\alpha)]^{-3}$	$[-\ln(1-\alpha)]^{4}$	(same as above) $n = 4$
21	P. -T. equation[c]	$\alpha(1-\alpha)$	$\ln\left[\frac{\alpha}{1-\alpha}\right]$	auto catalysis branch random nucleation
22	Mampel power law	$4\alpha^{\frac{3}{4}}$	$\alpha^{\frac{1}{4}}$	$n = 1/4$
23	Mampel power law	$3\alpha^{\frac{2}{3}}$	$\alpha^{\frac{1}{3}}$	$n = 1/3$
24	Mainpel power law	$2\alpha^{\frac{1}{2}}$	$\alpha^{\frac{1}{2}}$	$n = 1/2$
25	Mampel power law	1	$1-(1-\alpha)^{\frac{1}{1}} = \alpha$	phase boundary reaction, R_1, n=1
26	Mampel power law	$\frac{2}{3}\alpha^{-\frac{1}{2}}$	$\alpha^{\frac{3}{2}}$	$n = 3/2$
27	Mampel power law	$\frac{1}{2}\alpha^{-1}$	α^{2}	$n = 2$
28	Reaction order	$4(1-\alpha)^{\frac{3}{4}}$	$1-(1-\alpha)^{\frac{1}{4}}$	$n = 1/4$

Table 9. Continued

No.	Name of function	$f(\alpha)$	$G(\alpha)$	Mechanism
29	Contracting sphere (volume)	$3(1-\alpha)^{\frac{2}{3}}$	$1-(1-\alpha)^{\frac{1}{3}}$	phase boundary reaction, R_3
30	Contracting sphere (volume)	$(1-\alpha)^{\frac{2}{3}}$	$3\left[1-(1-\alpha)^{\frac{1}{3}}\right]$	$n = 1/3$, $n = 3$, (three-dimension)
31	Contracting cylinder (area)	$2(1-\alpha)^{\frac{1}{2}}$	$1-(1-\alpha)^{\frac{1}{2}}$	phase boundary reaction, R_2
32	Contracting cylinder (area)	$(1-\alpha)^{\frac{1}{2}}$	$2\left[1-(1-\alpha)^{\frac{1}{2}}\right]$	$n = 1/2$, $n = 2$, (two-dimension)
33	Reaction order	$\frac{1}{2}(1-\alpha)^{-1}$	$1-(1-\alpha)^2$	$n = 2$
34	Reaction order	$\frac{1}{3}(1-\alpha)^{-2}$	$1-(1-\alpha)^3$	$n = 3$
35	Reaction order	$\frac{1}{4}(1-\alpha)^{-3}$	$1-(1-\alpha)^4$	$n = 4$
36	Second order	$(1-\alpha)^2$	$(1-\alpha)^{-1}$	chemical reaction, F_2
37	Reaction order	$(1-\alpha)^2$	$(1-\alpha)^{-1}-1$	chemical reaction
38	2/3 order	$2(1-\alpha)^{\frac{3}{2}}$	$(1-\alpha)^{-\frac{1}{2}}$	chemical reaction
39	Exponent law	α	$\ln\alpha$	E_1, $n = 1$
40	Exponent law	$\frac{1}{2}\alpha$	$\ln\alpha^2$	$n = 2$
41	Third order	$\frac{1}{2}(1-\alpha)^3$	$(1-\alpha)^{-2}$	chemical reaction, F_3

a Ginstling-Brounshtein equation; b Zhuralev-Lesokin-Tempelman equation; c Prout-Tompkins equation

The differential Eqs.(10) and (11) and integral Eqs. (12)-(16) are cited to obtain the values of E_a, A and the most probable kinetic model function $[f(\alpha)]$ from a single non-isothermal DSC curve[20].

Differential equation [13]

$$\ln\left[\frac{d\alpha/dT}{f(\alpha)[E(T-T_0)/RT^2+1]}\right]=\ln\left(\frac{A}{\beta}\right)-\frac{E}{RT} \tag{10}$$

Achar-Brindley-Sharp-Wendworth equation [14]

$$\ln\frac{d\alpha}{f(\alpha)dT}=\ln\frac{A}{\beta}-\frac{E}{RT} \tag{11}$$

ordinary-integral equation [15]

$$\ln\left[\frac{G(\alpha)}{T^2\left(1-\frac{2RT}{E_a}\right)}\right]=\ln\left(\frac{AR}{\beta E_a}\right)-\frac{E_a}{RT} \tag{12}$$

MacCallum-Tanner equation [16]

$$\lg[G(\alpha)]=\lg\left(\frac{AE_a}{\beta R}\right)-0.4828E_a^{0.4357}-\frac{0.449+0.217E_a}{0.001}\frac{1}{T} \tag{13}$$

Satava-Sestak equation [17]

$$\lg[G(\alpha)]=\lg\left(\frac{AE_a}{\beta R}\right)-2.315-0.4567\frac{E_a}{RT} \tag{14}$$

Coats-Redfern equation [18]

$$\ln\left[G(\alpha)/T^2\right]=\ln\left(\frac{AR}{\beta E}\right)-\frac{E}{RT} \tag{15}$$

Madhusudanan-Krishnan-Ninan equation [19]

$$\ln\left[\frac{G(\alpha)}{T^{1.921503}}\right]=\left[\ln\frac{AE}{\beta R}+3.772050-1.921503\ln E\right]-0.120394(E/T) \tag{16}$$

Forty-one types of kinetic model function in Table 9 and the data, α , T, $\mathrm{d}\alpha/\mathrm{d}T$ and T_0 from TG-DTG curves under various linear heating rate β in Tables (3)-(7) are put into Eqs. (10) –(16) for calculation, respectively. The values of E_a, A, linear correlation coefficient (r) , standard mean square deviation (Q) and believable factor (d) (where d=(1-r)Q) were obtained by the linear least-squares[20].

The kinetic parameters, E and A and the probable kinetic model functions selected by the logical choice method and satisfying the ordinary range of the thermal decomposition kinetic parameters for energetic materials (E = 80 – 250 kJmol^{-1}, log A = 7 – 30 s^{-1}) together with their appropriate values of r, Q and d, obtained by jointly employing seven methods [Eqs.(10)-(16)] are presented in Table 10. The most probable kinetic model function of the decomposition reaction is $f(\alpha)=(1-\alpha)^{0.71}$ Substituting $f(\alpha)$ with $(1-\alpha)^{0.71}$, E with 154.41 kJ·mol^{-1} and A with $10^{11.22}$s^{-1} in Eq. (17),

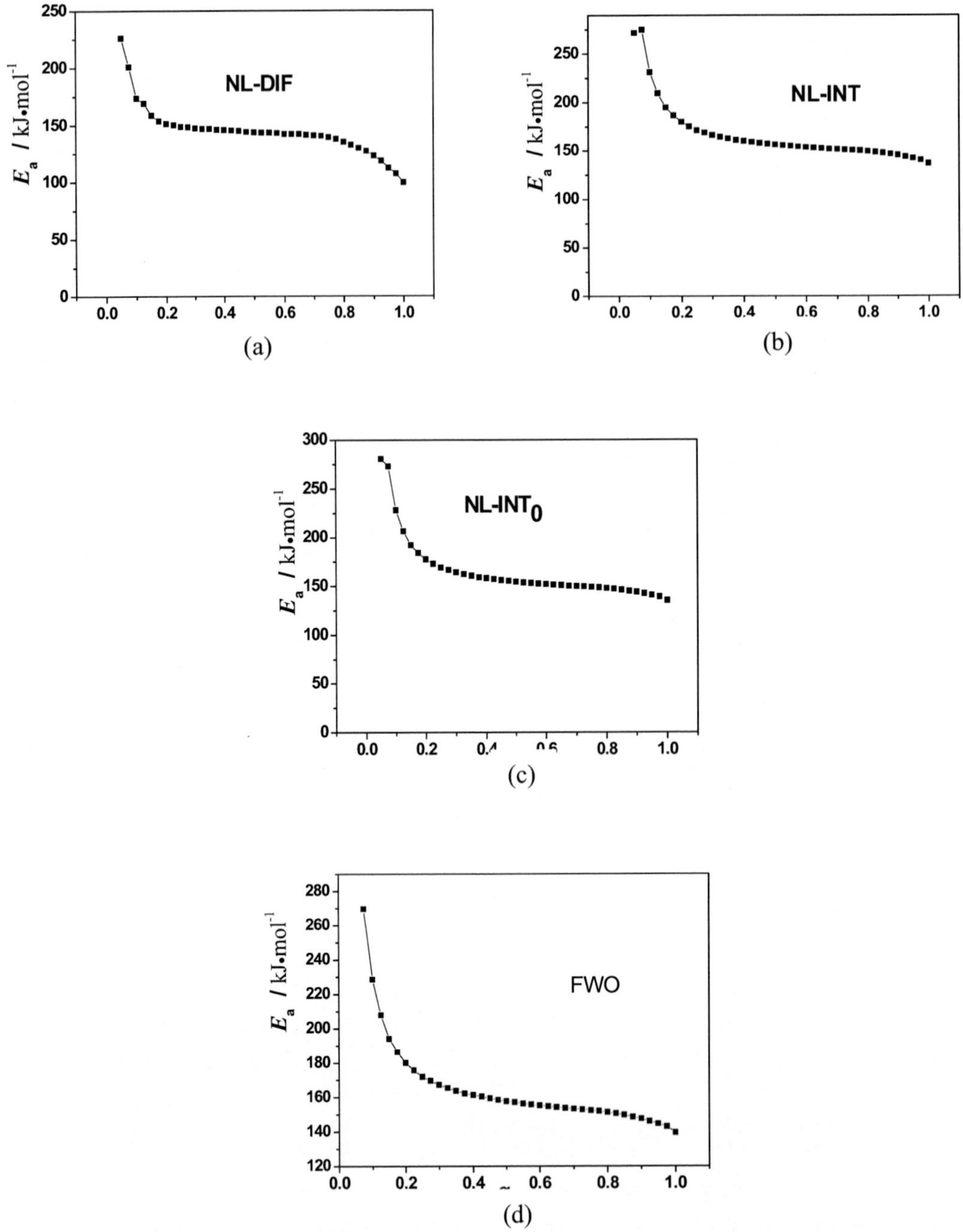

Figure .9 $E_{\alpha} \sim \alpha$ curves for the decomposition of LLM-105 by NL-DIF (a), NL-INT (b), NL-INT$_0$(c) and FWO (d) methods

$$\mathrm{d}\alpha / \mathrm{d}T = \frac{A}{\beta} f(\alpha) \mathrm{e}^{-E/RT} \tag{17}$$

We can now establish the kinetic equation of the exothermic decomposition process of LLM-105 as follows:

$$\frac{\mathrm{d}\alpha}{\mathrm{d}T}=\frac{10^{11.22}}{\beta}(1-\alpha)^{0.71}\exp\left(-1.857\times10^{4}/T\right)$$

The value (T_{p0}) of the peak temperature (T_p) corresponding to $\beta\rightarrow0$ obtained by Eq. (18) taken from Ref[12] is 307.47°C.

$$T_{pi}=T_{p0}+b\beta_i+c\beta_i^2+\mathrm{d}\beta_i^3,\ i=1,2,\ \cdots,5 \tag{18}$$

where *b, c* and d are coefficients.

The critical temperature of thermal explosion (T_b) obtained from Eq. (19) taken from [21] is 328.19 °C.

$$T_b=\frac{E_o-\sqrt{E_o^2-4E_oRT_{p0}}}{2R} \tag{19}$$

where *R* is the gas constant (8.314 J·mol^{-1}·K^{-1}), E_o is the value of *E* obtained by Ozawa's method.

Table 10. Kinetic parameters obtained by the data in Tables 3-7

No.	Equation	Function	E_a /(kJ·mol^{-1})	log(A/s^{-1})	r	Q	d•10^2
	$\beta=2.5$ ° C·min^{-1}						
1	(10)	$(1-\alpha)^{0.5}$	156.88	10.75	0.9865	0.2647	0.3569
2	(11)	$(1-\alpha)^{0.5}$	175.47	13.24	0.9900	0.2448	0.2452
3	(12)	$(1-\alpha)^{0.5}$	156.88	11.51	0.9991	0.0170	0.0015
4	(13)	$(1-\alpha)^{0.5}$	158.81	11.66	0.9992	0.0033	0.0003
5	(14)	$(1-\alpha)^{0.5}$	158.13	11.62	0.9992	0.0033	0.0003
6	(15)	$(1-\alpha)^{0.5}$	156.88	11.49	0.9991	0.0170	0.0015
7	(16)	$(1-\alpha)^{0.5}$	157.09	11.53	0.9991	0.0170	0.0015
	$\beta=5$ ° C·min^{-1}						
8	(10)	$(1-\alpha)^{0.5}$	148.58	10.10	0.9879	0.2276	0.2747
9	(11)	$(1-\alpha)^{0.5}$	169.42	12.70	0.9916	0.2035	0.1698
10	(12)	$(1-\alpha)^{0.5}$	150.92	11.01	0.9990	0.0196	0.0020
11	(13)	$(1-\alpha)^{0.5}$	152.96	11.16	0.9991	0.0038	0.0004
12	(14)	$(1-\alpha)^{0.5}$	152.61	11.16	0.9991	0.0038	0.0004
13	(15)	$(1-\alpha)^{0.5}$	150.92	10.98	0.9990	0.0196	0.0020
14	(16)	$(1-\alpha)^{0.5}$	151.14	11.03	0.9990	0.0196	0.0020
$\beta=10$ °C·min-1							
15	(10)	$(1-\alpha)^{0.5}$	139.83	9.12	0.9947	0.1108	0.0591
16	(11)	$(1-\alpha)^{0.5}$	152.73	11.08	0.9958	0.1026	0.0425
17	(12)	$(1-\alpha)^{1.0}$	155.94	11.49	0.9928	0.1863	0.1340

Table 10. Continued

No.	Equation	Function	Ea /(kJ·mol-1)	log(A/s-1)	r	Q	d•102
18	(13)	$(1-\alpha)^{1.0}$	158.25	11.67	0.9936	0.0355	0.0229
19	(14)	$(1-\alpha)^{1.0}$	157.59	11.63	0.9936	0.0355	0.0229
20	(15)	$(1-\alpha)^{1.0}$	155.94	11.46	0.9928	0.1863	0.1340
21	(16)	$(1-\alpha)^{1.0}$	156.16	11.51	0.9928	0.1864	0.1335
β = 15 °C·min-1							
22	(10)	$(1-\alpha)^{0.5}$	145.98	9.83	0.9828	0.3852	0.6644
23	(11)	$(1-\alpha)^{0.5}$	163.06	12.10	0.9871	0.3569	0.4604
24	(12)	$(1-\alpha)^{1.0}$	158.54	11.81	0.9926	0.1907	0.1402
25	(13)	$(1-\alpha)^{1.0}$	160.93	12.00	0.9934	0.0363	0.0239
26	(14)	$(1-\alpha)^{1.0}$	160.12	11.95	0.9934	0.0363	0.0239
No.	Equation	Function	E_a /(kJ·mol^{-1})	log(A/s^{-1})	r	Q	d•10^2
27	(15)	$(1-\alpha)^{1.0}$	158.54	11.79	0.9926	0.1907	0.1401
28	(16)	$(1-\alpha)^{1.0}$	158.77	11.83	0.9927	0.1908	0.1396
β = 20 ° C·min^{-1}							
29	(10)	$(1-\alpha)^{0.5}$	144.50	9.50	0.9869	0.3370	0.4406
30	(11)	$(1-\alpha)^{0.5}$	154.26	11.25	0.9890	0.3214	0.3528
31	(12)	$(1-\alpha)^{1.0}$	144.33	10.50	0.9899	0.2589	0.2619
32	(13)	$(1-\alpha)^{1.0}$	146.72	10.68	0.9910	0.4936	0.0442
33	(14)	$(1-\alpha)^{1.0}$	146.70	10.72	0.9910	0.0494	0.0442
34	(15)	$(1-\alpha)^{1.0}$	144.33	10.47	0.9899	0.2589	0.2619
35	(16)	$(1-\alpha)^{1.0}$	144.57	10.52	0.9899	0.2590	0.2608

Mean: n= 0.71; E_a= 154.41 kJ·mol^{-1}, log(A/s^{-1}) = 11.22

The entropy of activation ($\Delta S^{\neq}$), enthalpy of activation ($\Delta H^{\neq}$) and free energy of activation ($\Delta G^{\neq}$) corresponding to $T = T_{p0}$, E = 154.41kJ·mol^{-1} and $A = 10^{11.22}$s^{-1} obtained by Eqs.(20), (21) and (22) are -35.65 J·mol^{-1}·K^{-1}, 149.58 kJ·mol^{-1} and 170.28 kJ·mol^{-1}, respectively.

$$A = \frac{k_B T}{h} e^{\Delta S^{\neq}/R} \tag{20}$$

$$\Delta H = E - RT \tag{21}$$

$$\Delta G^{\neq} = \Delta H^{\neq} - T\Delta S^{\neq} \tag{22}$$

where, k_B is the Boltzmann constant and h the Planck constant.

3.4 Compatibility of LLM-105 with Components Used in Formulations

Vacuum stability test results from tests performed according to specifications of standardization method GJB 772A-97-501.2. e.g. 1:1 by weight, 100°C/40h [3] are given in Table 11. These data indicate that LLM-105 and contacted materials, cyclotrimethylenetrinitramine (RDX), N-nitro-dihydroxyethylamine-dinitrate (DINA), ammonium perchlorate (AP), lead 2,4-dihydrxybenzoate (β-Pb), and diethylphthalate (DEP) are compatible (gas volume generated by the mixture less those volumes generated by the individual components does not exceed 3 mL).

Table 11 Results of compatibility for LLM-105 and contacted materials

System	Weight /g	Gas evolved /mL	Net gas evolution /mL	Rating
LLM-105	5.0	0.17		
RDX	5.0	0.37		
50/50-RDX/LLM-105 mixture	2.5+2.5	0.23	-0.31	good compatibility
DINA	5.0	0.59		
50/50-DINA/LLM-105 mixture	2.5+2.5	1.95	1.19	fair compatibility
AP	5.0	0.04		
50/50-AP/LLM-105 mixture	2.5+2.5	0.27	0.06	good compatibility
β-Pb salt	5.0	1.18		
50/50-β-Pb salt/LLM-105 mixture	2.5+2.5	1.28	-0.07	good compatibility
NC/NG	5.0	1.02		
50/50-(NC/NG)/LLM-105 mixture	2.5+2.5	1.30	0.11	good compatibility
DEP	5.0	0.05		
50/50-DEP/LLM-105 mixture	2.5+2.5	0.04	-0.18	good compatibility

3.5 The Quantum Chemical Investigation

The initial geometry of ANPZ was drawn with Chem3D software and input to Gaussian-03W [22] program and optimized using DFT-B3LYP/6-311++G** method. The optimized geometry of ANPZ was added with an oxygen atom in N4 to get the structure of LLM105. Then the obtained geometry was optimized at the DFT-B3LYP level by the Berny method [23, 24] with 6-311++G** basis set. The NMR chemical shift calculations were performed using Hartree-Fock NMR calculations with 6-311+G (2d, p) basis set at geometries optimized with the B3LYP method and the 6-311++G** basis set. The scaled IR frequencies were calibrated by multiplying 0.96. All calculations were carried out on a Lenovo P4 computer with Gaussian-03W program.

Optimized Geometries

The stable structures of ANPZ and LLM-105 were obtained as shown in Fig. 10. After stationary points were located, vibrational frequencies were calculated in order to ascertain that the structure obtained corresponds to a minimum on the potential energy surface. In the result of frequency calculation, there is no imaginary frequency. It shows that our optimized geometry is the stable one. Table 12 shows the calculated bond lengths and bond angles.

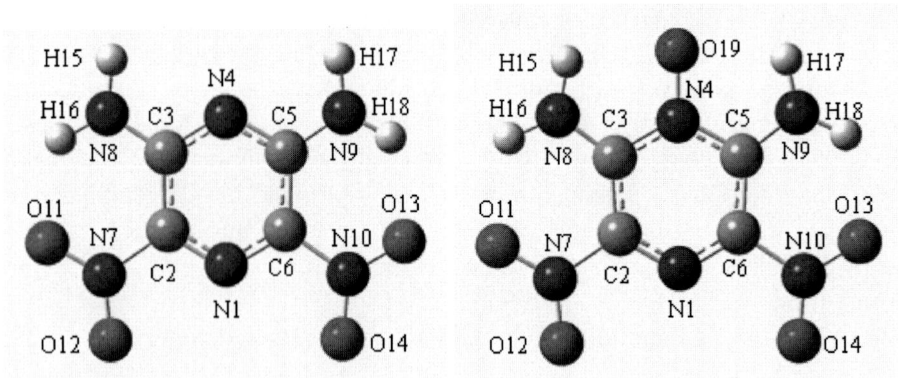

Fig. 10 The atomic number of ANPZ and LLM-105

Table 12 Some selected optimized geometric parameters for LLM-105

Bond lengths (Å)	ANPZ	LLM-105	Bond angle (°)	ANPZ	LLM-105
$R_{(1,2)}$	1.3051	1.3073	$A_{(2,1,6)}$	120.44	120.00
$R_{(1,6)}$	1.3050	1.3073	$A_{(3,4,5)}$	120.64	122.05
$R_{(2,3)}$	1.4401	1.4209	$A_{(2,3,4)}$	118.55	116.02
$R_{(3,4)}$	1.3366	1.3777	$A_{(4,5,6)}$	118.55	116.03
$R_{(4,5)}$	1.3365	1.3777	$A_{(1,2,3)}$	120.91	122.96
$R_{(5,6)}$	1.4401	1.4209	$A_{(1,6,5)}$	120.91	122.95
$R_{(2,7)}$	1.4720	1.4742	$A_{(4,3,8)}$	117.18	114.51
$R_{(3,8)}$	1.3390	1.3284	$A_{(2,3,8)}$	124.28	129.47
$R_{(5,9)}$	1.3390	1.3283	$A_{(4,5,9)}$	117.18	114.51
$R_{(6,10)}$	1.4721	1.4742	$A_{(6,5,9)}$	124.28	129.46
$R_{(7,11)}$	1.2418	1.2394	$A_{(1,2,7)}$	115.92	116.31
$R_{(7,12)}$	1.2122	1.2105	$A_{(3,2,7)}$	123.17	120.74
$R_{(10,13)}$	1.2418	1.2394	$A_{(1,6,10)}$	115.92	116.31
$R_{(10,14)}$	1.2122	1.2105	$A_{(5,6,10)}$	123.17	120.75
$R_{(8,15)}$	1.0078	1.0140	$A_{(2,7,11)}$	116.64	116.41
$R_{(8,16)}$	1.0101	1.0089	$A_{(2,7,12)}$	118.94	118.63
$R_{(9,17)}$	1.0078	1.0140	$A_{(6,10,13)}$	116.64	116.41
$R_{(9,18)}$	1.0101	1.0088	$A_{(6,10,14)}$	118.94	118.63
$R_{(4,19)}$		1.3008	$A_{(3,8,15)}$	118.07	115.95
Dihedral angle/°			$A_{(3,8,16)}$	119.39	119.11
$D_{(8,3,2,4)}$	180.00	180.00	$A_{(5,9,17)}$	118.07	115.94
$D_{(7,2,3,1)}$	180.00	180.00	$A_{(5,9,18)}$	119.39	119.11
$D_{(9,5,4,6)}$	180.00	180.00	$A_{(17,9,18)}$	122.54	124.95
$D_{(10,6,1,5)}$	180.00	180.00	$A_{(15,8,16)}$	122.54	124.94
$D_{(3,4,19,5)}$		179.98	$A_{(11,7,12)}$	124.42	124.96
			$A_{(13,10,14)}$	124.43	124.96

From Table 12 and Fig. 10, one can see that the whole atoms of ANPZ and LLM-105 were almost coplanar and their structures were symmetrical about a line formed by atoms of N(1) and N(4). When an oxygen atom of O(19) was add to the heteroatomic ring, the bond lengths and angles change a lot. The mean bond lengths of the six member ring increase by 9.7 pm, among these bonds the bond lengths of N(1)-C(2) and N(1)-C(6) increased by 2.2 pm and those of C(3)-N(4) and N(4)-C(5) increased largely by 41.1 pm while those of decreased by 19.2 pm. The bond lengths of C-N(-NO_2)

increased by 2.1 pm and those of C-N(-NH_2) decreased by 10.6 pm. The bond lengths of C-N(-NO_2) are longer than those of C-N(-NH_2) and the longest bond of LLM-105 is C-N(-NO_2), therefore the bond of C-N(-NO_2) will be broken first when the compound is heated to some certain temperature [25]. With the changes of the bond lengths of LLM-105 from ANPZ, the bond angles of C(3)-N(4)-C(5) increased by 1.41°which indicated an increased cyclical tensile force here. The bond angles of N(1)-C(2)-C(3) and N(1)-C(6)-C(5) increased by 2.05°while those of C(2)-C(3)-N(4) and N(4)-C(5)-C(6) decreased by 2.53°. From the above analysis we can concluded that all these changes from ANPZ to LLM-105 were caused by the O(19) to N(4) and the atom of O(19) has much influence on the properties of LLM-105.

IR Frequencies and NMR Chemical Shifts

Based on the simple harmonic oscillator analysis, 3*N*-6 (the number of atom is 19) frequencies and intensities of 51 frequencies were obtained, 34 of which more than 700 cm^{-1} were listed in Table 13. The scaled IR frequencies were calibrated by multiplying 0.96[26]. For the complexity of vibrational mode, it is difficult to assign all bands, so we have only analyzed some typical vibrational modes comparing with those of the experimental. The theoretical value at 796 cm^{-1} caused by C-N stretching vibration was in agreement with that of the experimental (815 cm^{-1}). It is found that the stretching vibrations of six-member ring appear at 1415, 1413 and 1340 cm^{-1} and those of the experimental were at 1490, 1454 and 1379 cm^{-1}. The asymmetric characteristic absorption peaks of C-NO_2 were found to be at 1552cm^{-1}, relevant to those of the experimental of 1565 cm^{-1}. The experimental stretching vibrations of the nitrogen in the ring with oxygen were at 890 cm^{-1} which were found to be 905 cm^{-1} in the theoretical calculation. The theoretical absorption peaks at 1284^{-1} was composed of the stretching vibration of N(4)-O(19) and the symmetric stretching vibration of –NO_2. The flexural oscillation of –NH_2 were found to be at 1603, 3378, 3381, 3523 and 3525 cm^{-1} which were consistent with the experimental values of 1647, 3226, 3228, 3403 and 3422 cm^{-1}.

Table 13 The scaled IR frequencies (cm^{-1}) for LLM-105

υ/ cm^{-1}	I / km·mol^{-1}	Exp./ cm^{-1}	υ/ cm^{-1}	I / km·mol^{-1}	Exp./ cm^{-1}
525	0.1		1232	490.8	
532	1.4		1234	110.5	
581			1284	206.0	
618	0.2		1333	234.9	
659	1.8		1341	34.2	1379
671			1413	4.5	1454
679	4.5		1415	34.5	1490
682	9.6		1458	197.2	
691	5.1		1472	14.5	
705			1538	90.5	
709	14.5		1552	268.7	1565
796	75.3	815	1584	724.8	
863	74.8		1604	319.7	1647
905		890	3378	160.0	3226
1052	52.9		3381	113.4	3282
1083	12.9		3523	200.3	3403
1164	3.7		3525	139.2	3422

Table 14 ^{13}C NMR chemical shifts (ppm)

NMR chemical shifts	Cal./ ppm	Exp. / ppm
$\delta_{(C2\text{-}N7)}$	127.4007	125.7136
$\delta_{(C3\text{-}N8)}$	161.1181	151.2849
$\delta_{(C5\text{-}N9)}$	161.1223	144.7482
$\delta_{(C6\text{-}N10)}$	127.3986	125.0115
$\delta_{(H15)}$	7.9925	8.774
$\delta_{(H16)}$	8.5094	9.049
$\delta_{(H17)}$	7.9938	8.774
$\delta_{(H18)}$	8.5078	9.049

Through the NMR calculation, we got the data of 4 chemical shifts of ^{13}C and 4 chemical shifts of ^{2}H which were listed in Table 14. Compared with the corresponding experimental values, the calculated results derivate slightly because the theoretical properties were calculated based on the gas phase.

Atomic Charges

The atomic NBO charges of ANPZ and LLM-105 were listed in Table 15

Table 15 Net atomic charges on atoms of LLM-105 at B3LYP/6-311++G level**

Atoms	Charges (e) ANPZ	Charges (e) LLM-105	Charges (e) ANPZ	Charges (e) LLM-105
N(1)	0.1574	0.1018	-0.3208	-0.3423
C(2)	-0.2586	-0.1107	0.2013	0.2097
C(3)	0.3631	0.3288	0.4150	0.3897
N(4)	-0.2979	-0.4442	-0.5458	-0.0124
C(5)	0.3631	0.3289	0.4150	0.3897
C(6)	-0.2586	-0.1108	0.2014	0.2097
N(7)	-0.3553	-0.3429	0.4700	0.4734
N(8)	-0.3432	-0.3294	-0.7259	-0.7244
N(9)	-0.3432	-0.3294	-0.7259	-0.7244
N(10)	-0.3553	-0.3429	0.4700	0.4734
O(11)	-0.0287	-0.0131	-0.4424	-0.4353
O(12)	0.0619	0.0646	-0.3222	-0.3139
O(13)	-0.0287	-0.0131	-0.4424	-0.4352
O(14)	0.0619	0.0646	-0.3222	-0.3139
H(15)	0.2937	0.3077	0.4027	0.4363
H(16)	0.3374	0.3487	0.4347	0.4408
H(17)	0.2937	0.3077	0.4027	0.43625
H(18)	0.3374	0.3487	0.4347	0.4408
O(19)		-0.1650		-0.5979

From Table 15, one can see that the charges of the nitrogen atoms of N(1), N(4), N(8) and N(9) are negative while those of N(7) and N(10) atoms in the nitro-group are positive for their connected atom of oxygen. Compared from ANPZ, the charges of LLM-105 on all the C atoms are positive and the charges on C connected with $-NH_2$ increase and the thickness of electron atmosphere on C (connected with $-NO_2$) decreased. The charges of O(19) are -0.5979 e and those on N(4) are -0.0124 e, greatly increased by 0.5334 e compared with the corresponding atom of ANPZ due to the strong electrophilic effect of the O(19) atom. From the total atomic calculation of $-NO_2$ and $-NH_2$, the total atomic charges of $-NO_2$ increased by 0.0188 e and those of $-NH_2$ increased by 0.0412 e. Therefore we can conclude that O(19) changes the whole electron cloud of ANPZ to form LLM-105.

Population Analysis

The population analysis of ANPZ and LLM-105 were listed in Table 16. From Table 16, we can see that the largest overlap contribution are C(2)-C(3) and C(5)-C(6) with the antibonding molecule orbital. Inspection to population analysis for LLM-105 of the six-member ring shows that among the ring atoms, N(1)–C(2), N(1)–C(6), C(2)–N(4), C(2)–C(5), N(1)-C(5), N(1)-C(6), C(3)-C(6) and N(4)-C(6) are bonding molecule orbitals and C(2)–C(3), C(3)–N(4), N(4)–C(5), C(5)–C(6), C(2)–C(6), N(1)-N(4) and C(3)-C(5) are antibonding molecule orbitals. The orbitals of C(3)–N(4) and N(4)–C(5) for LLM-105 changed to be antibonding compared with those of ANPZ.

Table 16 Population analysis of LLM-105

Bonds	Population (LLM-105)	Population (ANPZ)
N(1)–C(2)	0.4805	0.2137
N(1)–C(6)	0.4802	0.2141
C(2)–C(3)	-10.7988	-21.8748
C(3)–N(4)	0.3020	-0.7976
N(4)–C(5)	0.3020	-0.7976
C(5)–C(6)	-10.7983	-21.8782
C(2)–N(4)	0.0965	0.7747
C(2)–C(5)	4.7845	11.0645
C(2)–C(6)	-7.0691	-10.3139
N(1)-N(4)	0.0791	-0.0399
N(1)-C(5)	-0.1452	0.1442
N(1)-C(6)	0.4802	0.2141
C(3)-C(5)	-5.2205	-15.8825
C(3)-C(6)	4.7845	11.0644
N(4)-C(6)	0.0965	0.7749
C(2)–N(7)	-1.1999	-1.3803
C(3)–N(8)	-0.1575	0.7909
C(5)–N(9)	-0.1575	0.7906
C(6)–N(10)	-1.1996	-1.3806
N(7)–O(11)	-0.1151	-0.0949
N(7)–O(12)	0.2342	0.2505
N(10)–O(13)	-0.1151	-0.0949
N(10)–O(14)	0.2342	0.2505
N(4)–O(19)		0.2170
N(4)–N(8)	-0.2388	-0.3184
N(4)–N(9)	-0.2388	-0.3183
N(8)–O(19)		-0.0715
N(9)–O(19)		-0.0022

Note: - denotes antibonding molecule orbital.

Molecular Total Energies, Frontier Orbital Energy Levels and its Gap

Based on the composition of the molecule and the basis set we selected, there are 51 occupied molecule orbitals and 284 unoccupied molecule orbitals for ANPZ while those for LLM-105 are 55 and 303 respectively. Form Table 17 we can see that the total energy of LLM-105 is lower than that of ANPZ which indicates that LLM-105 is more stable than ANPZ. The frontier energy gap of ANPZ is lower than that of LLM-105, which indicates that it needs more energy for the electron transition from HOMO to LUMO. Therefore LLM-105 is more stable than ANPZ as indicated by the total energy analysis. The frontier orbitals of LLM-105 were shown at Fig. 11.

Table 17 Molecular total energies, frontier orbital energy levels and its gap (in Hartree)

Compounds	E_{total}	E_{HOMO}	E_{LUMO}	□E [a]
ANPZ	-784.2928944	-0.27620	-0.12897	0.14723
LLM-105	-859.4764263	-0.26690	-0.00919	0.25771

a—□E= E_{LUMO}-E_{HOMO}

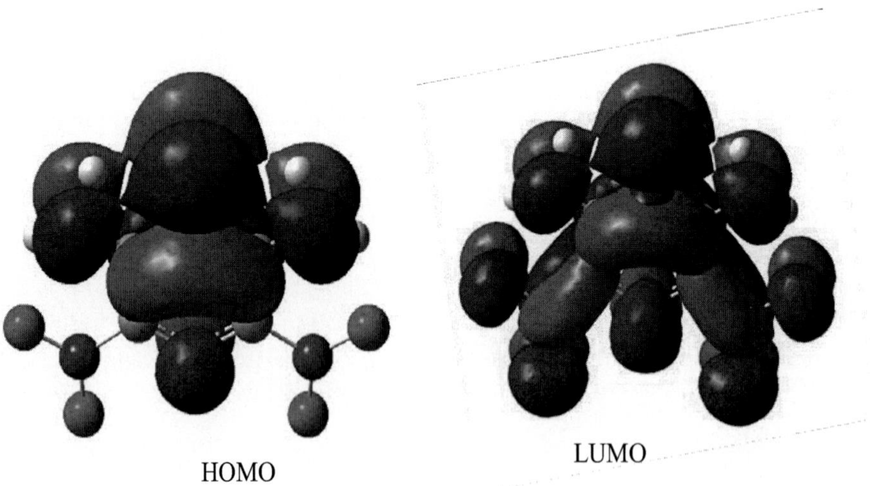

Fig. 11 HOMO and LUMO of LLM-105

Natural Bound Orbital Analysis

Table 18 summarizes some selected second-order perturbative estimates of "donor-accepter" (bond-antibond) interactions in the NBO basis for LLM-105. The stabilization energies are proportional to the NBO interacting intensities and the bigger the energy is, the stronger the interaction will be. From the NBO interactions from the table, one can see that the main NBO interacting are bonds as donor and antibonds as acceptor, especially the bonds within the six-member ring. The largest stable energy was 127.32 kJ mol^{-1} among C(2)–C(3) as donor and N(1)–C(6) as accepter, indicating a strong bond action with C(2), C(3), N(1) and C(6) atoms. The antibond N(4)–O(19) acts as accepters by σ electron with C(2)–C3 and C(5)–C(6) as donors with the equal stable energy of 14.31 kJ mol^{-1} which was in well accordance with the symmetric structure.

Table 18 Selected natural bond orbital interacting and the corresponding stable energy *E* for LLM-105[a]

Donor NBO (i)	Accepter NBO (j)	E / kJ mol^{-1}
BD(1)N1–C2	BD*(1)C2–C3	11.51
BD(1)N1–C2	BD*(1)C6–N10	13.56
BD(1)N1–C6	BD*(1)C2–N7	13.56
BD(1)N1–C6	BD*(1)C5–C6	11.51
BD(2)N1–C6	BD*(2)C2–C3	67.66
BD(2)N1–C6	BD*(2)N4–C5	91.34
BD(2)N1–C6	BD*(2)N10–O14	58.03
BD(1)C2–C3	BD*(1)N1–C2	12.59
BD(1)C2–C3	BD*(1)C3–N8	14.48
BD(1)C2–C3	BD*(1)N4–O19	14.31
BD(2)C2–C3	BD*(2)N1–C6	127.32
BD(2)C2–C3	BD*(2)N4–C5	36.61
BD(2)C2–C3	BD*(2)N7–O12	105.73
BD(1)C3–N8	BD*(1)C2–C3	11.25
BD(2)N4–C5	BD*(2)N1–C6	32.89
BD(2)N4–C5	BD*(2)C2–C3	90.96
BD(1)C5–C6	BD*(1)N1–C6	12.59
BD(1)C5–C6	BD*(1)N4–O19	14.31
BD(1)C5–C6	BD*(1)C5–N9	14.48
BD(1)C5–N9	BD*(1)C5–C6	11.25
BD(2)N7–O12	BD*(2)C2–C3	15.52
BD(2)N7–O12	BD*(2)N7–O12	29.25
BD(1)N8–H15	BD*(1)C2–C3	23.77
BD(1)N8–H16	BD*(1)C3–N4	19.04
BD(1)N9–H17	BD*(1)C5–C6	23.77
BD(1)N9–H18	BD*(1)N4–C5	19.04
BD(2)N10–O14	BD*(2)N1–C6	15.69
BD(2)N10–O14	BD*(2)N1–O14	29.25

[a] BD * means antibond, (1) denotes σ bond and (2) denotes π. Only the stable energies over 10 kJ mol^{-1} are listed.

ACKNOWLEDGMENTS

This work was supported by the National Natural Science Foundation of China (20573098). We are also grateful for Prof. Song Jirong (Northwest University, Xi'an) for the theoretical investigation with G03W.

REFERENCES

[1] J. Kerth, W. Kuglstatter. *32nd In Annual conference of ICT*, pp.166-1-166-11, Karlsrouke, 2001.

[2] T. D. Tran, P. F. Pagoria, D. M. Hoffman, J. L. Cutting, R. S. Lee, R. S. Simpson. *12th In Detonation Symposium,* pp.45-1-45-16, San Diego, 2002.

[3] GJB 772A-97-501.2, China, 1997.

[4] Yang Xuwu, Chen Sanping, Gao Shengli., Li Huanyong, Shi Qizhen. *Instrumentation Science and Technology.*,*30*(3) (2002), 311-321.

[5] Liu J. R., Yang X. W., Hou Y. D., GAO S. L., *Thermochim. Acta,* 329, (1999), 123.

[6] M.M.Popov. *Thermometry and Calorimetry.:* Moscow University Publishing House, Moscow, 1954, p.382.

[7] J.D. Cox. *J.Chem. Thermodynamic., 10*(1978) 903.

[8] H. E.Kissinger, *Anal. Chem. 29*(11)(1957) 1702 -1706.

[9] P. Budrugeac, *J. of Thermal Analysis and Calorimetry*, 68(2002) 131-139.

[10] T.Ozawa, Bull. *Chem. Soc. Jpn. 38*(1)(1965) 1881-1886.

[11] S. Vyazovkin, D. Dollimore, *J. Chem. Information Comp. Sci.*, 36(1996)42.

[12] Hu Rongzu,Shi Qizhen. *Thermal Analysis Kinetics,*Science Press, Beijing, 2001, p.127.

[13]] Bagchi, T. P., Sen, K. P.. *Thermochim, Acta*, 1981, 51(2-3): 175-189.

[14] B. N. Achar, G. W. Brindley, *J. H. Sharp. Proc. Int. Clay Conf.*, Jerusalen, 1966, 1:67, J. H. Sharp, S. A. Wendworth, *Anal. Chem.,* 1969, 41(14): 2060-2062.

[15] Yang Zhengquan, Hu Rongzu, Liang Yanjun, Li Xiangdong, *Acta Phys. Chim. Sin.*, Vol. 2, No.1 (1986)13.

[16] J. R. Mac Callum, J. Tanner, *Eur. Polymer J.,* 1968, 4: 333.

[17] V. Satava, J. Setak, *J. Them. Anal.,* 1975, 8(3): 477-489.

[18] A. W. Coats, J. P. Redfern. *Nature,* 1964, 201(4914): 68-69.

[19] P. M. Madhusudanan, K. Krishnan, K. N. Ninan, *Thermochim. Acta,* 1986, 97: 189-201.

[20] Hu Rongzu, Yang Zhengquan, Liang Yanjun. *Thermochim. Acta,* 123(1988) 135-151.

[21] Zhang Tonglai, Hu Rongzu, Xie Yi, Li Fuping. *Thermochim. Acta 244*(1994) 171 -176.

[22] M. J. Frisch, G. W. Trucks, Schlegel H B, et al. Gaussian 03, Revision A.1[M], Gaussian, Inc., Pittsburgh PA, 2003.

[23] Fletcher R, Powell M J D, *Comput. J.* 1963,6,163.

[24] Schlegel H B, *J. Comput. Chem.* 1982,3,214.

[25] Xiao H M, Structures and properties of energetic compounds[M] . *National Defence Industry Press*, 2004.

[26] Wong M W, *Chemical Physics Letters*, 1996, 256(4-5),391.

In: Thermochemistry and Advances in Chemistry Research ISBN: 978-1-60692-376-4
Ed: Anton G. Mirskiy

Chapter 6

THERMOCHEMICAL PROPERTIES FROM AB INITIO CALCULATIONS – G3MP2B3 AND CCSD(T)-CBS (W1U) – : FREE RADICALS OF IMPORTANCE: $^{\bullet}C_3H_3$ (PROPARGYL) AND $^{\bullet}C_4H_3$ AND THE REACTIONS *n*-$^{\bullet}C_4H_3 + C_2H_2 \rightarrow {}^{\bullet}C_6H_5$ AND $^{\bullet}C_6H_5 + 2\ C_2H_2 \rightarrow C_{10}H_8 + H^{\bullet}$

Rudolf Janoschek** and Walter M.F. Fabian
Institut für Chemie, Karl-Franzens-Universität, Heinrichstraße 28, A-8010 Graz, Austria

ABSTRACT

We applied computational electronic energy methods, G3MP2B3 and CCSD(T)-CBS (W1U), for which we have the MAD (Mean Absolute Deviation) between calculated and experimental enthalpies of formation, $\Delta_f H^o(298.15)$, for sets of well known radicals. For G3MP2B3, the MAD is 3.9 kJ mol^{-1}, and for CCSD(T)-CBS (W1U), it is 2.3 kJ mol^{-1}. The former method can be applied to systems with up to 20 C-atoms or more; but for the latter, it is limited to systems up to 4 – 5 C-atoms. The systems $^{\bullet}C_2H$, C_2H_2, $^{\bullet}C_3H_3$, C_3H_4, *n*-$^{\bullet}C_4H_3$, *i*-$^{\bullet}C_4H_3$, and C_4H_4 were calculated for both computational methods. The systems $^{\bullet}C_6H_5$ (phenyl), C_6H_6, $^{\bullet}C_8H_7$ (styryl), $^{\bullet}C_{10}H_9$ (4-phenyl-1,3-butadien-1-yl and 2-(1,3-butadien-1-yl)phenyl) transition structures and intermediates, and $C_{10}H_8$ (naphthalene) were calculated only by G3MP2B3. The properties are energy minimum structures as well as transition structures for the title reactions, harmonic vibrational wave numbers ν, moments of inertia I_A, I_B, and I_C, heat capacities $C^o_p(T)$, standard entropies $S^o(T)$, thermal energy contents $H^o(T) - H^o(0)$, standard enthalpies of formation $\Delta_f H^o(T)$, and atomic spin densities ρ. The temperature T is extended from 0 to 3000 K. Parent molecules of radicals were calculated to obtain bond dissociation energies $D^o(298.15)$.

---

* Email: rudolf.janoschek@uni-graz.at

INTRODUCTION

Small hydrocarbon radicals, e.g. $^{\bullet}C_3H_3$ and $^{\bullet}C_4H_3$, are of paramount importance in combustion science. They have been proposed to be involved in the formation of benzene in combustion flames, commonly considered as the first and rate-limiting step in the growth of polycyclic aromatic hydrocarbons (PAHs). Besides the health risk posed by many of these PAHs, they are widely considered as precursors to soot. Formation of soot is a persistent problem in fossil fuel combustion [1–5]. A number of pathways producing benzene in flames exist with the dominant one depending on fuel composition [2]. Of particular importance are the odd-carbon, e.g. the radical–radical combination of propargyl $^{\bullet}C_3H_3$ (eq. 1) [6], and even-carbon paths, e.g. the reaction of *n*-$^{\bullet}C_4H_3$ or *n*-$^{\bullet}C_4H_5$ with acetylene (eq.2,3) [7].

$$^{\bullet}C_3H_3 + {}^{\bullet}C_3H_3 \rightarrow C_6H_6 \ldots\ldots \tag{1}$$

$$n\text{-}^{\bullet}C_4H_3 + C_2H_2 \rightarrow {}^{\bullet}C_6H_5 \tag{2}$$

$$n\text{-}^{\bullet}C_4H_5 + C_2H_2 \rightarrow C_6H_6 + H \tag{3}$$

Most radical–radical reactions are insignificant in flames because their reactants do not exist in sufficiently high concentrations. However, propargyl $^{\bullet}C_3H_3$ as well as *i*-$^{\bullet}C_4H_3$ or *i*-$^{\bullet}C_4H_5$ as opposed to *n*-$^{\bullet}C_4H_3$ or *n*-$^{\bullet}C_4H_5$, belong to the class of so-called resonantly-stabilized radicals. Because of their increased stability, they can accumulate to relatively high concentrations [2]. Furthermore, besides reacting with acetylene, *n*-$^{\bullet}C_4H_3$ or *n*-$^{\bullet}C_4H_5$ easily can be converted to the respective *i*-isomers by H-atom assisted isomerization [6]:

$$n\text{-}^{\bullet}C_4H_5 + H \rightarrow i\text{-}^{\bullet}C_4H_5 + H \tag{4}$$

Only in low-temperature zones of the flame insufficient quantities of hydrogen atoms may be present to catalyze this conversion [8], thus, resulting in relatively high concentrations of the less stable *n*-isomers of $^{\bullet}C_4H_3$ or $^{\bullet}C_4H_5$. The presumed lower stability of *n*-isomers as well as the ease of the isomerization to the resonantly-stabilized *i*-isomers has been used to argue against the even-carbon pathway in the formation of aromatic hydrocarbons [6]. Exact knowledge of the thermochemical properties of the involved species, especially isomeric energy differences, is therefore essential for kinetic modeling of soot formation in flames. In view of this importance a number of computational studies concerning small hydrocarbon radicals have been published [8–17]. Strikingly, considerable discrepancies for these quantities exist. For instance, calculated energy differences between *n*-$^{\bullet}C_4H_3$ and *i*-$^{\bullet}C_4H_3$ are in the range 28.5 kJ mol^{-1} (diffusion Monte Carlo [17]) to 78.2 kJ mol^{-1} (BAC-MP4 [6]). An overview of relevant earlier calculations is provided in ref. [14]. More recent calculations by CBS-QCISD(T) [10] or the focal point extrapolation [14] appear to converge at $\Delta\Delta H(n\text{-}^{\bullet}C_4H_3 - i\text{-}^{\bullet}C_4H_3)$ (0 K) $\approx$ 50 kJ mol^{-1}. The same computational procedures yielded standard enthalpies of formation for *i*-$^{\bullet}C_4H_3$, $\Delta_f H^o(0\text{ K}) = 498.3$ [10] and 497.7 kJ mol^{-1} [14]. Recently, photoionization efficiency spectra (PIE) have revealed the presence of the resonantly-stabilized *i*-$^{\bullet}C_4H_3$ isomer in flames. The assignment was based on a comparison of the experimental PIE spectra with simulated curves resulting from calculated adiabatic ionization energies combined with Franck-Condon factor analysis [10]. Because of the lower sensitivity of the *n*-isomers only upper bounds for the n-$^{\bullet}C_4H_3$/ i-$^{\bullet}C_4H_3$ ratio could be estimated. It was found to be substantially higher than expected at thermal equilibrium and might further increase if the photoionization cross-section were lower [10]. Thus, although the amount of n-$^{\bullet}C_4H_3$ present in flames could not be established, even-carbon species might not completely be ruled out as contributors to the formation of aromatic hydrocarbons. The dynamics and energetics of i-$^{\bullet}C_4H_3$ were recently obtained by a crossed molecular beam study of the reaction between ethylene and C_2 molecules. From these experiments the standard enthalpy of formation at 0 K for the resonantly-stabilized i-$^{\bullet}C_4H_3$ radical was derived as $\Delta_f H^{o}(0) = 504 \pm 10$ kJ mol^{-1} [18]. Apparently, so far no thermochemical data for temperatures relevant in combustion have been determined for these important species.

Previously we have obtained thermochemical properties, including MADs (mean absolute deviation) between experimental and calculated standard enthalpies of formation, for a set of free radicals by the G3MP2B3 [19,20] procedure [21,22]. Herein we apply this method as well as the CCSD(T)-CBS (W1U) scheme [23] to a set of free radicals of importance in combustion, e.g., $^{\bullet}C_3H_3$ (propargyl), n-$^{\bullet}C_4H_3$, i-$^{\bullet}C_4H_3$, and the respective parent hydrocarbons. In addition, G3MP2B3 results for the reactions n-$^{\bullet}C_4H_3 + C_2H_2 \rightarrow {}^{\bullet}C_6H_5$ and $^{\bullet}C_6H_5 + 2\ C_2H_2 \rightarrow {}^{\bullet}C_{10}H_8 + H^{\bullet}$ will be presented. In both reactions, the phenyl radical $^{\bullet}C_6H_5$ is of central importance. Besides computational approaches, the role of the phenyl radical $^{\bullet}C_6H_5$ in the growth of PAHs has also been studied experimentally by vacuum ultraviolet (VUV) single photon ionization (SPI) time-of-flight mass spectrometry (TOFMS) [24].

A similar G3-type approach as described here, G3(MP2,CC)//B3LYP [25], combined with statistical theory has recently been applied to a computational study of the formation of indene, naphthalene, and other PAHs in combustion flames [26–28]. Density functional calculations, mostly B3LYP, on the same topic [29,30] as well as thermochemical properties of PAHs from G3MP2B3 computation also have been published [31].

METHOD OF CALCULATION

The method, referred to as G3MP2B3, is described in the following [19,20]. First, the geometries are calculated at the B3LYP/6-31G(d) level, unrestricted Hartree-Fock is default, as well as the complete set of vibrational wave numbers. The zero-point energies (ZPE) are scaled by 0.96. Second, a single-point quadratic configuration interaction calculation is performed in the frozen core (FC) approximation for the correlation calculation, QCISD(T,FC)/6-31G(d). Third, second-order Mœller-Plesset perturbation theory, MP2(FC), is performed with the basis sets 6-31G(d) and 6-311+G(2df,2p) on Li – Ne, and 6-311+G(3d2f,2p) on Na – Ar, summarized as G3MP2large. Fourth, a spin-orbit (SO) correction is included for atomic species. Fifth, the higher level correction (HLC) takes into account remaining deficiencies in the energy calculations. The HLC is simply a linear function of the number of valence electrons with α and β spin, $-An_\beta - B(n_\alpha - n_\beta)$, with $n_\alpha \geq n_\beta$. The values of A and B are 10.041 and 4.995 mE_h, respectively, for molecules, and 10.188 and 2.323 mE_h, respectively, for atoms; 1 mE_h = 2.625500 kJ mol^{-1}. In summary, the G3MP2B3 total energy is

$$E_0[\text{G3MP2B3}] = E[\text{QCISD(T)/6-31G(d)}] + E[\text{MP2/G3MP2large}] - E[\text{MP2/6-31G(d)}] + \Delta E(\text{SO}) + \Delta E(\text{HLC}) + \Delta E(\text{ZPE}) \quad (5)$$

The spin density is a by-product of the geometry calculation. This property is the difference of the densities of electrons with α- and β-spin at the nuclei.

The above-mentioned B3LYP/6-31G(d) calculations for the energy minimum structures and vibrational wave numbers are the basis for the thermochemical corrections to the electronic total energy E_0. These corrections can be calculated at specified states such as temperature T, pressure p, global scaling factor of vibrational wave numbers, and isotopes. The default values for thermochemistry analysis are $T = 298.15$ K, $p = 1$ atm, scaling factor of the set of vibrational wave numbers is 1.0, and the most abundant isotopes (H = 1, C = 12, etc.). Standard rigid-rotor harmonic oscillator (RR-HO) partition function expressions are used. The influence of hindered rotation on thermodynamic functions can be studied if necessary. In addition to the electronic total energy E_0, heat capacities $C^{\circ}_{p}(T)$, entropies $S^{\circ}(T)$, and thermal energy contents $H^{\circ}(T) - H^{\circ}(0)$ are calculated. Standard enthalpies of formation $\Delta_f H^{\circ}(T)$ have been calculated via enthalpies of atomization. The mean absolute deviation (MAD) between G3MP2B3 calculated and reliable experimental $\Delta_f H^{\circ}(298.15)$ values by the previous set of 32 radicals is 3.91 kJ mol^{-1} [21].

For systems smaller than benzene, more expensive calculations can be performed using complete basis sets (CBS) and computations based on CCSD(T) (coupled cluster). One of these methods is known by the acronym W1 [23] and is modified for radicals where UCCSD (Unrestricted open shell, U is default) corresponds to W1U and is implemented in the programs of GAUSSIAN 03 [32]. The protocol for the various steps of W1U is described in the following. First, the geometry as well as the complete set of harmonic vibrational wave numbers are calculated at the B3LYP/cc-pVTZ level if only first first-row atoms are present. The zero-point energy ZPE is scaled for the total energy as well as for thermochemical properties by 0.985. Second, the following single-point calculations are carried out: CCSD(T)/cc-pAVDZ, CCSD(T)/cc-pAVTZ, and CCSD/cc-pAVQZ. The SCF component of total energy is extrapolated using $A + B/C^{l}$ from the above SCF components ($l = 2$, 3, and 4, respectively). The CCSD valence correlation component is obtained from applying $A + B/l^{\beta}$ to CCSD/cc-pAVTZ and CCSD/cc-pAVQZ valence correlation energies ($l = 3$ and 4, respectively; $\beta = 3.22$). The (T) valence correlation component is obtained from applying $A + B/l^{\beta}$ to CCSD(T)/cc-pAVDZ and CCSD(T)/cc-pAVTZ values for the (T) contribution. Third, core correlation contributions are obtained from CCSD(T)/MTsmall. Fourth, scalar relativistic corrections are calculated with the previous step. Fifth, spin-orbit (SO) coupling effects are considered manually from experimental values. SO splittings (cm^{-1}): C (16.4); energy corrections (cm^{-1}): C (–29.6); 1 kJ mol^{-1} = 83.5935 cm^{-1}. In summary, the W1U total energy E_0 may be expressed as the sum of the following terms:

$$E_0[\text{CCSD(T)-CBS(W1U)}] = E[\text{SCF-CBS}] + \Delta E_{\text{val.-correl.}}[\text{CCSD-CBS}] + \Delta E_{\text{val.-correl.}}[\text{CCSD(T)-CBS}] + \Delta E_{\text{core-correl.}}[\text{CCSD(T)/MTsmall}] + \Delta E_{\text{relativ.}}[\text{CCSD(T)/MTsmall}] + \Delta E(\text{ZPE}) + \Delta E_{\text{exp.}}[\text{SO}] \quad (6)$$

MT corresponds to “Martin-Taylor small basis set”. W1U resulted in a MAD of 2.3 kJ mol^{-1} for the enthalpies of formation [22]. All computations have been performed using the GAUSSIAN 03 computer program [32].

RESULTS AND DISCUSSION

The Radicals $^{\bullet}C_3H_3$, $^{\bullet}C_4H_3$, $^{\bullet}C_6H_5$, $^{\bullet}C_8H_7$ and the Reactions $^{\bullet}C_3H_3 + {}^{\bullet}C_3H_3 \rightarrow C_6H_6$

A number of computational studies at various levels of sophistication have been published on several putative intermediates of soot formation [6–17, 26–31]. Specifically, rather high-level computations, including summaries of previous computational and/or experimental studies, have been published on the thermochemistry of disputed species, e.g. resonantly stabilized *i*-$^{\bullet}C_4H_3$ and *i*-$^{\bullet}C_4H_5$ vs. non-stabilized *n*-$^{\bullet}C_4H_3$ and *n*-$^{\bullet}C_4H_5$ [10,14]. Similarly, standard enthalpies of formation at 0 K, $\Delta_fH^o(0)$ were computed for the propargyl radical and its precursors propyne and allene [15]. To facilitate comparison, we first present thermochemical properties of relevant neutral and radical species calculated at the same level of theory (G3MP2B3 and, partly, CCSD(T)-CBS-W1U). These results, namely heat capacity $C^o_p(298.15)$, entropy S^o (J mol^{-1} K^{-1}), and standard enthalpy of formation $\Delta_fH^o(298.15)$ (kJ mol^{-1}), are summarized in Table 1. Available experimental data [33–45] are also given there.

With the exception of butatriene calculated and experimental heat capacities agree to within a few kJ mol^{-1}. Entropies are also remarkably well reproduced by the calculations, Table 1. Consequently, the following discussion will be restricted to standard enthalpies of formation as the most important (and critical) thermodynamic quantity.

Although generally the agreement between the results for $\Delta_fH^o(298.15)$ obtained with the two computational procedures as well as with the corresponding experimental data is very good, there are a few discrepancies which need further consideration. For instance, $\Delta_fH^o(298.15)$ of propyne obtained by both computational procedures essentially coincide and are close to the experimental value, Table 1. A larger deviation is found for the isomeric allene molecule, especially with G3MP2B3. Lacking experimental data for the higher cumulenes, no definite conclusion whether there are special problems associated with the G3MP2B3 procedure in the correct description of systems with cumulated double bonds or poly-ynes, as found for the B3LYP density functional [14], can be drawn. Apparently, there is only one *estimate* for the enthalpy of formation of butatriene $H_2C{=}C{=}C{=}CH_2$ available, $\Delta_fH^o \approx 298.3$ kJ mol^{-1} [46]. Compared with the calculations, this value is too low by 15 (G3MP2B3) and 22 kJ mol^{-1} (W1U). Clearly, more reliable experimental results are highly desirable. In contrast to propyne and allene, W1U results in a significantly higher $\Delta_fH^o(298.15)$ for the propargyl radical, $\Delta_fH^o(298.15) = 351.6$ kJ mol^{-1} as compared with 346.4 (G3MP2B3) and ≈ 340 kJ mol^{-1} from experiment. Standard enthalpies of formation at 0 K resulting from the focal point extrapolation scheme of Allen and co-workers [47–49] are even higher, $\Delta_fH^o(0) = 198.4$ (allene), 193.8 (propyne), and 354.6 kJ mol^{-1} (propargyl) [15]. The calculated thermal corrections to Δ_fH^o are small for propargyl, $\Delta_fH^o(0) = 348.7$ (G3MP2B3) and 353.9 kJ mol^{-1} (W1U) but significant for allene, $\Delta_fH^o(0) = 191.4$ (G3MP2B3) and 194.4 kJ mol^{-1} (W1U) as well as propyne, $\Delta_fH^o(0) = 189.3$ (G3MP2B3) and

189.7 kJ mol^{-1} (W1U). Applying similar thermal correction of –8 (allene) and –7 kJ mol^{-1}(propyne) to $\Delta_f H^o(0)$ obtained by the focal point extrapolation scheme results in calculated $\Delta_f H^o(298.15)$ values of 190.4 (allene) and 186.8 kJ mol^{-1} (propyne). In view of the good agreement with the respective experimental data, the ≈10 kJ mol^{-1} overestimation of $\Delta_f H^o(298.15)$ (propargyl) found by CCSD(T)-CBS-W1U and the focal point extrapolation calls for further computational and experimental work. Accurate thermochemical data for this radical are highly needed because of the its central role in the chemistry of PAH formation.

Table 1 Thermochemical properties heat capacity $C^o_p(298.15)$ (J mol^{-1} K^{-1}), entropy $S^o(298.15)$ (J mol^{-1} K^{-1}), and standard enthalpy of formation $\Delta_f H^o(298.15)$ (kJ mol^{-1}) for calculations and experimental systems.

System	Method	C^o_p	S^o	$\Delta_f H^o$
$H^\cdot$	G3MP2B3	20.8	114.6	218.0
	CCSD(T)-CBS-W1U	20.8	114.6	218.0
	Experiment	20.8 [33][a]	114.7 [33][a]	218.0 [33][a]
$^\cdot C{\equiv}CH$	G3MP2B3	40.0	217.6	562.1
	CCSD(T)-CBS-W1U	43.5	216.9	569.7
	Experiment	37.1 [34]	209.7 [35]	567.4 ± 2.1 [36][b]
				566.1 ± 2.9 [33][c]
$HC{\equiv}CH$	G3MP2B3	44.8	201.7	225.0
	CCSD(T)-CBS-W1U	43.2	200.1	226.9
	Experiment	44.0 [33][d]	200.9 [33][d]	227.4 [33][d]
$CH_3{-}C{\equiv}CH$	G3MP2B3	62.0	248.9	182.4
	CCSD(T)-CBS-W1U	60.3	247.4	182.6
	Experiment	60.7 [40]		185.4 ± 0.88 [39]
$H_2C{=}C{=}CH_2$	G3MP2B3	58.9	243.0	183.8
	CCSD(T)-CBS-W1U	58.3	242.8	186.8
	Experiment	59.0 [40]		190.5 [36][e]
$H_2C^\cdot{-}C{\equiv}CH$	G3MP2B3	64.4	256.5	346.4
	CCSD(T)-CBS-W1U	62.1	253.8	351.3
	Experiment			339.0 ± 4.0 [38]
				340.6 ± 8.4 [33][c]
$HC^\cdot{=}CH{-}CH$	G3MP2B3	64.3	270.1	672.6

C_6H_6 (Triplet)	G3MP2B3	106.8	314.9	445.0
C_6H_6, benzene	G3MP2B3	83.8	270.1	76.9
	Experiment	82.4 [33][g]	269.2 [33][g]	82.9 [33][g]
$H_2C{=}CH{-}C{\equiv}CH$	G3MP2B3	73.7	278.8	283.3
	Experiment	73.2 [41,45]		295 [42,44]
$HC^{\bullet}{=}CH{-}C{\equiv}CH$ (**1**)	G3MP2B3	73.1	284.3	539.4
	CCSD(T)-CBS-W1U	71.4	282.8	547.6
$H_2C{=}C{=}C{=}CH_2$	G3MP2B3	73.5	266.8	313.8
	CCSD(T)-CBS-W1U	71.0	276.9	320.2
	Experiment	81.3 [43,45]		
$HC^{\bullet}{=}C{=}C{=}CH_2$	G3MP2B3	77.8	294.2	501.9
	CCSD(T)-CBS-W1U			502.6
$(HC{\equiv}C{-}CH_2)_2$	G3MP2B3	112.7	338.7	414.5
$^{\bullet}C_6H_5$ (**2**)	G3MP2B3	107.2	337.0	610.5
$^{\bullet}C_6H_5$, phenyl	G3MP2B3	82.1	289.9	334.8 (341.1 [31])
	Experiment		290.3 [37]	330.1 ± 3.3 [36][e]
				338 ± 3 [33][f]
				339.7 ± 2 [31]
$^{\bullet}C_8H_7$ (**3**)	G3MP2B3	121.9	347.8	402.5
	Experiment			393.5 ± 7 [31]
				387.0 [36][h]
$^{\bullet}C_{10}H_9$ (**4**)	G3MP2B3	155.9	407.1	496.3
$^{\bullet}C_{10}H_9$ (**5**)	G3MP2B3	144.0	366.9	244.6
$C_{10}H_8$, naphthalene	G3MP2B3	134.8	335.3	134.9
	Experiment	131.9 [33][g]	333.2 [33][g]	150.6 [33][g]

[a]p. **5**-76; [b]p. **5**-23; [c]p. **9**-70; [d]p. **5**-65; [e]p. **9**-69; [f]p. **9**-70; [g]p. **5**-67; [h]p. **9**-70.

The recombination of propargyl is of fundamental importance for the formation of benzene [2,6], equ. (1). An extensive computational investigation of the C_6H_6 potential energy surface with special emphasis on the formation of benzene from propargyl, has been published [9]. An important step in this radical-radical reaction is the formation of $HC{\equiv}C{-}CH_2{-}CH_2{-}C{\equiv}CH$ (hexa-1,5-diyne). Some possible pathways in the thermal rearrangement of hexa-1,5-diyne, based on computational [9] and experimental [50] investigations, are displayed in Scheme 1. Analysis of the reaction products from pyrolysis of hexa-1,5-diyne in a high-pressure single pulse shock tube by gas chromatography and matrix isolation IR spectroscopy led to the identification of six linear and two cyclic C_6H_6 species [50].

Relative energies as well as heats of formation and heats of hydrogenation for hexa-1,5-diyne have also been computed in studies related to benzene isomers [51] and conjugation effects in polyynes as compared with polyenes [52]. Using CCSD(T)/6-31G(d) calculations with MP2 corrections for basis set deficiencies and B3LYP/6-31G(d) thermal corrections for enthalpies, hexa-1,5-diyne was calculated to be 324.3 kJ mol^{-1} above benzene [51]. Several conformations for hexa-1,5-diyne are possible; the C_{2v} *cis*-rotamer which directly can lead to benzene actually is a transition state (τ(C2–C3–C4–C5) = 0°, ν = 162i cm^{-1}, $\Delta_f H^o(298.15)$ = 433.1 kJ mol^{-1}). The C_2 *gauche* form with τ(C2–C3–C4–C5) = 66°, however, is a true minimum, $\Delta_f H^o(298.15)$ = 414.5 kJ mol^{-1}. The experimental standard enthalpy of formation of hexa-1,5-diyne, most probably in the *trans* conformation, in the *crystal* is $\Delta_f H^o(298.15)$ = 384.2 kJ mol^{-1} [36]. Also, most probably for the *trans* conformation of hexa-1,5-diyne, an experimental value of $\Delta_f H^o(298.15)$ = 416.1 kJ mol^{-1} has been cited [52]. Concerted double 1,2 H–shift and cyclization of the *cis*-or *gauche*-rotamer of hexa-1,5-diyne to benzene proceeds *via* a second order saddle point ($\Delta_f H^o(298.15)$ = 770.8 kJ mol^{-1}, ν = 1490i and 1145i cm^{-1}, τ(C2–C3–C4–C5) = –18°), Figure 1. The barrier to this cyclization is just 78 kJ mol^{-1} with respect to two propargyl molecules, see Figure 1.

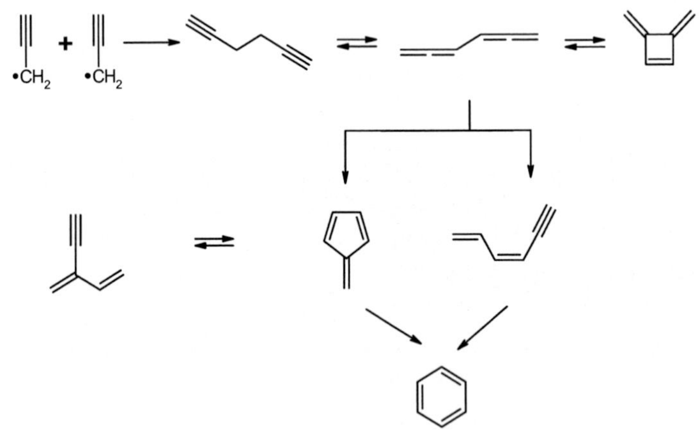

Scheme 1. Selected isomerization reactions of hexa-1,5-diyne [9,50].

Besides propargyl (prop-2-ynyl) other C_3H_3 isomers exist, for instance prop-1-ynyl, cycloprop-1-enyl, cycloprop-2-enyl, prop-1-en-1-yl-3-ylidene and prop-2-en-1-yl-3-ylidene [15]. Prop-1-en-1-yl-3-ylidene $HC^{\bullet}$=CH–CH is of some interest since, conceivably, it can combine to benzene in its triplet excited state. However, both prop-1-en-1-yl-3-ylidene and prop-2-en-1-yl-3-ylidene have been found to be kinetically unstable with respect to isomerization to the other isomers [53,54]. Furthermore, not only its energy is quite high, $\Delta_f H^o(298.15)$ = 672.6 kJ mol^{-1}, Table 1, but there is also a substantial barrier for its direct formation, $\Delta_f H^o(298.15)$ (TS) = 715.6 kJ mol^{-1}, via 1,2-H shift of propargyl. Triplet benzene is calculated to be 368 kJ mol^{-1} above its ground state. Experimentally, for gaseous benzene the $S_0 - T_1$ energy difference is 352.9 kJ mol^{-1} [55], yielding $\Delta_f H^o(298.15) \approx 436$ kJ mol^{-1} for triplet benzene – in good agreement with the calculated value of $\Delta_f H^o(298.15)$ = 445 kJ mol^{-1}, Table 1.

For vinyl acetylene (but-1-en-3-yne), $\Delta_f H^o(0)$ = 295.1 kJ mol^{-1} by the focal point extrapolation [14] and 289.9 kJ mol^{-1} by G3MP2B3. Thus, this latter method yields a thermal correction of –7 kJ mol^{-1}. Therefore, compared with experiment, calculated enthalpies of formation $\Delta_f H^o(298.15)$ of vinyl acetylene are too low by 7 (focal point extrapolation) and 12 kJ mol^{-1} (G3MP2B3). A surprisingly large discrepancy results for the standard enthalpy of formation of naphthalene, ca. 15 kJ mol^{-1} compared with a difference of 6 kJ mol^{-1} found for benzene, Table 1. Such a degradation of G3-type enthalpies of formation with molecular size has been noted previously and the experimental atomic value of the carbon atom has been proposed as a possible reason for this behavior [56].

In fact, from this study [56] as well as the Active Thermochemical Table (ATcT) approach [57] it was concluded that the *experimental* value for $\Delta_f H^o$(0, gas) of the carbon atom needs redetermination. The experimental standard enthalpy of formation of the phenyl radical shows a rather large spread, 330 – 340 kJ mol^{-1} [31,33,36]. The corresponding G3MP2B3 calculated value, $\Delta_f H^o(298.15)$ = 334.8 kJ mol^{-1}, Table 1, is exactly midway between these experimental data. A relatively large difference between calculations, $\Delta_f H^o(298.15)$ = 402.5 kJ mol^{-1} and experiments, $\Delta_f H^o(298.15)$ = 387.0 kJ mol^{-1} [36] and $\Delta_f H^o(298.15)$ = 393.5 kJ mol^{-1} [31], is also found for the styryl radical. However, the >5 kJ mol^{-1} difference between these two experimental values indicates that an experimental reinvestigation should be done.

Finally, it should be noted that an experimental estimate of the standard enthalpy of formation at 0 K for the *i*-$^{\bullet}C_4H_3$ radical, $\Delta_f H^o(0) = 504 \pm 10$ kJ mol^{-1} (crossed molecular beam experiments [18]) is in good agreement with calculated values, $\Delta_f H^o(0)$ = 498 kJ mol^{-1} [10,14]. The present G3MP2B3 and CCSD(T)-CBS-W1U calculations yield $\Delta_f H^o(0)$ = 502.1 and $\Delta_f H^o(0)$ = 502.8 kJ mol^{-1}, respectively. A combination of IR spectroscopic measurements with isodesmic reactions resulted in a slightly larger enthalpy of formation for the *i*-$^{\bullet}C_4H_3$ radical, $\Delta_f H^o$ = 512 kJ mol^{-1} [58]. The isomeric *n*-$^{\bullet}C_4H_3$ radical is calculated to be less stable than the *i*-$^{\bullet}C_4H_3$ radical by 37.5 (G3MP2B3) and 45.0 kJ mol^{-1} (CCSD(T)-CBS-W1U) at T = 298.15 K; the corresponding values at T = 0 K are 39.5 and 47.3 kJ mol^{-1}. Previous calculations at T = 0 K have resulted in a difference of $\Delta_f H^o(0)$ of 49.7 [14] and 50.2 kJ mol^{-1} [10].

In conclusion, the numerous comparisons between calculated and experimental data indicate the reliability of the computational approach. For larger systems where experimental thermochemical data are not available, calculations provide results of sufficient accuracy.

The Reactions *n*-$^{\bullet}C_4H_3 + C_2H_2 \rightarrow {}^{\bullet}C_6H_5$ and $^{\bullet}C_6H_5 + 2\ C_2H_2 \rightarrow C_{10}H_8 + H^{\bullet}$

The generally accepted opinion is that soot particles form *via* PAHs [1]. Therefore, in this section calculated results, G3MP2B3 relative enthalpies $\Delta H(T)$ of minima and transition structures for a prototype reaction sequence, equ. (7), leading to benzene and naphthalene as the simplest members of PAHs, will be described.

$$4\ C_2H_2 + {}^{\bullet}C_2H \rightarrow C_{10}H_8 + H^{\bullet} \qquad (7)$$

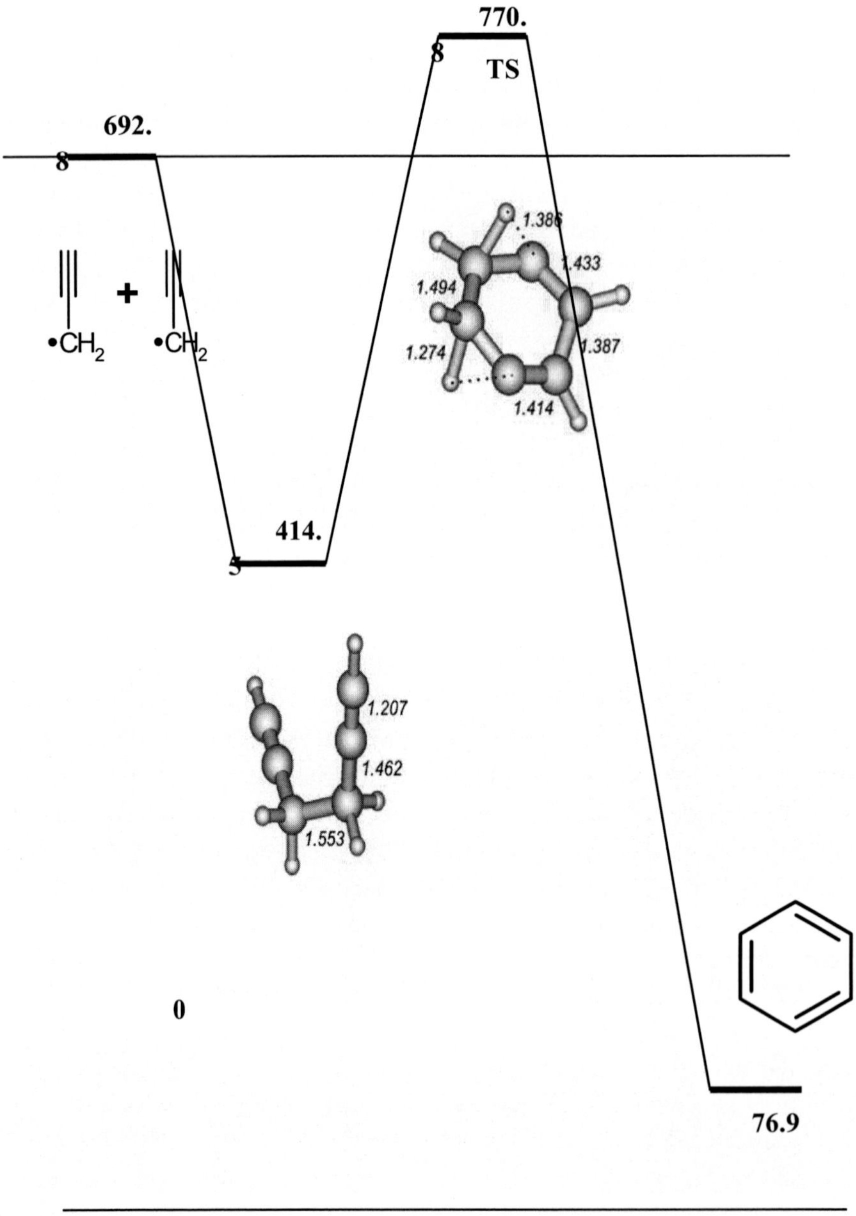

Figure 1. Schematic energy diagram for the recombination reaction of two propargyl radicals to benzene.

The relevant data for 11 stationary points along this reaction path at ten different temperatures ($T = 0 - 3000$ K) are summarized in Table 2.

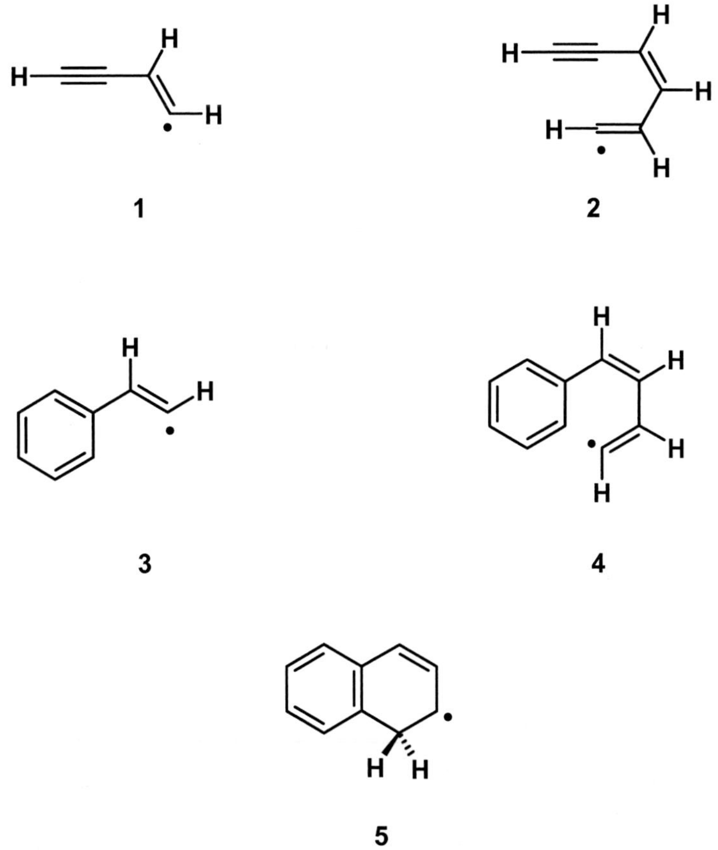

Scheme 2 Structures of radicals **1** – **5**.

The first steps in this reaction path are the formation of n-$^{\bullet}C_4H_3$ (1 in Scheme 2) by a radical-molecule combination reaction of acetylene with ethynyl followed by addition of n-$^{\bullet}C_4H_3$ to acetylene to give buta-1,3-dien-5-yn-1-yl 2 and cyclization of 2 to phenyl $^{\bullet}C_6H_5$. Only this last reaction involves crossing of a transition structure, stationary point #4 in Table 2, although the barrier height is quite small, 13.2 kJ mol^{-1} at $T = 298.15$ K, and even vanishes at higher temperatures, Table 2. Stepwise addition of two acetylenes to phenyl leads, *via* styryl 3, to 4-phenylbuta-1,3-dien-1-yl 4, $^{\bullet}C_{10}H_9$ which cyclizes *via* TS(4 $\rightarrow$ 5), stationary point #8 in Table 2, to 1,2-dihydronaphthalen-2-yl 5. The barrier for this cyclization is even lower than that of 2 but in contrast to this latter one, increases with increasing temperature, $\Delta\Delta H(4 \rightarrow 5)$ = 5.3 and 22.1 kJ mol^{-1} at $T = 0$ K and $T = 3000$ K, respectively. The final step, elimination of a hydrogen atom from 1,2-dihydronaphthalen-2-yl to naphthalene is slightly endothermic, $\Delta\Delta H(9 \rightarrow 11) = 104 - 113.5$ kJ mol^{-1}, and involves a comparably large barrier of $\Delta\Delta H(9 \rightarrow 10) = 162.0 - 173.0$ kJ mol^{-1}, Table 2. The overall reaction described by equ. (7) is strongly exothermic with $\Delta\Delta H(1 \rightarrow 11) \approx -1100$ kJ mol^{-1}.

Table 2. Relative enthalpies $\Delta H(T)$ (kJ mol^{-1}) of minima and transition structures

	T	0	298.15	300	500	1000	1500	2000	2500	3000
1	$4C_2H_2 + {}^{\cdot}C_2H$	0.0[a]	0.0[b]	0.0[c]	0.0[d]	0.0[e]	0.0[f]	0.0[g]	0.0[h]	0.0[i]
2	$3C_2H_2$ + n-$^{\cdot}C_4H_3$(**1**)	–240.7	–247.2	–247.2	–248.5	–246.4	–241.5	–235.8	–229.8	–223.7
3	$2C_2H_2$ + n-$^{\cdot}C_6H_5$(**2**)	–389.9	–401.5	–401.5	–404.2	–400.0	–390.0	–378.4	–366.3	–354.0
4	$2C_2H_2 + {}^{\cdot}C_6H_5$, TS(**2** → $^{\cdot}C_6H_5$, phenyl)	–376.6	–388.3	–388.3	–391.6	–390.5	–384.2	–376.6	–368.6	–360.4
5	$2C_2H_2 + {}^{\cdot}C_6H_5$, phenyl	–653.8	–670.9	–670.9	–677.8	–677.6	–668.7	–657.5	–645.6	–633.5
6	$C_2H_2 + {}^{\cdot}C_8H_7$(**3**)	–813.3	–834.5	–834.5	–841.8	–838.5	–824.0	–806.6	–788.5	–770.0
7	$^{\cdot}C_{10}H_9$(**4**)	–940.0	–965.6	–965.6	–974.4	–969.2	–949.6	–926.3	–902.0	–877.3
8	$^{\cdot}C_{10}H_9$, TS(**4** → **5**)	–945.3	–973.2	–973.2	–983.1	–979.5	–962.1	–941.7	–920.6	–899.4
9	$^{\cdot}C_{10}H_9$(**5**)	–1188.1	–1217.3	–1217.3	–1228.0	–1224.7	–1205.6	–1182.5	–1158.3	–1133.6
10	$^{\cdot}C_{10}H_9$, TS(**5** → $C_{10}H_8$, naphthalene + $H^{\cdot}$)	–1015.4	–1045.0	–1045.1	–1055.3	–1051.7	–1034.2	–1013.8	–992.8	–971.6
11	$C_{10}H_8 + H^{\cdot}$	–1084.1	–1109.0	–1109.0	–1117.7	–1111.8	–1092.2	–1069.7	–1046.6	–1023.4

[a]–385.320981, [b]–385.301137, [c]–385.301005, [d]–385.282247, [e]–385.226187, [f]–385.161986, [g]–385.092921, [h]–385.021047, [i]–384.947504; these total enthalpies are given in units of hartrees: $1H_h$=2625.5 kJmol^{-1}.

Dissociation energies

Dissociation energies $D^o(298.15)$ are obtained from G3MP2B3 and, partly, CCSD(T)-CBS-W1U, calculated standard enthalpies of formation $\Delta_f H^o(298.15)$ of parent molecules and radicals. The results are presented in Table 3 and compared with experimental data. In this comparison, it should be borne in mind that $D^o(298.15)$ values are difficult to measure in polyatomic molecules and, thus, are in many cases controversial and might be updated in the future [36]. In addition, some of the standard enthalpies of formation of the radicals $R^•$ used to derive dissociation energies according to

$$D^o(298.15)\,(\mathrm{H{-}R}) = \Delta_f H^o(298.15)\,(\mathrm{R}^•) + \Delta_f H^o(298.15)\,(\mathrm{H}^•) - \Delta_f H^o(298.15)\,(\mathrm{H{-}R}) \quad (8)$$

actually are calculated rather than experimental values, e.g. for n-$^•C_4H_3$ and i-$^•C_4H_3$.

Table 3. Parent molecules and radicals. Enthalpies of formation $\Delta_f H^o(298.15)$ and dissociation energies $D^o(298.15)$ calculated by G3MP2B3[a]

$\Delta_f H^o(298.15)$	$D^o(298.15)$	$\Delta_f H^o(298.15)$
HC≡CH	→	$^•$C≡CH + H$^•$
225.0	555.1	780.1
226.9[b]	560.8[b]	787.7[b]
(227.4)	(558.0)	(785.4)
CH_3–C≡CH	→	($H_2C^•$–C≡CH ↔ H_2C=C=$^•$CH) + H$^•$
182.4	382.1	564.5
182.6[b]	386.7[b]	569.3[b]
(185.4)	(371.6)	(557.0)
H_2C=C=CH_2	→	($H_2C^•$–C≡CH ↔ H_2C=C=$^•$CH) + H$^•$
183.8	380.7	564.5
186.8[b]	382.5[b]	569.3[b]
(190.5)	(366.5)	(557.0)
HC≡C–CH=CH_2	→	HC≡C–CH=CH$^•$ + H$^•$
283.3	474.1	757.4
		765.6[b]
H_2C=C=C=CH_2	→	(HC$^•$=C=C=CH_2 ↔ HC≡C–C$^•$=CH_2) + H$^•$
313.8	406.1	719.9
320.2[b]	400.4[b]	720.6[b]
C_6H_6	→	$^•C_6H_5$ + H$^•$
76.9	475.9	552.8
(82.9)	(465.2)	(548.1)
(82.9)	(473.1)	(556.0)
C_8H_8	→	$^•C_8H_7$ (3) + H$^•$
138.8	481.7	620.5
(147.9)	(457.1)	(605.0)
	(463.6)	(611.5 [31])
$^•C_{10}H_9$ (5)	→	$C_{10}H_8$ + H$^•$
244.6	108.3	352.9
		(368.6)

[a]G3MP2B3 calculated enthalpies of formation $\Delta_f H^o(298.15)$ of parent molecules and radicals; for the hydrogen atom, $\Delta_f H^o(298.15) = 218.0$ kJ mol^{-1} was used. Dissociation energies $D^o(298.15)$ were calculated as $\Delta\Delta_f H^o(298.15)$; all energies are given in kJ mol^{-1}. Experimental values are presented in parentheses and are taken, unless otherwise noted, from ref [36]. [b]CCSD(T)-CBS-W1U results.

The propargyl radical $^{\bullet}C_3H_3$, $H_2C^{\bullet}$–C≡CH ↔ H_2C=C=$^{\bullet}$CH, can be formed either from propyne CH_3–C≡CH or allene H_2C=C=CH_2. The respective dissociation energies are quite low, 386.7 and 382.5 kJ mol^{-1}, compared with that of, e.g. ethene (464.2 kJ mol^{-1} [36]). The calculated spin densities ρ indicate substantial contributions from both resonance structures with a preference for $H_2C^{\bullet}$–C≡CH, ρ($H_2C^{\bullet}$) = 0.8 and ρ(=$^{\bullet}$CH) = 0.5.

Conclusion

The rate-determining step of soot formation in a flame reportedly is the cyclization reaction leading to benzene that may occur either by homogeneous recombination of two propargyl radicals or by addition of C_2H_2 to n-$C_4H_3^{\bullet}$ corresponding to either an odd- or an even-carbon free radical path, or a combination thereof. Thermochemical properties obtained by quantum chemical calculations, mainly G3MP2B3 and partly CCSD(T)-CBS-W1U, for neutral molecules and radicals possibly involved in these pathways are presented. In addition, the consecutive step in the growth of polycyclic aromatic hydrocarbons as soot precursors, namely formation of naphthalene by the reaction $^{\bullet}C_6H_5 + 2\ C_2H_2 \rightarrow C_{10}H_8 + H^{\bullet}$, is described.

References

[1] Frenklach, M. *Phys. Chem. Chem. Phys.* 2002, *4*, 2028–2037.

[2] McEnally, C. S.; Pfefferle, L. D.; Atakan, B.; Kohse-Höinghaus, K. *Prog. Energ. Combust. Sci.* 2006, *32*, 247–294.

[3] Kohse-Höinghaus, K.; Atakan, B.; Lamprecht, A.; Alatorre, G. G.; Kamphus, M.; Kasper, T.; Liu, N.-N. *Phys. Chem. Chem. Phys.* 2002, *4*, 2056–2062.

[4] Richter, H.; Howard, J. B. *Phys. Chem. Chem. Phys.* 2002, *4*, 2038–2055.

[5] Appel, J.; Bockhorn, H.; Frenklach, M. *Combust. Flame* 2000, *121*, 122–136.

[6] Miller, J. A.; Melius, C. F. *Combust. Flame* 1992, *91*, 21–39.

[7] Wang, H.; Frenklach, M. *Combust. Flame* 1997, *110*, 173–221.

[8] Senosiain, J. P.; Miller, J. A. *J. Phys. Chem. A* 2007, *111*, 3740–3747.

[9] Miller, J. A.; Klippenstein, S. J. *J. Phys. Chem. A* 2003, *107*, 7783–7799.

[10] Hansen, N.; Klippenstein, S. J.; Taatjes, C. A.; Miller, J. A.; Wang, J.; Cool, T. A.; Yang, B.; Yang, R.; Wei, L. X.; Huang, C. Q.; Wang, J.; Qi, F.; Law, M. E.; Westmoreland, P. R. *J. Phys. Chem. A* 2006, *110*, 3670–3678.

[11] Harding, L. B.: Klippenstein, S. J.; Georgievskii, Y. *J. Phys. Chem. A* 2007, *111*, 3789–3801.

[12] Klippenstein, S. J.; Miller, J. A. *J. Phys. Chem. A* 2005, *109*, 4285–4295.

[13] Mebel, A. M.; Kislov, V. V.; Kaiser, R. I. *J. Chem. Phys.* 2006, *125*, 133113.

[14] Wheeler, S. E.; Allen, W. D.; Schaefer, H. F. *J. Chem. Phys.* 2004, *121*, 8800–8813.

[15] Wheeler, S. E.; Robertson, K. A.; Allen, W. D.; Schaefer, H. F.; Bomble, Y. J.; Stanton, J. F. *J. Phys. Chem. A* 2007, *111*, 3819–3830.

[16] Kollias, A. C.; Domin, D.; Hill, G.; Frenklach, M.; Golden, D. M.; Lester, W. A. *Int. J. Chem. Kinet.* 2005, *37*, 583–592.

[17] Krokidis, X.; Moriarty, N. W.; Lester Jr., W. A.; Frenklach, M. *Int. J. Chem. Kinet.* 2001, *33*, 808–820.

[18] Gu, X.; Guo, Y.; Zhang, F.; Mebel, A. M.; Kaiser, R. I. *Chem. Phys.* 2007, *335*, 95–108.

[19] Curtiss, L. A.; Raghavachari, K.; Redfern, P. C.; Rassolov, V.; Pople, J. A. *J. Chem. Phys.* 1998, *109*, 7764–7776.

[20] Baboul, A. G.; Curtiss, L. A.; Redfern, P. C.; Raghavachari, K. *J. Chem. Phys.* 1999, *110*, 7650–7657.

[21] Janoschek, R.; Rossi, M. J. *Int. J. Chem. Kinet.* 2002, *34*, 550–560.

[22] Janoschek, R.; Rossi, M. J. *Int. J. Chem. Kinet.* 2004, *36*, 661–686.

[23] Martin, J. M. L.; de Oliveira, G. *J. Chem. Phys.* 1999, *111*, 1843–1856.

[24] Shukla, B.; Susa, A.; Miyoshi, A.; Koshi, M. *J. Phys. Chem. A* 2008, *112*, 2362–2369.

[25] Curtiss, L. A.; Raghavachari, K.; Redfern, P. C.; Baboul, A. G.; Pople, J. A. *Chem. Phys. Lett.* 1999, *314*, 101–107.

[26] Kislov, V. V.; Mebel, A. M. *J. Phys. Chem. A* 2008, *112*, 700–716.

[27] Kislov, V. V.; Mebel, A. M.; Lin, S. H. *J. Phys. Chem. A* 2002, *106*, 6171–6182.

[28] Kislov, V. V.; Islamova, N. I.; Kolker, A. M.; Lin, S. H.; Mebel, A. M. *J. Chem. Theory Comput.* 2005, *1*, 908–924.

[29] Unterreiner, B. V.; Sierka, M.; Ahlrichs, R. *Phys. Chem. Chem. Phys.* 2004, *6*, 4377–4384.

[30] Aguilera-Iparraguirre, J.; Klopper, W. *J. Chem. Theory Comput.* 2007, *3*, 139–145.

[31] Blanquart, G.; Pitsch, H. *J. Phys. Chem. A* 2007, *111*, 6510–6520.

[32] Gaussian 03, Revision B.04. Frisch, M. J.; Trucks, G. W.; Schlegel, H. B.; Scuseria, G. E.; Robb, M. A.; Cheeseman, J. R.; Montgomery, J. A.; Vreven, T.; Kudin, K. N.; Burant, J. C.; Millam, J. M.; Iyengar, S. S.; Tomasi, J.; Barone, V.; Mennucci, B.; Cossi, M.; Scalmani, G.; Rega, N.; Petersson, G. A.; Nakatsuji, H.; Hada, M.; Ehara, M.; Toyota, K.; Fukuda, R.; Hasegawa, J.; Ishida, M.; Nakajima, T.; Honda, Y.; Kitao, O.; Nakai, H.; Klene, M.; Li, X.; Knox, J. E.; Hratchian, H. P.; Cross, J. B.; Adamo, C.; Jaramillo, J.; Gomperts, R.; Stratmann, R. E.; Yazyev, O.; Austin, A. J.; Cammi, R.; Pomelli, C.; Ochterski, J. W.; Ayala, P. Y.; Morokuma, K.; Voth, G. A.; Salvador, P.; Dannenberg, J. J.; Zakrzewski, V. G.; Dapprich, S.; Daniels, A. D.; Strain, M. C.; Farkas, O.; Malick, D. K.; Rabuck, A. D.; Raghavachari, K.; Foresman, J. B.; Ortiz, J. V.; Cui, Q.; Baboul, A. G.; Clifford, S.; Cioslowski, J.; Stefanov, B. B.; Liu, G.; Liashenko, A.; Piskorz, P.; Komaromi, I.; Martin, R. L.; Fox, D. J.; Keith, T.; AlLaham, M. A.; Peng, C. Y.; Nanayakkara, A.; Challacombe, M.; Gill, P. M. W.; Johnson, B.; Chen, W.; Wong, M. W.; Gonzalez, C.; Pople, J. A. Gaussian, Inc., Wallingford CT, 2004.

[33] [33] Kerr, J. A.; Stocker, D. W. In CRC Handbook of Chemistry and Physics, 82th ed.; Lide, D. R. (Ed.); CRC Press: Boca Raton, 2001.

[34] Chase, M. W.; Davies, C. A.; Downey, J. R.; Frurip, D. J.; McDonald, R. A.; Syverud, A. N. JANAF Thermochemical Tables *J. Phys. Chem. Ref. Data.* 1985, 14 (Suppl. 1).

[35] Gurvich, L. V.; Veyts, I. V.; Alcock, C. B. Thermodynamic Properties of Individual Substances: Hemisphere Publication: New York, 1991.

[36] [36] CRC Handbook of Chemistry and Physics, 87th ed.; Lide, D.R. (Ed.), CRC Taylor and Francis, Boca Raton (FL) 2006-2007.

[37] Benson, S. W. Thermochemical Kinetics, 2nd ed.; Wiley: New York, 1976.

[38] Tsang, W. In Energetics of Organic Free Radicals, Eds: Simoes, A.; Greenberg, A.; Liebman, J. F. Structure, Energetics and Reactivity in Chemistry Series (SEARCH series); Chapman and Hall: London, 1996.

[39] Wagman, D.D.; Kilpatrick, J.E.; Pitzer, K.S.; Rossini, F.D. *J. Res. Nat. Bur. Stand.* 1945, *35*, 467–496.

[40] Thermodynamics Research Center, Selected Values of Properties of Chemical Compounds. Thermodynamics Research Center, Texas A&M University, College Station, Texas, 1997.

[41] Stamm R.F. *J. Chem. Phys.* 1949, *17*, 104–105.

[42] Roth, W.R.; Adamczak, O.; Breuckmann, R.; Lennartz, H.-W.; Boese, R. *Chem. Ber.* 1991, *124*, 2499–2521.

[43] Joshi R.M. *J. Polym. Sci. A-2* 1970, *8*, 679–687.

[44] Afeefy, H.Y.; Liebman, J.F.; Stein, S.E. "Neutral Thermochemical Data" in Linstrom, P.J.; Mallard, W.G.; (Eds.) NIST Chemistry WebBook, NIST Standard Reference Database Number 69, June 2005, National Institute of Standards and Technology, Gaithersburg MD, 20899 (http://webbook.nist.gov).

[45] Glushko Thermocenter, Russian Academy of Sciences, Moscow "Entropy and Heat Capacity of Organic Compounds" in Linstrom, P.J.; Mallard, W.G.; (Eds.) NIST Chemistry WebBook, NIST Standard Reference Database Number 69, June 2005, National Institute of Standards and Technology, Gaithersburg MD, 20899 (http://webbook.nist.gov).

[46] Cowperthwaite, M.; Bauer, S.H. *J. Chem. Phys.* 1962, *36*, 1743–1753.

[47] East, A.L.L.; Allen, W.D. *J. Chem. Phys.* 1993, *99*, 4638–4650.

[48] Császár, A.G.; Allen, W.D.; Schaefer, H.F. *J. Chem. Phys.* 1998, *108*, 9751–9764.

[49] Schuurman, M.S.; Muir, S.R.; Allen, W.D.; Schaefer, H.F. *J. Chem. Phys.* 2004, *120*, 11586–11599.

[50] Tranter, R. S.; Tang, W.; Anderson, K. B.; Brezinsky, K. *J. Phys. Chem. A* 2004, *108*, 3406–3415.

[51] Dinadayalane, T. C.; Priyakumar, U. D.; Sastry, G. N. *J. Phys. Chem. A* 2004, *108*, 11433–11448.

[52] Rogers, D. W.; Matsunaga, N.; McLafferty, F. J.; Zavitsas, A. A.; Liebman, J. F. *J. Org. Chem.* 2004, *69*, 7143–7147.

[53] [53] Nguyen, T.L.; Mebel, A.M.; Kaiser, R.I. *J. Phys. Chem. A* 2001, *105*, 3284–3299.

[54] Vereecken, L.; Pierloot, K.; Peters, J. *J. Chem. Phys.* 1998, *108*, 1068–1080.

[55] J.B. Birks, in Organic Molecular Photophysics (J.B. Birks, Ed.), Vol. I, p. 25, J. Wiley, London 1973.

[56] Tasi, G.; Izsák, R.; Matisz, G.; Császár, A.G.; Kállay, M.; Ruscic, B.; Stanton, J.F. *ChemPhysChem* 2006, *7*, 1664–1667.

[57] Ruscic, B.; Pinzon, R. E.; Morton, M.L.; von Laszevski, G.; Bittner, S.J.; Nijsure, S.G.; Amin, K.A.; Minkoff, M.; Wagner, A.F. *J. Phys. Chem. A* 2004, *108*, 9979–9997.

[58] Ponomarev, D.; Takhistov, V. *Internet Electron. J. Mol. Des.* 2005, *4*, 367–380.

In: Thermochemistry and Advances in Chemistry Research ISBN: 978-1-60692-376-4
Ed: Anton G. Mirskiy

Chapter 7

IDEAL PROCESS CONCEPTUAL DESIGN BASED ON THE EXERGETIC ANALYSIS. NEW COUPLINGS OF THERMOCHEMICAL DIPOLES FOR COLD PRODUCTION FROM LOW GRADE HEAT

N. Le Pierrès, D. Stitou, N. Mazet and B. Spinner
PROMES (laboratoire PROcédés, Matériaux et Energie Solaire), UPR CNRS 521
Rambla de la Thermodynamique, Tecnosud, 66100 Perpignan (France).

ABSTRACT

The exergetic analysis of ideal processes gives parameters to be optimised to design processes converting or transforming inlet exergy as completely as possible. Thermochemical dipoles consist of two linked physico-chemical processes which produce or consume a working gas. These dipoles allow cold or heat production through the management of the physico-chemical processes involved. New couplings of such dipoles permits production of cold at low temperature (-22°C for deep-freezing purpose) from low grade heat (at 70°C, coming from thermal wastes or produced by solar collectors for example). In such systems, inlet exergy is completely used and no excess cold or energy is produced.

NOMENCLATURE

COP	**-coefficient of performance**
D	**-exergetic losses [J]**
d	**-relative exergy losses = D/E'**
e_w	**-exergy selectivity**
E'	**-inlet exergy [J]**

E'_{nu}	**-unused exergy [J]**
E''	**-outlet exergy [J]**
E''_{ex}	**-excess produced exergy [J]**
E''_{un}	**-untapped produced exergy [J]**
E''_{u}	**-useful produced exergy [J]**
P	**-Ppressure [Pa]**
Q	**-heat quantity [J]**
T	**-temperature [K]**
T_0	**-reference temperature [K]**
x_e	**-exergy conversion factor**
η	**-exergy yield**
θ	**-Carnot factor = $1-T_0/T$**
ΔH	**-heat of transformation [J/mol]**

Subscripts

C	**-cold**
h	**-hot**
i	**-ideal**

1. INTRODUCTION

When one designs new processes, one has to take into account environment and sustainable development questions and thus the type of energy used and of the energy produced involved in the processes. In the case of solar energy, either transformed into heat or converted into electricity, the question of the quality of the energy source is of particular importance. Indeed, this energy source is still not widely used and the cost of the solar collectors producing high temperature heat and convertors producing electricity is still high. The energy produced by the new process should be of the quantity and quality adapted to its use, of variable power and usable at the time when it is needed, independently from the discontinuous availability of the energy source.

In this study, focus will be directed toward the conversion of solar energy into heat. Nowadays, solar energy is only commonly used in domestic water heating systems and recent developments were made concerning solar cold production and particularly solar climatisation systems. Solar liquid-gas absorption systems were developed more than 50 years ago to transform solar heat into cold, but still present major drawbacks and constraints.

Indeed, this type of process works in a continuous way, thus with a continuous heat source. The storage of the needed heat or of the cold produced allows a good functioning of the process, but necessitates heavy and voluminous storage capacities. Moreover, the needed temperature of the heat source is high: from 90 to 140°C for simple or double effect systems [1], what imposes the use of expensive solar collectors, as parabolic concentrators or with vacuum tubes glazing. Problems are of the same kind concerning the solid-gas adsorption system (water-silicagel or water-zeolite for example), with however a lower temperature of the heat source, around 80°C in this case [2]. Adsorption and absorption systems allow to obtain cold at a minimum temperature of about -5°C.

Sorption based on the chemical reaction between a solid and a gas also allows the simple effect production of cold at around –5 to –10°C from heat at 80° to 90°C [3], and a solar prototype proved the effectiveness of that concept as soon as 1986 [4]. The reactive gas used is ammonia. Lower temperature levels were also obtained using this thermochemical technology for deep-freezing purpose, but from higher temperature levels of the heat source, at about 140 to 160°C [5].

The aim of this study is to determine the possibilities to design new process concepts to produce cold at low temperature (at about -30°C for storage of frozen food) from a heat source at about 70°C compatible with the use of flat simple glazed solar collectors. These processes separate the time of solar heat availability and the time of cold production.

Recent research work based on the exergetic analysis characterised classical and original processes. Studies were directed toward cold production systems with ideal coefficient of performance (COP) of 1, 2 and 3 [6], and toward systems producing simultaneously cold and heat at higher temperature level presenting the possibility of heat transport over long distances [7, 8].

The first goal of this work is to characterise exergetic methods allowing the design of processes using low grade heat to produce cold at low temperature: the quality and quantity of the involved heat should be distinguished. The second goal of this work comes from the definition of the exergetic efficiency, yield and effectiveness [9] based on the work of Brodyanski et al. [10]. These definitions take into account the concept of transiting exergy in the process : the determination of the ideal exergy conversion factor x_e [11] allows the optimisation of the ideal and practical exergetic yields of a process, as soon as its conception. This type of analysis will be used to determine all the possible processes.

This work thus takes place in the scientific goal of ranking and analysis of ideal processes, before their endoreversible analysis [12] and the studies of the irreversibilities distribution.

2. Analysis Basis

First, as represented on figure 1, considering an ideal process, we should separate from the inlet exergy E' [13]:

- unused exergy E'_{nu} that only transits through the ideal system,
- exergy that is converted or transformed in the process $E'-E'_{nu}$.

Then, the ideal exergy conversion factor x_e is defined as:

$$x_e = \frac{E' - E'_{nu}}{E'} \tag{1}$$

This criterion x_e should be maximised up to 1.

In real processes, destroyed exergy D, exergy produced in excess E''_{ex} and untapped exergy E''_{un} must be minimised, such that the useful produced exergy E''_u is maximal.

The exergetic yield η is defined as:

$$\eta = \frac{E''_u}{E'} = x_e - d - e_w \tag{2}$$

where $d = \frac{D}{E'}$ and the exergetic selectivity $e_w = \frac{E'_{ex} + E''_{un}}{E'}$

The exergetic yield applied to an ideal system (where d=0), is η_i :

$$\eta_i = x_e - e_w \tag{3}$$

η_i should be maximised, such that the ideal exergy conversion factor x_e approaches 1 and the exergetic selectivity e_w approaches 0.

The second basis of the analysis is the exergetic dipole, used for the study of heat transformation processes by Spinner et al [8]. An exergetic dipole results in the coupling, via an active gas transfer, of two different physico-chemical processes such as evaporation/condensation or the sorption of a gas on a liquid or a solid phase. These reactions and transformations are reversible and, depending on the direction of the process, either endothermic or exothermal. We will focus here on the gas-solid reaction, a monovariant process, and which can be coupled to a liquid-gas phase change, also a monovariant process, or to another gas-solid reaction involving the same fluid but a different reactive solid. Thermodynamic conditions applied to each pole of a closed dipole lead either to a useful heat or cold production phase, or to a regeneration phase. The two phases are independent temporally from each other.

In real processes, the heats ΔH of solid-gas reactions or transformations vary with the salts involved. Moreover, the liquid/gas heat of transformation is generally smaller than the solid/gas heat of reaction. In the following analysis of ideal processes, all heats of transformation will be considered equal. Real values will be taken into account only in the last paragraph, dealing with real processes.

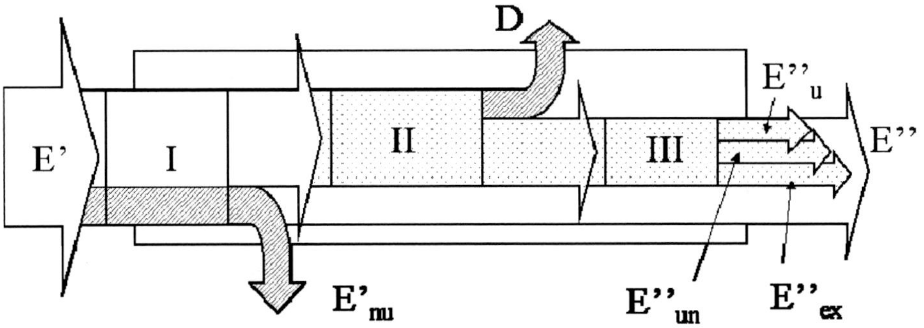

Figure 1. Exergies within (I) an ideal system, (II) a real system and (III) with the different output exergies

Transfers from a pole to the other are characterised by the quantity and quality of heat transferred. The quantity of heat transferred is proportional to the heat of the involved transformation or reaction. The quality of the heat transferred depends on the temperature levels T_h, T_0 or T_c imposed during the different phases. T_h is the regeneration temperature, the highest temperature level involved. T_c is the cold production temperature. T_0 is the reference or ambient temperature. If the heat released by the process is at T_0, the corresponding Carnot factor equal zero and thus the unused and untapped exergies E'_{nu} and E''_{un} equal zero. Thus, x_e equals 1 following equation (1). Moreover, if T_c is the cold temperature level which is necessary to the user, the excess exergy production E''_{ex} is zero and consequently, following equations (2) and (3) in the ideal case, the selectivity e_w equals 0 and the exergetic yield η_i is maximum.

The dipole COP is defined as :

$$\textbf{COP}= \frac{\Delta H_1}{\Delta H_2} \tag{4}$$

for the poles 1 and 2 of the basic dipole represented figure 2, where pole 1 is the one where cold production occurs during one phase of the dipole functioning and 2 is the one that has to be heated for regeneration during the other phase.

In the ideal case, using (2) as exergies E' and E''_u in the case of this dipole, equal respectively $\Delta H_2 . \theta_h$ and $\Delta H_1 . \theta_c$, the COP_i is proportional to the exergetic yield η_i:

$$\textbf{COP}_i = \frac{\Delta H_1}{\Delta H_2} = \eta_i \frac{\theta_h}{|\theta_c|} \tag{5}$$

where θ is the carnot factor $\theta = 1 - \frac{T_0}{T}$.

From equation (5), we can see that, for an ideal system with an exergetic yield of 1 and a COP of 1 producing cold at -22°C, T_h should be at least of 51°C. It is supposed that the process enthalpies ΔH_1 and ΔH_2 involved are equal, as precised above, and the reference temperature T_0 of 10°C.

Endoreversible thermodynamics analysis show the importance of external irreversibilities on real performances of such processes [12]. For a real process, the exergetic yield could be of up to 0.5, that means the necessary heat source temperature should be of 107°C for a system COP of 1 to produce cold at -22°C. A COP_i value of 1 thus cannot be the targeted performance, if the available high temperature level is of only of 70°C. Consequently, we will look for the design of ideal machines of COP_i lower than 1, and more precisely of COP_i fixed to ½. The ideal high temperature level necessary in this case is of 302K (29°C) for the process where T_0=10°C. This high temperature is then compatible with the practical available high temperature of 70°C when considering a reasonable exergetic efficiency of 0.5.

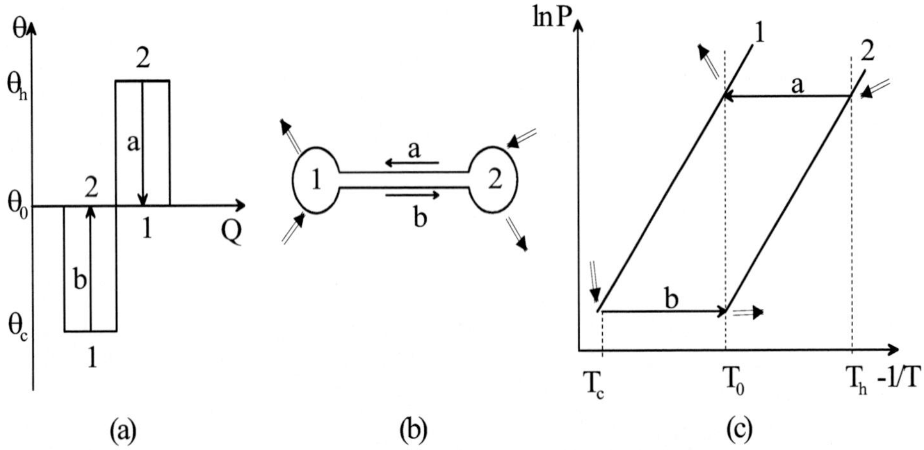

Figure 2. (a) Representation in the θ = f(Q) diagram of exergies Qθ involved in an ideal process producing cold at θc from heat at θh. (b) The dipole constituted of physico-chemical processes (1) and (2) produces cold at θc in direction (b) and is regenerated in direction (a).(c) Representation in Clausius-Clapeyron diagram of the pressure and temperature conditions in the two phases

3. Choice of Ideal Dipole Couplings

Following (4) and (5), to get a COP_i value of ½ with an exergetic yield of 1, one has to introduce twice as much heat into the process as the cold quantity produced. This allows to have values of Carnot factor θ_h twice smaller than θ_c and a functioning of the thermochemical dipole as presented on figure 3a. For an exergetic yield of 1, inlet exergies (downward arrow) and outlet exergies (upward arrow), represented by the area θ.Q on the diagram, are equal.

The two dipole regenerations can happen simultaneously or successively. Three physico-chemical processes are necessary to ensure the desired functioning, as shown by the three equilibrium lines on figure 3b. To avoid having to build twice pole number 2, it is possible to use valves between poles 1, 2 and 3, allowing a successive functioning of the dipoles with only three poles (figure 3c).

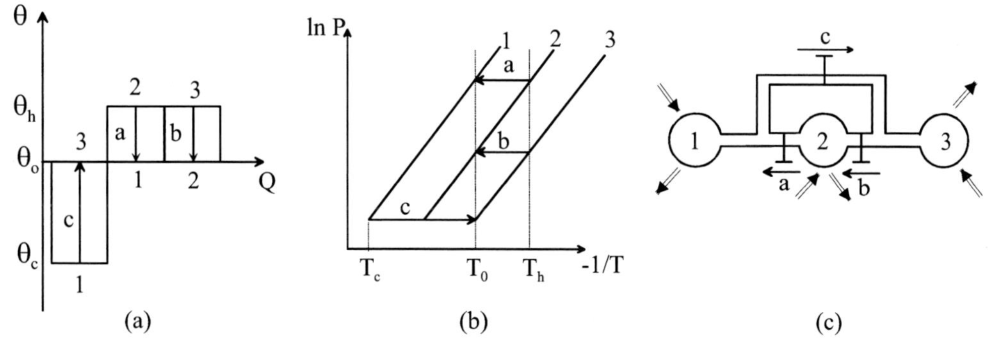

Figure 3. Representation in the θ = f(Q) diagram (a), in Clausius-Clapeyron diagram (b), and of the dipole coupling (c) in case (A)

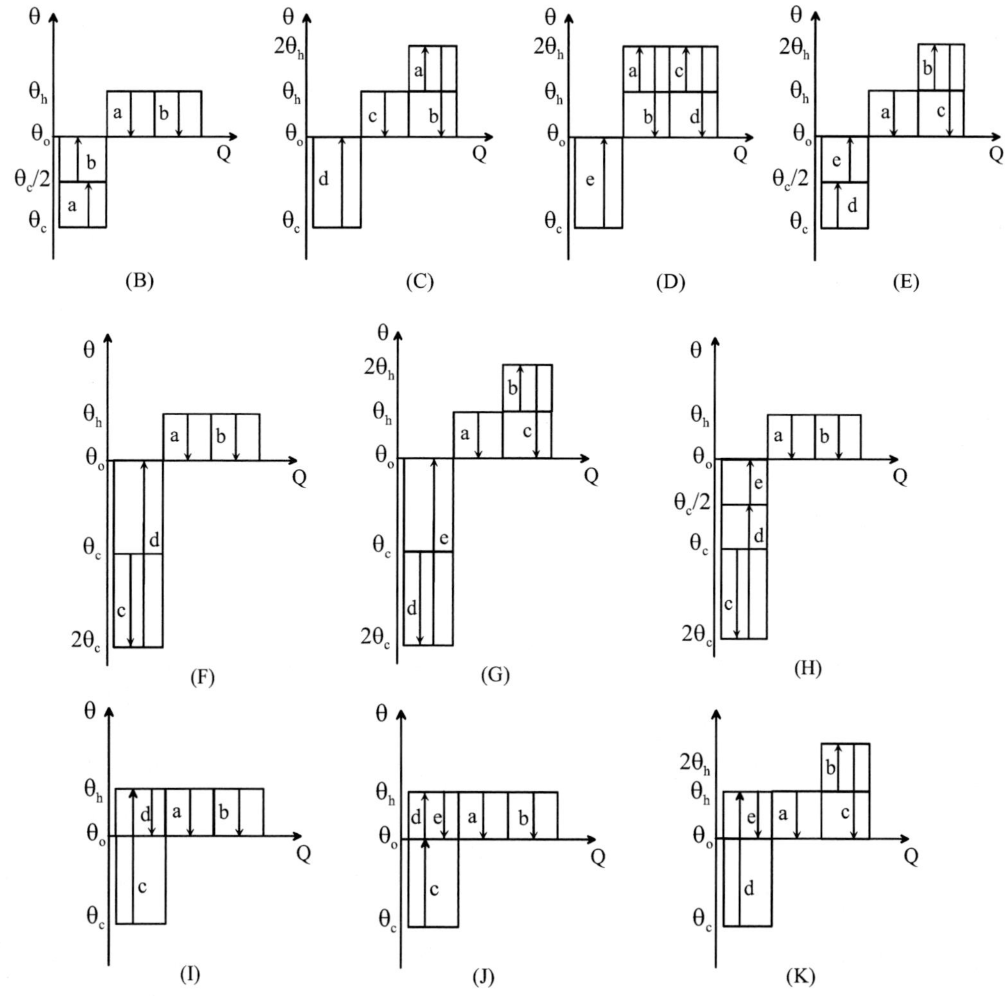

Figure 4. θ = f(Q) diagrams of diferent dipole couplings (cases (B) to (K)) for cold production at θc from heat at θh = ½ |θc|

Coupling (B) on figure 4 is particularly interesting, as it only involves two physico-chemical processes 1 and 2 and two dipoles working:

- without thermal connections during the regeneration phase,
- with one thermal transfer during cold production.

Temperature and pressure levels and the functioning principle of that system is presented on figure 5. Figure 6 represents the Clausius-Clapeyron diagrams corresponding to couplings (C) to (K).

On Clausius-Clapeyron diagrams here, equilibrium lines of the different processes are supposed parallel. In practical cases, the involved processes have different heat of reactions and thus different slopes on the diagram. Calling (I) the condensation/evaporation heat of ammonia (23 kJ/mol), (II) the reaction heat of $BaCl_2$ (0-8)NH_3 (37 kJ/mol of NH_3), (III) the reaction heat of $SrCl_2$ (0-7)NH_3 (42 kJ/mol of NH_3), (IV) the reaction heat of $MnCl_2$ (2-6)NH_3 (49 kJ/NH_3) and (V) the reaction heat of $NiCl_2$ (2-6)NH_3 (62 kJ/mol of NH_3), the performances of the different dipole couplings with autothermal connections are presented on table 1. The COP values presented here depend only on the heat of reactions. In practice, they also depend on the reactants specific heats, on the mass of metal parts used and of the heat and gas exchangers needed. The study of these parameters has already been performed [14].

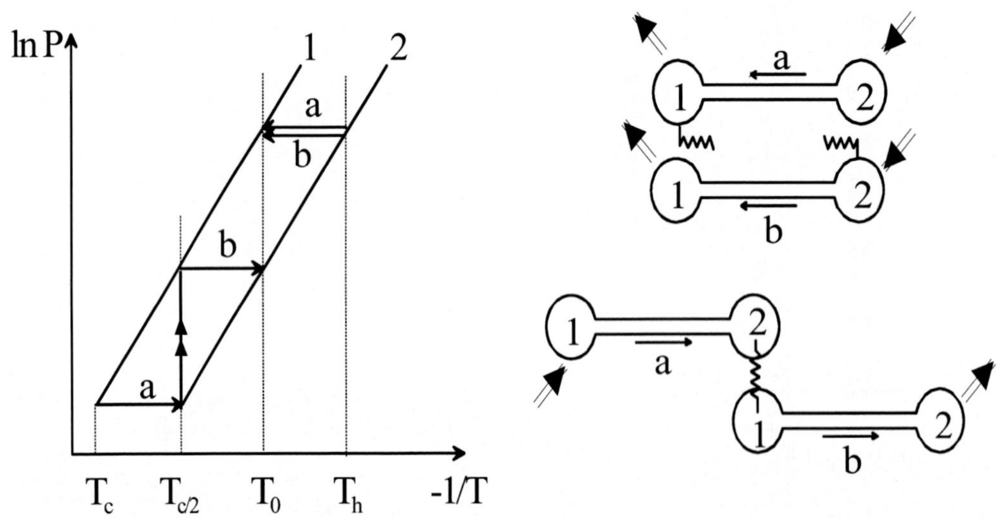

Figure 5. Clausius-Clapeyron diagram of case (B) involving two dipoles and two physico-chemical processes (1) and (2), and realisation of the coupling.

The final choice between all the different possible couplings can be the result of a compromise between the COP and the simplest coupling to realise, as the cases (A) and (B) presented before, that present either gaseous connections through valves or autothermal connection between two poles. Possibilities of application of dipole couplings for long distance heat transport have been recently presented [7].

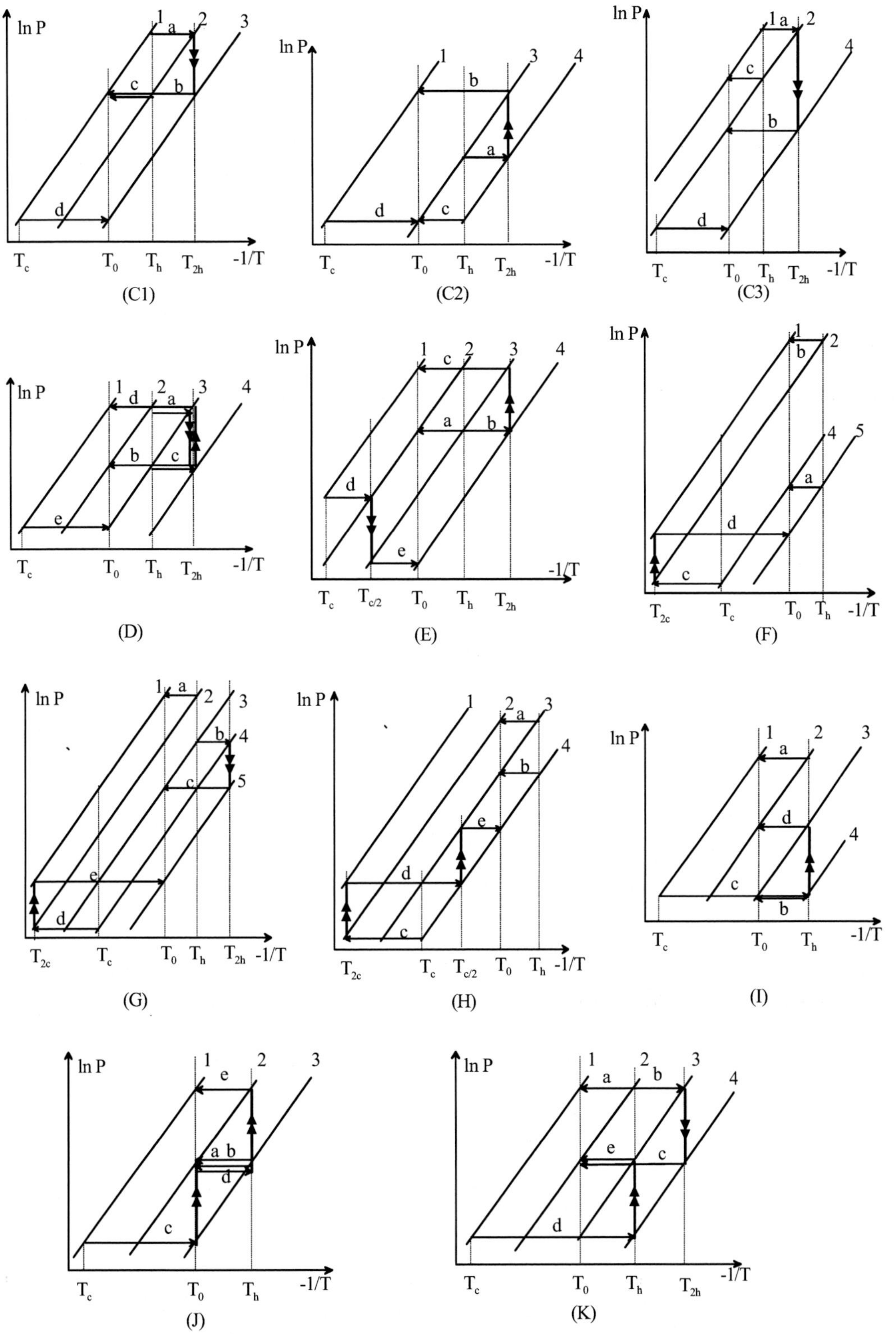

Figure 6. Clausius-Clapeyron diagram of cases (C) to (K) presented on figure 3

Table 1. COP of the different dipole couplings depending on the physico-chemical processes heat of reaction or transformation

Case	Autothermal connections	couples	COP	couples	COP	couples	COP	couples	COP
A, D	0	I/ II III	0.39	II/ III IV	0.40	III/ IV V	0.38		
B, K	1	I/ II II	0.31	II/ III III	0.44	III/ IV IV	0.43	IV/ V V	0.39
C1	1	I/ I II	0.38	II /II III	0.47	III/ III IV	0.46	IV/ IV V	0.44
C2	1	I/ III IV	0.25	II/ IV V	0.33				
C3	1	II/ I II	0.61	III/ II III	0.53	IV/ III IV	0.54	V/ IV V	0.56
E, J	2	I/ III III	0.27	II/ IV IV	0.37	III/ V V	0.34		
F	1	IV/ II V	0.5						
G	2	IV/ II III	0.62	V/ III IV	0.68				
H	2	IV/ II III	0.62	V/ III IV	0.68				
I	1	I/ IV II	0.27	II/ V III	0.35				

4. Conclusion

The exergetic analysis of heats involved in thermochemical dipoles, here for cold production purpose, allows distinguishing between quality and quantity of the inlet and outlet exergies. A low temperature level available at the inlet must be counterbalanced by a high quantity of heat, for production of cold at low temperature and at a given quantity. This study also allows a total use of the inlet exergy and no production of excess or unused exergy. $\theta = f(Q)$ diagrams allow the exhaustive analysis and the choice of the dipole couplings to be used. The theoretical study presented here is followed by the dimensioning of the process corresponding to coupling (B) presented figure 3 and 4 [14]. The construction and experimentation of a prototype linking coupling (B) to simple flat plate solar collectors should prove the possibility of deep-freezing cold production from solar low grade heat.

Acknowledgments

This work was supported by the French governmental agency for energy management and environment ADEME and the 'ENERGIE' program of the research ministry – CNRS – DGA through PRI 'Froid Solaire' 6.1.

References

[1] Alefeld G., Radermacher A., Heat conversion systems. *FL CRC Press Inc.,* Boca Raton, Florida, 1994.

[2] Meunier F., La sorption solide: une alternative aux CFCs, Proceedings of the Symposium *'Le froid à sorption solide',* Paris, France, Nov. 18-20, 1992, p 44-52.

[3] Crozat G., Spinner B., Amouroux M., Systèmes de gestion de l'énergie thermique basés sur des réactions solide-gaz, Proceedings of 'pompes à chaleur chimiques de hautes performances', *Perpignan,* France, Sept. 1988, p 310-319.

[4] Balat M., Crozat G., Testing of a prototype of a solar powered solid-absorption refrigeration system, *Proceedings of the ISES Solar World Congress,* Hamburg, Germany, Sept. 13-18, 1987.

[5] Stitou D., Spinner B., Neveu P., Patent FR 02 09392, Installation et procédé pour la production de froid par un système à sorption renversable, 24/07/2002 - CNRS.

[6] Stitou D., Spinner B., Sorin M., An approach for structural design of cooling sorption processes, *I. J. Heat and Technology* 23 (1), june 2005.

[7] Spinner B., Stitou D., Mazet N., New sorption cycles for heat and/or cold production adapted for long distance heat transmission, *Proceedings of IMECE2002, ASME Int. Mechanical Eng. Congress and Exposition,* New Orleans, Louisiana, Nov. 17-22, 2002.

[8] Spinner B., Stitou D., Sorin M., New processes for heat transformation: exergetic analysis of cold production, heat upgrading, with long distance transportation from a discontinuous heat source at low temperature, Paper 345, *Proceeding of the 53rd Canadian Chem. Eng. Conference, Hamilton*, Canada, Oct. 26-29, 2003.

[9] Le Goff P., Cachot T., Lasalle A., Rahbar M.S., Proceedings of JETC IV conference, *Nancy, France,* Sept. 1995, p. 354-362.

[10] Brodyanski V.M., Sorin M., Le Goff P., The efficiency of industrial processes: exergy analysis and optimisation, *Elsevier Science,* New York, 1994.

[11] Sorin V.M., Spinner B., Stitou D., Thermodynamic techniques for the conceptual design of thermochemical refrigerators using two salt materials, *Chem. Eng. Sci.* 57 (2002) 4243-4251.

[12] Stitou D., Spinner B., A new characterization of real energy conversion processes: A contribution of finite size thermodynamics, Paper 348, *Proceedings of the 53rd Canadian Chem. Eng. Conference*, Hamilton, Canada, Oct. 26-29, 2003.

[13] Stitou D., Spinner B., Sorin A., A new advanced exergy analysis of sorption systems: characterisation of transiting exergy and exergy yield concept, *Proceedings of the Int. Sorption Heat Pump Conference'99*, Munich, Germany, March 24-26, 1999.

[14] Le Pierrès N., Stitou D., Mazet N., *Modelling and performances of a deep freezing process using low grade solar heat*, submitted to Energy, March 2005.

In: Thermochemistry and Advances in Chemistry Research ISBN: 978-1-60692-376-4
Ed: Anton G. Mirskiy

Chapter 8

CRYSTALLOGRAPHIC AND STRUCTURAL ANALYSIS OF IRON HETEROOCTA- AND HETEROOLIGONUCLEAR CLUSTERS

***Milan Melnik*[1]*, Ján Garaj*[1] *and Clive Holloway*[2]**
[1] Department of Chemical Theory of Drugs, Faculty of Pharmacy, Comenius University, 832 32 Bratislava, Slovak Republic,
[2]Department of Chemistry, York University, 4700 Keele str. North York, M3J 1P3. Ontario, Canada.

ABSTRACT

This review includes one hundred and thirty heteroocta – and heterooligonuclear iron clusters of known crystal structures. The heterometal atoms include both transition and non – transition metals, including both hard (class A) and soft (class B) examples with one example of actinide (uranium). The clusters range from heterooctametallic of up to sixty five metal atoms.

The shortest Fe–M (non – transition) and Fe–M (transition) bonds are 2.471(2) Å for Fe–Ge and 2.540(4) Å for Fe-Pt. The shortest M-M and M'-M bonds are 2.456(1) Å for Hg-Hg and 2.346(2) (4)Å for Co-Ge. The shortest Fe-Fe bond is 2.517(3) Å. There are some relationships between the structural parameters which are discussed.

ABBREVIATIONS

Ac	acetate
Bu	butyl
Bu^t	terc-butyl
Bu_2O	dibutylester
Cat	catecholate
$catCl_4$	tetrachlorocatecholate
$catPn_2$	3,6- dipropylcatecholate
$C_5H_4CH_2NMe_2$	N,N-dimethylaminoethylcyclopentadien

C_5H_4COO	cyclopentadien-carboxylate
Chp	6-chloro-2-pyridinate
C_5H_4Me	methylcyclopentadienyl
$CH(Me)NCH_2NH_2$	N,N-dimethylaminoethylenediamine
$C_5H_4NMe_2$	N.N-dimethylaminocyclopentadienyl
$CH(Ph_2)_3$	tris(diphenylphosphino)methylene
$p\text{-}ClC_6H_4S$	p-chlorothiobenzoate
C_2O_4	oxalate
cp	cyclopentadienyl
cp*	pentamethylcyclopentadienyl
dmf	dimethylformamide
ert	erythritolate(3-)
ert H	erythritolate(2-)
Et	ethyl
EtS	thioethyl
m	monoclinic
Me	methyl
mea	mercaptoacetate
$m\text{-}MeC_6H_4S$	meta-methylthiophenolate
$N(CH_2CH_2O)_3$	tris(ethoxyl)amine
N_2H_4	hydrazine
or	orthorhombic
PEt_3	triethylphosphine
Ph	phenyl
$PhCO_2$	benzoate
$PMePh_2$	diphenylmethylphosphine
PPh_4	tetraphenylphosphonium
PPh_2Et	diphenylethylphosphine
PPr_3	tripropylphosphine
PPh_3	triphenylphosphine
PhS	thiophenyl
pz	pyrazole
tg	tetragonal
thf	tetrahydrofurane
tla	thioacetolate
tmen	N,N,N',N'-tetramethylethylendiamine
tol	toluene
tr	triclinic
triphos	1,1,1-tris(diphenylphosphinomethyl)ethane

INTRODUCTION

The chemistry of iron is an important area, particularly in the field of chemistry, catalysis and theory. The last forty years have seen many studies of the reaction of various metal salts with iron salts. In many cases compounds have been obtained for which crystallographic and structural data could be obtained. Previous reviews have covered heterobinuclear [1], heterotri - [2], heterotetra - [3], heteropenta - [4], heterohexa – and heteroheptanuclear [5] iron clusters.

This review presents a brief survey of the crystal and structural data for heteroocta – and heterooligonuclear iron clusters. The structures are organized and discussed with a view to obtaining a better understanding of the factor common to each type of structure. As such it represents the first comprehensive overview of over one hundred and thirty heteroocta – to heterooligonuclear iron clusters for which structures have been established by X-ray crystallographic techniques.

HETEROOCTANUCLEAR IRON CLUSTERS

Sixty five heterooctanuclear clusters can be divided into six categories: Fe_6M_2 (38 examples), Fe_4M_4 (18 examples), Fe_3M_5, (3 examples), Fe_2M_6 (2 examples), $Fe_4M_2M_2'$ (3 examples) and $Fe_2Ge_2Co_4$ and their structural data are given in Tables 1 and 2.

Fe_6M_2

Structural parameters of the clusters are summarized in Table 1. X-ray analysis of brown $[(MeO)_6(\mu\text{-}MeO)_6(\mu_6\text{-}O)Fe_6(\mu_3\text{-}OMe)_6\{Na(MeOH)_3\}_2]$ shows [6] that the central μ_6-O atom is approximately in the centre of an octahedron formed by the six Fe(III) atoms. The iron atoms themselves are linked to each other by six μ-OMe and six μ_3-OMe bridges. A distorted octahedral coordination sphere of the Fe(III) atoms is completed by six terminal methoxide ligands. Each of the two Na(I) atoms, reguired for charge balance , is bound to three μ_3-OMe of the $OFe_6(OMe)_6$ moiety and to three additional methanol molecules leading to an idealized C_3 – coordination geometry (NaO_6).

Structure of orange $(CO)_8Fe_2GeFe_2(\mu\text{-}CO)(CO)_6GeFe_2(CO)_8$ [7] consists of four linked $GeFe_2$ triangles. The central Fe-Fe bond of 2.634(1) Å is shorter than the external Fe-Fe bonds at 2.789(1) and 2.794(1) Å. The Fe-Ge bond distances fall into two groups, those to the central iron atoms average 2.431(1) Å, while those to the peripheral iron atoms average 2.405(1) Å. The dihedral angles between the triangles with the common Ge apex are 87.5(1)^o at Ge(1) and 88.8(1)^o and Ge(2), which are close to the expected perpendicular arrangement. However, the dihedral angle between the triangles sharing a common Fe(3)-Fe(4) edge is very wide at 132.3(1)^o. The Ge...Ge separation is 3.74 Å.

Structure of orange $(CO)_8Fe_2SnFe_2(\mu\text{-}CO)(CO)_6SnFe_2(CO)_6$ [7] is essentially isostructural with the Ge analogue [7] . The Fe-Fe and eight Fe-Sn bonds are as expected by comparism with Ge analogue, allowing for the larger size of Sn (1.44 Å compared to Ge (1.22 Å). Thus the inner Fe-Sn bonds (av. 2.575(4) Å) and outer Fe-Sn bonds (av. 2.547(4) Å are both about 0.14 Å longer than the Fe-Ge bonds in the germanium analogue. The outer Fe-Fe bond has expanded by 0.05 Å (av. 2.860 Å) and the inner by 0.10 Å (2.734(5) Å) on going from Sn to Ge cluster.

In red Fe_6Sn_2 compound [8] two $(\eta^5\text{-cp})_2(CO)_7Fe_3Sn(\mu\text{-}CO)$ moieties are bridged by two isocarbonyls, in such that one cis-$Fe(CO)_4$ based CO group from each $[Fe(CO)_4\{SnFe_2(CO)_2(\eta^5\text{-cp})\}]$ unit is joined to the other $[Fe(CO)_4\{SnFe_2(CO)_2(\eta^5\text{-cp})\}]$ unit via an O-Sn linkage, resulting in an eight – membered Sn-O-C-Fe-Sn-O-C-Fe ring. There are three pairs of Fe-Sn bonds with the mean value of 2.538(1) Å.

The $[(CO)_8Fe_2SbFe_2(CO)_6SbFe_2(CO)_8]$ which exists in two isomeric forms, monoclinic dark red [9] and orthorhombic deep red [10] are isostructural, differ mostly by degree of distortion. The antimony – iron skeleton possesses approximately C_{2v} symmetry, the C_2 axis lying on the midpoint of the Fe(3)-Fe(4) axis. The structure consists of a central Fe_2Sb_2 butterfly arrangement. Each Sb atom is involved in a Fe_2Sb ring, thereby achieving a μ_4 – spiro geometry. The butterfly dihedral angles are 109.3° [9] and 109.1(1)° [10] at the Fe-Fe cross. The naked Sb atoms serve as bridges between the inner $Fe_2(CO)_6$ groups (Fe-Fe bonds are 2.857(2) Å [9] and 2.841.1(1)Å [10]) and the outer $Fe_2(CO)_8$ groups (av. Fe-Fe bonds are 2.843(2) in the former and 2.831(11) Å in the latter), with outer Sb-Sb bonds somewhat longer (av. 2.557(2) [9] and 2.559(1) Å [10]) than the inner set (av. 2.494(2) and 2.499(1) Å, respectively).

The asymmetric unit of the red Fe_6In_2 compound [12] consists of the ½ of a Et_4N^+ cation and ¼ of the $[In_2Fe_6(CO)_{24}]$ anion. The $[Et_4N]^+$ group lies on twofold axis, while complex anion is situated about a centre of 2/m symmetry. The anion $[In_2Fe_6(CO)_{24}]^{2-}$, consists of a central In_2Fe_2 parallelogram with each In atom linked to an additional $Fe_2(CO)_6(\mu\text{-}CO)_2$ group. The In…In and Fe…Fe' separations across the ring are long (3.369(2) and 4.222 Å) and considered nonbonding and Fe-In-Fe angle is 102.82(4)° Each In atom exhibits a distorted tetrahedral coordination ($InFe_4$). The Fe-Fe bond in $Fe_2(CO)_6(\mu\text{-}CO)_2$ groups is equal 2.616(2) Å.

Structure of black Fe_6Tl_2 compound shows [13] that the asymmetric unit consists of $[ppn]^+$ cations and $[Tl_2Fe_6(CO)_{24}]^{2-}$ anions. The anion is composed of a Tl_2Fe_2 planar parallelogram with $d_{Tl\text{-}Fe}$ = 2.746(2) and 2.762(2) Å situated on a crystallographic inversion center, with the thallium atoms each bound to one additional $Fe_2(\mu\text{-}CO)_2(CO)_6$ group with $d_{Tl\text{-}Fe}$ = 2.746(2) and 2.765(2) Å. The thalium atoms are four-coordinated, severely distorted tetrahedral ($TlFe_4$).

Structure of black Fe_6V_2 consists [14] of well separated $[Et_4N]^+$ cations and $[\{(EtS)_3(\mu_3\text{-}S)Fe_3(\mu_3\text{-}S)_3V\}_2(\mu\text{-}SEt)_3]^{3-}$ anion. The structure of the anion is set out in Fig. 1. In this crystal, the cluster has an imposed C_3 axis on which are located atoms S(1,1') and V, V'and a perpendicular mirror plane containing atoms S(20,20',20''), corresponding to molecular D_{3h} symmetry. Two cubane – type $\{VFe_3S_4\}$cores are triply connected by μ-SEt groups with a V…V separation of 3.601(3) Å. The V-Fe and Fe-Fe bond distances are 2.575(2) and 2.687(1) Å.

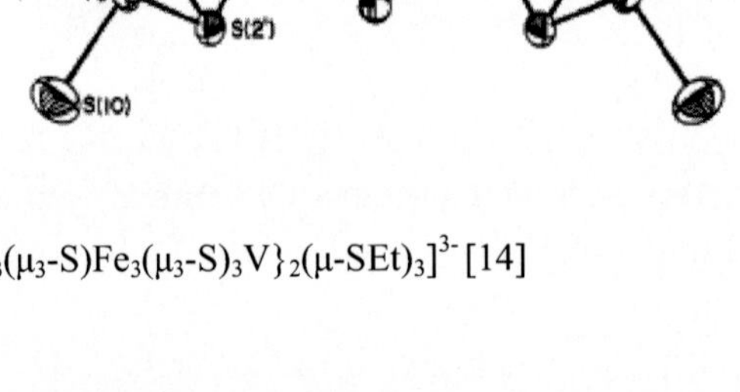

Figure 1. Structure of $[\{(EtS)_3(\mu_3\text{-}S)Fe_3(\mu_3\text{-}S)_3V\}_2(\mu\text{-}SEt)_3]^{3-}$ [14]

In orange red Fe_6Cr_2 compound [15] the carboxylato groups of four acetate anions bridge the Cr_2^{4+} moiety in a paddle wheel fashion. On the axial positions, two acetate acid ligands are coordinated via their carboxylic C=O groups. The carboxylic OH group forms a hydrogen bond of the one of the O atoms of the bridging carboxylates. The molecule is almost perfectly eclipsed although the O –Cr – Cr- O chains have very small torsion angles, 0.94(8) and 3.12(8)°. The Cr-Cr bond distance is 2.3070(8) Å. The $Fe(CO)_2(\eta\text{-cp})(\mu\text{-acH})$ molecules are axially bond to both ends of the Cr_2 unit. The remaining four $Fe(CO)_2(\eta\text{-cp})$ moieties are equatorially bond to $Cr_2(\mu\text{-ac})_4$ unit in such a manner that the C atom of CH_2 group of acetate is coordinated to the iron atom .

There are five black Fe_6Mo_2 clusters [16-20] which contain $[\{(L_1)_3Fe_3(\mu_3\text{-E})_4Mo\}_2(\mu\text{-}EL_2)_3]^{3-}$ anion, (L_1 = EtS, E = S and L_2 = EtS [16]; L_1 = EtS, E = S and L_2 = S(x1) and EtS(x2) [16]; L_1 = PhS (x 1.5) and Cl(x 1.5); E = S and L_2 = OMe [17]; L_1 = Cl, E = S and L_2 = S-tolyl-2 [18, 19]; L_1 = EtS, E = Se and L_2 = EtS [20]). The anions in all five structures are of the "double – cubane" type containing trigonally distorted $MoFe_3E_4$ sub clusters, which are linked through the Mo atoms. Three E-groups serve bridges between the two Mo atoms. The mean cube face diagonal Mo-Fe distance elongated in the order: 2.723(2) Å [16] < 2.726(5) Å [17] < 2.730(2) Å [16] < 2.741 Å [18, 19] < 2.793(2) Å [20].

There are another eight black double – cubane clusters $[Mo_2Fe_6S_8(SR)_6(cat)_2]^{4-}$ [21, 22], $[Mo_2\ Fe_6S_8Cl_8(C_2O_4)]^{4-}$ [23], $[Mo_2Fe_6S_8Cl_6\ (L)_2]^{4-}$ (L = tla or maa) [24], $[Mo_2Fe_6S_8Cl_6(L)]^{4-}$ (L = pz or N_2H_4) [25] and $[Mo_2Fe_6S_9Cl_4.(catCl_4)_2(N_2H_4)]^{4-}$ [26]. The structures of the centrosymmetric anions are that of double bridged, SEt [21], SCl_6H_4Cl-p [22], C_2O_4 [23], tla or maa [24], pz or N_2H_4 [25], double – cubane clusters . In [26] the anion is located on a crystallographic mirror plane that contains the μ-S and the side-on, bridging N_2H_4 ligand and bisects the Mo – Mo and Fe – Fe intercube vectors. The mean Mo-Fe and Fe-Fe bond distances in the series are 2.727 and 2.722 Å.

Structure of other black Fe_6Mo_2 clusters [23, 27, 28] contain a pentaanions $[Mo_2Fe_6S_9Cl_4(C_2O_4)_2(CN)]^{5-}$ [23], $[Mo_2Fe_6S_8(PhS)_9]^{5-}$ [27]and $[Mo_2Fe_6S_9(catCl_4)(OH)]^{5-}$ [28]. The first pentaanion [23] is a doubly-bridged doubly-cubane with a CN^- bridging end- -on the two Mo atoms and a S^{2-} bridging two Fe atoms. Among the outstanding structural features in the pentaanion [23] are included long Mo…Mo and Fe…Fe (intercube) distances of 5.267(5) and 3.443(7) Å respectively, and a Fe – μS – Fe angle of 104.1(4)°. In the second pentaanion [27] is a triply-bridged double cubane with S^{2-} anions end - on the two Mo atoms (Mo…Mo, 3.813(2) Å. The third one [28] contains two S^{2-} bridging ligands and bisects the Mo…Mo and Fe…Fe intercube sectors. The Mo…Mo and Fe…Fe (intercube) distances are 4.248(5) and 3.35(1) Å, respectively. The mean Mo-Fe and Fe-Fe bond distances in the pentaanion clusters are 2.755 and 2.730 Å.

Black $(NEt_5)_6[\{Cl_2Fe_3(\mu_3\text{-}S)_4Mo(\eta^2\text{–}catCl_4)\}_2(\mu\text{-}S)_2]$ [28] is only example which contains hexaanion. X-ray analysis of the cluster shows the hexaanion as a bis (μ-sulfido) doubly bridged doubly cubane with two homometallic M – μS – M bridges (M = Mo or Fe). The Mo…Mo and Fe…Fe (intercube) distances are 4.926(8) and 3.33(21) Å. The mean Mo-Fe and Fe-Fe bond distances are 2.75(2) and 2.73(2) Å.

The mean Mo-Fe bond distance elongated in the sequence: 2.727 Å (tetraanions) < 2.742 Å (trianions) < 2.75 Å (pentaanions and hexaanion). The mean Fe-Fe bond weaker in the sequence: 2.71 Å (tri-) < 2.72 Å (tetra-) < 2.73 Å (penta – and hexaanion).

There are six black Fe_6Mo_2 clusters [29,30]. Three of them contain trianion of the composition $[X_6Fe_6(\mu_3\text{-}S)_6\{Mo(CO)_3\}_2]^{3-}$ (X = Cl or Br [29], p–OMePhO [30]). Remaining three contain tetraanion, $[X_6Fe_6(\mu\text{-}S)_6\{Mo(CO)_3\}_2]^{4-}$ (X = Cl [29], p-COMePhO [30]). The six complex anions are quite similar, and all contain the Fe_6S_6 prismane core. The Fe_3S_3 hexagonal cross section of this core have a "chair" conformation and serve as tridentate ligands for $Mo(CO)_3$ structural subunits. The centrosymmetric octanuclear $Mo_2Fe_6S_6$ core, which forms following addition of the $Mo(CO)_3$ fragments to a Fe_6S_6 unit, shows a distorted Mo_2F_6 cube with a $\mu_3 - S^{2-}$ ligand occupying a position above the center of each of the six faces. The mean Mo-Fe and Fe-Fe bond bond distances are 2.941 and 2.763 Å.

Structure of black Fe_6W_2 cluster [30, 31] contains a trianionic $[(\eta^1\text{-}MePhO)_6Fe_6(\mu_3\text{–}S)_6\{W(CO)_3\}_2]^{3-}$ cluster. The trianionic cluster has similar structure to that of the molybdenum analogous [30]. The mean W-Fe and Fe-Fe bond distances are 2.960 and 2.791 Å.

Structure of another black Fe_6W_2 cluster [32] also contains a trianionic $[\{(EtS)_3Fe_3(\mu_3 - S)_4W\}_2(\mu\text{-}SEt)_3]^{3-}$ cluster. The structure of centrosymmetric anion is that of a double bridged by two SEt groups, double – cubane type. The mean W-Fe and Fe-Fe bond distances are 2.718 and 2.702 Å. The W…W separation is 3.674(3) Å.

Structure of black Fe_6Ni_2 cluster [33] consists of well separated $PhCH_2NEt_2^+$ cations and $[I_6Fe_6(\mu_4\text{-}S)_6Ni_2.(PMePh_2)_2]^{2-}$ anion. The structure of the anion (Fig. 2) contains a cubic $[Fe_6Ni_2S_6]$ cluster with the metal atoms located unambiguously by means of the ligation. As expected, the phosphanes are coordinated to the electron - richer nickel atoms, due to the shorter Fe-Ni (av. 2.645 Å) compared with Fe-Fe (av. 2.719 Å) bond distances. The Fe_6N_2 cube is slightly distorted (M – M - M between 88.0 and 93.5°).

X-ray analysis of dark green Fe_6Re_2 cluster shows [34] that there are well separated NEt_4^+ cations and $[(EtS)_3Fe_3(\mu_3\text{-}S)_4Re_2(\mu\text{-}SEt)_3]^{3-}$ anion. The anion consists two terminal cubane – type $Re(\mu_3\text{-}S)_4Fe_3(SEt)_3$ subclusters bridged by three ethanethiolate groups through the Re atoms. The subclusters are related by an imposed centre of symmetry. The Fe-Re bond distances of 2.739(1) Å is about 0.024 Å longest than the Fe-Fe bond distance (2.715(1) Å . The Re…Re separation of 3.543(1) Å excluded a direct bond.

2.2 Fe_4M_4

Crystallographic and structural data for Fe_4M_4 clusters are summarized in Table 2A. There are eleven clusters (M = non-transition) and seven (M = transition) metal atoms.

Three yellow Fe_4Ga_4 clusters [35] are isostructural contain a Ge_4E_4 (E = S, Se or Te) heterocubane core with an alternating arrangement of gallium and chalcogen atoms. The gallium atoms are coordinated nearly tetrahedrally by three chalcogen atoms and {Fe(CO)(cp)} group. The mean Fe-Ga bond distance elongated with an increases covalent radius of E with the values: 2.337(1) Å (E = S) < 2.347(1) Å (Se) < 2.353(2) Å (Te). The mean Ga – E bond distance elongated with covalent radii of E in the order: 2.375 Å (E = S, 1.02 Å) < 2.501 Å (Se, 1.16Å) < 2.700 Å (Te, 1.36 Å), as expected.

Two orange Fe_4In_4 clusters [36] have a similar structure to those of Fe_4Ga_4 clusters [35]. Each indium atom is a tetrahedrally coordinated. The mean Fe-In as well as In-E bond distances elongated with a covalent radius of E in the orders: 2.493(1) Å (E = S) < 2.504(2) Å (E = Se); and 2.555(3) (E = S) < 2.674(2) Å (E = Se), respectively.

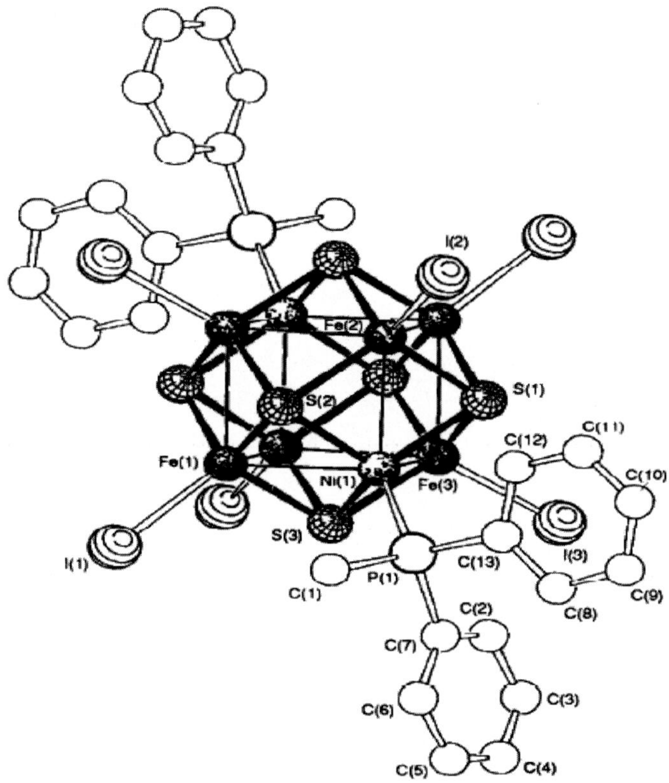

Figure 2. Structure of [I6Fe6(μ4-S)6Ni2(PMePh2)2]2- [33]

Structure of Se_4Sn_4 clusters of the composition [{(cp)$(CO)_2$Fe}$_4Sn_4(\mu\text{-}E)_6$] (yellow E = S [37] and orange E = Se [38]) confirms the semichalcogenide as a tetramer bearing the usual adamantane type. Selenium clusters contains two crystallographically independent molecules differ mostly by degree of distortion. The mean Fe-Sn bond distances are 2.506 Å [37], 2.490 Å (molecule 1) and 2.508 Å (molecule 2) [38] . Each tin atom is tetrahedrally coordinated (SnE_3Fe). The mean Sn-E bond distances, 2.416 Å (E = S) and 2.542 Å (E = Se) follow covalent radius of E.

X-ray analysis of red orange {(cp)Fe(C_5H_4COO)Sn(Bu_2O)}$_4$ [39] adopts an oxo – bridging dimer structure mode with four anisobidentate bridging carboxylate groups as a ligands and has planar tortuous ladder geometry (Fig. 3). The atom Sn1, Sn2, Sn1A, Sn2A, O5 and O5A are coplanar to within ± 0.0238 Å. The atoms Sn1, O5A, O5, Sn2A are coplanar to within ± 0.018 Å. The dihedral angle between the central and terminal four atoms ring is 3.2^{o}. The ester oxygen atoms as a bridge – oxygen coordination atom have a normal linkage (Sn-O 2.121(10) ~ 2.228(12) Å), simultaneously coordinating to an adjacent tin atom with long - range interaction (distances 2.740 and 3.122 Å), while each of the carbonyl oxygen atoms, as terminal – oxygen coordination atoms, coordinates to the same tin atom with long – range interaction (2.710 and 2.888 Å). Each iron atom is sandwiched.

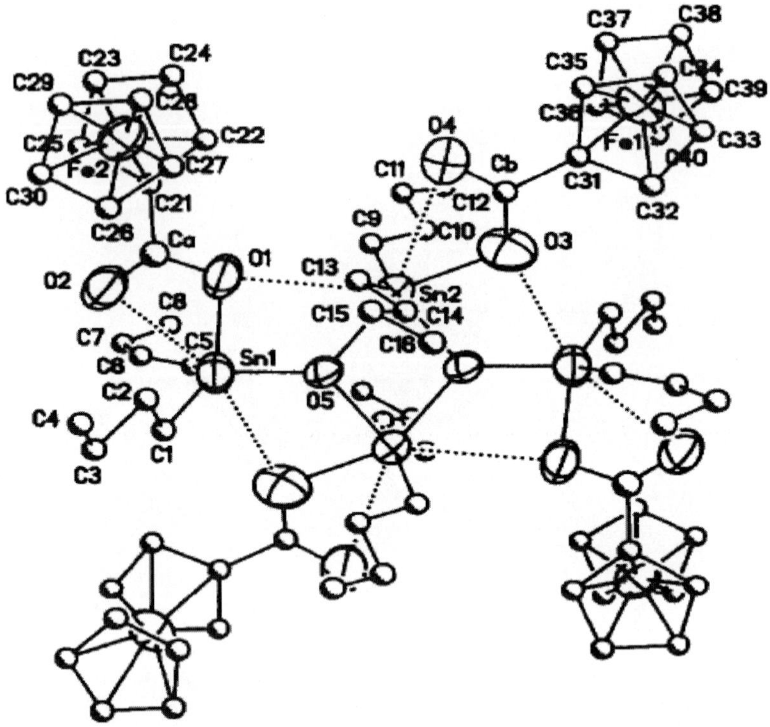

Figure3. Structure of {(cp)Fe)C_5H_4COO)Sn($Bu_2O)_2\}_4$ [39]

Structure of brown – red Fe_4Bi_4 cluster [40] contains well separated NEt_4^+ cations and $[(CO)_4\ FeBi_4\ \{Fe(CO)_3\}_3]^{2-}$ anion. The dianion can be described as a hybrid Zintl ion – metal carbonyl, with three $Fe(CO)_3$ units capping faces of a Bi_4 tetrahedron. An additional $Fe(CO)_4$ fragment is apically bonded to that unique bismuth atom which is bonded to all three $Fe(CO)_3$ units. The Fe-Bi bond distances range from 2.699(6) to 2.753(6) Å (av. 2.725 Å). The dianion cluster as a whole has approximate C_{3v} symmetry.

Yellow orange Fe_4Zn_4 cluster [41] possesses C_1 symmetry with a distorted cubic Zn_4O_4 core made up of interpenetrating tetrahedral of zinc and oxygen atoms with Zn-O bond distances in the range 2.048(5) – 2.099(4) Å. Each oxygen atom of the core is bonded to a methyl group, and each zinc is bonded to the iron atom of a $Fe(CO)_2$(cp) group. The Zn-Fe bond lengths range from 2.344(2) to 2.354(2) Å (2.350 Å average). The cp $Fe(CO)_2$ groups assume a gauche rotational conformation (about the Zn-Fe bond) relative to the three core oxygen atoms adjacent to that bond.

Molecular structure of pale yellow $[(CO)_4FeCd]_4$ [42] consist of a nearly planar, centrosymmetric, eight – membered ring of alternating Cd and cis–$Fe(CO)_4$ units. The geometry deviates from D_{4h} square symmetry principally in that two opposite Fe – Cd – Fe angles are 170.25(5)° and two are 189.85(5)°. The Cd-Fe bond distances are within experimental error all equal at 2.562(2) Å.

The core molecular structure of $[\{cpFe(C_5H_4CH_2NMe_2)Cu\}(\mu_4\text{-}O)(\mu_4\text{-}Cl)_6]$ [43] consists of a central oxo group that is tetrahedrally coordinated to four Cu(II) centers. Each Cu(II) carries monodentate (cp)Fe($C_5H_4CH_2$ NMe_2) ligand and is bridged to three other Cu(II) centers by chlorine ligands. The coordination sphere about copper centers is trigonal bipyramidal [$CuCl_3ON$], with three chlorine donors occupying the equatorial position and the axial sites defined by the central oxo group and the amine donor group of the {cp-$Fe(C_5H_4CH_2NMe_2$} moiety. The copper atom is displaced 0.17 Å from the Cl_3 plane in the direction of the amine donor. Each iron is sandwiched.

There are two Fe_4M_4 (M = Cu or Ag) clusters of the composition [$cpFe(C_5H_4NMe_2)M]_4$ [44] which are isostructural. Four M(I) atoms form plane square . Each cp $Fe(C_5H_4NMe_2)$ moiety serve as bridge between two M(I) atoms by amine donor. Unfortunately, structural parameters are not available.

Structure of pale yellow Fe_4Ag_4 clusters [45] contains well separated $NMeCH_2Ph^+$ cations and $[Ag_4\{\mu\text{-}Fe(CO)_4\}_4]^{4-}$ tetraanion. The molecular structure of the tetraanion consists of an Ag_4 square, spanned on all the edges, by $Fe(CO)_4$ groups. The idealized molecular symmetry is D_{4h}. The Ag-Fe bond distances range from 2.570(2) to 2.600(2) (av. 2.581 Å). The Ag – Ag interactions are distributed over the range 3.036 – 3.334(1) Å.

X-ray analysis of black Fe_4Mo_4 clusters shows [46] that there are well separated PPh_4^+ cations and $[\{(CO)_6Fe_2(\mu_3\text{-}S)_2\}_2\ \{Mo(O)(\mu\text{-}O)_2Mo(O)_2\}(\mu\text{-}OMe)_2]^{2-}$ dianion. The centrosymmetric anion is comprised of two cis-Mo_2O_4 moietes bridged by two methoxide ligands and chelated at the terminal Mo atoms by one $Fe_2S_2(CO)_6$ unit per Mo. The Mo-Mo bond distance is 2.592(1) Å. The Fe-Fe bond distance in $Fe_2S\ _2(CO)_6$ unit is 2.503(8) Å. The Fe_2S dihedral angle is 79.2°.

The $[(\mu_3\text{-}S)_4Fe_4\{(\mu\text{-}NC)W(CO)_4\}_4]^{2-}$ dianion [47] is highly symmetrical a crystallographic two fold axis passes through the centers of two opposing faces of the Fe_4S_4 cube. The mean Fe-Fe bond distance is 2.733 Å. Each CN^- group serve as bridge between (μ_3-S)$_4$Fe$_4$ cube and

$W(CO)_4$ groups, with the mean Fe-N and W-C bond distances of 1.915 and 2.143 Å, respectively.

Black $[I_4Fe_4(\mu_4\text{-}S)_6Ni_4(PMePh_2)_4]$ 2 MeCN [33]contains cubic {M_8S_6} cluster. The cube is slightly distorted (M – M – M between 87.7 – 92.5°). The overall oxidation state of the metals is +2 and hence the number of metallic valence electrons is 108. As expected the phosphanes are coordinated to the electron –richer nickel atoms. The mean Fe-Ni bond distance is 2.650 Å.

2.3 Fe_3M_5

There are three Fe_3M_5 (M = Hg [48], Ni [49] or Ru [50]) clusters and their isostructural parameters are gathered in Table 2B. Structure of yellow $[\{(CO)_4Fe\}_3Hg_5(\mu\text{-}PBu^t_2)_2(\mu\text{-}Br)_2]$ [48] has pseudosymmetry C_{2v}. The structure consists of $Hg_3\ Fe(CO)_4)_2$ [Hg(3)-Hg(5), Fe(1), Fe(2)] and $Fe(CO)_4(Hg)_2$ [Hg(1), Hg(2), Fe(3)] fragments, connected by two PBu^t_2 groups [P(1), P(2)]. There is wide variety of Fe – Hg – Fe/P bond angles.: Fe(1) – Hg(3) – P(1) 145(1)°, Fe(2) – Hg(5) – P(2) 152.4(1)°; Fe(3) – Hg(1) – P(2) 159.1°; Fe(2) – Hg(4) – Fe(1) 169.9(1)°; and Fe(3) – Hg(2) – P(1) 176.8(1).

Structure of black Fe_3Ni_5 cluster [49] contains well separated $Et_3NCH_2Ph^+$ cations and $[(I)_3Fe(\mu_4\text{-}S)_6Ni_5I_5]^{4-}$ tetraanion. The cluster anion with approximate O_h symmetry has crystallographic C_{4h} symmetry. The eight cubically arranged metals atoms together with circumscribing S_6 octahedtron form a rhombododecahedron. The metal – metal distances are 2.681 Å (mean value).

Structure of brown $[(CO)_7Fe_3(\mu_6\text{-}C)(\mu_5\text{-}C)Ru_5(\mu\text{-}PPh_2)_2(CO)_{10}]$ $CHCl_3$ [50] containing a square – pyramidal $Fe_2Ru_3(\mu_5\text{-}C)$ cluster fused to an octahedral $Fe_3Ru_3(\mu_6\text{-}C)$ cluster via a shared Fe_2Ru face (Fig. 4). The two carbide atoms, C(A) and C(B), are found in the center of octahedral and the base of the square pyramidal clusters, respectively C(B) being 1.15(2) Å below the Fe_2Ru_2 plane. Opposite Ru(1)Ru(2) and Ru(3) Ru(4) edges are bridged by P(1) and P(2), respectively, of the two PPh_2 groups. The "outer" Fe-Ru, Fe-Fe and Ru-Ru bond distances of 2.85, 2.75 and 2.78 Å; are longer than the "inner", Fe-Ru, 2.69 Å and Fe-Fe 2.65 Å.

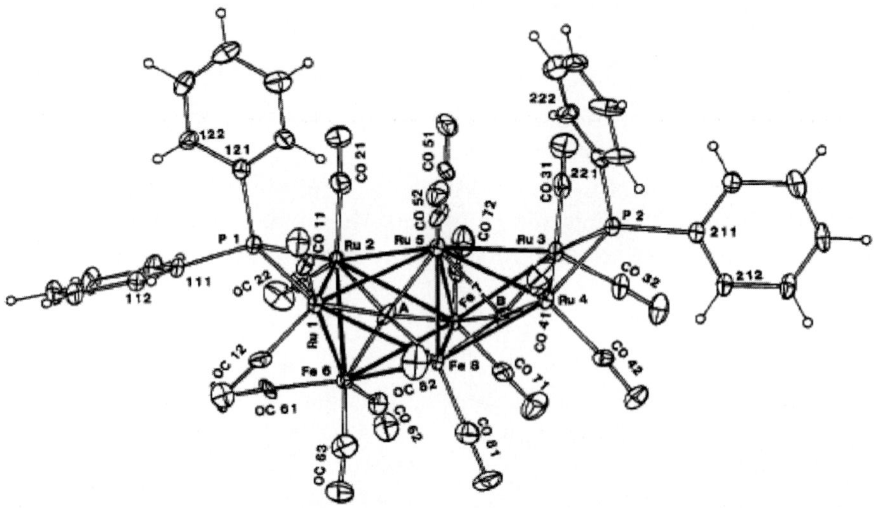

Figure 4. Structure of $[(CO)_7Fe_3(\mu_6\text{-}C)(\mu_5\text{-}C)Ru_5(\mu\text{-}PPh_2)_2(CO)_{10}$ [50]

2.4 Fe_2M_6

There are two Fe_2M_6 (M = Sn [51], Ru [52]) clusters (Table 2C). Structure of orange – red Fe_2Sn_6 cluster [51] contains the $Sn_6(\mu_3\text{-}O)_4(\mu_3\text{-}OSiMe_3)_4$ core. Each $Fe(CO)_4$ group is coordinated to the tin(II) atom (Fe-Sn 2.463(1) Å). Two Sn(II) atoms are pentacoordinated (SnO_4Fe). Remaining the tin(II) atoms are tetrahedrally coordinated (SnO_4).

Structure of Fe_2Ru_6 cluster [52] comprises two C_2 ligands and an octametallic framework. There are two crystallographically independent molecules differ mostly by degree of distortion. Every edge of the Ru_4 square (Ru 1-4) is bridged by iron or ruthenium and each C_2 moiety above and below the Ru_4 square interacts with a boat – like hexanuclear array. Of the six Ru atoms, Ru5 and Ru6 in the apical position lie closer to the C_2 ligands. The Fe–Ru metal bond distances range from 2.687(4) to 2.855(3) Å and Ru-Ru bond distances from 2.746(3) to 2.857(4) Å.

2.5 $Fe_4M_2M_2'$

There are three $Fe_4M_2M_2'$ (M = Sn and M' Cr [8]; M = Hg and M' Co [53]; M = Cr and M'= Co [54] clusters and their structural parameters are summarized in Table 2D. Structure of red $[cp_2(CO)_2Fe_2SnCr(CO)_4(\mu\text{-}CO)]_2$ [8] is isocarbonyl – bridged dimer such that one cis – unit is joined to the other $\{Cr(CO)_n(SnFecp_2)\}$ unit via an O-Sn linkage, resulting in an eight membered Sn – O – C - Cr – Sn – O – C - Cr ring. The mean Fe-Sn bond distance of 2.560 Å is about 0.064 Å shorter than those of Sn-Cr bond distances (av. 2.624 Å).

Molecule of $[(CO)_2Co(\mu_3\text{-}Cl)(\mu\text{-}CO)Fe_2(CO)_6Hg(\mu\text{-}S)]_2$ [53] has C_{2h} symmetry. Two $\{(CO)_2Co(\mu\text{-}Cl)(\mu\text{-}CO)Fe_2(CO)_6Hg\}$ fragments are connected by two sulphide atoms. The Fe_2Co atoms are bridged with μ_3-Cl atom to form a trigonal pyramidal cluster. There are metal-metal bonds with mean values of 2.575 Å (Fe-Co), 2.631 Å (Fe-Hg) and 2.855 Å (Fe-Fe').

Structure of brown $Fe_4Cr_2Co_2$ cluster [54] is shows in Fig. 5. The central fragment is a distorted cube $Cr_2Co_2(\mu_3\text{-}S)_4$ and its Cr_2Co_2 metal core has the structure of a butterfly with one Cr-Cr bond (2.772(3) Å, four Cr-Co (2.704(2) – 2.739(2) Å) and a non – bonded Co…Co (3.160(1) Å) separation. The dihedral angle between the Cr_2Co rings is 85.11°. One Co atom is bonded to the $(S)_2Fe_2(CO)_6$ fragment (Fe-Fe 2.575(2) Å via two Co-S σ-bonds and one Fe(1)-Co(1) bond 2.645(2) Å, while the second Fe(2)…Co(1) is absent (3.266(1) Å. The second Co atom is bonded to the $SFe_2(CO)_6$ fragment by the S → Co (2.194(4) Å) and two Co → Fe (2.571(2) Å and 2.584(2)Å) dative bonds.

2.6 $Fe_2Ge_2Co_4$

X-ray analysis of dark orange $Fe_2Ge_2Co_4$ cluster shows [55] it to have a metal framework made up of four linked GeM_2 triangles with two spiro Ge atoms. Each of the Co-Co (2.491(5) and 2.528(5) Å) and Fe-Fe (2.678(5) Å) bonds are bridged by CO ligands, and each transition metal carries three terminal CO groups (Table 2E). Ignoring irregularities the overall symmetry is approximately C_2, with two-fold axis coincident with the μ-CO across the

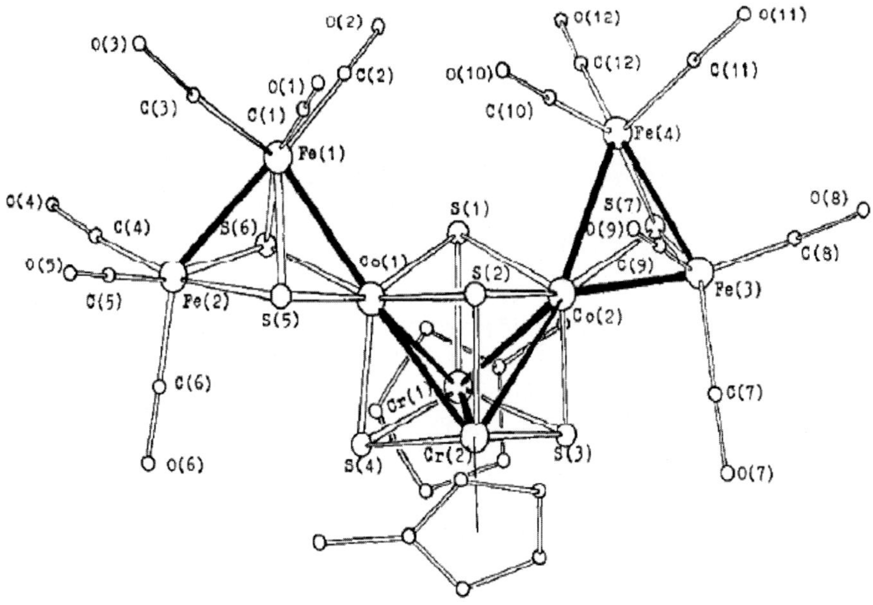

Figure 5. Structure of $cp_2Cr_2(\mu_3\text{-}S)_4Co_2\{(\mu_3\text{-}S)_2Fe_2(CO)_6\}\{(\mu_3\text{-}S)Fe_2(CO)_6\}$ [54]

Fe-Fe bond. The central position of the molecule is clearly related to $\{Fe_2(CO)_9\}$ with two of the bridging CO groups replaced by bridging $Ge[Co_2(CO)_7]$ units.

In the Ge_2Fe_2 unit the edge – linked triangles are unsymmetrical so that Ge(1) lies closer to Fe(1) while Ge(2) lies closer to Fe(2); the individual Ge-Fe bond distances range from 2.365(4) to 2.437(4) Å. The dihedral angle between the two $GeFe_2$ triangles is only 111°, with a non-bonded Ge…Ge distance of 3.302 Å.

In the two $\{GeCo_2(CO)_7\}$ ends in the complex the Ge(1)Co(1)Co(2) triangle is only slightly unsymmetrical (Ge(1)-Co(1) 2.283(5) Å; Ge(1)-Co(2)2.351(4) Å, Co(1)-Co(2) 2.528(5) Å); However the Ge(2)Co(3)Co(4) triangle is strongly distorted, with the Ge-Co bonds (Ge(2)-Co(3) 2.309(5), Ge(2)-Co(4) 2.399 Å) and Co(3)-Co(4) of 2.491(5) Å. The dihedral angles between the triangles sharing a common apex at the spiro –Ge atoms also differ (84 and 88°).

Inspection of data in Tables1 and 2 reveals that there are six categories of the clusters (Fe_6M_2 38 examples), Fe_4M_4 (18 examples), Fe_3M_5 (3 examples), Fe_2M_6 (2 examples), $Fe_4M_2M'_2$ (3 examples), and $Fe_2Ge_2Co_4$. The heterometals include the non-transition and transition metals.

Table 3A. Summary of Fe-M Bond Distances (< 3.0 Å) in Heterooctameric Iron Clusters

Fe – M	Shortest [ref]	Longest [ref]	Average
Fe – Ga	2.337(2) [35]	2.353(2) [35]	2.346
Fe – Zn	2.344(2) [41]	2.354(2) [41]	2.350
Fe – Ge	2.365(4) [55]	2.452(1) [7]	2.450
Fe – Sn	2.463(1) [51]	2.612(7) [51]	2.540
Fe – Cd	2.599(2) [42]	2.562(2) [42]	2.562
Fe – Sb	2.471(1) [9]	2.702(2) [11]	2.579
Fe – Hg	2.627(2) [8]	2.635(2) [8]	2.631
Fe – In	2.493(2) [36]	2.708(2) [12]	2.652
Fe – Bi	2.699(2) [40]	2.753(2) [40]	2.752
Fe – Tl	2.746(1) [13]	2.765(2) [13]	2.755
Fe – Ag	2.570(2) [45]	2.600(2) [45]	2.581
Fe – Co	2.568(4) [53]	2.645(2) [54]	2.590
Fe – Ni	2.627(1) [33]	2.685(1) [49]	2.658
Fe – Re	2.739(1) [34]		
Fe – V	2.752(1) [14]		
Fe – Mo	2.503(8) [46]	2.99(3) [29]	2.768
Fe – Ru	2.644(2) [50]	2.913(2) [50]	2.789
Fe – W	2.718(2) [31]	2.990(2) [31]	2.919

A summary of the Fe-M bond distances in the heterooctameric derivatives is given in Table 3A. The Fe-M (non-transitions) and Fe-M (transition) bonds are given in the order of elongated average Fe-M bond lengths. The shortest Fe-M (non-transition) and Fe-M (transition bonds) are 2.337(2) Å for Fe-Ga [35] and 2.503(8) Å for Fe-Mo [46].

The summary of the M-M and M-M' bond distances are given in Table 3B, where the shortest are 2.307(1) Å for Cr-Cr [15] and 2.351(4) Å for Ge-Co [55]. The shortest Fe-Fe bond is 2.546(3) Å [54].

Table 3B. Summary of M-M and M-M'' Bond Distances (< 3.0 Å) in Heterooctanuclear Iron Clusters

M – M	Shortest [ref]	Longest [ref]	Average
Co – Co	2.491(5) [55]	2.528(5) [55]	2.510
Cr – Cr	2.307(3) [15]	2.772(3) [54]	2.539
Mo – Mo	2.592(1) [46]		
W – W	2.674(3) [32]		
Ni - Ni	2.677(1) [49]	2.685(1) [49]	2.681
Fe – Fe	2.546(3) [54]	2.862(7) [7]	2.734
Ru – Ru	2.733(2) [50]	2.946(2) [50]	2.832
Ge – Co	2.351(4) [55]	2.399(4) [55]	2.360
Sn – Cr	2.642(1) [8]		
Cr – Co	2.704(2) [54]	2.739(2) [54]	2.717

The $[(Co)_8Fe_2SbFe_2(CO)_6SbFe_2(CO)_8]$ exists in two isomeric forms, monoclinic [9] and orthorhombic [10]. Two clusters, $\{(cp)(CO)_2Fe\}_4Sn_4(\mu\text{-}Se)_6$ [38] and $cp^*{}_2Fe_2(\mu_6\text{-}\eta^2\text{-}C_2)_2(\mu\text{-}CO)_3Ru_6(CO)_{14}$ [52] contain two crystallographically independent molecules. The isomers as well as the independent molecules differing only by degree of distortion in the M-L bond distances and L – M – L angles. Such examples are typical of the general class of distortion isomerism [56].

HETERONONANUCLEAR IRON CLUSTERS

There are twenty-two heterononanuclear iron clusters and their structural parameters are given in Table 4. There are seven categories of the heterononanuclear derivatives Fe_8M(x2), Fe_7M_2(x9), Fe_6M_3(x3), Fe_4M_5(x3), Fe_3M_6(x2), FeM_8(x2) and $Fe_2Co_5Ge_2$(x1).

Fe_8M

Two Fe_8M (M = Cs [57], or Au [58]) are only examples of such clusters. In the structure of amber $[Fe_8\{\mu_3\text{-}\eta^4\text{-}N(CH_2CH_2O)_8\}_8Cs]Cl.2CH_2Cl_2$ [57], the eight iron atoms of the approximately centrosymmetric are located in the corners of a nearly regular octagon. The Fe…Fe separations are 3.143 and 3.162 Å and the impair angles of the octagon range from 133.33 to 135.17°. The diameter of the octagon, defines, as the mean distance between opposite iron atoms is 8.224 Å. The distorted octahedral arrangement of the iron atoms is composed of one N donor, one μ_2- and two μ_2- and μ_3- oxygen donors. Three sets of eight of a total of twenty four oxygen atoms are located in the corners of three pairs of a square faces that are rotated 45° relative to each other. The square faces are located parallel and atom equidistant in pairs, and one plane lies above and one plane lies below the octagon generated by the iron atoms, these structural elements share a common S_8 axis. The Cs(I) lies directly above the center of the cation. The eight μ_3-O atoms form a quadratic antiprism coordination sphere around the Cs(I) atom.

Structure of dark orange Fe_8Au cluster [58] consist of $[Au(PMePh_2)_2]^+$ cation and$[\{(CO)_{12}(\mu\text{-}H).Fe_4(\mu_5\text{-}BH)\}_2Au]^-$ anion. Complex anion contains two $\{HFe_4(CO)_{12}BH\}$ fragment fused in "face-to face" manner via a single gold(I) atom which bridges one Fe wingtip –B edge of each cluster unit. The geometry of each butterfly fragment varies little. The internal dihedral angle of the Fe_4 skeleton is 116.6(3)°. The two clusters subunits are twisted with respect to one another with a spiro twist angle of 30.9(5)° at the gold(I) atom. The Fe-Au bond distance is 2.615(1) Å. The Fe-Fe bond distances range from 2.630(2) to 2.668(2) Å (av. 2.651 Å). In complex cation $[Au(PMePh_2)]^+$, the geometry is unexceptional with linear coordination at the Au(I) center, < P – Au – P = 180.0(1)°.

Fe_7M_2

There are nine Fe_7M_2 clusters (M = Ge [7] , Mo [20, 25-62], W [61], or Re [34]). In dark brown Fe_7Ge_2 cluster [7] two μ-Ge$\{Fe_2(CO)_8\}$ units bridging two of the edges of the central Fe_3 triangle, whole the third edge is bridged by a nearly symmetrically placed CO (Fig. 6). The central Ge_2Fe_5 skeleton is essentially planar, with the outer $GeFe_2$ triangles virtually perpendicular to this plane. None of the eight Fe-Ge bonds differ significantly from the mean value of 2.40 Å. The Fe-Fe bond distances range from 2.650(9) to 2.845(9) Å.

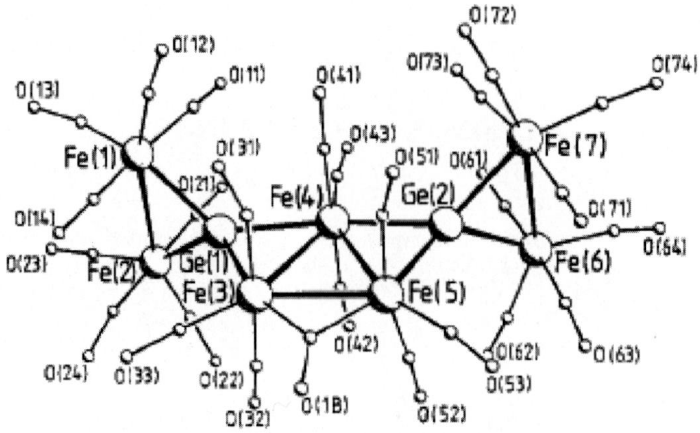

Figure 6. Structure of $(CO)_6Fe_2GeFe_3(\mu\text{-}CO)(CO)_9GeFe_2(CO)_6$ [59]

There are eight Fe_7M_2 (M 'Mo [20, 59-60], W [61], or Re [34]) clusters which contain $[\{(RS)_3Fe_3(\mu_3\text{-}X)_4M(\mu\text{-}SR)_3\}_2Fe]^{n-}$ (X = Se [20] or S [34, 59-61]; n = 2 [34], 3 [20, 59-61] or 4 [34, 59-61] anion. The anions have an imposed center of symmetry and are of the "double - cubane" type. Each anion contains two $MFe_3X_4(SR)_3$ clusters with trigonally distorted MFe_3S_4 cores which are connected through their M atoms by $M(SR)_3Fe(SR)_3M$ bridging units. The dimensions of the cores do not exhibit any chemically significant differences. The primary dimensional differences inside in the $Fe(SR)_6$ bridge subunits which possess distorted trigonal – antiprismatic coordination. These differences were satisfactorily interpreted in term of high spin Fe(II) in [34, 59-61] and Fe(III)(probably low-spin) in [20, 34, 61]. The mean Fe(II)-M bond distance elongated with the covalent radius of M (in parenthesis) in the order 2.725 Å (Mo, 1.45Å) < 2.728 Å (W, 1.46Å) < 2.758Å (Re, 1.59 Å). The mean Fe(II)-Fe(II) bond distance follows the trend: 2.692 Å (in clusters where M is Mo) < 2.695Å (W) < 2.707 Å (Re).

Structure of dark yellow $[\{(CO)_9Fe_3(\mu_3\text{–}S)_3Mo(CO)_2(\mu_4\text{-}S)\}_2Fe]$ [62] is shows in Fig. 7. The structure consists of two fragments $(CO)_9Fe_3(S)_4Mo(CO)_2$ in which each atom has three terminal CO groups, and each molybdenum atom is bonded to two CO groups. Each of the $Fe(CO)_3$ units is bonded to the molybdenum atom via the bridging sulphur atoms and three short Fe-Mo bonds (mean 2.762(2) Å). The molybdenum atoms in the tetranuclear fragments are bonded with each other via the central Fe by two elongated Mo-Fe (2.900(2) Å) bonds. It is also bonded to the six sulphur atoms by four short Fe-μ_4S bonds (mean 2.268(3) Å and two long Fe-μ_4S bonds (mean 2.529(2) Å). There are no Fe-Fe bonds in the cluster.

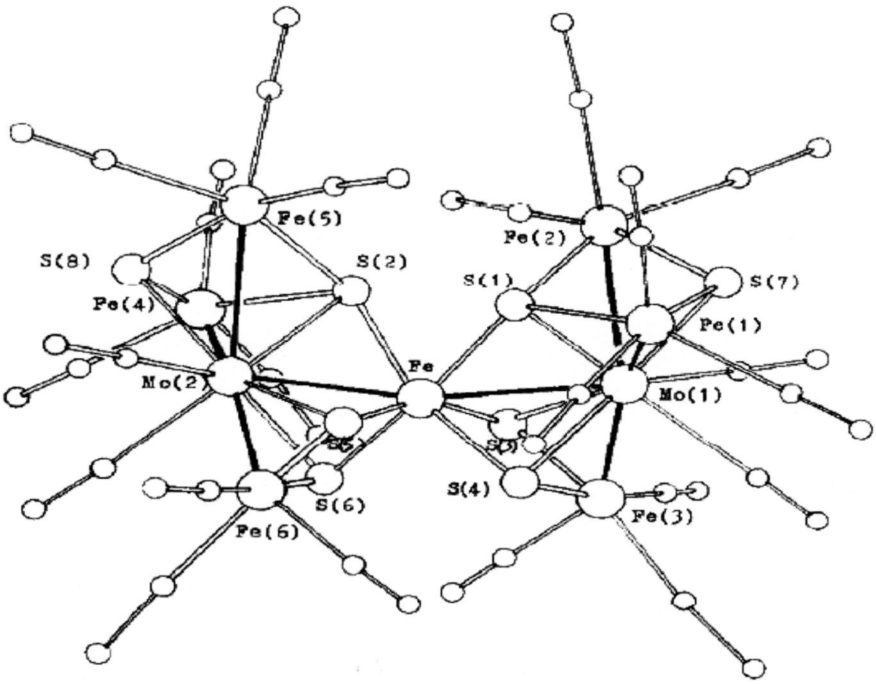

Fig.ure 7. Structure of $[\{CO)_9Fe_3(\mu_3\text{-}S)_3Mo(CO)_2(\mu_4\text{-}S)\}_2Fe]$ [62]

3.3 Fe_6M_3

There are three Fe_6M_3 (M = Bi [63.64] or Au [65]) clusters (Table 5). Structure of dark green [{(L)(CO)$_2$Fe}$_2$ Bi(μ-Cl)]$_3$ L = C_5H_4Me [64] containing a planar six – membered Bi_3Cl_3 ring in which each bismuth is bonded to two chlorine atoms and two (L) $Fe(CO)_2$ fragments. The mean internuclear Bi…Bi separation is 3.95 Å. The mean Fe-Bi bind distance is 2.684 Å.

In dark green Fe_6Au_3 cluster [65] three Fe_2 Au units are linked by the three – legged propeller – like structure of the triphos ligand. Interestingly, the gold atoms are placed at the corners of an almost equilateral triangle. The mean Fe-Au bond distance of 2.689 Å is about 0.083 Å longer than the mean Fe-Fe bond distance (2.606 Å).

Table 5. Summary of Fe-M, M-M and M-M' Bond Distances (< (3.0 Å) in Heterononanuclear Iron Clusters

Fe - M	Shortest [ref]	Longest [ref]	Average
- Ge	2.35(1) [71]	2.415(8) [7]	2.384
- Cu	2.387(3) [66]	2.491(2) [66]	2.450
- Co	2.62(1) [71]	2.64(1) [7]	2.63
- Ag	2.585(1) [45]	2.727(1) [45]	2.671
- Bi	2.666(3) [63]	2.710(2) [64]	2.684
- Au	2.600(6) [65]	2.922(2) [68]	2.717
- Re	2.680(4) [34]	2.776(4) [34]	2.728
- W	2.709(6) [61]	2.738(6) [61]	2.728
- Mo	2.715(1) [60]	2.901(2) [62]	2.763
- Cr	2.575(5) [69]	2.819(5) [69]	2.788
Cu – Cu	2.527(2) [66]	2.691(2) [66]	2.582
Ir – Ir	2.698(2) [70]	2.890(2) [70]	2.734
Cr – Cr	2.774(5) [69]	2.839(5) [69]	2.804
Fe – Fe	2.60(1) [70]	2.845(9) [7]	2.808
Ag – Ag	2.795(1) [45]	2.980(1) [67]	2.878
Ge – Co	2.31(1) [71]	2.39(1) [71]	2.35

3.4 Fe_4M_5

There are two yellow Fe_4M_5 (M = Cu [66] or Ag [45]) clusters. Structure of $[\{(CO)_4Fe\}_4Cu_5]^{3-}$ [66] showed the metal atoms to be arranged in a nearly perfect 3x3 rhombus with $Fe(CO)_4$ groups occupying the four corners and copper atoms located in the remaining position. The metal atoms thus form a small section of a close-packed surface. There are two different types of iron and copper atoms in the clusters, the central Cu atom having six nearest neighbors in the plane. The central Cu is 2.527(2) Å from the copper atom on the adjacent edge, whereas the edge lying coppers are 2.691(4) Å from the copper atom on the adjacent edge. The two iron atoms, which defined the short diagonal of the rhombus bridge three coppers and are, located 2.410(3) Å from the Cu's on the edge and 2.491(3) Å from the central copper atom. The remaining two iron atoms bridge two Cu's with a Fe-Cu bond distance of 2.39Å.

Structure of $[Ag_5\{\mu\text{-}Fe(CO)_4\}_2\{\mu\text{-}Fe(CO)_4\}_2]^{3-}$ [45] contains a planar cluster of nine metal atoms packed in a rhombus. The iron atoms define the vertices and the silver atoms are placed almost midway along the edges and in the center. Both the idealized and crystallographic symmetry of the anion is D_{2h}. The silver atoms define a bow tie with four equivalent atoms at the corners and unique central atoms. The latter is six – coordinated to four silver and two iron atoms. (Ag(1)–Ag(2) 2.975(1) Å and Ag(1)-Fe 2.727(1) Å. The corner atoms are four – coordinated with two Ag…Ag and two Ag…Fe interactions. The Ag…Ag distance bridged by the $Fe(CO)_4$ group of 3.017(1) Å is longer than the values above. The $FeCO)_4$ groups form two independent pairs, both form to a precise C_{2v} symmetry. The Fe(1) and Fe(1)' are three connected to the silver atoms resulting in the distorted pentagonal-bipyramid (FeC_4Ag_3), The Fe(2) and Fe(2)' are two connected and realize octahedral like coordination (FeC_4Ag_2). The actual values Ag-Fe are 2.585(1) Å for the six – Fe(2) coordination and 2.650(1) and 2.727(1) Å for the seven coordination Fe(1).

3.5 Fe_3M_6

There are three Fe_3M_6 (M = Ag [67], Au [68] or Cr [69]) clusters (Table 5). The molecule of dark red $[\{(CO)_4Fe\}_3Ag_6\{\mu\text{-}\eta^3\text{-}CH(PPh_2)_3\}].2Me_2(CO)$ [67] has no crystallographically imposed symmetry, but the skeletal atoms have approximately C_3 symmetry (Fig. 8). The silver atoms define a distorted octahedron with the trigonal $CH(PPh_2)_3$ ligand coordinated to three silver atoms on the face of the octahedron. These silver atoms {Ag(I), Ag(2) and Ag(3)} are separated by 3.288(1) Å. The remaining Ag…Ag distances lie in the range 2.817(1) – 3.065(1) Å (mean 2.909(1) Å. The three $Fe(CO)_4$ fragments cap three of the faces of the octahedron in a symmetrical fashion. The Fe-Ag bond distances lie in a narrow range 2.66(1) – 2.720(1) Å. In view of the long Ag…Ag distances associated with the silver atoms coordinated to the tripodal ligand, the cluster geometry can be described as a triangle of vertex linked Ag_3Fe tetrahedra.

Structure of yellow $[\{Fe(C_5H_4PPh_2\}_3\{(\mu_3\text{-}S)Au_3\}_2].8CHCl_3$ [68] consists of an S_2Au_6 core with Au-Au distances vary from 2.905(2) to 3.272(2) Å. The shortest contacts, Au(1)-Au(4) 2.905(2) Å, Au(1) – Au(2) 2.922(2), and Au(4) – Au(5) 2.920(2) Å, involve gold atoms of the $[Au_2dppf]^{2+}$ fragment that are bonded to the same sulphur atom. The shortest Fe…Au separation is 4.284 Å.

In black Fe_3Cr_6 cluster [69] two tetrahedra $cp_3Cr_3(\mu_3\text{-}S_4)Fe$ (Cr-Cr 2.774(5) – 2.824(4) Å, Fe-Cr 2.779(4) – 2.805(3) Å) are held together by $(\mu_3\text{-}S)Fe(CO)_3$ fragment, from which μ_3-S as well as iron atom serve as a bridges with Fe-Fe bond distances of 2.649(2) and 2.679(3) Å.

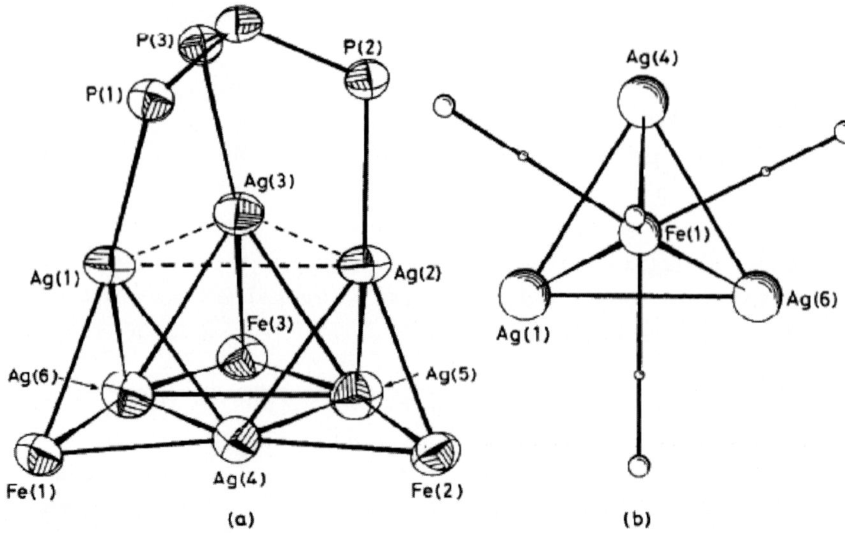

Figure 8. Structure of skeletal geometry of [{(CO)$_4$Fe}$_3$Ag$_6${(CH(Ph$_2$)$_3$}] in (a). For reason of clarity the phenyl rings and carbonyls have been omited. In (b) the geometry ofa single Ag$_3$Fe(CO)$_4$. [67]

3.6 FeM$_8$

Brown FeIr$_8$ cluster [70] is only example of this class and its structure is in Fig 9. The structure consists of two Ir$_4$ tetrahedra linked by [Fe(cp){P$_3$CH$_2$(CMe$_2$)CBu^t}] via P(1) to Ir(1) and both P(2), that is not adjacent to P(1), and the CH$_2$ group to Ir(6). The metal-metal bond distances in the Ir$_4$(CO)$_{11}$ unit range from 2.698(2) to 2.748(2) Å) (average 2.719 Å).

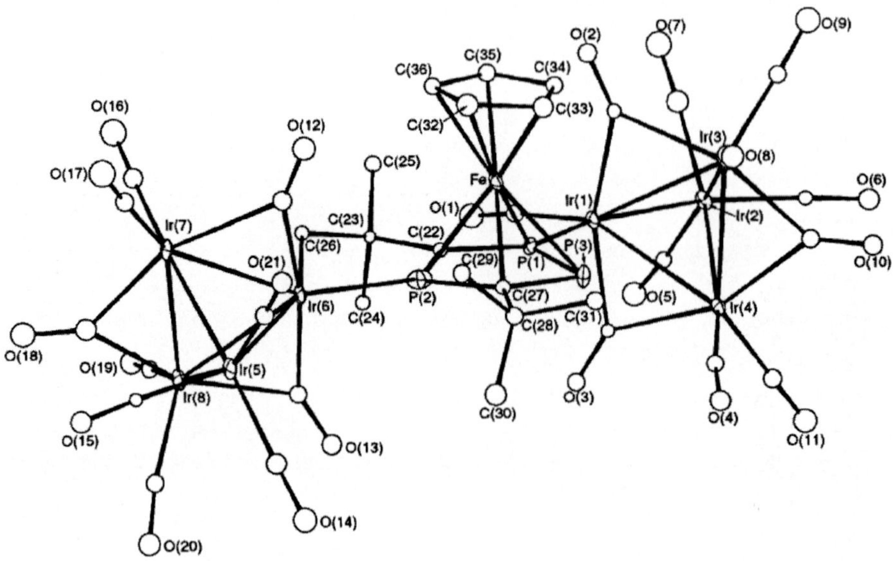

Figure 9. Structure of [{H(μ-CO)3Ir4(CO)3}Fe(cp){PCH2(CMe)2CBut}.Ir(μ-CO)3(CO)8] [70].

3.7 $Fe_2Ge_2Co_5$

The unit cell of $Fe_2Ge_2Co_5$ cluster [71] contains $[(CO)_6(\mu\text{-}CO)Fe_2\{Co(CO)_3\}.Ge\{Co(CO)_3\}Ge\{Co(CO)_3\}_3]^{2-}$ anions and two independent $[NEt_4]^+$ cations at crystallographic inversion centers. The anion can be viewed as a trans $L_2Co(CO)_3$ species with $GeCo_3(CO)_9$ and $GeCoFe_2(CO)_{10}$ groups coordinated to the planar $Co(CO)_3$ unit via the germanium atoms. The Ge-Co bonds to the central Co(1) are 2.30 Å (Table 5).

A summary of the Fe-M, M-M and M-M' bond distances in the heterononameric derivatives is given in Table 5. The metal – metal bonds are given in the order of elongated average metal – metal bond lengths. The shortest Fe-M (non-transition) and Fe-M (transition) bonds are 2.35(1) Å for Fe-Ge [71] and 2.387(3) Å for Fe-Cu [66].

The average Fe-M bond lengths in the heterononametallic iron derivatives are somewhat shorter than those found in the heterooctameric iron derivatives when M is Ge, Bi or Re; but opposite is true when M is Co, W or Mo. The mean value Fe-M for the former are: 2.35 vs 2.405 Å (M = Ge), 2.684 vs 2725 Å (Bi); 2.728 vs 2.739Å (Re); and 2.630 vs 2.590 Å (Co), 2.728 vs 2.919 Å (W) and 2.763 vs 2.768 Å (Mo).

Heterodeca – And Heterooligonuclear Iron Clusters

Crystallographic and structural data for heterodeca - and heterooligonuclear iron clusters are gathered in Table 6.

4.1 Heterodecanuclear Iron Clusters

There are four categories of the heterodecanuclear clusters, Fe_8M_2 (M = V [72] or Ru [73]), Fe_6Ru_4 [74], Fe_4Pt_6 [75] and $Fe_2Ge_4Co_4$ [76], Table 6A). Structure of black Fe_8V_2 cluster [72] contains two cubane $[Fe_4(\mu_3\text{-}S)_3(\mu\text{-}S)_3V(PEt_3)_4]^+$ fragments which are connected by a single sulphur with the mean Fe-S (bridge) bond distance of 2.192(2) Å and Fe-S-Fe bridge angle of 134.0(1)°. The Fe…Fe separations of 4.03 Å excluded bond.

Structure of black Fe_8Ru_2 is shown in Fig. 10. The structure is composed of a $\{Ru(SPMePh_2)(MeCN)_4\}^{2+}$ fragments bridged via μ_5 – sulphur owing to Ru(II) coordination the rhombododecahedral $\{Fe_8S_6\}^{4+}$ core exhibits a tetragonal distortion, i.e. an elongation of the cube parallel to the Ru-Ru axis. The mean values of the Fe-Fe bond distances parallel and perpendicular to this axis are 2.819 and 2.671 Å, respectively.

Structure of red-orange Fe_6Ru_4 cluster [74] has C_2 symmetry and consists of four Ru atoms bridged by a central μ_4-S. Two of the Ru-Ru vectors are bridged by single symmetrical SPhFecp bridges and the other two are each bridged symmetrically by two SPhFecp moieties.

Structure of Fe_4Pt_6 cluster [75] has an overall idealized D_{2h} symmetry. The pseudo – C_2 axis of two trapezoidal $\{Fe_2Pt_3(CO)_{11}\}^-$ units to give rise to a tetrahedral core of Pt atoms. The mean Fe-Pt and Pt-Pt bond distances are 2.568 and 2.733 Å, respectively.

Molecule of orange $Fe_2Ge_4Co_4$ cluster [76] lies on a crystallographic two-fold axis coincident with the bridging CO group and bisecting the Fe-Fe bond. It has a chain of six linked GeM_2 triangles. The molecule contains central Fe_2Ge_2 core. (Fe-Fe 2.641(4) Å) and two outer Ge_2Co_2 parts (Co-Co 2.636(3) Å). The mean Ge-Co and Ge-Ge bond distances are 2.365(3) Å and 2.471(2) Å, respectively. The Ge…Ge separations are 3.132 Å.

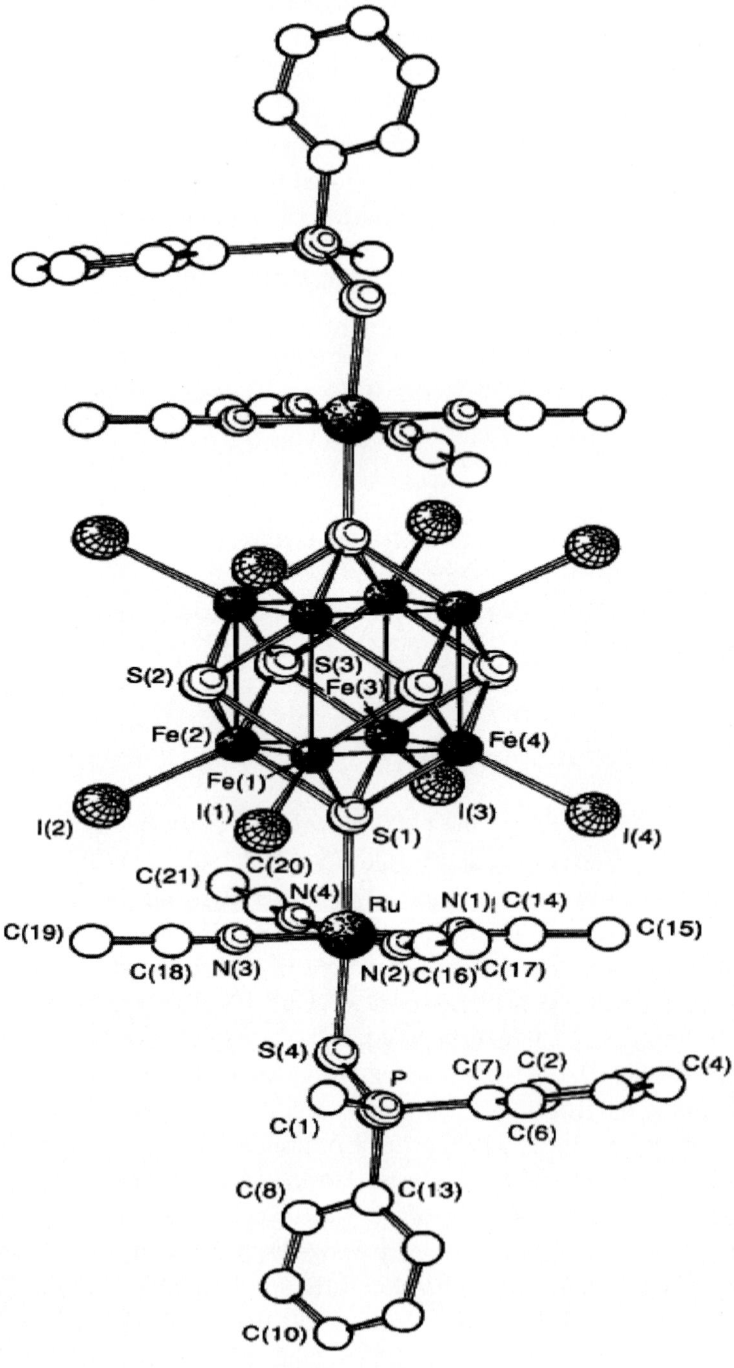

Figure 10. Structure of $[\{(Ph_2MePS)(MeCN)_4Ru(\mu_5\text{-}S)Fe_4Ir_4\}_2(\mu_4\text{-}S)_4]$ [73].

Heteroundecanuclear Iron Clusters

There are four such clusters, Fe_8Hg_3[77] Fe_6Cu_5 [78], $FeOs_{10}$ [79] and Fe_8Au_2K [80] for which structural data are available (Table 6). In brown [{$(CO)_{10}$(μ-COMe)Fe_3HgFe. $(CO)_4\}_2$Hg] [77] the eleven metal atoms are connected in this manner.

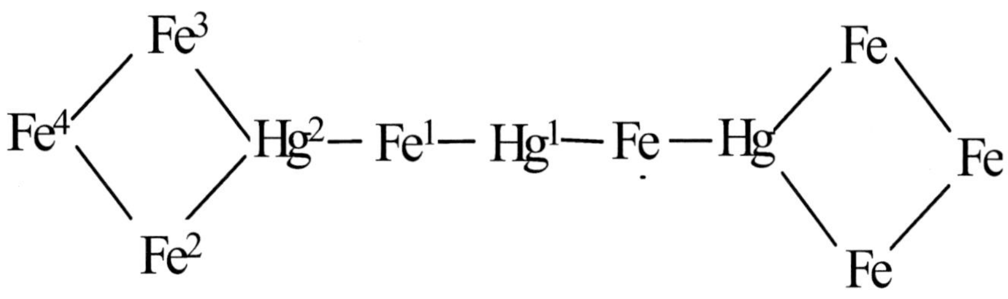

The central Hg(1) atom lies on a crystallographic inversion centre, so that Fe(1)-Hg(1)-Fe(1') angles is exactly 180°. The cluster has exact C_i symmetry. The butterfly dihedral angle Hg(2)-Fe(2)-Fe(3)-Fe(4) is 127.9(1)°. Two Fe_3(μ-COMe)$(CO)_{10}$ units are linked by a "zig-zag" Hg_3Fe_2 chain. The Fe-Hg bond distance range from 2.544(3) – 2.706(3) Å (av. 2.625 Å) and Fe-Fe bonds from 2.670(4) – 2.691(3) Å (av. 2.680Å). The narrow Hg(2)-Fe(1)-Hg(1) angle of 78.9(1)° results in a Hg…Hg separation of 3.257(1)Å, excluded direct bond.

Structure of dark red Fe_6Cu_5 cluster [78] contains well separated $[NEt_4]^+$ cation, $[(CO)_{12}Fe_6(\mu\text{-}CO)_6(\mu_4\text{-}S)_6Cu_5(PPh_3)_2]^-$ anion and MeCN molecule. Structure of the complex anion is shown in Fig. 11. The cluster consists of a central Cu_5 in trigonal bipyramidal arrangement with the equatorial plane Cu(1)Cu(2)Cu(3), and three puckering "butterfly" type Fe_2S_2 units chelating to the copper atoms of the central Cu_5 along the pseudo – D_3 axis through the Cu(4) and Cu(5) atoms to the trigonal bipyramidal resulting in pseudo – D_3 symmetry. There are fifteen M-M bonds, including six Cu-Fe of 2.710 Å, three Fe-Fe bonds of 2.525 Å and six Cu-Cu of 2.604 Å.

Structure of orange red $FeOs_{10}$ cluster [79] reveals a bridging ferrocenylphosphine unit axially and singly linking two $\{Os_5(\mu_5\text{-}C)(CO)_{14}\}$ moieties. While the Os(basal)-Os(basal) bonds range from 2.831(3) to 2.940(3) Å, the Os(basal)-Os(apical) bonds are much more narrow with the values of 2.819(3) and 2.844(3) Å.

In black Fe_8Au_2K cluster [80] two "butterfly clusters" $\{(\mu\text{-}CO)(CO)_{11}(\mu_3\text{-}CO)Fe_4Au(PEt_3)\}$ are connected by {K(18-crown-6)} unit. The potassium is coordinated to the six oxygen of the crown ether in a planar array and also axially to two carbonyl oxygen of two "butterfly" clusters. The Fe-Fe bond distances in butterfly range from 2.623(1) to 2.654(1) Å (av. 2.642 Å).

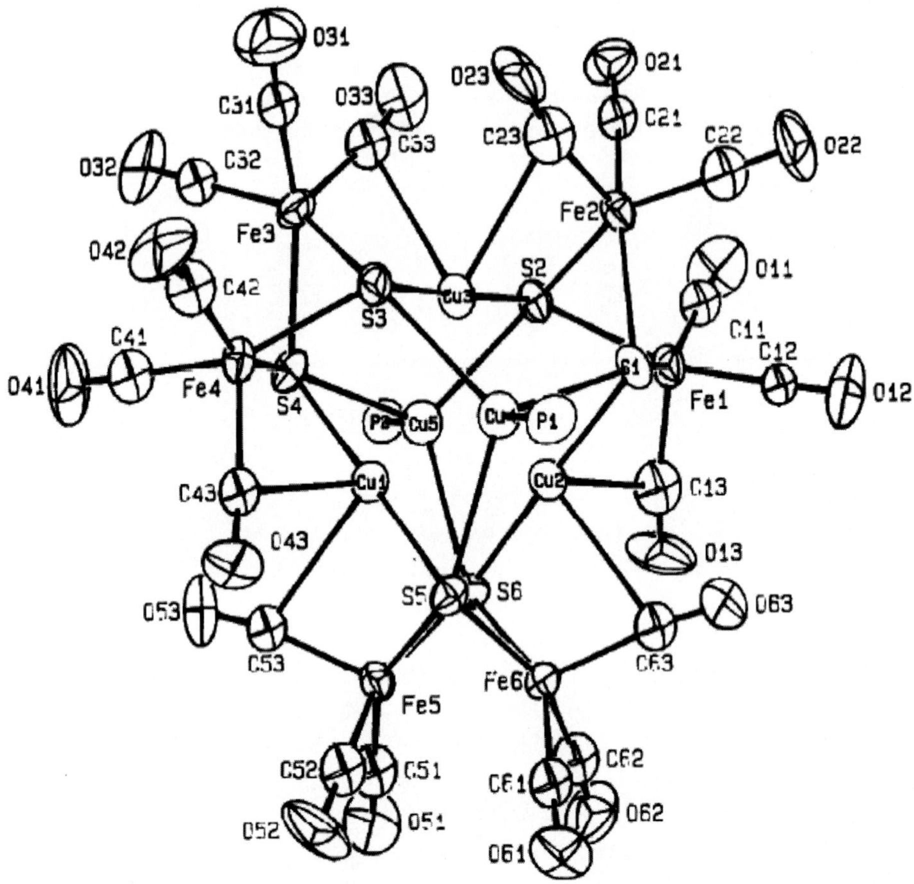

Figure 11. Structure of $[(\mu CO)_{12}Fe_6(\mu\text{-}CO)_6(\mu_4\text{-}S)_6Cu_5(PPh_3)_2]^-$ [78].

Heterododecanuclear Iron Clusters

There are seven examples, which contain twelve metal atoms, $Fe_{10}Na_2$ [81], Fe_6M_6 (M = Mo [82] or Pd [83]), Fe_5Hg_7 [84] (x2) or Mn [85] and Fe_4M_8 (M = Hg [48]) and their structural data are gathered in Table 6.

Structure of $Fe_{10}Na_2$ cluster [81] is shows in Fig 12. The structure corresponds to a decanuclear iron complex copped by two sodium atoms. The metal atoms are held together by a combination of μ_4-oxo groups each bridging three iron atoms and one sodium atom, μ_3 –oxo and μ_3 – hydroxo units bridging exclusively iron centers, 1,1',3-bridging and 1,3 – bridging benzoates and chp ligands bridging in a μ – fashion through the exocyclic O atoms. Terminally bound water molecules are attached to Fe 4 and Fe 4a, and acetone molecules are bound to the Na atoms. The iron – oxo core can be considered to contains two distorted Fe_6O_6 hexagonal prisms sharing one "square" face containing Fe4, O4, Fe4a and O4a. The shortest Fe-Fe distance is 2.974(2) Å.

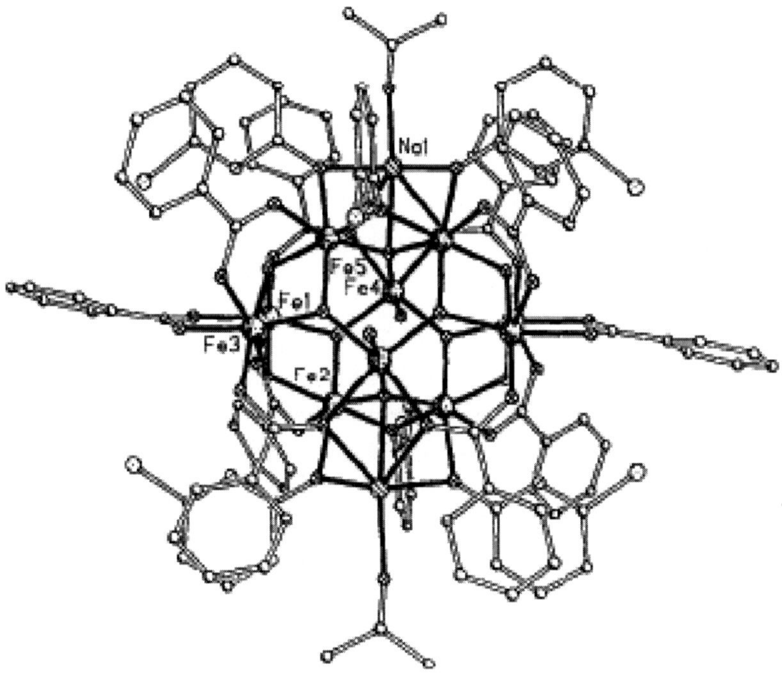

Figure 12. Structure of $[Fe_{10}Na_2(\mu_4\text{-}O)_2(\mu_3\text{-}O)_4(\mu_3\text{-}OH)_4(\mu\text{-}\eta^2\text{-}PhCO_2)_{10}(\mu\text{-}chp)_6(H_2O)_2(Me_2CO)_2]$ [81].

In black Fe_6Mo_6 cluster [82] three $Mo_2Fe_2S_4$ cubane skeletons are linked by three $\mu\text{-}\eta^2\text{-}S_4(2\text{-})$ ligands. Each $S_4(2\text{-})$ ligand bridges two Fe atoms of different cubanes, which is situated perpendicular to the Fe-Fe vector with the intercubane Fe-Fe bond of 2.610(3) Å. The molecule has crystallographic D_{3h} symmetry. A C_3 axis runs through the center of molecule, and all six iron atoms and six cubane S atoms, lie in a mirror plane. Within each cubane core, the metal – metal distances are: 2.760(2) Å (Mo-Mo), 2.801(2) Å (Fe-Mo) and 3.325(2) Å (Fe...Fe).

In brown Fe_6Pd_6 cluster [83] the metal skeleton consists of a trigonal antiprismatic array of six palladium atoms with the six lateral faces capped by six iron metals, of the 24 – CO groups, 12 are terminally bonded, two per iron atom, six are doubly bridging the six Fe-Pd bonds connected the two Fe_3Pd_3 fragments, and six are triply bridging the six $FePd_3$ triangles of the two monolayers. The mean Fe-Pd and Pd-Pd bond distances are 2.654(1) and 2.879(2) Å. The Fe...Fe separation of 3.604 Å excluded direct bond.

Structure of orange Fe_5Hg_7 cluster [84] consists of an Hg_3Fe_2 (Hg5-Fe4-Hg6-Fe5-Hg7) and an Hg_4Fe_3 fragment (Hg1-Fe1-Hg2-Fe2-Hg3-Fe3-Hg4). The Fe – Hg chains are linked together by three sulfur atoms from the Bu^tS groups and a chlorine atom. The Fe-Hg bond distances range from 2.505(4) to 2.672(4) Å (av. 2.573 Å).

In black Fe_4Mn_8 cluster [85] twelve μ_3-oxo groups each bridging two manganese atoms and one iron atom and sixteen acetates bridging one manganese and one iron atom. The Mn-Mn bond distances range from 2.773(3) to 2.938(4) Å. The Fe...Fe separations (3.357(3) – 3.478(3) Å) excluded direct bond.

Heterooligonuclear Iron Clusters

Structural data for these clusters are summarized in Table 6. Structure of black $Fe_{12}Na_2$ cluster [86] consists of well separated $[NEt_4]^+$ cations and $[\{MeS)_2(\mu\text{-}S)(\mu_3\text{-}S)_2Fe_6(\mu_3\text{-}S)_5\ (\mu_6\text{-}S)\}_2\ Na_2]^{6-}$ anion. Structure of the anion can be described in terms of two hexanuclear iron-sulphide – thiolate units there are connected by two sodium atoms. The sodium atoms are located in within trigonal prisms formed by opposite S_3 triangles of two symmetry-related Fe_6S_9 subunits. The Fe-Fe bond distance range from 2.707(3) to 2.767(3) Å (av. 2.735 Å).

Structure of red Fe_4Li_{10} cluster [87] consists of discrete molecules (Fig. 13) possessing exact C_2 symmetry, each containing four ferrocene moieties, ten lithium atoms, two tmen units, and two ethoxy groups. Each of the five symmetry independent Li atoms achieves distorted tetrahedral coordination in a different way (Table 6D). The Li-Li bond distance range from 2.50(3) to 2.78(3) Å. Each ferrocene moiety exhibits eclipsed cyclopentadienyl rings which are nearly parallel, the dihedral angles between normal to mean planes being 4.3 and 4.8° for the Fe(1) and Fe(2) ferrocene groups, respectively. The Fe-Li bond distances are 2.72(2) and 2.75(2) Å.

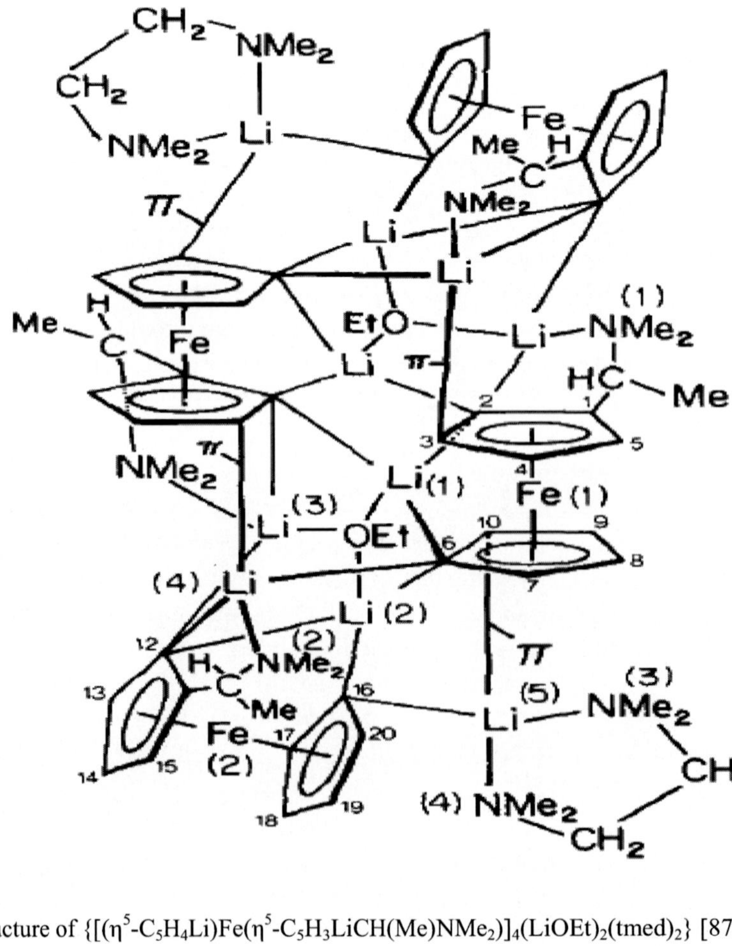

Figure 13. Structure of $\{[(\eta^5\text{-}C_5H_4Li)Fe(\eta^5\text{-}C_5H_3LiCH(Me)NMe_2)]_4(LiOEt)_2(tmed)_2\}$ [87].

Structure of $Fe_2Mo_{12}U$ complex [88] contains well separated NMe_4^+ cations, $[(H_2O)_6Fe_2UMo_{12}O_{42}]^{2-}$ anion and 18 water molecules. The metal – metal separations Fe-M are 3.595(2) Å (M = Mo) and 5.4863(3) Å (M = U). The Mo...Mo distances range from 3.189(2) to 3.907(2) Å and Mo...U distances are 3.508(3) and 3.539(3) Å, excluded direct bonds.

There are two dark brown $Fe_{16}M$ (M = Mn or Co) clusters [89], which are isostructural. The structure may be formally constructed by fusion of two twisted, pentacapped trigonal prisms of $\{Fe_{11}O_6(OH)_6(O_2CPh)_{15}\}$ molecules encapsulating a M(II) atom. The metal are bridged by four types of oxo and hydroxo ligands, μ_4-O, planar μ_3-O, and μ- and μ_3-OH. Six μ_4- oxide atoms are bound to the central M(II) atom. An interesting structural feature of the clusters, is the presence of two corner-sharing $\{\{Fe_3O_3(OH)\}_2\ M\}^{6+}$ cubes. The octahedral coordination spheres of the iron atoms are completed by 20 $PhCOO^-$ groups, 18 of which are bidentate and bridging while 2 are monodentate.

For red $Na_{12}Sn_4Fe_3(citrate)_9.38\ H_2O$ cluster [90] only unit cell dimension data are available.

Structure of black $Fe_{18}Na_2$ cluster [91] consists of discrete $[Pr_4N]^+$ cations, and two Na(I) cations weakly associated with the cluster $[Na_2Fe_{18}S_{30}]^{8-}$. Structure of cluster anion (Fig. 14) is constructed by the fusion of 24 nonplanar Fe_2S_2 rhombs in edge - and corner – sharing modes such that there are 20 μ-S, 8 μ_3-S and 2 μ_4-S, every FeS_4 unit tetrahedral, and the 18 Fe atoms are essentially planar. Two sodium (I) atoms are bound to interior sulfur atoms of the cluster whose framework can be considered to arise from the sequential connection of two linear Fe_3S_4 and two Fe_6S_9 units. The cluster is a mixed-valence species, having 14 Fe(III) and 4 Fe(II), or a mean Fe oxidation state of 2.784. The Fe-Fe bond distance range from 2.672(3) to 2.783(3) Å (av. 2.719 Å).

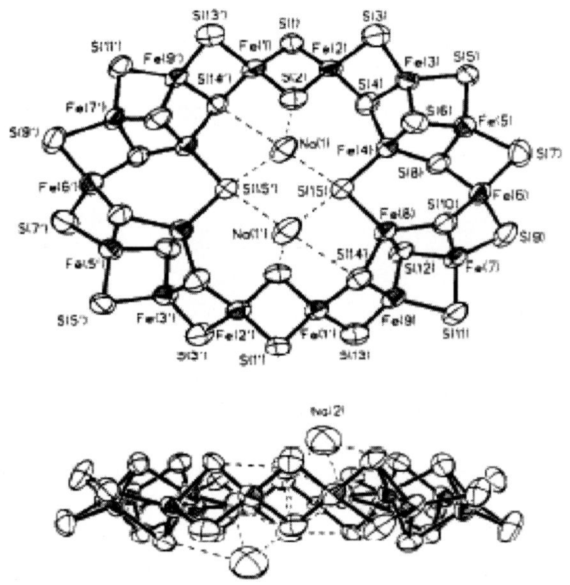

Figure 14. Structure of $[(\mu\text{-S})_{14}Fe_{18}(\mu_3\text{-S})_{14}(\mu_4\text{-S})_2Na_2]$ [91]. Upper planar projection emphasiring centrosymmetry. Lower side-on view – rotated 90° from the upper view.

Structure of another black $Fe_{18}Na_2$ cluster [92] consists of discrete $[Bu_4N]^+$ cations, two Na(I) atoms and $[Na_2Fe_{18}S_{30}]^{8-}$ cluster anion. Cluster anion is constructer by the fusion of 22 nonplanar Fe_2S_2 rhombs in edge – and vertex – sharing modes such that there are 18 μ-S and 12 μ_3-S atoms binding 18 nearly coplanar iron atoms in tetrahedral FeS_4 sites. Two sodium atoms are bound to interior sulfur atoms of the cluster whose framework can be considered to arise from the sequential connection of two Fe_2S_2 and four cuboidal Fe_3S_4 fragments and two tetrahedral FeS_4 sites. The Fe-Fe bond distance range from 2.678(3) to 2.771(3) Å (av. 2.724 Å).

Structure of black Fe_8Ag_{13} cluster [93] consists of discrete $\{PhCH_2NMc_3\}^+$ cations, $[\{(CO)_4Fe\}_8Ag_{13}]^{4-}$ anion and two molecule of acetone. View of the $Ag_{13}Fe_8$ cluster core is shows in Fig. 15. The cluster anion contains a centered cubooctahedron of silver atoms, i.e., a fragment of the cubic face centered structure of the bulk metal. All the triangular faces of the cubooctahedron are capped by $Fe(CO)_4$ fragments adopting C_{3v} ligand arrangement. The Fe-Ag bond distances range from 2.695(1) to 2.772(1) Å (av. 2.737 Å). The Ag-Ag distances range from 2.826(1) to 3.108(1) Å (av. 2.906(1) Å).

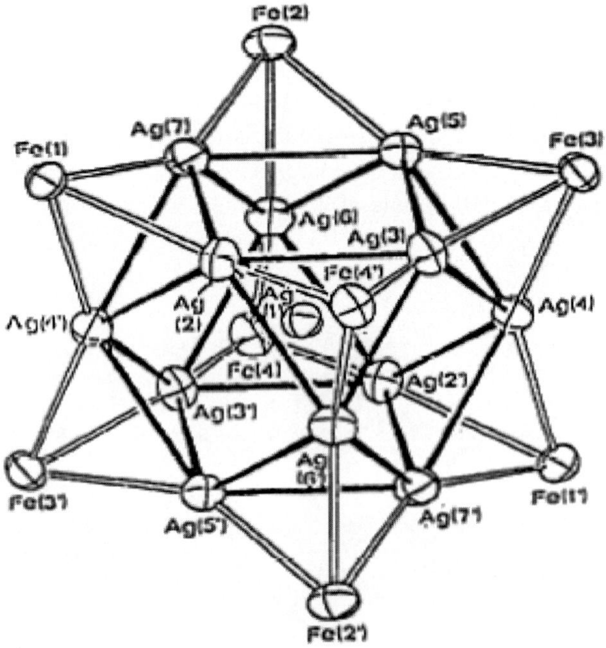

Figure 15. View of the $Ag_{13}Fe_8$ cluster core [93].

Structure of brown $Fe_{14}Ca_{12}$ cluster [94] consists of $[Ca_{12}(H_2O)_{24}Fe_{14}(\mu_3\text{-}O)_{12}(\mu\text{-}OH)_4(OH)_{12}(ert)_4(ertH)_4]^{6+}$, $HCOO^-$ anions and eight water molecules. The S_4 symmetric hexacations contain an $Fe_{14}O_{12}(OH)_{16}$ core that is surrounded by erythritolato ligands and diaquacalcium ions. The calcium ions are bound to all of the hydroxo ligands. The inner core of the oxo cluster is formed by ten iron centers, which are hexacoordinated by μ_3-oxo, hydroxo, and erythrilato ligands. Four additional iron atoms are attached to the core through two oxo/hydroxo ligands each; these iron atoms are also coordinated to one erythritolato ligand and one terminal hydroxo ligand and thus are five-coordinate with a trigonal – bipyramidal arrangement.

Structure of the discrete $[Na_9Fe_{20}Se_{38}]^{9-}$ cluster [92] is depicted in Fig. 16. The cluster contains no terminal ligands and thus is necessary cyclic. An imposed C_2 axis lies perpendicular to the plane of the rhomb Fe_2(3,3')Se_2(6,6'). The structure consists of three $Fe_6(\mu\text{-}Se)_{10}$ chains, Fe(1)-Fe(1'), Fe(4)-Fe(7'), and Fe(7)-Fe(4'). Terminal atoms Fe(1,4,7) are each included in a bridgehead rhomb by μ-Se(1,7,13)bridges, the three rhombs being fused along the common edge Fe(10) – Se(9) to form tetranuclear fragment [Fe_4(1,4,7,10)-μ-Se_3(1,7,13)-μ_4-Se(19)]. All iron atoms are tetrahedrally coordinated ($FeSe_4$). The Fe-Fe distances range from 2.744(1) to 2.887(1) Å. Within the cluster cavity are nine sodium atoms (Na(0), Na(1-4), Na(1'-4'), encapsulated in the manner of a cryptand and also coordinated to ethanol solvate molecule.

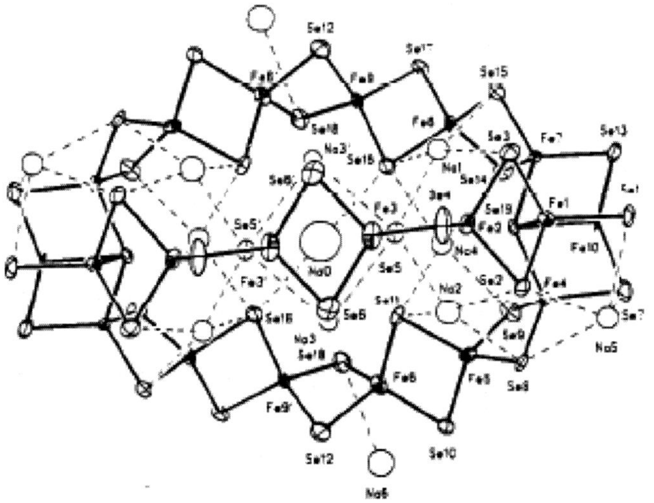

Figure 16. Structure of $[Na_9Fe_{20}Se_{38}]^{9-}$ [92].

Structure of orange $Fe_{18}Hg_{14}$ cluster [84] contains the central $Hg_{14}Fe_{12}$ core and six periphery $FeHg_2$ units (Hgl- Fel- Hg2, Hg3- Fel'-Hg, Hg5-Fe3-Hg6). The iron atoms Fel, Fe2, Fe3, Fel', Fe2' and Fe3' are coordinated in a distorted octahedral fashion by four CO ligands and two cis – positioned Hg atoms, Fe-Hg 2.523(3) (av.) and Hg – Fe – Hg angle range from 77.96(9) to 79.62(9)°. The shortest Hg-Hg distances are 3.182(2) Å. Each iron atom in central $Hg_{14}Fe_{12}$ core has distorted tetrahedral environments (FeS_2Br_2). The central core of the cluster contains six μ_3-S, six μ-Br, two μ-S, six SBu^t and twelve Br^- ligands (Table 6D).

Structure of brown $Fe_{26}Hg_{39}$ cluster [84] is built up from two $Hg_{19}Fe_{13}$ clusters that are linked together by a bridging central Hgl atom, which is situated on an inversion center. A characteristic of this cluster is an almost planar Fe_3S_3 six – membered ring (S7, Fe8, S6, Fel0, S8, Fe9), in the center of the $Hg_{19}Fe_{13}$ fragment.

If the $Hg_{19}Fe_{13}$ cluster is described starting from the six – membered ring, than three S-Hg-Fe-Hg-S-Hg chains (e.g. S6 - Hg14 - Fel1- Hg15 - S9 - Hg20) connect one sulfur atom from the six – membered ring to at terminal Hg20 atom and likewise three S – Hg – Fe – Hg

– Br – Fe – S – Hg chains (e.g. S6 – Hg 3 – Fe2 – Hg2 – Br2 – Fel – S1 – Hgl) link the Fe_3S_3 ring to the bridging Hgl atom. The first Hg3 atom is linked through a bridging S3 to a further Hg_2Fe unit (Hg4 – Fe3 – Hg5) that lies in the plane of the six – membered ring. The Fe-Hg bond distance range from 2.510(2) to 2.563(2) Å.

Structure of orange $Na_6[H_6Mo_{57}Fe_6O_{183}(NO)_6(H_2O)_{18}]$. 91 H_2O [95] consists of three 17 – molybdate subunits connected by six FeO_6 octahedra and three $\{Mo_2O_9\}$ groups, each of which are linked by one bridging oxygen atom two bridging water molecules, and the subunits related to each other by a three – fold rotation axis, the cluster center sits on a special position with 6m2 crystallographic symmetry and each individual submit has internal C_{2v} symmetry (Fig. 17). In the center cavity there are 12 water molecules, which are situated around the six inversion axis and are coordinated to six molybdate and six iron groups, respectively. Moreover, these 12 water molecules are linked to each other by hydrogen bonding to form two six – membered rings.

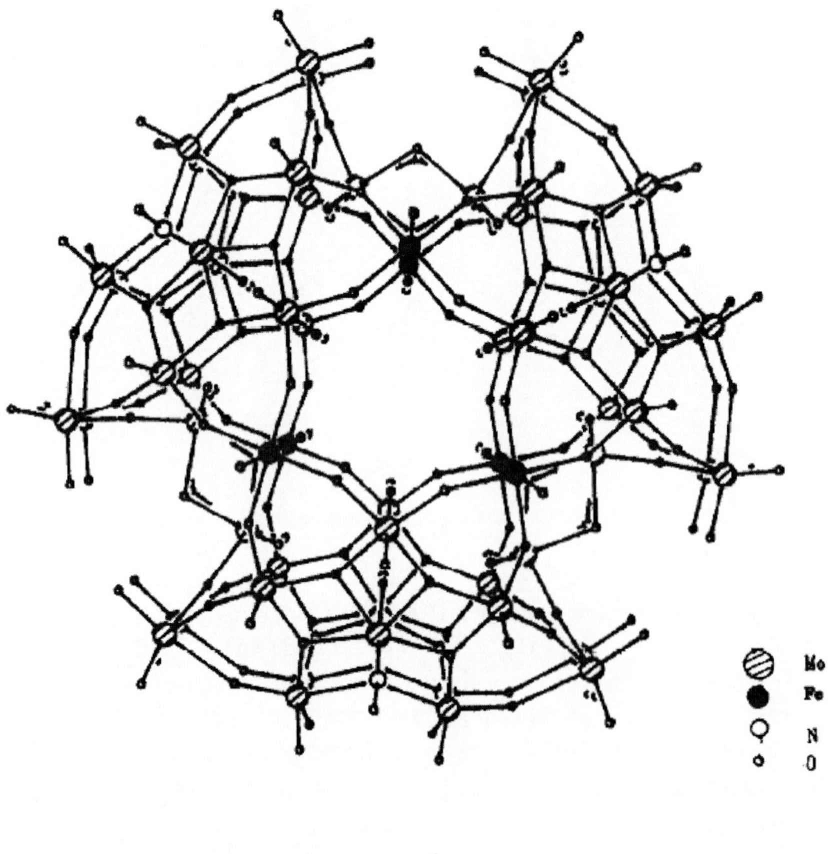

Figure 17. Structure of $[H_6Mo_{57}Fe_6O_{183}(NO)_6(H_2O)_{18}]^{6-}$ [95].

The 17 – molybdate submit consists of one compact group of five predominantly edge-sharing MoO_6 octahedra and two $MoO_6(NO)$ pentagonal bipyramids viewed as an $\{Mo_7O_{24}(NO)_2\}$ unit of C_{2v} symmetry. The $\{Mo_7O_{24}(NO)_2\}$ unit is circled by a ring of 10 corner-sharing distorted MoO_6 octahedra to form a 17 – molybdate units.

Inspection of the data in Table 6 reveals that there are thirty four clusters with several variations: Fe_8M_2 (M = V [72] or Ru [73]), Fe_6Ru_4 [74], Fe_4Pt_6 [75], $Fe_2Ge_4Co_4$ [76], Fe_8Hg_3

[77], Fe_6Cu_5 [78], $FeOs_{10}$ [79], Fe_8Au_2K [80], $Fe_{10}Na_2$ [81], Fe_6M_6 (M = Mo or Pd [83]), Fe_5Hg_7 [84], Fe_4M_8 (M = Hg [48] (x2) or Mn [85]), $Fe_{12}Na_2$ [86], Fe_4Li_{10} [87], Fe_4Hg_{10} [48], $Fe_2Mo_{12}U$ [88], Fe_6Hg_{10} [48], $Fe_{16}M$ (M = Mn or Co [89]), Fe_7Hg_{12} [48], Fe_5Hg_{14} [48], $Fe_3Na_{12}Sn_4$ [90], $Fe_{18}Na_2$ [91, 92], Fe_8Ag_{13} [93], $Fe_{14}Ca_{12}$ [94], $Fe_{20}Na_9$ [92], $Fe_{18}Hg_{14}$ [84], Fe_6Mo_{57} [95] and $Fe_{26}Hg_{39}$ [84].

Table 7. Summary of Fe-M, M-M and M-M' Bond Distance (< 3.0 Å) in Heterodeca – and Heterooligonuclear Iron Clusters

Fe – M	Shortest [ref]	Longest [ref]	Average
- Ge	2.471(2) [76]		
- Pt	2.540(4) [75]	2.597(4) [75]	2.568
- Hg	2.505(4) [84]	2.715(4) [48]	2.584
- Pd	2.609(1) [83]	2.699(1) [83]	2.654
- Au	2.666(1) [80]		
- Cu	2.689(5) [78]	2.734(5) [78]	2.710
- Ag	2.695(1) [93]	2.762(1) [93]	2.731
- Li	2.72(2) [87]	2.75(2) [87]	2.735
- Mo	2.800(2) [82]	2.802(2) [82]	2.801
- Os	2.819(3) [79]	2.940(3) [79]	2.871
Hg - Hg	2.456(1) [84]	2.546(1) [84]	2.501
Co – Co	2.636(2) [76		
Li - Li	2.50(3) [87]	2.78(3) [87]	2.64
Cu – Cu	2.574(3) [78]	2.823(3) [78]	2.679
Fe – Fe	2.517(3) [78]	2.974(2) [81]	2.714
Pt – Pt	2.677(1) [75]	2.790(1) [78]	2.733
Mn – Mn	2.773/4) [85]	2.938(4) [85]	2.844
Os – Os	2.819(3) [79]	2.940(3) [79]	2.871
Pd – Pd	2.810(2) [83]	2.948(2) [83]	2.874
Ag – Ag	2.826(1) [93]	2.999(1) [93]	2.907
Ge – Co	2.346(2) [76]	2.482(2) [76]	2.415

Examination of the metal – metal bond distances (Table 7) shows, that the shortest Fe-M bond distances (M = non transition and transition) are 2.471(2) Å (M = Ge [76]) and 2.540(4) Å (M = Pt [75]), respectively. The shortest Fe-Fe bond distance is 2.517(3) Å [78] and M-M bond distance is 2.456(1) Å (Hg-Hg).

5. CONCLUSIONS

There are one hundred and thirty heteroocta - and heterooligonuclear iron clusters analyzed in this review. Of these 65 are octanuclears, 21 nanonuclears, 5 decanuclears, 7 dodecanuclears, 4 tetradecanuclears, 1 hexadecanuclear, 2 heptadecanuclears, 3 nonadeca-nuclears, 2 eicosinuclears, Fe_8Ag_{13}, $Fe_{14}Ca_{12}$, $Fe_{20}Na_9$, $Fe_{18}Hg_{14}$, Fe_6Mo_{57}, $Fe_{26}Hg_{39}$ (each x 1).

One cluster, $[(CO)_8Fe_2SbFe_2(CO)_6SbFe_2(CO)_8]$ exists in two isomeric forms, monoclinic [9] and orthorhombic [10] with similar stereochemistry, differing by degree of distortion. Another two clusters, $\{(cp)(CO)_2Fe\}_4Sn_4.(\mu\text{- Se})_6$ [36] and $cp_2^*Fe_2(\mu_6\text{-}\eta^2\text{- C} - C)_2(\mu\text{-CO})_3Ru_6(CO)_{14}$ [52] contain two crystallographically independent molecules within the same crystal, differing only by degree of distortion. The isomers as well as independent molecules are typical examples of distortion isomerism [56].

The heterometal atoms include the transition and non-transition metals, with one example actinide $Fe_2Mo_{12}U$ [88]. There is a much richer chemistry along with the transition metals than those with the non-transition elements, with some examples, which contain both non-transition as well as transition metals (x5) [8, 53, 55, 71, 76].

A summaries of the metal – metal bond distances below 3.0 Å (Tables 3, 5 and 7) show that although iron is classified as a soft, or class B, metal atom, it can be seen that it is found bonded to variety of other metal atoms of both soft and hard (class A) variety. This is also observed in the other heterometallic iron derivatives [1-5], illustrating a rich diversity in the chemistry of iron amongst its 2 300 or so structurally determined heterometallic compounds.

This review together with its precursors [1-5] has surveyed nearly 2 300 heterometallic iron compounds serving to focus attention to the rich stereochemistry of iron and to encourage more extensive studies on this topic.

Correlation has been found to assist the investigation in a variety of individual, biological and environmental topic. During the collection of the data it was found out that some data were unattainable, some transparent and some very difficult to reach. In few cases incorrect information was published in the literature. Where possible, these faults were corrected in this review.

A survey for heteropolynuclear iron compounds is now in progress. It is our hope that this review will summarize common structural features and further stimulate activity in areas of particular interest.

6. ACKNOWLEDGMENTS

The authors wish to thank those who gave permission for reproduction of original figures, the Faculty of Pure and Applied Science of York University, and the Ministry of Education of the Slovak Republic, Grant Agency of the Slovakia (Grant No. 1/0353/08) and Technology Assistance Agency under the contract No. APVT-20-005504.

7. References

[1] M. Melník, Z. Baloghová, I. Ondrejkovičová, V. Vančová and C.E. Holloway, Rev. *Inorg. Chem.*, 26 (2006) 1-365

[2] M. Melník, I. Ondrejkovičová, D., Mikloš, P. Segľa and C.E. Holloway, *Rev. Inorg. Chem.*, 27(2007) 67-317

[3] M. Melník, I. Ondrejkovičová, M. R. Sundberg and C.E. Holloway, *Rev. Inorg, Chem.*, in press.

[4] M. Melník, I. Omdrejkovičová, M. Koman and C.E. Holloway, *Rev. Inorg. Chem.*, in press.

[5] M. Menik, I. Ondrejkovičová, J. Garaj and C.E. Holloway, *Rev. Inorg. Chem.*, in press.

[6] Hegetschweiler, H.W. Schmalle. H.M. Streit, V. Gramlich, H.U. Hund and I. Erni, *Inorg. Chem.* 31 (1992) 1299.

[7] S.G. Anema, K.M. Mackay and B.K. Nicholson, *Inorg. Chem.*, 28 (1989) 3158.

[8] P.B. Hitchcock, M.F. Lappert and M.J. McGeary, *Ogranometallics*, 9 (1990) 884.

[9] A.M. Arif, A.H. Cowley and M. Pakulski, J. Chem. Soc., *Chem. Commun.*, (1987) 622.

[10] A.L. Rheingold, S.J. Geib, M. Shieh and K.H. Whitmire, *Inorg. Chem.*, 26 (1987) 463.

[11] S. Luo and K.H. Whitmire, J. *Ogranomet. Chem.*, 376 (1989) 297.

[12] J.M. Cassidy and K.H. Whitmire, *Acta Crystallogr.*, Sect. C, 46 (1990) 1781.

[13] J.M. Cassidy and K.H. Whitmire, *Inorg. Chem.*, 28 (1989) 1432.

[14] W. Cen, S.C. Lee, J. Li, F.M. MacDonnell and R.H. Holm, *J. Am. Chem. Soc.*, 115 (1993) 9515.

[15] F.A. Cotton and G. Schmid, *Inorg. Chim. Acta*, 254 (1977) 235.

[16] T.E. Wolff, J.M. Berg, K.O. Hodgson, R.B. Frankel and R.H. Holm, *J. Am. Chem. Soc.*, 101 (1979) 4140.

[17] J.Q. Xu, X.G. Zhang, H. Cai, N.H. Hu, Ch. P. Wei, G.Y. Jang and T.G. Wang, *Chem. Commun.*, (1996) 757.

[18] B.S. Kang, J.H. Cai, D.X. Wu, H.Q. Liu, Q.T. Liu, L.H. Weng and J.X. Lu, *Acta Chim. Sin.*, 47 (1989) 744.

[19] B.S. Kong, H.Q. Liu, J.H. Cai, L.G. Huang, Q.T. Liu, D. Wu, L.H. Weng and J.X. Lu, *Transition Met. Chem.*, 14 (1989) 427.

[20] M.A. Greaney, C.L. Coyle, R.S. Pilato and E.J. Stiefel, *Inorg. Chim. Acta*, 189 (1991) 81.

[21] W.H. Armstrong and R.H. Holm, *J. Amer. Chem. Soc.*, 103 (1981) 6246; W.H. Armstrong, P.K. Mascharak and R.H. Holm, *J. Amer. Chem. Soc.*, 104 (1982) 4373.

[22] R.E. Palermo, R. Singh, J.K. Bashkin and R.H. Holm, *J. Amer. Chem. Soc.*, 106 (1984) 2600.

[23] K.D. Demadis and D. Coucouvanis, *Inorg. Chem.*, 34 (1995) 436.

[24] K.D. Demadis and D Coucouvanis, *Inorg. Chem.*, 34 (1995) 1995.

[25] P.E. Mosier, Ch.G. Kim and D. Coucouvanis, *Inorg. Chem.*, 32 (1993) 2620.

[26] P.R. Challen, S.M. Koo, C.G. Kim, W.R. Dunham and D. Coucouvanis, *J. Amer. Chem. Soc.*, 112 (1990) 8606.

[27] G. Christou, P.K. Mascharak, W.H. Armstrong, G.C. Papaefthymiou, R.B. Frankel and R.H. Holm, J. *Amer. Chem. Soc.*, 104 (1982) 2820.

[28] D. Coucouvanis, P.R. Challen , S.M. Koo, W.M. Davis, W. Butler and W.R. Dunham, *Inorg. Chem.*, 23 (1989) 4181.

[29] M.G. Kanatzidis and D. Coucouvanis, J. Amer. Chem. Soc., 108 (1986) 337; D. Coucouvanis, A. Salifoglou, M.G. Kanatzidis, A. Simopoulos and A. Kostikas, ibid, 109 (1987) 3807; D. Coucouvanis, A. Salifoglou, M.G. Kanatzidis, W.R. Dunham, A. Simopoulos and A. Kostikas, *Inorg. Chem.*, 27 (1988) 4066.

[30] S.A. Al-Ahmad, A. Salifoglou, M.G. Kanatzidis, W.R. Dunham and D. Coucouvanis, *Inorg. Chem.*, 29 (1990) 927.
[31] A. Salifoglou, M.G. Kanatzidis and D Coucouvanis, *J. Chem. Soc., Chem. Commun.* (1986) 559.
[32] G. Christou, C.D. Garner, R.M. Miller and T.J. King, *J. Inorg. Biochem.*, 11 (1979) 349.
[33] Ch. Junghans, W. Seak and S. Pohl, *J. Chem.. Soc., Chem. Commun.*, (1994) 2327.
[34] S. Ciurli, M. Carrie and R.H. Holm, *Inorg. Chem.*, 29 (1990) 3493.
[35] U. App and K. Merzweiler, *Z. Anorg. Allg. Chem.*, 623 (1997) 478.
[36] K. Merzweiler, F. Rudolph and L. Brands, *Z. Naturforsch.*, 47b (1992) 470.
[37] R.M. Harker, M.F. Mahon and K.C. Molloy, *Main Group Met. Chem.*, 19 (1996) 29.
[38] K. Merrweiler and L. Weisse, *Z. Naturforsch.* 45b (1990) 971.
[39] Jiaxum Tao, Wenjing Xiao and Qingchuan Yang, *J. Organomet. Chem.*, 531 (1997) 223.
[40] K.H. Whitmire , T.A. Albright, S.K. Kang, M.R. Churchill and J.C. Fettinger, *Inorg. Chem.*, 25 (1986) 2799.
[41] J.M. Burlitch, S.E. Hayes and G.E. Whitwell II, *Organometallics*, 1 (1982) 1074.
[42] R.D. Ernst, T.J. Marks and J.A. Ibers. *J. Amer. Chem. Soc.*, 99 (1977) 2090.
[43] M.A El-Sayed, A. Ali, G. Davies, S. Larsen and J. Zubieta, *Inorg. Chim. Acta*, 194 (1992) 139.
[44] S.K. Mojseev, N.N. Mileschonkova and F.M. Spiridonov, *Koord. Khim.*, 13 (1987) 1589.
[45] V.G. Albano, F. Azzaroni, M.C. Iapalucci, G. Longoni, M. Monari, S. Mulley, D.M. Proserpio and A. Sironi, *Inorg. Chem.,* 33 (1994) 5320.
[46] D.E. Barbir, R.E. Bryan, M. Sabat, K.S. Bose, and B.A. Averill, *Inorg. Chem.*, 35 (1995) 4635.
[47] N. Zhu, J. Pebler and H. Vahrenkamp, Angew. *Chem. Int. Ed. Engl.*, 35 (1996) 894.
[48] J. Eisenmann, G. Fenske and F. Hezel, *Z. Anorg. Allg. Chem.*, 624 (1998) 1095.
[49] W. Saak and S. Pohl, Angew. Chem., 103 (1991) 869; *Angew. Chem. Int. Ed. Engl.*, 30 (1991) 881.
[50] Ch.J. Adams M.L. Bruce, B.W. Skelton and A.H. White, *Inorg. Chem.*, 31 (1992) 3336.
[51] L.R. Sita, R. Xi, A.P.A. Yap, L.M.Liable-Sande and, A.L. Rheigold, *J. Amer. Chem. Soc.*, 119 (1997) 756.
[52] M. Akita, S. Sugimoto, M. Tanaka and Y. Moro-oka, *J. Amer. Chem. Soc.*, 114 (1992) 7581.
[53] Feng-Ying Jiao, Yuan-Pi Yin, Quan-Fu Zhore, Ji-Chun Xu and Liang-Ren Huang, *Acta Chim. Sin.*, 50 (1992) 788.
[54] A.S. Katugin, A.A. Pasinskii, I.L. Erimenko, B. Orazsakhatov, A.S. Aliev, Yu.T. Struchkov and A.I. Yanovsky, *J. Organomet. Chem.*, 382 (1990) 423.
[55] S.G. Anema, K.M. Mackay, L.C. McLeod, B.K. Nickolson and J.M. Whitaker, Angew. *Chem. Int. Ed. Engl.* 25 (1986) 759; S.G. Anema, J.A. Audett, K.M. Mackay and B.K. Nicholson, *J. Chem. Soc.,* Dalton Trans., (1988) 2629.
[56] M. Melník, *Coord. Chem. Rev.*, 47 (1982) 239.
[57] R.W. Saalfrank, I. Bernt, E. Uhler and F. Hampel, Angew. *Chem. Int. Ed. Engl.*, 36 (1997) 2482.
[58] C.E. Housecroft, A.L. Rheingold and M.S. Shongwe, *J.Chem. Soc.*, Chem. Commun., (1998) 1630; S.M. Draper, C.E. Househeroft, J.E. Rees, M.S. Shongwe, B.S. Haggerty and A.L. Rheingold, *Organometallics*, 11 (1992) 2356.
[59] Bei-Sheng , Jing-Hua Cai, Chang-Neng Cheng and Jia-Xi Lu, *Acta Chim. Sin.*, 44 (1986) 781.

[60] Bei-Sheng Kang, Liang-Ren Huang, Jing-Hua Cai, Yu Yang and Jia-Xi Lu, *Acta Chim. Sin.*, 45 (1987) 1152.
[61] T.E. Wolff, J.M. Berg, P.P. Power, K.O. Hodgson and R.H. Holm, *Inorg. Chem.*, 19 (1980) 430.
[62] I.L. Eremenko, A.A. Pasynskii, A.S. Abdulaev, A.S. Aliev, B. Oraszakhatov, S.A. Sleptsova, A.I. Nekhaev, V.E. Sklova and Yu.T. Struchkov, *J. Organomet. Chem.*, 365 (1989) 297.
[63] W. Clegg, N.A. Compton, R.J. Errington and N.C. Norman, Polyhedron, 6 (1987) 2031; *J. Chem. Soc.*, Dalton Trans., (1988) 1671
[64] J.M. Wallis, G. Müller and H. Schmibauer, *J. Organomet. Chem.*, 325 (1987) 159.
[65] M. Ferrer, A. Julia, O. Rossel, M. Seco, M.A. Pellinghelli and A. Tiripicchio, *Organometallics*, 16 (1977) 3715.
[66] G. Doyle, K.A. Eriksen and D. Van Engen, *J. Amer. Chem. Soc.*, 108 (1986) 445.
[67] C.E. Briant, R.G. Smith AND d.m.p. Mingos, *J. Chem.. Soc., Chem. Commun.*, (1984) 586.
[68] F. Canales, M.C. Gimeno, A. Laguna and P.G. Jones, *J. Amer. Chem. Soc.*, 118 (1996) 4839.
[69] S.E. Nefedov, A.A. Pasynskii, I.L. Eremenko, E.E. Stomakhina, A.I. Yanovsky and Yu.T. Struchkov, *J. Organomet. Chem.*, 405 (1991) 287. S.E. Nefedov, I.L. Eremenko, A.A. Pasynskii E.E. Stomakhina , S.A. Sleptsova, A.I. Yanovsky, M.Yu. Antipin and Yu.T. Struchkov, *Metallorg.* Khim., 3 (1990) 951.
[70] M.H.A. Benvenutti, P.B. Hichcock, J.F. Nixon and M.D. Vargas, *J. Chem. Soc.*, Dalton Trans., (1996) 739.
[71] S.G. Anema, G.C. Baris, K.M. Mackay and B.K. Nicholson, *J. Organomet. Chem.*, 441 (1992) 35.
[72] J. Huang, S. Mukeyee, B.M. Segal, H. Akashi , J. Zhou and R.H. Holm, *J. Amer. Chem. Soc.*, 119 (1997) 8662.
[73] S. Pohl, W. Barklage, W. Saak and U. Gitz, *J. Chem. Soc., Chem. Commun.*, (1993) 1251.
[74] W.R. Cullen, S.J. Rettig and T.C. Zheng, *Polyhedron*, 14 (1995) 2653.
[75] G. Longoni, M. Manassero and M. Sansoni, *J. Amer. Chem. Soc.*, 102 (1980) 7973.
[76] S.G. Anema, K.M. Mackay and B.K. Nicholson, *J. Chem. Soc.,* Dalton Trans., (1995) 3853.
[77] A. Bianchini and L.J. Farrugia, *Organometallics*, 11 (1992) 540.
[78] B. Zhuang, B. Pan, L. Huang and P. Yu, *Inorg. Chim.* Acta, 227 (1994) 119.
[79] J. Wing-Sze Hui and Wing-Tak Wong, *J. Chem. Soc.*, Dalton Trans., (1997) 2445.
[80] C.P. Horwitz, E.M. Holt, C.P. Brock and D.F. Shriver, *J. Amer. Chem. Soc.*, 107 (1985) 8163.
[81] C. Bernelli, S. Parsons, G.A. Solan and R.E.P. Wimpenny, Angew. *Chem. Int. Ed. Engl.,* 35 (1996) 1825.
[82] H. Kawaguchi, K Yamada, S. Ohniski and K. Tatsumi, *J. Amer. Chem. Soc.*, 119 (1997) 10871.
[83] G. Longoni, M. Manassero and M. Sansoni, *J. Amer. Chem. Soc.*, 102 (1980) 3242.
[84] D. Fenske and M. Bettenhausen, Angew. *Chem. Int. Ed. Engl.*, 37 (1998) 1291.
[85] R.R Schake, H.L. Tsai, N. de Vries, R.J. Webb, K. Folting, D.N. Hendrikson and G. Christou, *J. Chem. Soc., Chem. Commun.*, (1992) 181.
[86] H. Strasdeit, B. Krebs and G. Henkel, *Inorg. Chem.*, 23 (1984) 1816.
[87] I.R. Butler, W.R. Cullen, J. Reglinski and S.J. Rettig, *J. Organomet. Chem.*, 249 (1983) 183.

[88] M.A. Petruhina, V.N. Molchanov, I.V. Tatianina and E.A. Torchenkova, *Kristallografiya*, 35 (1990) 386.
[89] W. Micklitz and S.J. Lippard, *J. Amer. Chem. Soc.*, 111 (1989) 6856.
[90] B. Binder, *J. Inorg. Nucl. Chem.*, 41 (1979) 257.
[91] J.F. You, B.S. Snyder and R.H. Holm, *J. Amer. Chem. Soc.*, 110 (1988) 6589; J.F.You, B.S. Snyder, G.C. Papaefthymiou and R.H. Holm, ibid, 112 (1990) 1067.
[92] J.F. You and R.M. Holm, *Inorg. Chem.*, 30 (1991) 1431; J.F. You, G.C. Papaefthymiou and R.H. Holm, *J. Amer. Chem. Soc.*, 114 (1992) 2697.
[93] V.G. Albano, L. Grossi, G. Longoni, M. Monari, S. Mulley and A. Sironi, *J. Amer. Chem. Soc.*, 114 (1992) 5708.
[94] J. Burger and P. Klüfers, Angew. *Chem. Int. Ed. Engl.*, 36 (1997) 776.
[95] G.Q. Huang, S.W. Zhang and M.Ch. Shao, *Polyhedron*, 12 (1993) 2067.

In: Thermochemistry and Advances in Chemistry Research ISBN: 978-1-60692-376-4
Ed: Anton G. Mirskiy

Chapter 9

PHOTOLUMIENESCENCE FROM RARE EARTH IONS DOPED NANOCRYSTALS

Zuoling Fu[1,2] and Siyuan Zhang[1]

[1]Laboratory of Rare Earth Chemistry and Physics, Changchun Institute of Applied Chemistry, Chinese Academy Sciences, Changchun 130022, China

[2]Department of Physics, Pukyong National University, Busan 608-737, South Korea

ABSTRACT

Research and development of nanoscale luminescent and laser materials are part of the rapidly advancing nanoscience and nanotechnology. Besides well-known quantum dots, lanthanide-doped oxide nanocrystals form a new promising class of nanophosphors in view of various application because of unique spectroscopic properties related to two types of electronic transitions: 4f-4f and 4f-5d transitions. The subjects of this chapter are the synthesis of rare earth ions doped nanocrystals and their spectral properties. We begin this chapter with some introductory comments on the nature of spectroscopy of rare earth ions, and then discuss an important issue that also needs to be addressed is whether the rare earth ions are really being incorporated in the host lattice in the synthesis of rare earth ions doped nanocrystals. A crucial test for the incorporation is an excitation spectrum of the rare earth emission. Citing examples suggest that some of the difficulties encountered in nanodoping are often attributed to “self-purification”. Finally, we will discuss the effect of nanostructure on the optical properties of rare earth ions. Specifically, the various mechanisms of spectral properties of rare earth ions from bulk to nanocrystals are detailed. The luminescence properties of rare earth doped nanocrystals differ from corresponding bulk materials in several ways: (i) spectroscopic changes, as a broadening and shift(related to the 5d energy level change) of the luminescence bands in nanocrystals, resulting from a structural disorder intrinsic to the small size; (ii) a lower luminescence quantum yield related to the presence of chemical species adsorbed at the surface, especially the OH groups; (iii) a shift towards higher values of the optimum doping concentration, due to the alteration of energy transfers rate from luminescence centers to quenching centers; (iv) new luminescent peak occurs due to the new site occupation of rare earth ion on the surface or another phase; (v) the change of photoluminescence decay time due to surface-bound centers. All the results are emphasis on the photoluminescence from the rare earth-doped oxide nanocrystals in the last 15 years.

INTRODUCTION

Inorganic compounds doped with rare earth ions form an important class of phosphors for many luminescence applications [1-3], including phosphors for fluorescent lighting, display monitors, X-ray imaging[4-6], scintillators[7], lasers[8], and amplifiers for fiber-optic communication[9]. In the last years growing interest for lanthanide(also known as rare earth)-doped nanoparticles has arisen because of the observed modification of their significant properties as compared to the bulk materials [10-13]. The motivation to study nanostructures of lanthanide-doped insulating materials are concerned because the spectral and dynamic properties of these technologically important materials change when the reduced dimensions affect the chemical and physical properties of the host. In a similar manner, the lanthanide dopant can serve as a sensitive probe of the chemistry and structure of its host. In addition, preparing nanoscale host materials can change the physical properties of the host, which affects the luminescence and dynamics of an optically active dopant. Investigating and understanding these materials will be important for optimizing their emissive properties for technological applications in future communication and display devices. Several groups have reported particle-size-dependent phenomena in lanthanide-doped nanoparticles that affect emission lifetime [14,15], quantum efficiency[15,16], and concentration quenching[17]. Some effects have been attributed to quantum-confinement effects when the particle diameter is less than 5 nm[15]. In both Tb^{3+}: ZnS and Mn^{2+}: ZnS, luminescence lifetimes become shorter by several orders of magnitude [11,15]. These size-dependent effects are postulated to result from mixing of the s-p electrons of the host with the d-f valence electrons of the activator due to quantum confinement, causing the normally forbidden d-d or f-f transitions to become allowed[15]. Although research in nanostructured semiconductors is beginning to produce a clear theoretical picture of these materials, very little theoretical work has addressed the size-dependence of localized emissive ions in nanostructured hosts. What is clear is that the current theories and experimental database are inadequate to understand the optical properties of localized luminescent centers in nanostructures. Therefore, investigation and study on nanosized physical properties, enriching and developing the prepared methods of nanostructures, which have important significance for basic research and exploring application field of new material.

To maintain a concise and focused review, this chapter does not review advances in nanocrystalline semiconductors, which are being studied extensively for their unique size-dependent quantum-confinement effects[18-20], nor does it include the active research in lanthanide-doped one-dimensional oxide nanowires[21,22] or core-shell nanostructure[23,24]. The subjects of this chapter are the synthesis of rare earth ions doped nanocrystals and their spectral properties. We begin this chapter with the preparation of rare earth-doped semiconductor or oxidic insulator, and then discuss an important issue that also needs to be addressed is whether the rare earth ions are really being incorporated in the host lattice in the synthesis of rare earth ions doped semiconductor nanocrystals. Finally, we will discuss the effect of nanostructure on the optical properties of rare earth ions after some introductory comments on the nature of spectroscopy of rare earth ions are given. Specifically, the various mechanisms of spectral properties of rare earth ions from in bulk to in corresponding oxidic insulator nanocrystals are detailed.

2. General Synthesis Of Rare Earth Doped Nanocrystals

There are a large number of methods for forming discrete metal oxide nanoparticles, including common precipitation methods [25, 26], such as microemulsion technique[27], sol-gel processing[28, 29], electrospray and spray pyrolysis[30-32], combustion synthesis[33], high energy mechanical milling[34], self-assembly[35, 36], and hydrothermal or solvothermal synthesis[37, 38]. The surfaces of nanoparticles can be changed by chemical reactions or by dispersing the particles in a polymer or glass matrix. There are also a variety of synthesis and processing methods to prepare metal oxide nanocrystals in matrixes. In these synthesis processes, there are two urgent questions that must be solved. One is that smaller and monodisperse nanoparticles can be obtained, which shows the nanosize effect on the structure and properties of the materials; another is that shape and morphology can be controlled, which plays an important role in nanotechnology as self-assembled device structures become more accessible.

The first report on doped nanocrystals that showed all features such as increased band gap, shift of excitation and emission spectra of nanosized semiconductor was reported by Bhargava et al. [11] in 1994. They synthesized manganese doped nanocrystals of zinc sulfide. The nanomaterials had external photoluminescence quantum efficiency of 18% and it ushered the era of nanophosphors. In addition, it was argued that the lifetime of the Mn^{2+} emission in nanocrystalline ZnS was shortened from milliseconds to nanoseconds due to quantum size effects. However, it was later shown that lifetime shortening in nanocrystalline $ZnS:Mn^{2+}$ does not occur [39, 40]. The Mn^{2+} emission of nanocrystalline $ZnS:Mn^{2+}$ has a "normal" millisecond lifetime. Despite the absence of lifetime shortening, numerous papers on the luminescence of semiconductor nanoparticles doped with 3d transition metal (TM) ions or rare earth (RE) ions have appeared and in various articles it was concluded that both the TM- and RE-doped ZnS nanocrystals "form a new class of luminescent materials"[11, 41, 42]. Many literatures have reports on both oxidic insulators and II-VI semiconductor nanoparticles doped with RE ions. Based on the survey made in the literature on nanocrystalline RE-doped oxides, it is clear that it is very well possible to incorporate RE ions in oxidic nanoparticles such as Y_2O_3[16, 43, 44], YVO_4[45], and Gd_2O_3[46], due to the very similar chemical properties of all rare earth ions. Quantum size effects are not expected in these oxides since the Bohr radius of the exciton in insulating oxides, like Y_2O_3, YVO_4, and Gd_2O_3, is very small. The exciton Bohr radius of II-VI semiconductors is larger (e.g., 2.5 nm for CdS) resulting in pronounced quantum confinement effects for nanoparticles of about 2.5 nm and smaller. Therefore, a possible influence of quantum size effects on the luminescence properties of RE ions is only expected in II-VI semiconductor nanocrystals. A number of papers reported on the luminescence of nanocrystalline II-VI semiconductors doped with Tb^{3+}[15, 25, 47], Sm^{2+}[25], Eu^{3+}/Eu^{2+}[26, 27, 49], or Er^{3+}[50]. Papers on the luminescence of nanocrystalline $ZnS:Tb^{3+}$ [15, 25, 47] reported emission spectra characteristic for $^5D_4$-^7F_n transitions of the Tb^{3+} ion. Still, it is known that RE ions can be incorporated in bulk II-VI semiconductors [15]. Efficient luminescence from intraconfigurational 4f-4f transitions has been observed for RE ions in bulk ZnS upon excitation over the band gap [51-53]. However, high temperatures (900- 1200°C[54]) are needed to accomplish this. The reported synthesis conditions for the preparation of RE-doped nanocrystalline semiconductors are very mild compared to the conditions used for bulk RE-doped semiconductors. Therefore, it is not evident that RE ions are incorporated in ZnS or CdS nanoparticles. Bol et al.[55] found that a crucial test for the incorporation is an excitation spectrum of the Eu^{3+} emission; if the ZnS absorption band is present in the excitation spectrum, energy transfer from the ZnS semiconductor nanoparticle to the Eu^{3+} ion occurs and it is clear that Eu^{3+} ions are incorporated in the ZnS nanoparticles. If the excitation spectrum does not show the ZnS absorption band, it can be concluded that the Eu^{3+} emission does not originate from Eu^{3+} in ZnS nanoparticles.

Recently, this area has been the subject of several excellent review articles [55, 56, 57]. A wide variety of characterization techniques to ensure dopant incorporation into the host semiconductor lattice occurred, such as X-ray diffraction, Raman spectroscopy, electron paramagnetic resonance spectroscopy (EPR), electronic absorption spectroscopy, X-Ray photoelectron spectroscopy (XPS), extended X-ray absorption fine structure (EXAFS) and luminescence spectroscopy [56]. Norris et al.[57] also reviewed advances in the chemical synthesis of doped nanocrystals, in the theoretical understanding of the fundamental mechanisms that control doping, and in the creation of highly conducting nanocrystalline films. Three doping mechanisms have been proposed, as summarized in Figure 1. Turnbull argued that crystallites tend to be pure simply because they contain very few atoms [58]. Turnbull model is purely statistical and does not propose any new physics for small crystals (Figure 1A). Moreover, it cannot explain the issues with Mn doping of II-VI nanocrystals. Mn is sufficiently soluble in the bulk that even very small nanocrystals should still contain many impurities. An alternative explanation is that nanocrystals undergo "self-purification", that is, the impurity solubility is much lower than in the bulk [59]. Self-purification model implies that nanocrystals are hard to dope for thermodynamic reasons and that impurities are expelled (Figure 1B). However, a fundamental assumption in this view is that the nanocrystal is in thermodynamic equilibrium with its environment, including chemical equilibrium with reservoirs of impurity and nanocrystal constituents. The system then minimizes its Gibbs free energy. This principle underlies quantitative concepts such as the impurity formation energy and the solubility limit, as well as more heuristic ideas such as impurity size. For the assumption of thermodynamic equilibrium to be justified, however, atomic exchanges between the system and its reservoirs must be possible. This means that impurity atoms must be able to diffuse readily through the nanocrystal.

However, high temperatures (above 1000°C) are needed to diffuse most dopants rapidly. If thermodynamic equilibrium is not established, then kinetic factors such as activation barriers will control doping (Figure 1C). The impurity must adsorb on the surface of the nanocrystal and then be covered by additional material [61, 62]. For this to occur, the surface must be favorable for impurity binding. The nature of the surface depends on the crystal structure and nanocrystal shape, and hence these two factors affect doping. Moreover, in colloidal synthesis, surfactants must not bind to the surface or the impurity too strongly. Lastly, additional material must adhere so that, in the absence of diffusion, the impurity becomes trapped. These criteria have been used to explain Mn doping in II-VI nanocrystals [62].

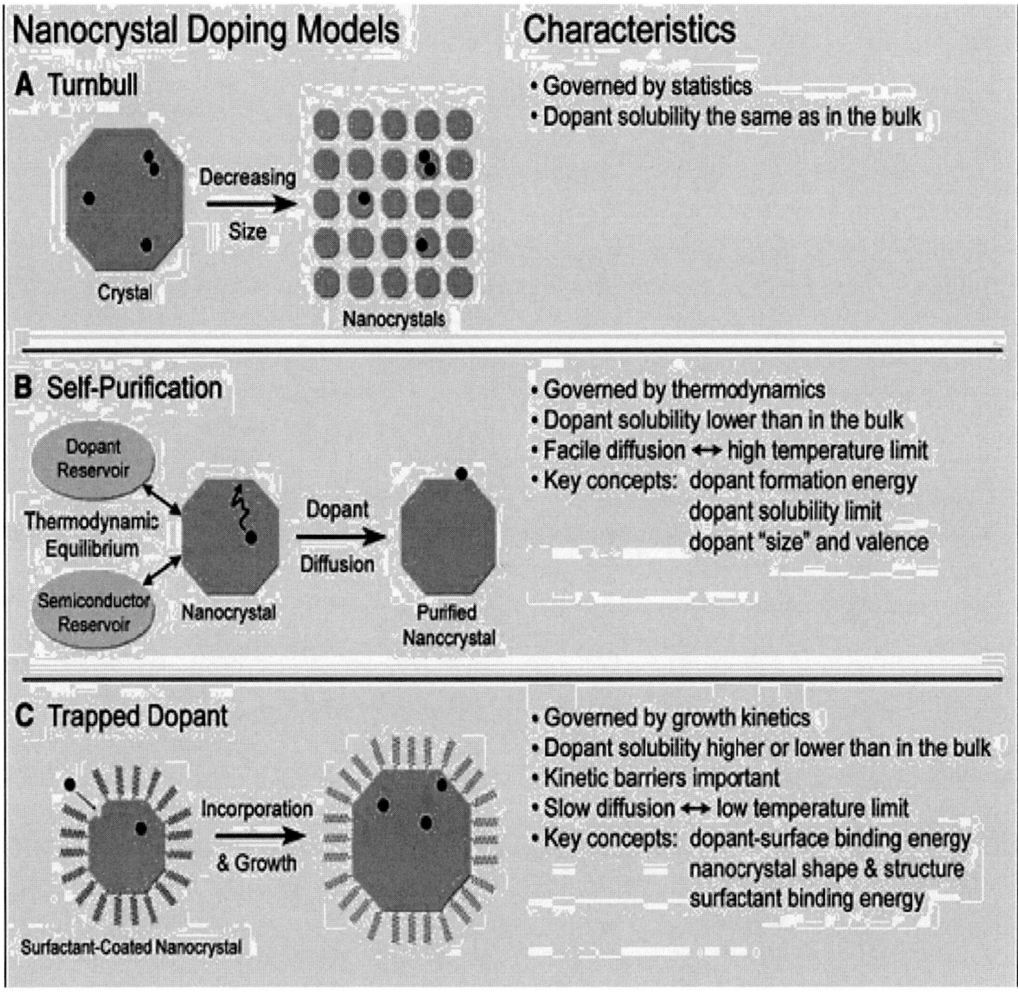

Figure 1. Schematic and characteristics of three models used to explain doping in semiconductor nanocrystals. (A) Turnbull model, in which the number of impurities per nanocrystal decreases with crystallite size based solely on statistics [Key reference: 58]. (B) Self-purification model, in which dopant solubility is lower in the nanocrystal than in the bulk semiconductor [Key reference: 59, 60]. (C) Trapped dopant model, in which kinetic factors govern the doping process [Key reference: 61, 62]. (Adapted from ref. [57])

Despite the above content mentioned refer to the TM-doped semiconductor nanocrystals, it has guide significance for RE-doped semiconductor. Moreover, other evidence is that an acceptable percentage different in ion radii between doped and substituted ions must not exceed 30% [63].

Doping enhances the properties of nanocrystals by providing another means to control their remarkable electronic, optical, transport, and magnetic properties. In the next section, we will discuss the change of optical properties in rare earth doped nanocrystals. Particularly, the various mechanisms of spectral properties of RE ions from bulk to corresponding oxidic insulator nanocrystals are analyzed in detail.

3. OPTICAL PROPERTY

3.1 General Introduction To The Nature Of Spectroscopy Of Rare Earth Ions [64] (4fn, 4f-5d And Charge-Transfer Transitions)

The rare earth ions are characterized by an incompletely filled 4f shell. The 4f orbital lies inside the ion and is shielded from the surroundings by the filled $5s^2$ and $5p^6$ orbits, therefore the influence of the host lattice on the optical transitions within the $4f^N$ configuration is small(but essential). Emission transitions yield, therefore, sharp lines in the spectra such as $Pr^{3+}(4f^2)$, $Sm^{3+}(4f^5)$, $Eu^{3+}(4f^6)$, $Gd^{3+}(4f^7)$, $Tb^{3+}(4f^8)$ and $Dy^{3+}(4f^9)$. The unique properties of the 4f electrons that have localized states and exhibit weak coupling to ligand electrons and lattice vibrations have been applied for the fluorescent lamps, solid-state lasers and optical amplifiers in fiber optics, etc.

Rare earth ion emission is not necessarily sharp line emission. With energy levels at more than 30,000 cm^{-1} above the ground state of the $4f^N$ configurations, there are 5d, 6s, 6p orbits in the rare earth ion electronic structure. The allowed optical transitions of the rare earth ions are inter-configuration and consist of two different types, one is $4f^N$-$4f^{N-1}5d$ transition; the other is charge-transfer transitions ($4f^N$-$4f^{N+1}L^{-1}$, where L = ligand). Both are the allowed transitions, and appear in the spectra as broad absorption bands such as trivalent ions Ce^{3+}, Pr^{3+}, Tb^{3+}, and divalent ions Eu^{2+}, Sm^{2+}, Yb^{2+}, which show 4f-5d absorption bands. The 5d states exhibit less localized nature and stronger coupling to lattice vibrations, therefore, the electronic transitions between the $4f^N$ and $4f^{N-1}5d$ states are expected to be characteristically very different from transitions within the $4f^N$ configuration, and their intensities up to 10,000 times stronger than the strongest $4f^N$-$4f^N$ transitions. Charge-transfer transitions are found for rare earth ions which like to reduced, 4f-5d transitions for ions which like to be oxidized. The tetravalent rare earth ions (Ce^{4+}, Pr^{4+}, Tb^{4+}) show charge-transfer absorption bands. Recently, an unusual blue phosphor Sr_2CeO_4 has been developed using the combinatorial techniques by Danielson et al. [65]. In this oxide, the Ce ion is in the +4 oxidation state and the 4f shell is empty. The emission has been assigned to Ce^{4+} charge transfer (CT) luminescence. Until now, the Sr_2CeO_4 lattice is the only host lattice in which CT luminescence of Ce^{4+} has been observed at room temperature [66].

3.2 Spectral Diferences Of Rare Earth Ions Between The Bulk And Nanocrystals

It is well known that the luminescent spectra of lanthanide ions in crystals arise mainly from two types of electronic transitions: $4f-4f$ and $4f-5d$ transition. The former generally shows sharp line emission, while the latter have broad band character, and the crystalline environment has greater effect on interconfigurational transitions of Ln^{3+} than on intraconfigurational transitions. This is because the excited states such as $5d$ are not shielded from the surrounding electronic shells and the 5d-electron has a strong interaction with the neighboring anion ligands in the compound. In addition, the $5d$ energy level has closer relation to the conduction band of compound, when the grain size become small, the band gap may become wide, then the $5d$ energy level also change correspondingly, i.e., he $5d$ energy level will raise.

It is also known that the CT band is related to stability of the electron of the surrounding O^{2-}, i. e., the charge-transfer transition is sensitive to ligand environment (the bonding energy between the central ion and the ligand ions) [67]. If the bonding energy between the central ion and the ligand ions increases, the highest occupied molecular orbital (HOMO) will locate at the lower energy position, i.e., the valence band descend, then, the charge transfer transition will occur at higher energy.

A lot of papers reported on the spectral differences of rare earth ions between the bulk and nanocrystals. The luminescence properties of rare earth doped nanocrystals differ from corresponding bulk materials in several ways:

[1] spectroscopic changes, as a broadening and shift of the luminescence bands in nanocrystals, resulting from a structural disorder intrinsic to the small size.

The spectral band undergoes to higher energies, which is called a blue shift; it is the other way round—which is called a red shift. The shift of the luminescence bands is mainly related to the 5d energy level change. In present, many literatures [68-73] have reported that the charge-transfer band, absorption band of host and f-d transition in absorption or excitation spectra show blue or red shift with the decrease of particle size in rare earth doped nanocrystals. Zhang et al.[68]prepared nanocrystalline Y_2O_3:Eu by a combustion reaction. For samples with particle sizes of 80, 40, and 5 nm, the CT absorption peaks are centered at 239, 243, and 250 nm, respectively. It is attributed that the lattice constant of a nanocrystal is larger than that of a commercial bulk sample, or there is a larger Eu–O distance in a nanosample than in the normal one. A red shift of the Eu–O charge transfer has also been observed in YBO_3:Eu[69], and Gd_2O_3: Eu[70]and explained by an increase of the Eu–O bond length (the lattice constant becomes larger) in nanoparticles induced by an increase of the volume cell. Compared with that in bulk materials, the lattice constants in corresponding nanocrystals become larger or smaller; the spectra bands show blue or red shift with the decrease of the particle size. Maybe, these phenomenons occurred in the same nanocrystals but these were prepared by different authors. The following is the case. Another blue shift was also reported by Igarashi et al. [71], who gave a 5-nm shift of the CT band in nanocrystalline Y_2O_3: Eu. Such a change of crystallinity increases the ionicity of Eu–O bonds (the lattice constants become smaller) to shift the CT excitation peak towards the higher-energy side.

Konrad et al. [72] synthesized nanocrystalline yttria powder by chemical vapor method. A blue shift and a broadening of the absorption edge, both depending on the particle size were found. Spectroscopic effects have also been essentially studied in Eu^{3+}-doped oxide nanoparticles by Tissue et al.[73], as europium ions are good probes for their local environment. Besides small changes in the relative intensities and in the positions of excitation and emission bands, a systematic increase of the bandwidths for fine 4f–4f transitions has been noted decreasing the size below 50 nm. This inhomogeneous broadening has been essentially attributed to the existence of a structural disorder within nanometric particles related to the high surface/volume ratio.

[2] a lower luminescence quantum yield related to the presence of chemical species adsorbed at the surface, especially the OH groups.

The surface of small particles, where the coordination of atoms differs from the bulk and where different chemical species may be adsorbed, is suspected to be the most important source of alteration of the luminescence efficiency. YVO_4:Eu nanoparticles are well-known to be quenched by hydroxyl groups[74-77] since the energy of the $^5D_0$–$^7F_6$ transition nearly corresponds to the third harmonic of the vibration mode of the OH oscillator. This is the case for terbium ions, where the $^5D_4$ emitting level is about 5 times the OH vibration. This effect has been evidenced by transferring nanoparticles into D_2O, the Tb quantum yield reaches 61%(instead of 42% for Tb emission for monazite nanoparticles in aprotic solvents) [78]. This makes the probability of relaxation about 200 times lower than in presence of non-deuterated OH groups[79]. These experimental results confirm that surface OH groups are efficient quenchers of the Eu^{3+} excited states in red nanophosphors[78]. Similarly, the existence of the residual OH groups also explained the low emission efficiency for nanocrystalline YPO_4:Eu prepared by sol-gel method[80]. However, as expected, these yields are lower than for corresponding bulk materials (between 70– 90%).The decrease of the luminescence efficiency does not seem to be directly correlated to the decrease of the crystallite size, but depends much more on the surface chemistry [45, 74, 78]. In particular, the luminescence is drastically quenched ($<$ 10%) for IR emitting lanthanides such as Nd, Yb and Er ions[81-83] due to the existence of the residual OH groups on surface of nanoparticles. In addition, because the increase of local environment difference between ions in nanoparticles, the optimum concentrations of doping lanthanide ions are raised.

[3] A shift towards higher values of the optimum doping concentration, due to the alteration of energy transfers rate from luminescence centers to quenching centers.

Table 1 Optimum europium concentrations for different oxide nanoparticles and comparison with the corresponding bulk materials.

	Bulk	Nanocrystals	Size	Ref.
Y_2O_3:Eu	6%	8%	20nm	85
		13%	40nm	68
		18%	5nm	68
Y_2SiO_5:Eu	20%	60%	50nm	86
YBO_3:Eu	10%	22%	20nm	87
		20%	40nm	88
YVO_4:Eu	5%	20%	7nm	76

The increase of the optimum concentration has not been systematically explained, but seems to result from the alteration of energy transfers within the nanoparticles. The quenching concentration of Y_2O_3:Eu phosphors prepared by conventional synthesis is 6mol % Eu [84], but the quenching concentration was apparently 14mol % Eu for corresponding nanocrystals[43]. It is thought that the interface effects of the nanoscale materials may hinder the energy transfer and, as a result, the quenching concentration increases because the quenching concentration is related to the energy transfer between the Eu ions. The shift of the optimum doping concentration to high values for other materials was summarized in Table 1.

[4] New luminescent peak occurs due to the new site occupation of rare earth ion on the surface or another phase.

Song et al. [89] reported that a new broad excitation line at 579.9 nm was observed in nanocrystalline $Y_2O_3:Eu^{3+}$. The 579.9nm line was attributed to the luminescence of Eu^{3+} ions near the surface of the nanocrystals. The surface ions are unstable and symmetry in the near surface becomes lower due to numerous surface defects. Igarashi et al. [71] also reported that the lattice distortion of the nanocrystals was seven times larger than that of the micrometer particles. Lattice distortion may lead to the site symmetry down and the degeneration of the energy level. The Eu_2O_3 and Tb_2O_3 nanocrystals[90] in the range of sizes from 2nm up to 5nm exhibit new luminescence peaks at 620 nm and 548 nm, which correspond to modulation of the $^7F_2$ transition of the Eu^{3+} ion and modulation of the $^7F_5$ transition of the Tb^{3+} ion, respectively. In nanocrystalline Lu_2O_3: Eu^{3+} with the particle size at 12nm[91], some additional emission lines located around 580nm and 585.7nm were also observed. The lines are completely absent for the sintered ceramic sample (having largest crystallites). An appearance of the extra emission lines when the crystallites get smaller seems to indicate that the emissions may have their origin at the Eu^{3+} ions located in the vicinity of the crystallites surface, where the environment is typically significantly distorted.

[5] The change of photoluminescence decay time due to surface-bound centers.

Important modifications have also been noted for emitting lifetimes, which are generally shorter in nanoparticles. This has been associated with the increase of non-radiative transfers. In cubic $Y_2O_3:Eu^{3+}$[73], for the samples with particle size larger than 10 nm, the decay is single exponential process, with lifetime values of 1.7, 1.39, 1.28, and 1.08 ms for 3μm, 80nm, 40nm, and 10nm samples, respectively. However, the luminescent decay curve of 5-nm $Y_2O_3:Eu^{3+}$ could only be fitted by a double exponential, with τ_1 = 1.04 ms and τ_2 = 0.35 ms, which corresponds to the decay of a conventional emission and the lifetime of Eu^{3+} in the $^5D_0$ state on the surface, respectively. A shorter lifetime for the $^5D_0$-$^7F_2$ emission in YPO_4:Eu nanocrystals derived from wet chemical routes was also observed due to including residual OH groups in soft chemistry derived materials[80]. However, in some cases, this effect can be counterbalanced by the influence of the low refractive index of the dispersion medium which decreases the average index and increases the lifetime. As a consequence, longer lifetimes have been observed for dispersed nanoparticles[92].

In next section we will address issues of the spectral various mechanisms of rare earth ions in bulk and nanocrystals, new features appear in these materials and this will be the emphasis of the present discussion. Three examples of the rare earth ions doped nanocrystals systems will be described in detail: alkaline earth oxide, yttrium oxide, and strontium yttrium oxide with different rare earth-doped sites.

3.3 Various Mechanisms Of Rare Earth Ions In Bulk And Nanocrystals

3.3.1 Photoluminescence Excitation Spectra

3.3.1.1 $SrAl_2O_4$: Ln (Ln = Ce^{3+}, Pr^{3+}, Tb^{3+}) Nanocrystals

The $SrAl_2O_4$: Ln (Ln = Ce^{3+}, Pr^{3+}, Tb^{3+}) nanocrystals were synthesized by the combustion method [93]. The refined crystallographic unit cell parameters were obtained and a decreasing trend of unit cell parameter values in nanocrystals could clearly be observed compared with the bulk materials (Table 2).

Table 2. Unit Cell Parameters for $Sr_{1-x}Al_2O_4$: Ln_x (x = 0.01, Ln= Ce^{3+}, Pr^{3+}, Tb^{3+}) Bulk and Nanocrystalline Samples[93]

Sample	a(Å)	b(Å)	c(Å)	β(°)
JCPDS std no. 34-379	8.447	8.816	5.163	93.42
$SrAl_2O_4$: Ce^{3+} (Bulk)	8.431±0.0007	8.816±0.0069	5.155±0.0125	93.41±0.0014
$SrAl_2O_4$:Ce^{3+} (Nano)	8.421±0.0006	8.807±0.0054	5.159±0.0030	93.36±0.0601
$SrAl_2O_4$: Pr^{3+} (Bulk)	8.432±0.0009	8.817±0.0012	5.155±0.0007	93.40±0.0109
$SrAl_2O_4$: Pr^{3+} (Nano)	8.428±0.0018	8.808±0.0073	5.150±0.0019	93.38±0.0237
$SrAl_2O_4$: Tb^{3+} (Bulk)	8.441±0.0007	8.816±0.0011	5.156±0.0124	93.43±0.0013
$SrAl_2O_4$: Tb^{3+} (Nano)	8.421±0.0007	8.808±0.0006	5.148±0.0097	93.29±0.0058

The photoluminescence excitation spectra in bulk and nanocrystalline $SrAl_2O_4$: Ln (Ln = Ce^{3+}, Pr^{3+}, Tb^{3+}) were shown in Figure 2, and it is found that the excitation peaks of 5d energy levels of Ln^{3+}(Ln = Ce^{3+}, Pr^{3+}, Tb^{3+}) show red shift in nanocrystals in contrast to that in bulk crystals. The order of the degree of red shift for nano $SrAl_2O_4$: Ln (Ln = Ce^{3+}, Pr^{3+}, Tb^{3+}) crystals is $Pr^{3+} > Ce^{3+} > Tb^{3+}$.

The mechanism of spectra and energy changes is investigated according to the optical spectrum simulations. All the interactions for the $4f^{N-1}5d$ configuration were shown in Figure 3. It is shown that the energy level change results from three interactions for Ln^{3+}: the shift of the energy centroid (E_c) of the $5d$ orbital; the Coulomb interactions between $4f$ and $4f$ electrons ($E_{Coul}(ff)$) and between $4f$ and $5d$ electrons ($E_{Coul}(fd)$); the effect of the crystal field ($E_{cf}(d)$).

It is assumed that $E_{\exp}$ denotes the lowest energy level of $5d$ state in experimental spectra of crystal, then, the following equation can be obtained from the energy diagram: $E^b_{\exp} = E^b_c - E^b_{Coul}(ff) \pm E^b_{Coul}(fd) - E^b_{cf}(d)$, (Plus is suitable for N>7; Minus is suitable for N ≤ 7). According to Refs [94-96], two factors, h_e and F_c depending on the crystals were introduced.

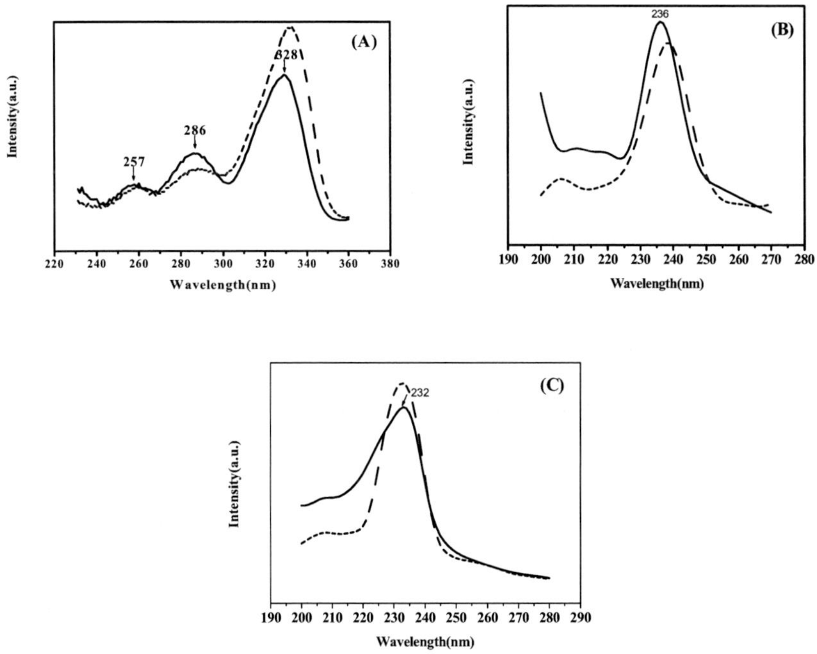

Figure 2. Photoluminescence excitation spectra of bulk (solid line) and nanocrystalline (dot line) $SrAl_2O_4$: Ce^{3+} (A), Pr^{3+} (B) and Tb^{3+}(C) at the room temperature. (Adapted from ref. [93]).

When the particle size becomes small, the unit cell volume accordingly becomes small. One can notice that h_e of nanocrystal has a decreasing trend compared with the bulk counterpart according to theoretic calculation. Decrease of h_e leads to increase of E_c ($E_c = A + Be^{-kh_e}$), which indicates that the central energy E_c in nanocrystals has an ascending trend; $E_{Coul}(ff)$ and $E_{Coul}(fd)$ also increase. At the same time, the bond length Sr-O become shorter in nanocrystals, which results in increase of the crystal field strength around the rare earth ions, Consequently, $E_{cf}(d)$ will increase in nanocrystals. Finally, the following equation can be obtained in nanocrystals:

$$E_{\exp}^{n} = E_{c}^{b} + \delta E_{c} - E_{Coul}^{n}(ff) \pm E_{Coul}^{b}(fd) \pm \delta E_{Coul}(fd) - E_{cf}^{b}(d) - \delta E_{cf}(d),$$

(Plus is suitable for N>7; Minus is suitable for N $\leq$ 7). Where the footnote with n indicates the parameter values in nanocrystals. δ denotes increment in nanocrystals. Since $4f$ electron is less influenced by crystal environment, it is considered $E_{Coul}^{n}(ff) = E_{Coul}^{b}(ff)$, then, we can obtain,

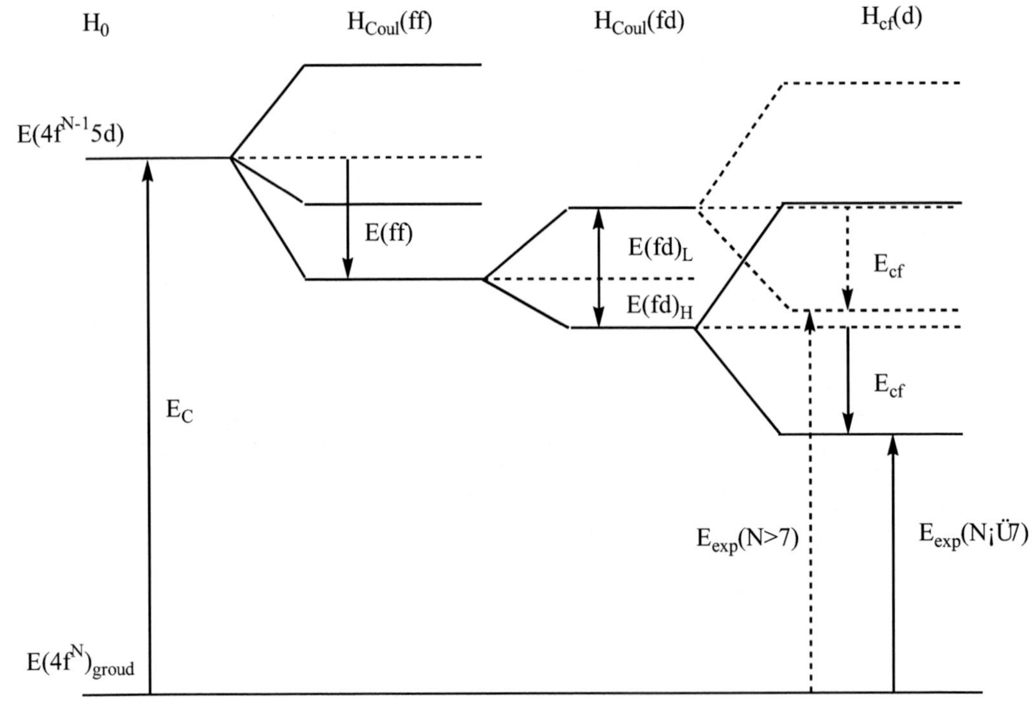

Figure 3. A schematic energy diagram of the $4f^{N-1}5d$ configuration of Ln^{3+} in any crystal A. (Adapted from ref. [93]).

$$\Delta E = E^{n}_{\exp} - E^{b}_{\exp} = \delta E_{c} \pm \delta E_{Coul}(fd) - \delta E_{cf}(d)$$

According the above equation, if the luminescent mechanism of Ce^{3+} in any compound A is known, the luminescent behaviors of other trivalent lanthanide ions in the same host can be predicted through above mentioned physical analysis. For example, the excitation spectra for $SrAl_2O_4$: Ln (Ln = Ce^{3+}, Pr^{3+}, Tb^{3+}) are chosen as an example for our investigation (see Figure 2). For Ce^{3+} ion, the energy change $\Delta E = \delta E_{c} - \delta E_{cf}(d)$ because of $E_{Coul}(fd) = 0$. The excitation spectra in nano $SrAl_2O_4$: Ce^{3+} crystal show apparent red-shift in comparison with that of the bulk crystal (from 328nm to 332nm), which indicates that the descent of energy level caused by crystal-field splitting is more than the ascent of central energy level of 5d orbits. For Pr^{3+} ion, $\Delta E = \delta E_{c} - \delta E_{Coul}(fd)_{H} - \delta E_{cf}(d)$, therefore, it can be concluded that the excitation spectra in nano $SrAl_2O_4$: Pr^{3+} crystal should show red-shift in comparison with that of the bulk crystal, and the degree of the red-shift is larger than that of Ce^{3+} ion due to the contribution of $\delta E_{Coul}(fd)_{H}$. In experimental spectra, the lowest 5d level in nano $SrAl_2O_4$: Ce^{3+} crystal shifts downwards by 368cm^{-1}(from 328nm to 332nm); the lowest 5d level in nano $SrAl_2O_4$: Pr^{3+} crystal shifts downwards by 532cm^{-1}(from 236nm to 239nm). The experimental results are consistent with the predicted results. For Tb^{3+} ion, $\Delta E = \delta E_{c} + \delta E_{Coul}(fd)_{L} - \delta E_{cf}(d)$, because $\delta E_{Coul}(fd)_{L}$ makes the energy level shift upwards, it can be predicted that the degree of the red-shift in nano $SrAl_2O_4$: Tb^{3+} must be smaller than that of Ce^{3+}. In experimental spectra, the lowest $5d$ -level excitation band in nano $SrAl_2O_4$: Tb^{3+} has slightly red-shift about 185cm^{-1}. Finally, it is concluded that the order of the degree of red-shift for nano $SrAl_2O_4$: Ln (Ln= Ce^{3+}, Pr^{3+}, Tb^{3+}) crystals is Pr^{3+}> Ce^{3+}> Tb^{3+}, which is in good agreement with the experimental results. By contrast, it is found that although the order of magnitude of the spectral change is small, the analysis reflects a profound physical mechanism and it can be applicable to the spectral prediction of other rare earth doped nano-materials.

3.3.1.2 Y_2O_3: Eu^{3+} Nanocrystals

In this case, Y_2O_3:Eu^{3+} nanocrystals were prepared by combustion synthesis[99]. The results of X-ray diffraction (XRD) showed that the materials have a reduced and distorted cubic lattice in contrast to the bulk materials(Table 3). A blue-shift of the charge transfer (CT) band in excitation spectra was also observed (Figure 4).

Table 4 Unit Cell Parameters for Y_2O_3:Eu^{3+} Bulk and Nanocrystalline Samples[97]

Sample	a(Å)	Std.error for a (Å)	Cell volume (Å^3)
JCPDS std no. 43-1036	10.604		1192.4
Bulk Sample	10.606	±0.0004	1193.12
Nano Sample	10.580	±0.0044	1184.38

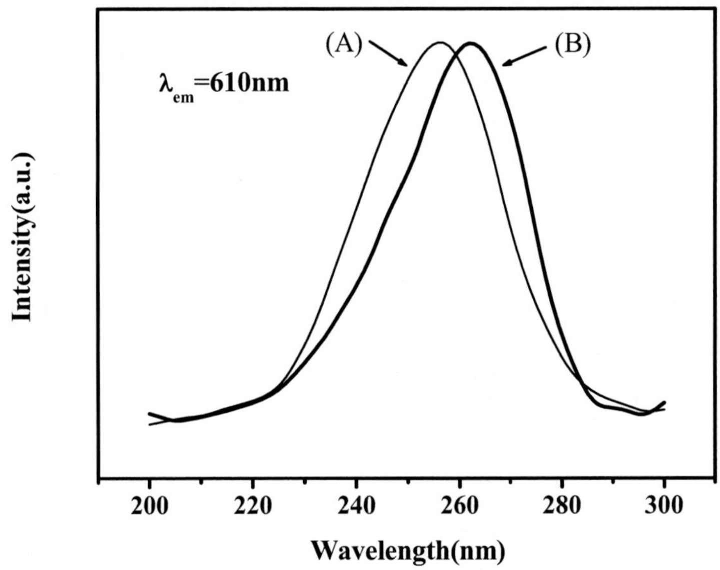

Figure 4. Excitation spectra for the nanocrystalline Y_2O_3:Eu^{3+} (A) and bulk Y_2O_3:Eu^{3+} (B). (Adapted from ref.[97])

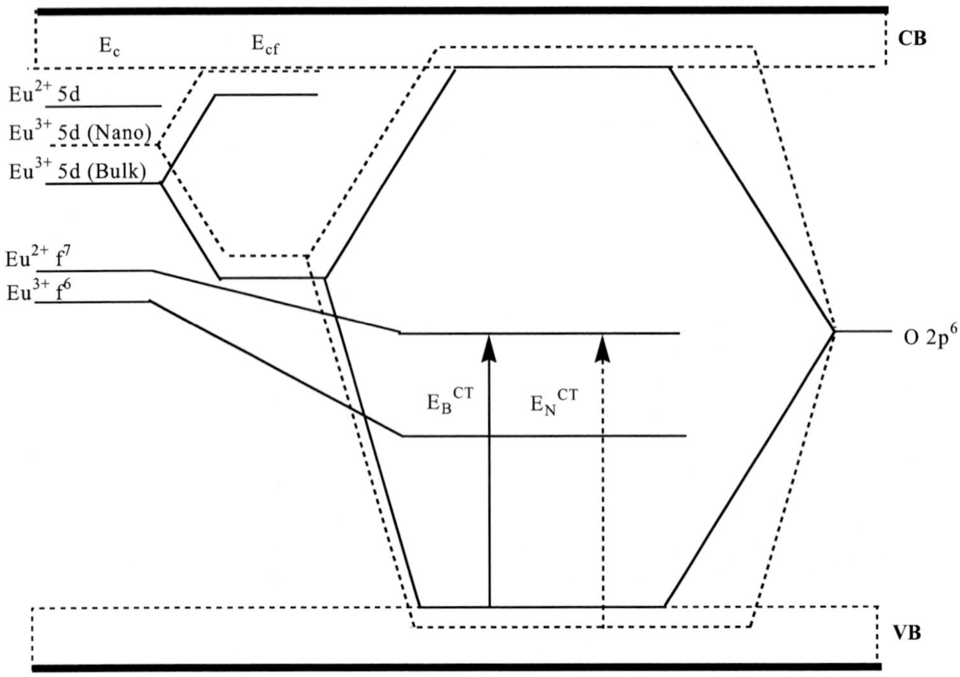

Figure 5. A schematic energy diagram of charge transfer band in Y2O3:Eu3+. The solid arrow indicates the charge-transfer (CT) transition of Eu^{3+} substituting Y site in bulk (E_B^{CT}). The dashed arrow indicates the charge-transfer transition of Eu^{3+} substituting Y site in nanocrystal (E_N^{CT}).(Adapted from ref.[97])

The charge-transfer (CT) band in the energy need to the energy transfer an electron from the valence band (VB) to a trivalent lanthanide. The final state in the transition is the ground state of the divalent lanthanide ion [98]. A schematic energy diagram of CT band taken Y_2O_3: Eu^{3+} as samples is shown in Figure 5[97]. The 5d energy level of rare-earth ions has closer relation to the conduction band (CB) of compound. In addition, it is considered that the energy level of f^N configuration is unchanged in bulk and nanocrystal since f^N are shielded from the surrounding electronic shell. The experimental result shows that the CT band in nanocrystal is blue-shifted compared with the bulk crystal, suggesting that the energy centroid (E_c) of the 5d orbital (Eu^{3+}) and the crystal-field splitting ($E_{cf}(d)$) have important effect on CT band.

3.3.2 Photoluminescence Emission Spectra

3.3.2.1 $SrAl_2O_4$: Eu^{2+} Nanocrystals

The electric dipole-allowed transition, $4f^7$-$4f^6$5d, originating from bivalent europium doped strontium aluminate has profound dependence on the particle size. Figure 6 shows the excitation and emission spectra at room temperature of $SrAl_2O_4$: 2mol%Eu phosphor prepared by solid-state reaction method (Bulk) and solution combustion (Nano), respectively [99]. By comparison, it is easily to notice that the position of the emission band in nanocrystals shifts to higher energies compared with the bulk materials (from 513nm to 494nm). It should be also noted that the band edge of excitation spectra also had a slightly blueshift from 460nm of bulk to 455nm of nanocrystals. Moreover, a new band at 404nm is also detected in nanocrystals, which is ascribed to the emission of the higher excited state energy level of Eu^{2+} ions. It is found that fluorescent lifetime of nanocrystal is shorter than that of bulk crystal (from 762 ns of bulk sample to 200 ns of 18.7 nm particles) maybe due to non-radiation relaxation caused by surface defects and hydrated or carbonated species that act as quenching centers.

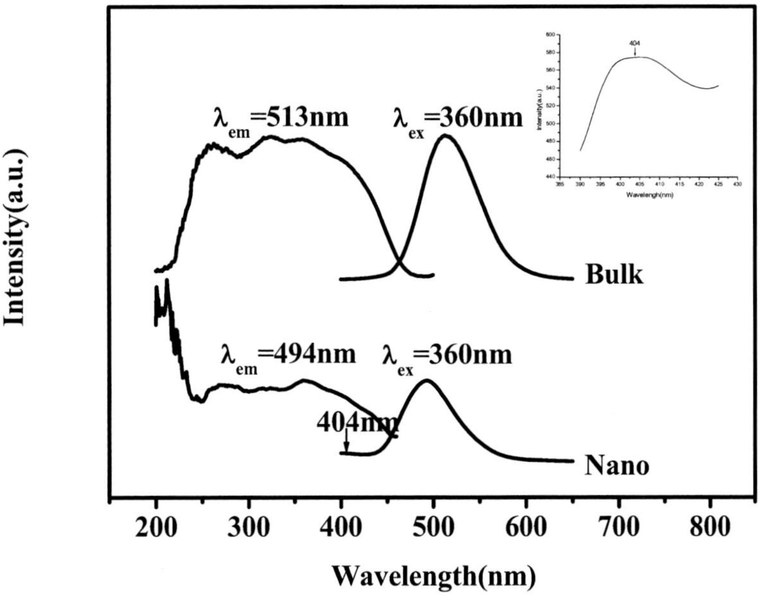

Figure 6. Photoluminescence emission and excitation spectra of bulk and nanocrystalline $SrAl_2O_4:Eu^{2+}$ samples (2%). In the inset the new band at 404 nm is shown. (Adapted from ref. [99])

The first-principles band calculation has been performed using CASTEP code [100]. This is well-tested method that can be applied to a great variety of materials. The minimum forbidden gap of bulk material is calculated about 4.479 eV, while that of nano-material is 4.519 eV. The band gap of nano-$SrAl_2O_4$ increases by 0.040 eV than that of bulk $SrAl_2O_4$. While the experimental result is that the change of energy between the bulk and nano-material is 0.03 eV (the shift of the band edge). Consequently, it is concluded that the width of band gap becomes large when the lattice parameters become small, and the calculated results is in reasonable agreement with the experimental results. In addition, if the unit cell parameters of nanocrystals have been known, this method can be used to predict the changes of energy gap for different particle sizes.

3.3.3 Site-Selected Spectroscopy

3.3.3.1 Bulk And Nanocrystalline SrY_2O_4: Eu^{3+}

It is well known that the Eu^{3+} ion is sensitive to the surrounding environment, and the effect of the crystal-field will cause shifts and splittings of crystal field levels. Therefore, the laser-selective excitation experiments and high resolution emission spectra are a good method to clarify the site occupation of Eu^{3+} doped in SrY_2O_4[101].

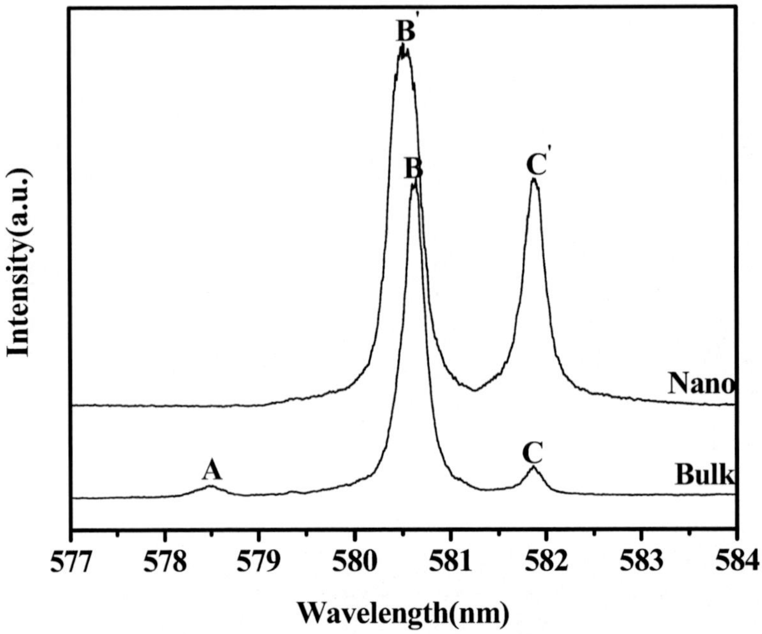

Figure 7. Laser-selective excitation spectra of bulk and nanocrystalline $SrY_2O_4:Eu^{3+}$ were measured by monitoring at 588.58nm. (Adapted from ref. [101])

It is also known that the initial and final energy state involved in $^5D_0$-$^7F_0$ transition is nondegenerate, only a single transition is expected. Laser-selective excitation spectra of bulk $SrY_2O_4:Eu^{3+}$ clearly shows three distinct $^7F_0$ -$^5D_0$ lines at 578.49, 580.63 and 581.86nm in Figure 7, which are labeled A, B and C, respectively. The different positions confirm the existence of at least three sites over which the Eu^{3+} dopant is distributed. For nanocrystalline $SrY_2O_4:Eu^{3+}$, the excitation spectrum shows two distinct $^7F_0$ -$^5D_0$ lines at 580.53 and 581.88nm, which are labeled B′ and C′, respectively. By contrast, it is clearly seen that the spectral line widths in nanocrystal is significantly broader than those in bulk material, indicating a higher disorder of the Eu^{3+} crystalline environment. In addition, it is suggested that it is energetically not easy for Eu^{3+} to substitute Sr^{2+} in comparative lower temperature due to the different valence states and difference of the ion sizes between Sr^{2+} and Eu^{3+}. Therefore, in nanocrystalline $SrY_2O_4:Eu^{3+}$, there are two different sites, i.e., Y(1) and Y(2) site, for Eu^{3+} replacing Y sites. Figure 8 shows the high-resolved emission spectra of bulk SrY_2O_4: Eu^{3+} upon excitation in the A-C bands. One can observe that there is distinct difference in three emission spectra, which also confirmed there are three different Eu sites, i.e., Sr site, Y(1) site and Y(2) site, respectively. Finally, it is also found that when the particle size becomes small, the change of d energy level of rare earth ions is obvious while f energy level of rare earth ions almost unchanged.

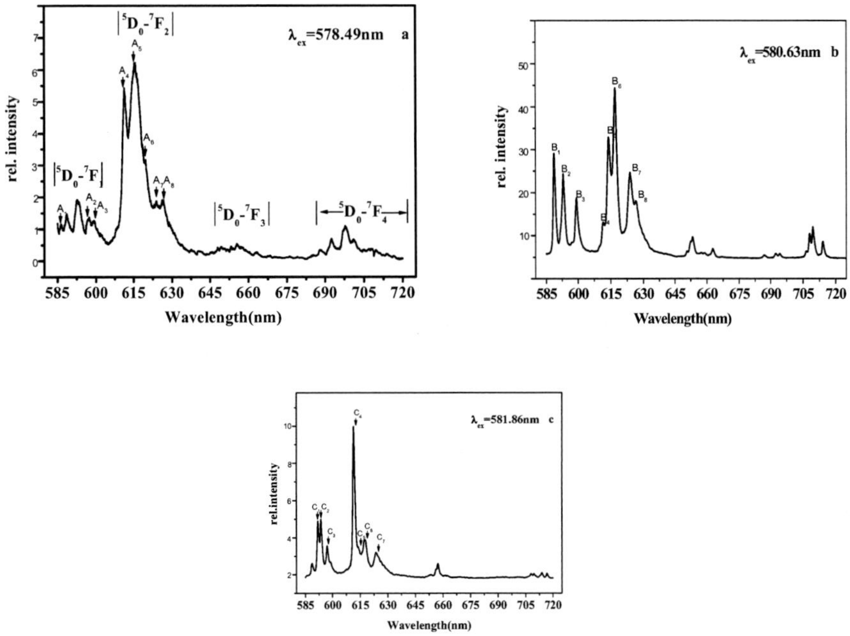

Figure 8. The high-resolved emission spectra of bulk $SrY_2O_4:Eu^{3+}$ for the excitation at 578.49(A), 580.63(B) and 581.86nm(C), respectively. (Adapted from ref. [101])

Conclusion

The purpose of this chapter has been to summarize the subject of luminescent properties from rare earth ions in nanocrystals and show the difference between the bulk and nanocrystals wihin the framework of examples of work done during the past 15 years. It is found that the energy change related to 5d level of rare earth ions is obvious, while energy level of $4f^N$ configuration of rare earth ions almost unchanged with the particle size decrease. The various mechanisms of spectral properties of rare earth ions from in bulk to in nanocrystals are analyzed in detail by taking the rare earth-doped oxidic insulator nanocrystals as samples. Doping enhances the properties of nanocrystals by providing another means to control their remarkable electronic, optical, transport, and magnetic properties. Therefore, the question should be considered when rare earth doped nano-semiconductors are to be made and characterized, how do we know when nanocrystals have been successfully doped? In addition, this chapter is certainly not complete in its coverage. It does not treat all of the work done in this area in the last fifteen year. For example, it does not deal with the extensive amount of research done on rare earth doped semiconductors luminescence. It also suffers from an emphasis on the author's own work, a weakness which is difficult to avoid because of its familiarity. Hopefully the extended bibliography will partially make amends for this bias.

REFERENCES

[1] Fouassier, C. (1997). Development of highly radiation-resistant aramid fiber for concrete reinforcing. *Curr. Opin. Solid State Mater. Sci., 2*(2), 231-235.

[2] Blasse, G. (1995). Luminescent materials: is there still news? *J. Alloys Compd., 225*(1-2), 529-533.

[3] Blasse, G., Grabmaier B. C. (1994). *Luminescent Materials*; Springer-Verlag: Berlin.

[4] Ronda, C. R. (1997). Recent achievements in research on phosphors for lamps and displays. *J. Lumin., 72-74*, 49-54.

[5] Hase, T., Kano, T., Nakazawa, E., Yamamoto, H. (1990) Phosphor materials for cathode-ray tubes. *Adv. Electronics Electron Phys., 79*, 271-373.

[6] Blasse, G. (1989). New luminescent materials. *Chem. Mater., 1*, 294-301.

[7] Blasse, G. (1994). Scintillator materials. *Chem. Mater., 6*, 1465-1475.

[8] Kaminskii, A. A., Weber, M. J. (1996) *Crystalline Lasers: Physical Processes and Operating Schemes*; CRC Press: Boca Raton, FL.

[9] Barber, D. B., Pollock, C. R., Beecroft, L. L., Ober, C. K. (1997). Amplification by optical composites. *Optics Lett., 22*, 1247-1249.

[10] Yoffe, A. D. (1993). Low-dimensional systems: quantum size effects and electronic properties of semiconductor microcrystallites (zero-dimensional systems) and some quasi-two-dimensional system. *Adv. Phys., 42*, 173-266.

[11] Bhargava, R. N., Gllagher, D., Hong, X., Nurmikko, A. (1994). Optical properties of manganese-doped nanocrystals of ZnS. *Phys. Rev. Lett.*, 72, 416-419.

[12] Willianms, D. K., Bihari, B., Tissue, B. M., Machale, J. M. (1998). Preparation and fluorescence spectroscopy of bulk monoclinic Eu^{3+}: Y_2O_3 and comparison to Eu^{3+}: Y_2O_3 nanocrysrals. *J. Phys. Chem. B., 102*, 916-920.

[13] Qi, Z., Shi, C., Zhang, W., Hu, T. (2002). Local structure and luminescence of nanocrystalline Y_2O_3: Eu. *Appl. Phys. Lett., 81*, 2857-2859.

[14] Tissue, B. M.; Bihari, B. (1998). Lanthanide luminescence as a probe of nanocrystalline materials. *J. Fluor., 8*(4), 289-294.

[15] Bhargava, R. N. (1996). Doped nanocrystalline materials-Physics and applications. *J. Lumin., 70*, 85-94.

[16] Goldburt, E. T., Kulkarni, B., Bhargava, R. N., Taylor, J., Libera, M. (1997). Size dependent efficiency in Tb doped Y_2O_3 nanocrystalline phosphor. *J. Lumin., 72/74*, 190-192.

[17] Tao, Y., Zhao, G., Ju, X., Shao, X., Zhang, W., Xia, S. (1996). EXAFS studies of luminescence centres in Eu^{3+} doped nanoscale phosphors. *Mater. Lett., 28*, 137-140.

[18] Khairutdinov, R. F. (1997). Physical chemistry of nanocrystalline semiconductors. *Colloid J., 59*, 535-548.

[19] Brus, L. E., Efros, Al. L., Itoh, T. (Eds.). (1996). Special issue on spectroscopy of isolated and assembled semiconductor nanocrystals, *J. Lumin.***,** 70.

[20] Reed, M. A., Seabaugh A. C. (1994). Prospects for semiconductor quantum devices. *Adv. Chem. Ser., 240*, 15-42.

[21] Song, H. W., Yu, L. X., Yang, L. M., Lu, S. Z. (2005). Luminescent properties of rare earth ions in one dimensional oxide nanowire. *J. Nanosci. Nanotech., 5,* 1519-1531.

[22] Song, H. W., Pan, G. H., Bai, X., Li, S. W., Yu, H. Q., Zhang, H.(2008). One-dimensional rare earth compounds and complexes: preparation and improved photoluminescence properties. *J. Nanosci. Nanotech., 8,* 1316-1325.

[23] Yu, M., Wang, H., Lin, C. K., Li, G. Z., Lin, J. (2006). Sol–gel synthesis and photoluminescence properties of spherical SiO_2@$LaPO_4$:Ce^{3+}/Tb^{3+} particles with a core–shell structure. *Nanotechnology., 17,* 3245-3252.

[24] Wang, H., Yu, M., Lin, C. K., Lin, J. (2006). Core–shell structured SiO_2@YVO_4:Dy^{3+}/Sm^{3+}phosphor particles: sol-gel preparation and characterization. *J. Colloid Interface Sci., 300,* 176-182.

[25] Ihara, M., Igarashi, T., Kusunoki, T., Ohno, K. (2000). Preparation and characterization of rare earth activators doped nanocrystal phosphors. *J. Electrochem. Soc.*, *147*, 2355-2357.

[26] Chen, W., Malm, J-. O., Zwiller, V., Huang, Y., Liu, S., Wallenberg, R., Bovin, J-. O., Samuelson, L. (2000). Energy structure and fluorescence of Eu^{2+} in ZnS:Eu nanoparticles. *Phys. Rev. B.*, *61*, 11021-11024.

[27] Xu, S. J., Chua, S. J., Liu, B., Gan, L. M., Chew, C. H., Xu, Q. Q. (1998). Luminescence characteristics of impurities-activated ZnS nanocrystals prepared in microemulsion with hydrothermal treatment. *Appl. Phys. Lett.*, *73*, 478-480.

[28] Levy, D., Esquivias, L. (1995). Sol-gel processing of optical and electrooptical materials. *Adv. Mater.*, *7*, 120-128.

[29] Gutzov, S., Bredol, M., Wasgestian, F. (1998). Cathodoluminescence study of europiumdoped zirconia and cassiterite powders. *J. Phys. Chem. Solids.*, *59*, 69-74.

[30] Park, D. G., Burlitch, J. M. (1996). Electro-spray synthesis of titanium oxide nano-particles. *J. Sol-Gel Sci. Technol.*, *6*, 235-249.

[31] Jain, S., Skamser, D. J., Kodas, T. T. (1997). Morphology of single-component particles produced by spray pyrolysis. *Aerosol Sci. Technol.*, *27*, 575-590.

[32] Jain, S., Skamser, D. J., Kodas, T. T. (1993). Aerosol processing of materials. *Aerosol Sci. Technol.*, *19*, 411-452.

[33] Shea, L. E., McKittrick, J., Lopez, O. A., Sluzky, E. (1996). Synthesis of red-emitting, small particle size luminescent oxide using an optimized combustion process. *J. Am. Ceram. Soc.*, *79*, 3257-3260.

[34] Matteazzi, P., Alcala, M. (1997). Mechanomaking of Fe/Al_2O_3 and FeCr/Al_2O_3 nanocomposites powders fabrication. *Mater. Sci. Eng. A*, *230*, 161-170.

[35] Giannelis, E. P. (1992). Advances in the processing and properties of $YBa_2Cu_4O_8$. *JOM* (March), *44*, 28-30.

[36] Kawa, M., Fre´chet, J. M. (1998). Self-assembled lanthanide-cored dendrimer complexes: enhancement of the luminescence properties of lanthanide ions through site-isolation and antenna effects. *Chem. Mater.*, *10*, 286-296.

[37] Dutta, P. K., Asiaie, R., Akbar, S. A., Zhu, W. (1994). Hydrothermal synthesis and dielectric properties of tetragonal $BaTiO_3$. *Chem. Mater.*, *6*, 1542-1548.

[38] Kumar, S., Messing, G. L., White, W. B. (1993). Metal orgamic resin derived barim titanate: I, formation of barium titanate oxycarbonate intermediate. *J. Am. Ceram. Soc.*, *76*, 617-624.

[39] Murase, N., Jagannathan, R., Kanematsu, Y., Watanabe, M., Kurita, A., Hirata, K., Yazawa, T., Kushida, T. (1999). Fluorescence and EPR characteristics of Mn^{2+} -doped

ZnS nanocrystals prepared by aqueous colloidal method. *J. Phys. Chem. B.*, *103*, 754-760.

[40] [40] Bol, A. A., Meijerink, A. (1998). Long-lived Mn^{2+} emission in nanocrystalline $ZnS:Mn^{2+}$. *Phys. Rev. B.*, *58*, R15997-R16000.

[41] [41] Yang, P., Lu, M.; Xu, D., Yuan, D., Zhou, G. (2001). Photoluminescence properties of ZnS nanoparticles co-doped with Pb^{2+} and Cu^{2+}. *Chem. Phys. Lett.*, *336*, 76-80.

[42] Xu, S. J., Chua, S. J.; Liu, B., Gan, L. M., Chew, C. H., Xu, Q. Q. (1998). Luminescence characteristics of impurities-activated ZnS nanocrystals prepared in microemulsion with hydrothermal treatment. *Appl. Phys. Lett.*, *73*, 478-480.

[43] Ye, T., Guiwen, Z., Weiping, Z., Shangda, X. (1997). Combustion synthesis and photoluminescence of nanocrystalline Y_2O_3: Eu phosphors. *Mater. Res. Bull.*, *32*, 501-506.

[44] [44] Soo, Y. L., Huang, S. W., Ming, Z. H., Kao, Y. H., Smith, G. C., Goldburt, E., Hodel, R., Kulkarni, B., Veliadis, J. V. D., Bhargava, R. N. (1998). X-ray excited luminescence and local structures in Tb-doped Y_2O_3 nanocrystals. *J. Appl. Phys.*, *83*, 5404-5409.

[45] Riwotzki, K., Haase, M. (1998). Wet-chemical synthesis of doped colloidal nanoparticles: YVO_4:Ln (Ln = Eu, Sm, Dy). *J. Phys. Chem. B.*, *102*, 10129-10135.

[46] Sun, L., Yao, J., Liu, C., Liao, C., Yan, C. (2000). Rare earth activated nanosized oxide phosphors: synthesis and optical properties. *J. Lumin.*, *87-89*, 447-450.

[47] Kane, R. S., Cohen, R. E., Silbey, R. (1999). Synthesis of doped ZnS nanoclusters within block copolymer nanoreactors. *Chem. Mater.*, *11*, 90-93.

[48] Kushida, T., Kurita, A., Watanabe, M., Kanematsu, Y. (2000) .Optical properties of Sm-doped ZnS nanocrystals. *J. Lumin.*, *87-89*, 466-468.

[49]] Morita, M., Rau, D., Fujii, H., Minami, Y., Murakami, S., Baba, M., Yoshita, M., Akiyama, H. (2000). Photoluminescence of CdS : Mn^{2+} and Eu^{3+} nanoparticles dispersed in zirconia sol–gel films. *J. Lumin.*, *87-89*, 478-481.

[50] [50] Schmidt, T., Muller, G., Spanhel, L. (1998). Activation of 1.54 μm Er^{3+} fluorescence in concentrated II-VI semiconductor cluster environments. *Chem. Mater.*, *10*, 65-71.

[51] Hommel, D., Hartmann, H., Godlewski, M., Langer, J. M., Stapor, A. In *Rare Earth Spectroscopies*; Jezowska-Trzebiatowska, B., Legendziewicz, J., Strek, W., Eds.; World Scientific: Singapore, (1985), p 643.

[52] Zhang, H., Shen, Y. (1988). Luminescence of ZnS:Ho under excitation around the band edge (EBE). *J. Lumin.*, *40-41*, 401-402.

[53] Stapor, A., Godlewski, M., Przybylinska, H., Hommel, D. (1988). Novel excitation mechanism of rare earth intra-shell emission. *J. Lumin.*, *40-41*, 625-626.

[54] Yen, W. M., Shionoya, S., Eds., *Phosphor Handbook*, CRC Press: Boca Raton, 1999.

[55] Bol, A. A., Beek, R.V., Meijerink, A. (2002). On the incorporation of trivalent rare earth ions in II-VI semiconductor nanocrystals. *Chem. Mater.*, *14,* 1121-1126.

[56] [56] Bryan, J. D., Gamelin, D. R. (2005). Doped semiconductor nanocrystals: synthesis, characterization, physical properties, and applications. *Progr. Inorg. Chem., 54,* 47-126.

[57] Norris, D. J., Efros, A. L., Erwin, S. C. (2008). Doped nanocrystals. *Science., 319,* 1776-1779.

[58] Turnbull, D. (1950). Formating of crystal nuclei in liquid metal. *J. Appl. Phys., 21,* 1022-1028.

[59] Dalpian, G. M., Chelikowsky, J. R., (2006). Self-purification on semiconductor nanocrystals. *Phys. Rev. Lett., 96,* 226802-226804.

[60] Mikulec F. V., Kuno, M., Bennati, M., Hall, D. A., Griffin, R. G., Bawendi, M. G. (2000).Organometallic synthesis and spectroscopic characterization of manganese-doped CdSe nanocrystals. *J. Am. Chem. Soc., 122,* 2532-2540.

[61] Bryan, J. D., Gamelin, D. R., (2005). Doped semiconductor nanocrystals: synthesis, characterization, physical properties, and applications. *Prog. Inorg. Chem., 54,* 47-126.

[62] Erwin, S. C., Zu, L. J., Haftel, M. I., Efros, A. L., Kennedy, T. A., Norris, D. J. (2005). Doping semiconductor nanocrystals. *Nature., 436,* 91-94.

[63] Pires, A. M., Davolos, M. R. (2001). Luminescence of europium(III) and manganese(II) in barium and zinc orthosilicate. *Chem. Mater., 13,* 21-27.

[64] Blasse, G., Grabmaier, B. C. *Luminescent Materials;* Springer-Verlag: Berlin. 1994.

[65] Danielson, E., Devenney, M., Giaquinta, D. M., Golden, J. H., Haushalter, R.C., McFarland, E.W., Poojary, D. M., Reaves, C. M., Weinberg, W. H., Wu, X. D. (1998). A rare-earth phosphor containing one-dimensional chains identified through combinatorial methods. *Science., 279,* 837-839.

[66] Pieterson, L.V., Soverna, S., Meijerink, A. (2000). On the nature of the luminescence of Sr_2CeO_4. *J. Electrochem. Soc., 147,* 4688-4691.

[67] Blasse, G. (1966). On the Eu^{3+} fluorescence of mixed metal oxides. IV. the photoluminescent efficiency of Eu^{3+}-activated oxides. *J. Chem. Phys., 45,* 2356-2360.

[68] Zhang, W. W., Zhang, W. P., Xie, P. B., Yin, M., Chen, H. T., Jing, L., Zhang, Y. S., Lou, L. R., Xia, S. D. (2003). Optical properties of nanocrystalline Y_2O_3: Eu depending on the odding structure. *J. Colloid Interface Sci., 262,* 588-593.

[69] Wei, Z. G., Sun, L. D., Liao, C. S., Jiang, X. C., Yan, C. H. (2003). Luminescent properties for hexagonal YBO_3 : Eu nanocrystals in the vacuum ultraviolet region. *J. Appl. Phys., 93,* 9783-9788.

[70] Louis, C., Bazzi, R., Flores, M. A., Zheng, W., Lebbou, K., Tillement, O., Mercier, B., Dujardin C., Perriat, P. (2003). Synthesis and characterization of $Gd_2O_3:Eu^{3+}$ phosphor nanoparticles by a sol-lyophilization technique. *J. Solid State Chem., 173,* 335-341.

[71] Igarashi, T., Ihara, M., Kusunoki, T., Ohno, K., Isobe, T., Senna, M., (2000). Relationship between optical properties and crystallinity of nanometer Y_2O_3: Eu phosphor. *Appl. Phys. Lett., 76,* 1549-1551.

[72] Konrad, A., Herr, U., Tidecks, R.(2001). Luminescence of bulk and nanocrystalline cubic yttria. *J. Appl. Phys., 90(7),* 3516-3523.

[73] Williams, D. K., Yuan, Y., Tissue, B. (1999). Size dependence of luminescence spectra and dynamics of Eu^{3+}: Y_2O_3 nanocrysrals. *J. Lumin., 83–84*, 297-300.

[74] Huignard, A., Gacoin, T., Boilot, J. P. (2000). Synthesis and luminescence properties of colloidal YVO_4: Eu phosphors. *Chem. Mater., 12,* 1090-1094.

[75] Huignard, A., Buissette, V., Laurent, G., Gacoin, T., Boilot, J. P. (2002). Synthesis and characterizations of YVO_4: Eu colloids. *Chem. Mater., 14*, 2264-2269.

[76] Huignard, A., Buissette, V., Franville, A. C., Gacoin, T., Boilot, J. P. (2003). Emission processes in YVO_4: Eu nanoparticles. *J. Phys. Chem. B., 107*, 6754-6759.

[77] Buissette, V., Huignard, A., Gacoin, T., Boilot, J. P., Aschehoug, P., Viana, B. (2003). Luminescence properties of YVO_4:Ln (Ln=Nd, Yb, and Yb–Er) nanoparticles. *Surf. Sci.,532-535*, 444- 449.

[78] Riwotzki, K., Meyssamy, H., Schnablegger, H., Kornowski, A., Haase, M. (2001). Liquid-phase synthesis of colloids and redispersible powders of strongly luminescing $LaPO_4$: Ce, Tb nanocrystals. *Angew. Chem., Int. Ed., 40,* 573-576.

[79] Haas Y., Stein, G. (1972). Radiative and nonradiative pathways in solutions. excited states of the europium(III) ion. *J. Phys. Chem., 76,* 1093-1104.

[80] Nedelec, J. M., Avignant, D., Mahiou, R. (2002). Soft chemistry routes to YPO_4-based phosphors: Dependence of textural and optical properties on synthesis pathways, *Chem. Mater., 14,* 651-655.

[81] Huignard, A., Gacoin, T., Chaput, F., Boilot, J. P., Aschehoug, P., Viana, B., Synthesis and luminescence properties of colloidal lanthanide doped YVO_4, Materials Research Society, 2001 Spring Proceedings, *Volumes (664 – 685)* Symposium G4.5.

[82] Kumar, G. A., Chen, C. W., Riman, R., Chen, S., Smith, D., Ballato, J.(2005). Optical properties of a transparent $CaF_2:Er^{3+}$ fluoropolymer nanocomposite, *Appl. Phys. Lett., 86,* 241105-1-241105-3.

[83] Morita, M., Buddhudu, S., Rau, D., Murakami, S. (2004). Photoluminescence and excitation energy transfer of rare earth ions in nanoporous xerogel and sol-gel SiO_2 glasses, *Optical Spectra and Chemical Bonding in Transition Metal Complexes. 107,* 115-143.

[84] Ozawa, L., Forest, H., Jaffe, P. M., Ban, G. (1971). The effect of exciting wavelength on optimum activator concentration. *J. Electrochem. Soc., 118,*482-486.

[85] Li, Q., Gao, L., Yan, D.(1997). Effects of grain size on wavelength of $Y_2O_3:Eu^{3+}$ emission spectra, *Nanostruct. Mater., 8,* 825-831.

[86] Duan, C. K., Yin, M., Yan, K., Reid, M. (2000). Surface and size effects and energy transfer phenomenon on the luminescence of nanocrystalline $X_1–Y_2SiO_5:Eu^{3+}$. *J. Alloys Compd., 303–304,* 371-375.

[87] Wei, Z. G., Sun, L. D., Liao, C. S., Jiang, X. C., Yan, C. H. (2003). Luminescent properties for hexagonal YBO_3 : Eu nanocrystals in the vacuum ultraviolet region. *J. Appl. Phys., 93,* 9783-9788.

[88] Jiang, X. C., Yan, C. H., Sun, L. D., Wei, Z. G., Liao, C. S. (2003). Hydrothermal homogeneous urea precipitation of hexagonal $YBO_3:Eu^{3+}$ nanocrystals with improved luminescent properties. *J. Solid State Chem., 175,* 245-251.

[89] Peng, H. S., Song, H. W., Chen, B. J., Lu, S. Z., Huang, S. H. (2003). Spectral difference between nanocrystalline bulk $Y_2O_3:Eu^{3+}$. *Chem. Phys. Lett., 370,* 485–489.

[90] VMahajan, S., Dickerson, J. H. (2007). Synthesis of monodisperse sub-3 nm RE_2O_3 and $Gd_2O_3:RE^{3+}$ nanocrystals, *Nanotechnology.,18,* 325605-325610.

[91] Streka, W., Zych, E., Hreniak, D.(2002). Size effects on optical properties of Lu_2O_3: Eu^{3+} nanocrystallites, *J. Alloys Compd. , 344,* 332–336.

[92] Meltzer, R. S., Feofilov, S. P., Tissue B., Yuan, H. B. (1999). Dependence of fluorescence lifetimes of $Y_2O_3:Eu^{3+}$ nanoparticles on the surrounding medium. *Phys. Rev. B, 60,* R14012-R14015.

[93] Fu, Z. L., Zhou, S. H., Zhang, S. Z. (2005). Study on optical property of rare earth ions in nanocrystalline monoclinic $SrAl_2O_4$: Ln (Ln= Ce^{3+}, Pr^{3+}, Tb^{3+})."*J. Phys. Chem. B., 109,* 14396-14400.

[94] [94] Shi. J. S., Zhang, S.Y. (2003). Predicting the position of the spin-forbidden band for Tb^{3+} ions in crystal hosts. *J. Phys.: Condens. Matter.*, *15,* 4101-4107.

[95] Shi. J. S., Wu. Z. J., Zhou. S. H., Zhang. S. Y. (2003). Dependence of crystal field of 5d levels on hosts in the halide crystals. *Chem. Phys. Lett.*, *380,* 245-250.

[96] Shi. J. S., Zhang, S. Y. (2004). Barycenter of energy of lanthanide $4f^{N-1}5d$ configuration in inorganic crystals. *J. Phys. Chem. B.*, *108,* 18845–18849.

[97] Fu, Z. L. Zhou, S. H., Pan, T. Q., Zhang, S. Y. (2007). Preparation and luminescent properties of cubic Eu^{3+}:Y_2O_3 nanocrystals and comparison to bulk Eu^{3+}: Y_2O_3. *J. Lumin.*, *124,* 213-216.

[98] Dorenbos, P. (2005).The Eu^{3+} charge transfer energy and the relation with the band gap of compounds. *J. Lumin.*, *111,* 89-104.

[99] [99] Fu, Z. L., Zhou, S. H., Yu, Y. N., Zhang, S. Y. (2004). Combustion synthesis and luminescence properties of nanocrystalline monoclinic $SrAl_2O_4$:Eu^{2+}. *Chem. Phys. Lett.*, 395, 285-289.

[100] Segall, M. D., Lindan, P. L. D., Probert, M. J., Pickard, C. J., Hasnip, P. J., Clark, S. J., Payne, M. C. (2002). First-principles simulation: ideas, illustrations and the CASTEP code. *J. Phys.: Condens. Matter.*, *14,* 2717-2744.

[101] Fu, Z. L., Zhou, S. H., Zhang, S. Y. (2006). Preparation and optical properties of trivalent europium-doped bulk and nanocrystalline SrY_2O_4." *J. Opt. Soc. Am. B.*, *23,*1852-1858.

Reviewed by Jung Hyun Jeong, Professor, Department of Physics,Pukyong National University, Busan 608-737, South Korea

In: Thermochemistry and Advances in Chemistry Research ISBN: 978-1-60692-376-4
Ed: Anton G. Mirskiy

Chapter 10

LOW-TEMPERATURIC TEMPLATE SYNTHESIS INTO BIOPOLYMER-IMMOBILIZED MATRIX WITH USING OF CYANOCOMPLEXES OF *D*-ELEMENTS

O. V. Mikhailov*

Kazan State Technological University, K. Marx Street 68, 420015 Kazan, Russia

ABSTRACT

Template synthesis processes proceeding for metal ion – ligand synthon A – ligand synthon B triple systems into solutions and biopolymer-immobilized matrix, have been and analyzed with thermodynamical point of view. It has been shown that final results of these processes should be distinguished extremely considerably. The distinction predicted has been displayed by examples of template synthesis for some M(II)-,M(III) – (N,S)-,(N,O,S)-ambidentate ligand synthon – carbonyl-containing ligand synthon in water-ethanol solutions and metalhexaceanjferrate gelatin-immobilized matrix implantates.

INTRODUCTION

Template synthesis is comspicuous place in a number of synthetic methods of modern coordination and supramolecular chemistry [1-4]. In this method, metal ion having specific stereochemistry of coordination environment and electronic structure, is a peculiar "pattern" and ensures a formation of specific metalcomplexes, the synthesis of which under other conditions either has been difficult or cannot be realized at all, from corresponding starting chemical compounds (what are known as ligand synthons or ligsons). Such a synthesis, in principle, enables to obtain metalmacrocyclic and supramolecular coordination compounds, including even having very complex structure (for example, coordination compounds of various metal ions with phthalocyanines and porphirines) from comparatively simple

* E-mail: ovm@kstu.ru

fragments. However, such a synthesis leads to formation of chemical compounds which are far more complex composition and structure in compared with starting ligand synthones, and, according to general rule of chemical thermodynamic, it proceeds with considerable decrease of entropy of reactionary medium. As it is easily noted, in conformity with classic equation for isobaric chemical process $\Delta G = \Delta H - T\Delta S$ [where Δ, ΔH and ΔS is change of free energy (Gibbs's energy), enthalpy and entropy in the course of chemical reaction, respectively, T is a temperature in K], the possibility of realization of any process accompanied with entropy decrease, decreases with temperature growth. In this connection, next regrettable circumstance takes place: at relatively low temperatures, template processes owing to considerable values of activation energy have very small rates; at relatively large T values, when the rate of given processes may be acceptable with practical point of view, these reactions are thermodynamically forbidden. One way or another, such a state of things may be eased by force of high pressure to reactionary medium under relatively small temperature, and, what is more, during of extremely long time (up to several hours) [1,3]. That is why, «softening» of conditions of template synthesis, and, first of all, of ensuring of possibility of its realization in conditions what are known as standard ($\boldsymbol{T}$ = 298 K, $\boldsymbol{p}$ = 101325 Pa), is one of actual problems of modern synthetic coordination chemistry.

RESULTS

One of possible ways to decision of given problem, preliminary "regulation" of reactionary system (and, namely, "compulsory" decrease of entropy in it) which, as it is easily noticed, leads to that tangent of angle of slope of linear $\Delta G(T)$ dependence to abcissa axe decreases, may be. Indeed, because entropy is additive value, and, in the case of availability of such a preliminary "regulation", general equation $\Delta S = \Delta S' + \Delta S_{os}$, where ΔS is a change of entropy of reaction in an absence of "regulation" indicated, $\Delta S'$ is a change of of entropy of reaction in an availability of it, ΔS_{os} is a change of entropy in the course of the very "regulation", will takes place. As a result, the equation of $\Delta G'(T)$ dependence in the system where "compulsory" decrease of entropy indicated occurs, is described as

$$\boldsymbol{\Delta G' = \Delta H - T(\Delta S - \Delta S_{os})} \qquad (1)$$

and, because $\boldsymbol{\Delta S_{os}}$ is negative value, $\boldsymbol{\Delta S} < 0$, hence, $|\boldsymbol{\Delta S} - \boldsymbol{\Delta S_{os}}| < |\boldsymbol{\Delta S}|$. According to this, the range of temperature values in which given template process is thermodynamically possible, increases. This circumstance has been illustrated on the *Fig. 1.*

One of examples of reactionary mediums where preliminary "regulation" of reactionary system and "compulsory" decrease of entropy takes place, are biopolymer-immobilized matrix, for example thin-layer implantates on polyethyleneterephthalate film, containing as immobilized compounds hexacyanoferrates(II) of various transition metals, among their number $Co_2[Fe(CN)_6]$, $Ni_2[Fe(CN)_6]$, $Cu_2[Fe(CN)_6]$, $KFe[Fe(CN)_6]$ and $KCo[Fe(CN)_6]$. The most suitable biopolymer for creation of biopolymer-immobilized systems indicated is gelatin which very easily forms thin layers.

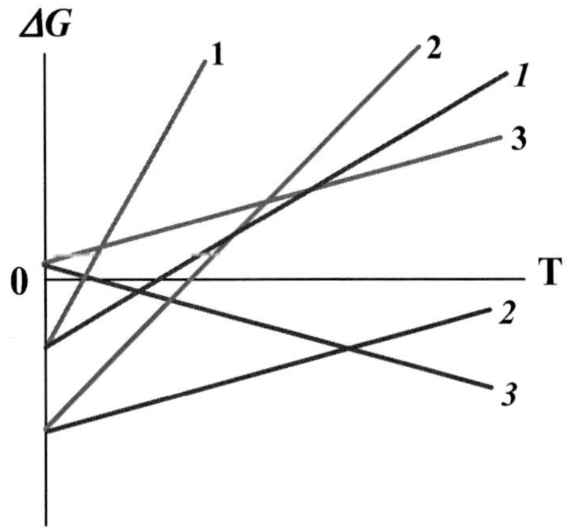

Figure 1. ***ΔG(T)*** dependences **(1,2,3)** и ***ΔG′(T)*** ones ***(1,2,3)*** for three various variants of template process: **1**, ***1*** and **2**, ***2*** – for such template processes having ***ΔH*** < 0 which, in principle, can be realized in the metal ion – ligand synthon A – ligand synthon B triple systems without "compulsory" decrease of entropy indicated, **3**, ***3*** – for template having ***ΔH*** > 0 which may be realized only in the case of "compulsory" decrease of entropy. Lower slope of straight lines of ***ΔG(T)*** and ***ΔG′(T)*** dependences to abscissa axe.

Moreover, according to experimental data, such thin-layer gelatin-immobilized matrix implantates (further MHF-GIM) are transparent as regards to electron flow, and, hence, have nano-structured level of organization of immobilized substance. Structure of such a gelatin-immobilized matrix has been shown on the *Fig. 2.* The overwhelming majority of template reaction found is known to be related to Shiff's condensation accompanied with intra-molecular formation of water at the expense of mobile hydrogen atoms of one ligand synthon and oxygen atoms of another ligand synthon [3]; in this connection, a probability of any such a template reaction is the higher, the more mobile hydrogen atoms in corresponding ligand synthon, that, in its turn, is connected with proton-donor ability of this chemical compound. Because gelatin molecules, as ampholyte and polyelectrolyte, acquire in alkaline medium the negative charge, proton-donor ability of any chemical compound immobilized in it, becomes more considerable in comparison with proton-donor ability of this compound in water solution; hence, template synthesis into MHF-GIM, in principle, must proceed more effectively than in solution or solid phase. And already in the end of XX century, it was noted in reviews [5,6] that character of «classic» complexing [which proceeds according to scheme (metal ion + ligand → complex)] as well as character of template synthesis [which proceeds according to scheme (metal ion + ligand synthones → complex)] into MHF-GIM for a number of (metal ion – ligand) and, correspondingly, (metal ion – ligand synthones) differs considerably from character for complexing and template synthesis in conditions what is known as «traditional» (that is to say in solution and solid phase). As suitable objects that give possibility more clearly to reveal such distinctions, are ambidentate ligand synthones having in its composition three or more number of donor centers and capable to coordinate to metal ion by various way in dependence of conditions of realization of complexing process. To a number of ligand synthones indicated, there are related such (N,S)- and (N,O,S)-donor organic compounds as ethanedithioamide-1,2 (L^{I}), thiocarbamoylmethaneamide (L^{II}), hydrazinomethanethiohydrazide

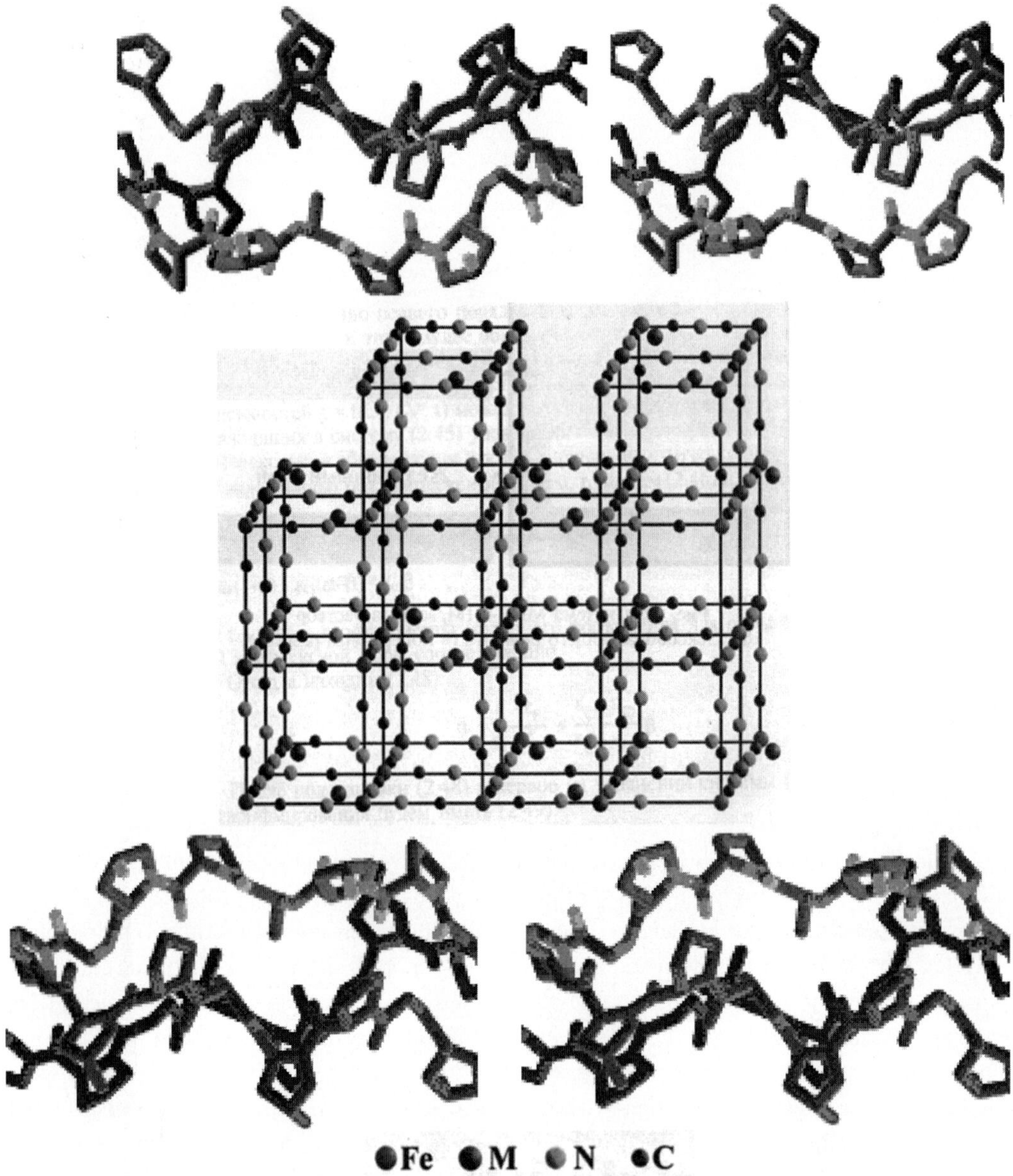

Figure 2. The structure of metalhexacyanoferrate(II) gelatin-immobilized matrix [7].

(L^{III}), propandithioamide-1,3 (L^{IV}) et al, capable to coordinate to metal ion through nitrogen and sulfur atoms; as attendant ligand synthones ensuring template "stitching" of metalchelate cycles in common closed contour, chemical compounds having carbonyl group C=O, for example methanal (L^{I}), ethanal (L^{II}), propanone (L^{III}), ethanedial (L^{IV}), buthanedione-2,3 (L^{V}) and penthanedione-2,4 (L^{VI}), can be used.

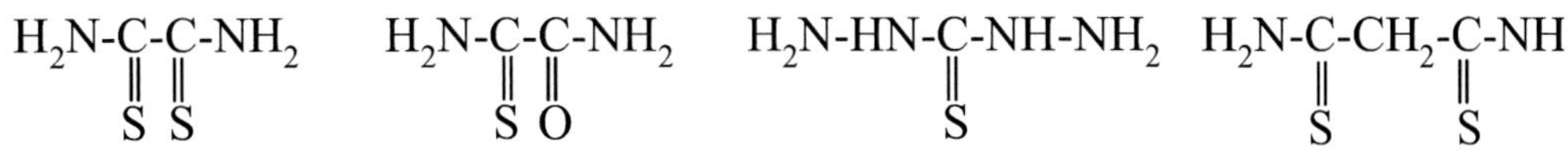

(L^{I}) (L^{II}) (L^{III}) (L^{IV})

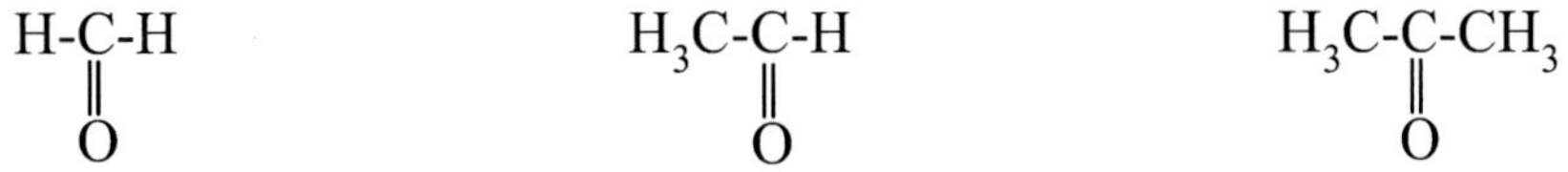

(L^{I}) (L^{II}) (L^{III})

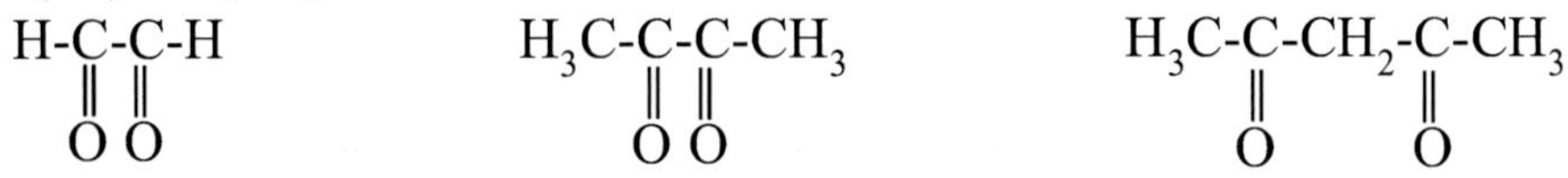

(L^{IV}) (L^{V}) (L^{VI})

The data concerning the final products of template synthesis in triple systems containing some of such ligand synthones, have been presented in the Table 1.

According to theoretical views indicated above, it may be expected that assortment of processes of template synthesis in a separately taken M(II) – (N,S)-containing ligand synthon – carbonyl-containing ligand synthon triple system in organizing reactionary medium with preliminary entropy decrease (and, among their number, into MHF-GIM) will be more considerable than assortment of template processes in «traditional» conditions where the aforesaid decrease of S values does not occur. It should be expected correspondingly to the above that a number of metalcomplexes formed as a result of realization of template processes in organizing medium must be more considerable than a number of metalcomlexes in «non-organizing» one. The data presented in the *Table 1* as it may be easily seen, are in full agreement with such a forecast, in the last resort, for triple systems containing Cu(II) as a "templet" metal ion; besides, in an availability in it of hydrazinomethanethiohydrazide, template synthesis in water-ethanol solution leads to formation of only one metalchelate whereas template synthesis into MHF-GIM, leads to two or even more their number; in an availability in it of ethanedithioamide-1,2 or propanedithioamide-1,3, template synthesis in water-ethanol solution does not occur (only *bis*-chelate copper(II) complexes with ligand synthons indicated) whereas template synthesis into MHF-GIM takes place [7-10].

At the present time, quantum-chemical calculations of template complexes formed in the M(II) – ethanedithioamide-1,2 – methanal (M= Co, Ni, Cu) triple systems, were carried out [11,12]. According to experimental data presented in *Table 1*, (N,N,S,S)-coordination of donor centers of template ligand to M(II) occurs in the complexes arising during as a result of template synthesis in the systems indicated. Template process proceeds here according to general equation (2)

Table 1. The products of complexing processes proceeding in the M(II)- (N,S)-containing ligand synthon- O-containing ligand synthon in the solutions or solid phase and in the $M_2[Fe(CN)_6]$- gelatin-immobilized matrix

Triple system	Products of complexing processes In solution or solid phase	In gelatin-immobilized matrix
1	2	3
Co(II)- (L^I)- (L^I)	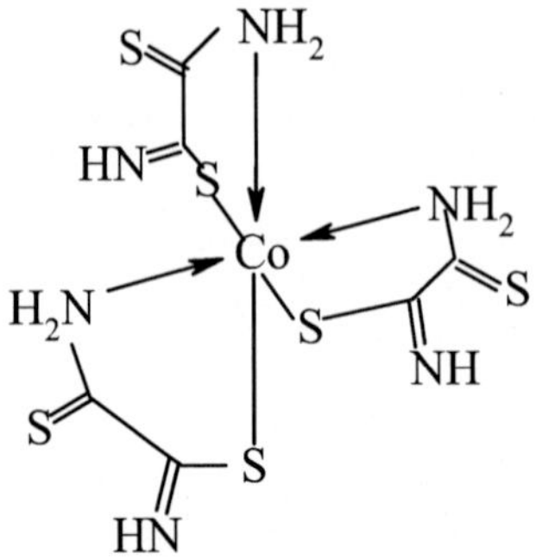	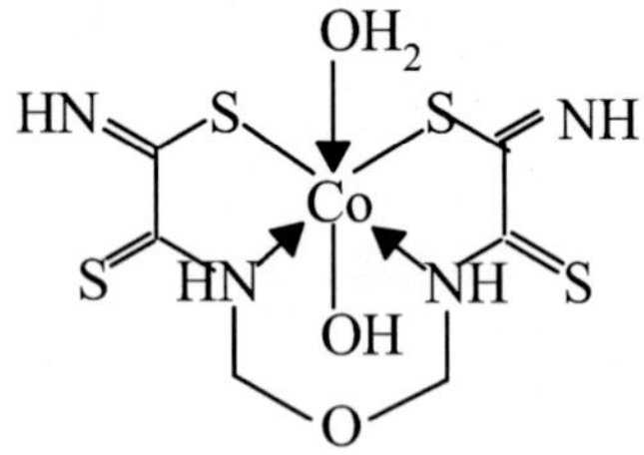
Co(II)- (L^I)- (L^{II})	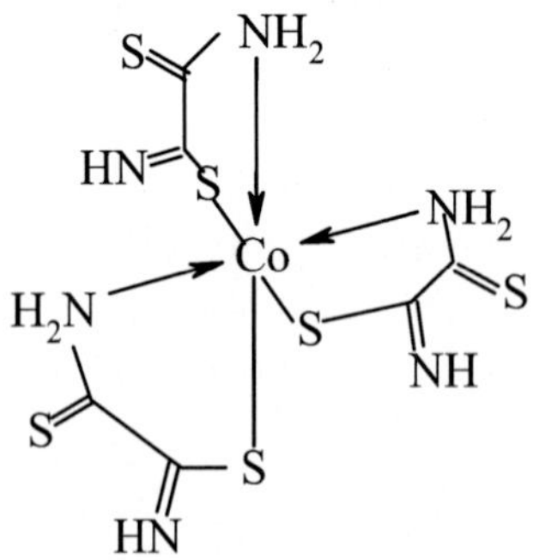	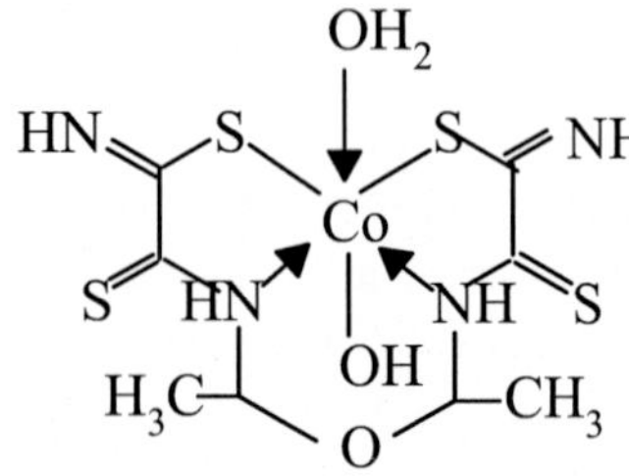
Co(II)- (L^I)- (L^{III})	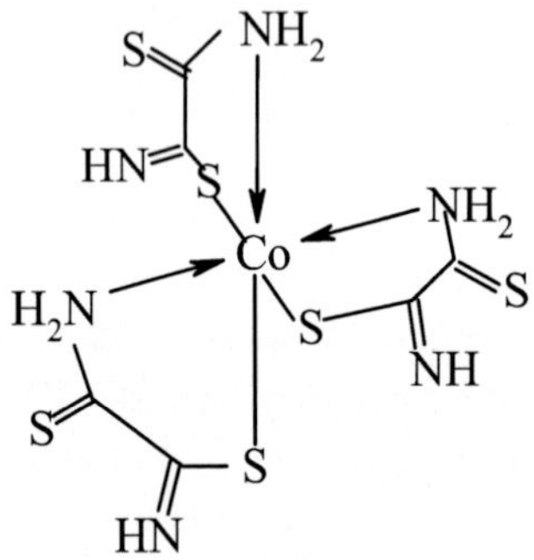	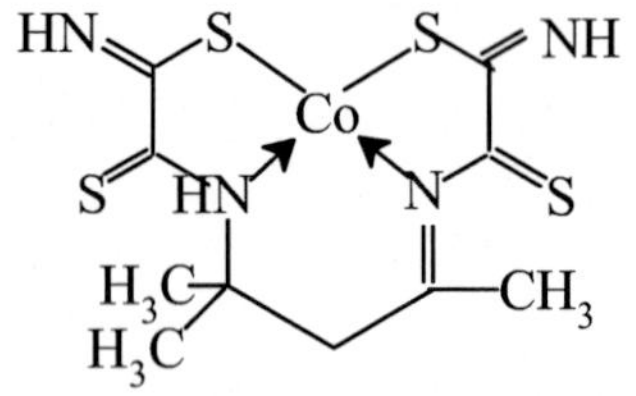
Co(II)- (L^I)- (L^{IV})	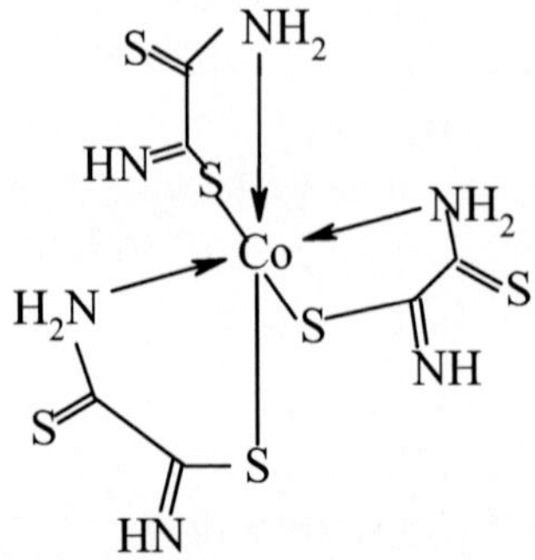	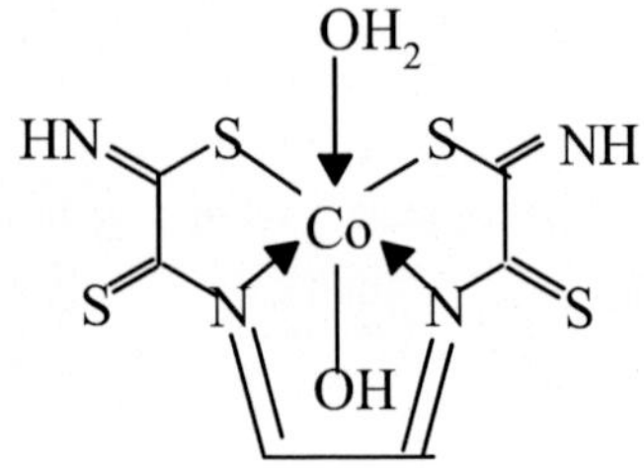

Table 1 (Continued).

Triple system	Products of complexing processes In solution or solid phase	In gelatin-immobilized matrix
1	2	3
Ni(II)- (L^{I})- (L^{I})	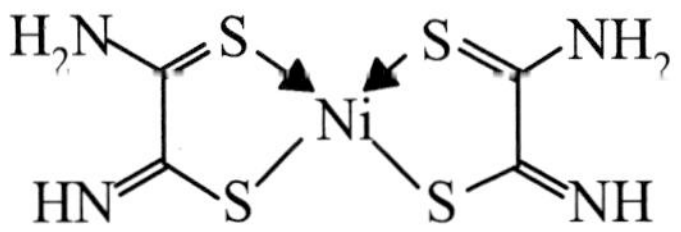	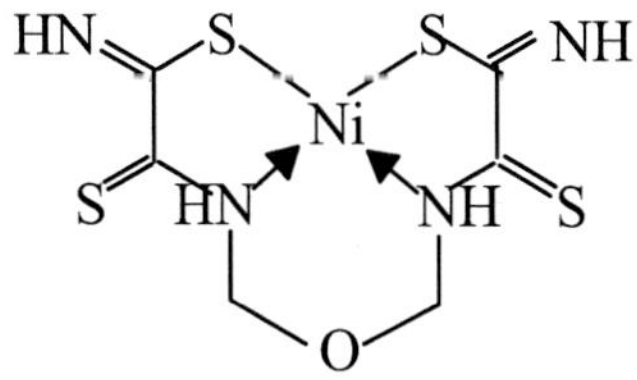
Ni(II)- (L^{I})- (L^{IV})	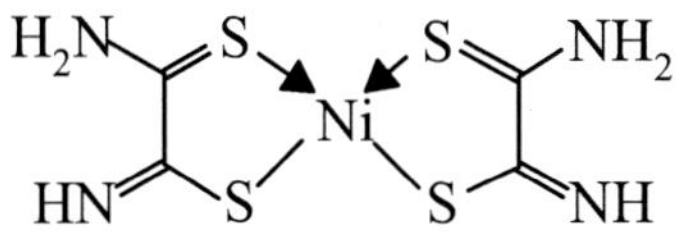	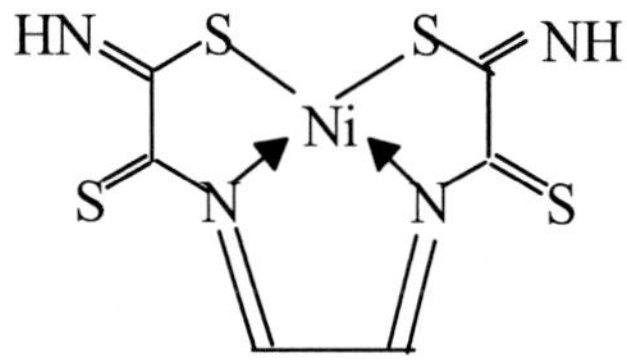
Ni(II)- (L^{I})- (L^{II})	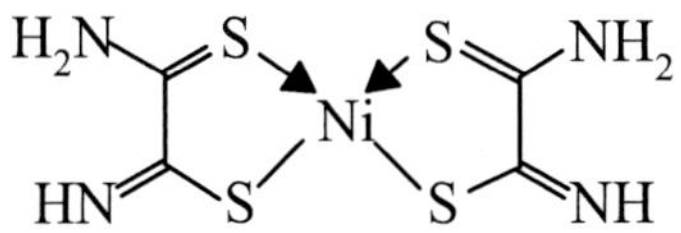	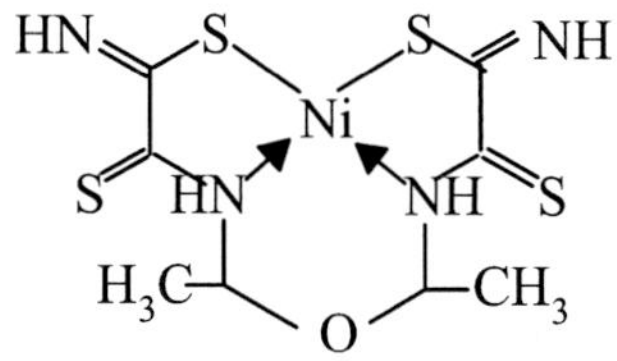
Ni(II)- (L^{I})- (L^{III})	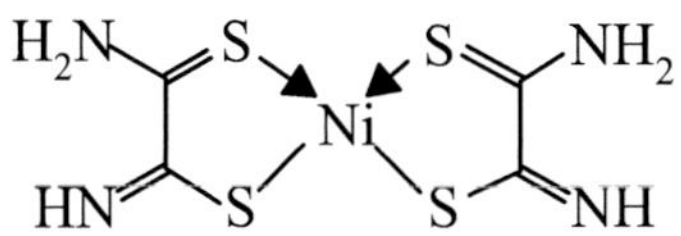	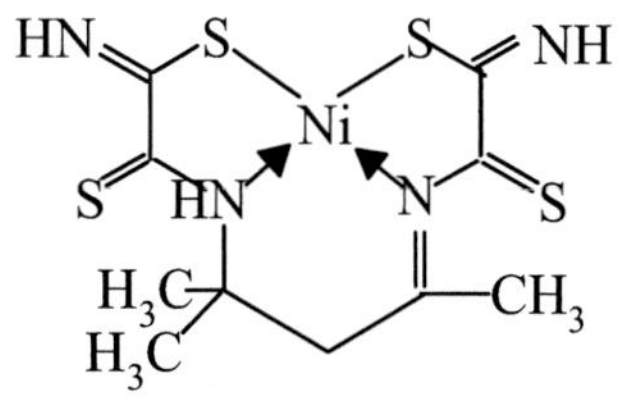
Ni(II)- (L^{I})- (L^{V})	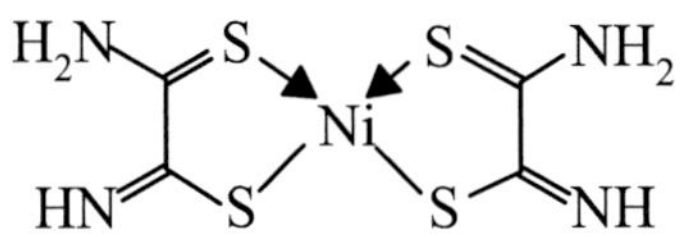	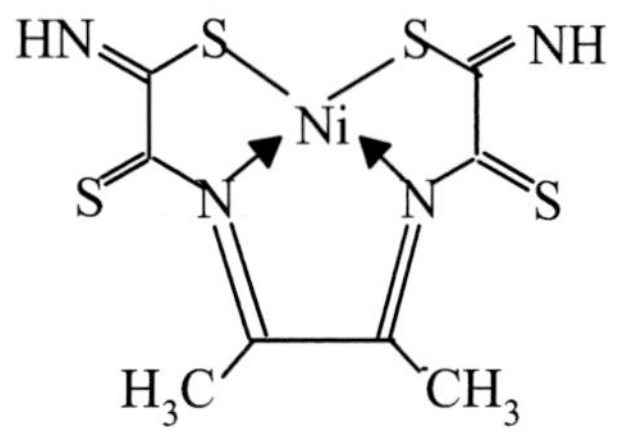
Ni(II)- (L^{IV})- (L^{I})	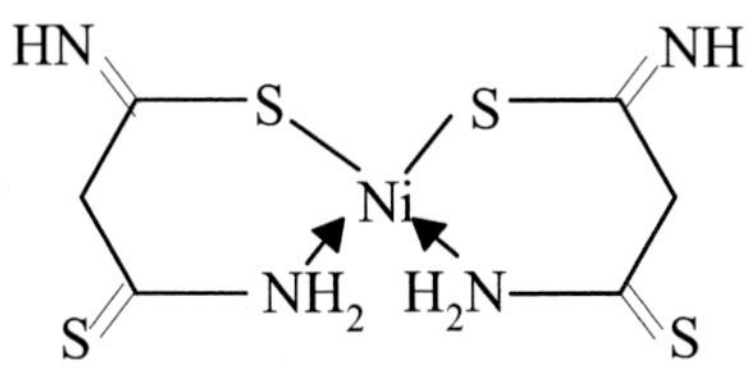	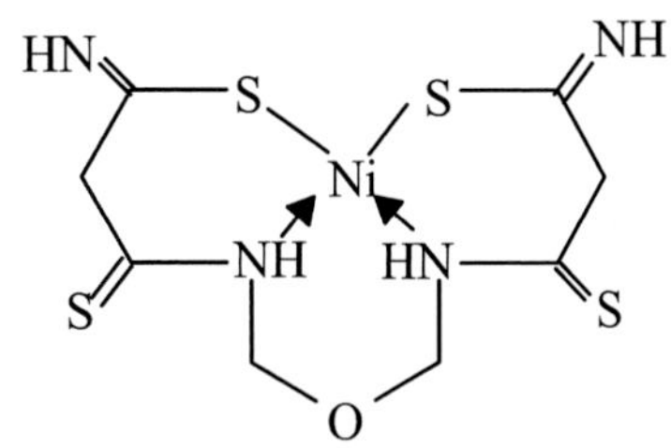

Table 1 (Continued).

Triple system	Products of complexing processes	
	In solution or solid phase	In gelatin-immobilized matrix
1	2	3
Ni(II)- (L^{IV})- (L^{V})	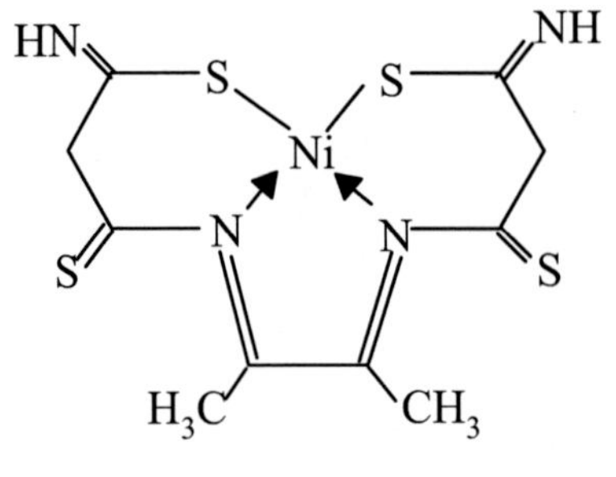	-
Ni(II)- (L^{III})- (L^{III})	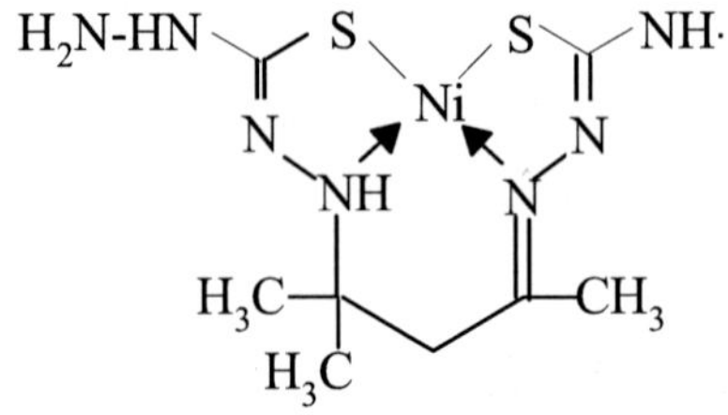	
	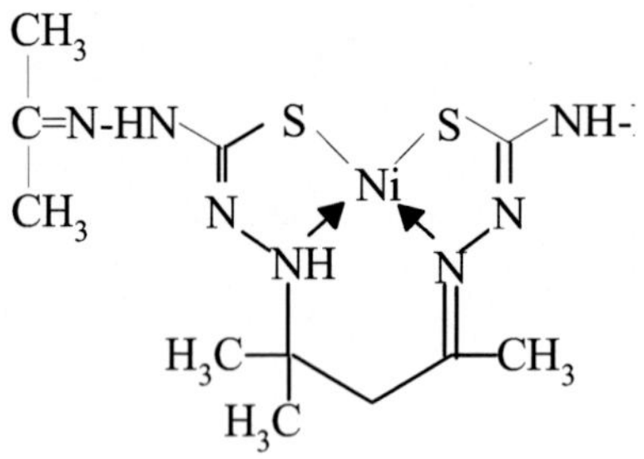	-
	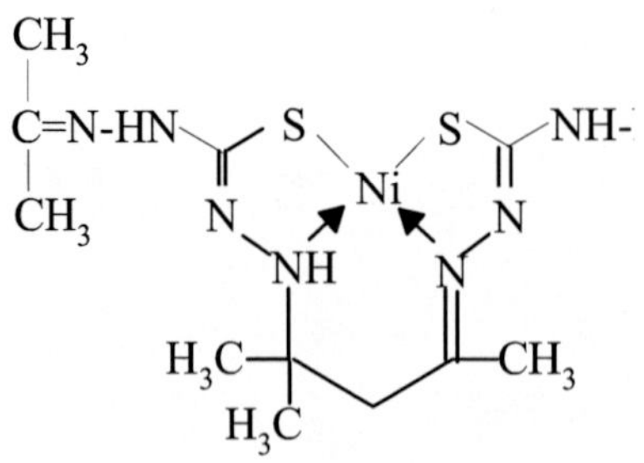	
Ni(II)- (L^{III})- (L^{V})	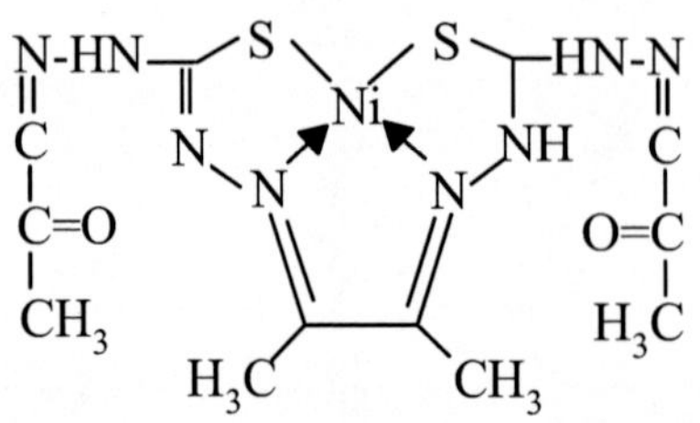	-

Table 1 (Continued).

Triple system	Products of complexing processes In solution or solid phase	In gelatin-immobilized matrix
1	2	3
Cu(II)- (L^I)- (L^I)	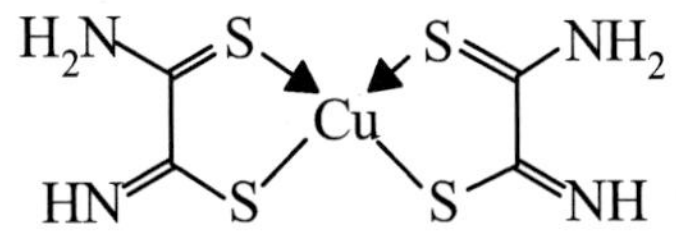	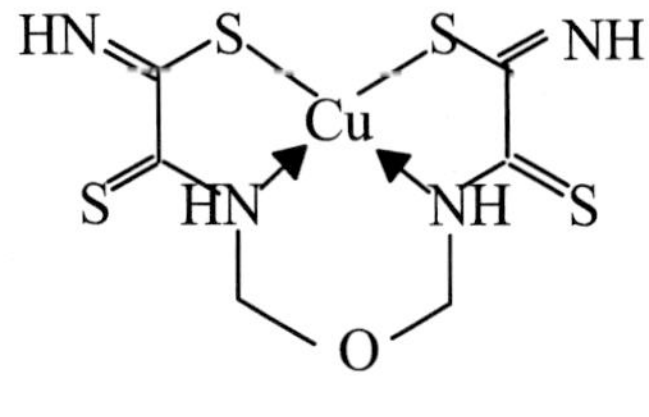
Cu(II)- (L^I)- (L^{IV})	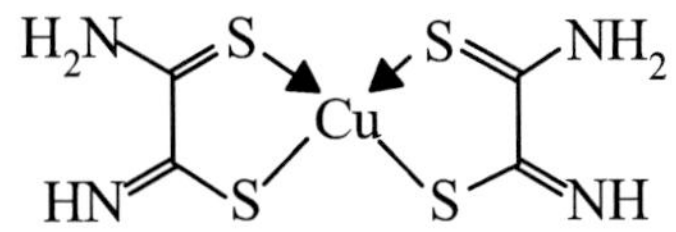	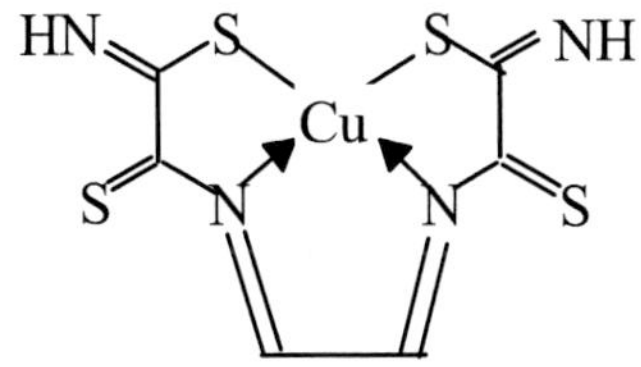
Cu(II)- (L^I)- (L^{II})	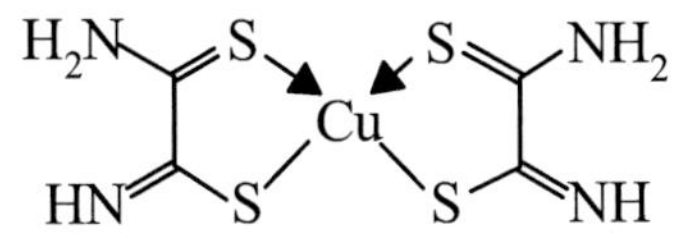	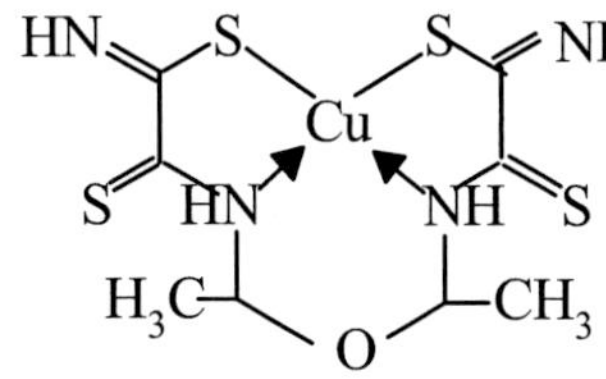
Cu(II)- (L^I)- (L^{III})	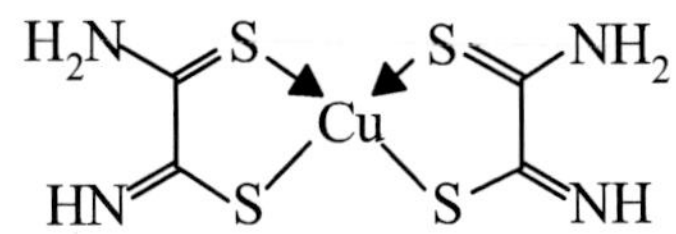	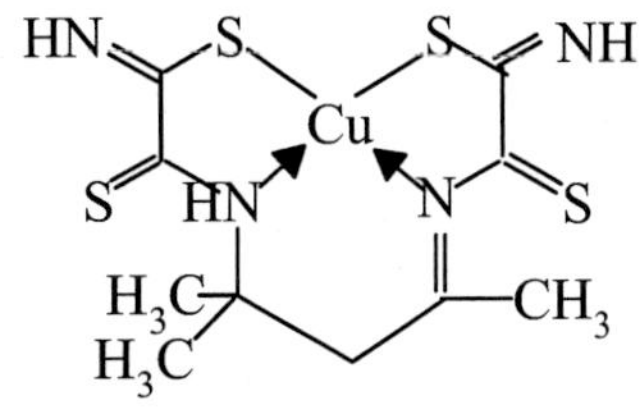
Cu(II)- (L^I)- (L^V)	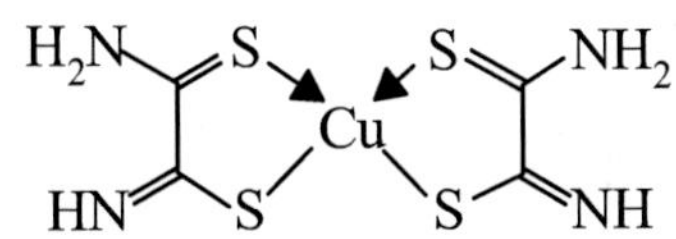	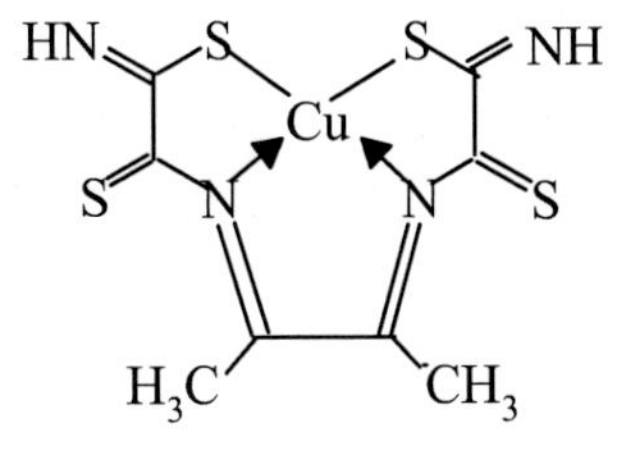
Cu(II)- (L^{III})- (L^{III})	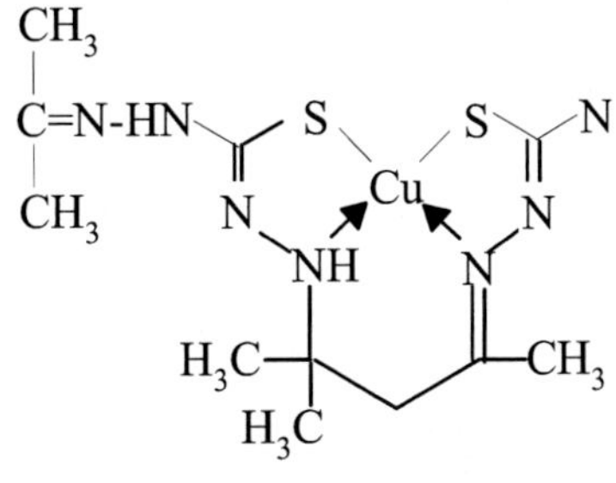	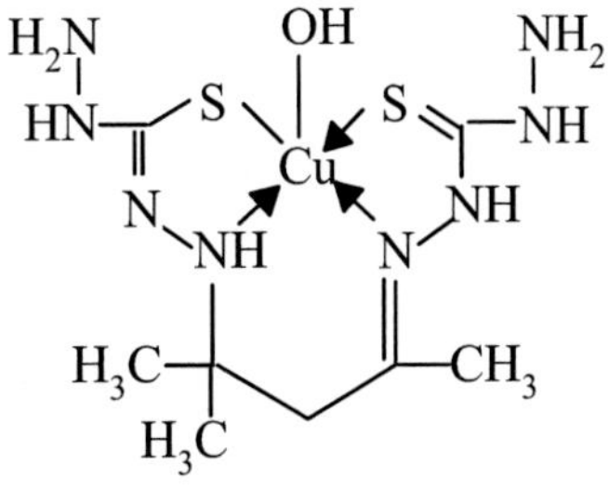

Table 1 (Continued).

Triple system	Products of complexing processes In solution or solid phase	 In gelatin-immobilized matrix
1	2	3
Cu(II)- (L^{III})- (L^{V})	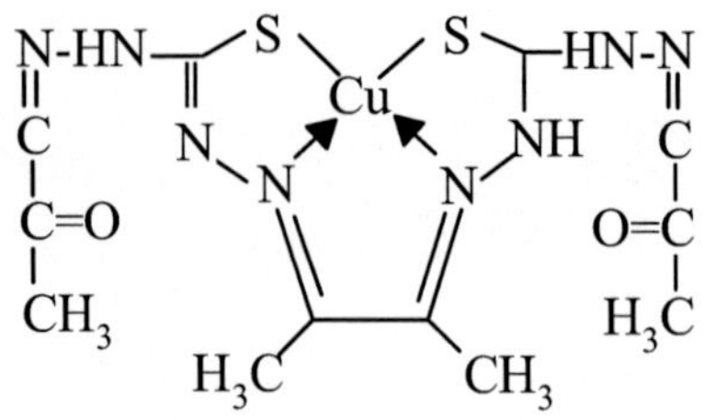	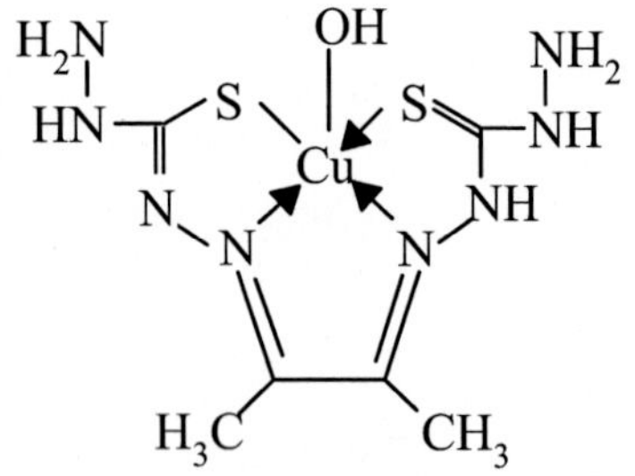 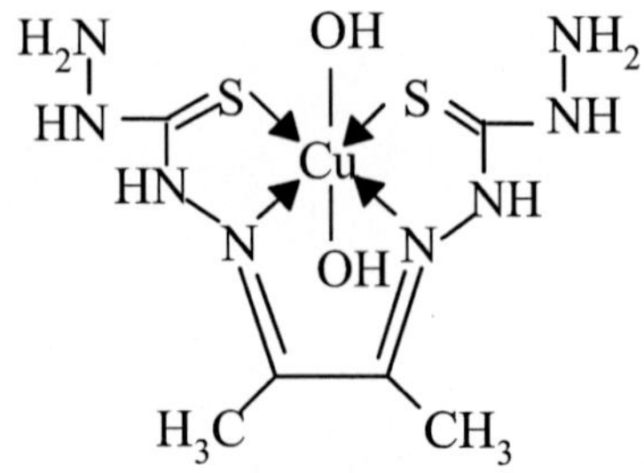 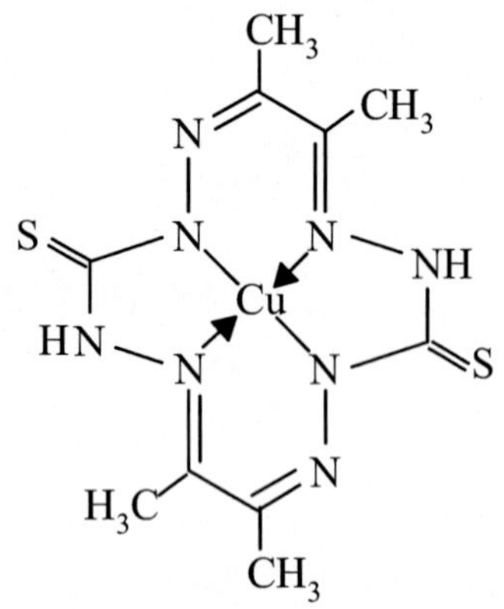
Cu(II)- (L^{IV})- (L^{I})	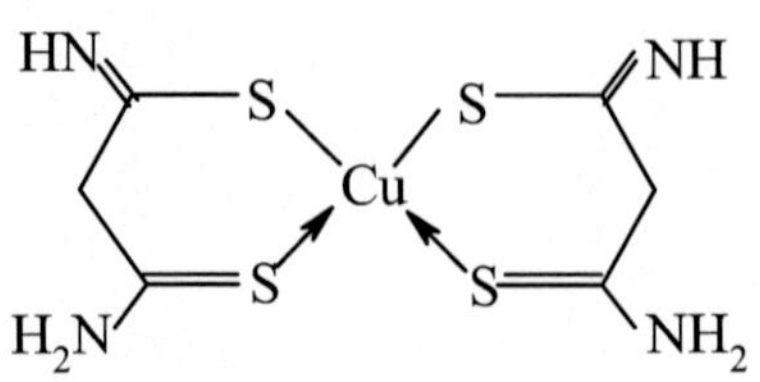	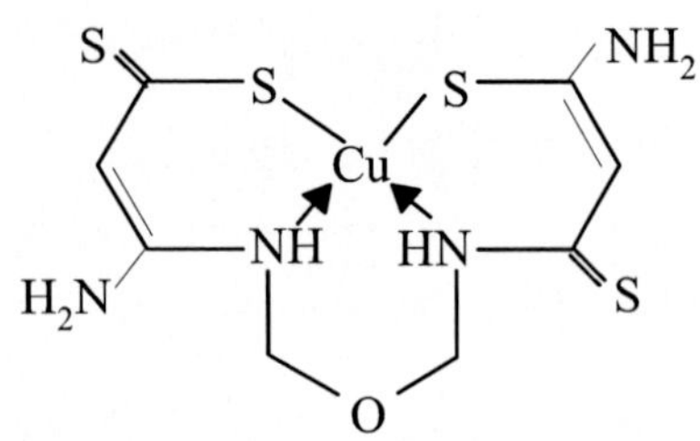

Table 1 (Continued).

Triple system	Products of complexing processes In solution or solid phase	In gelatin-immobilized matrix
1	2	3
Cu(II)- (L^{IV})- (L^{V})	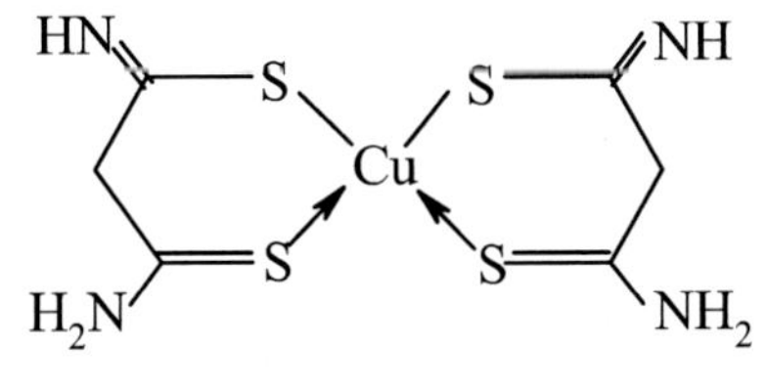	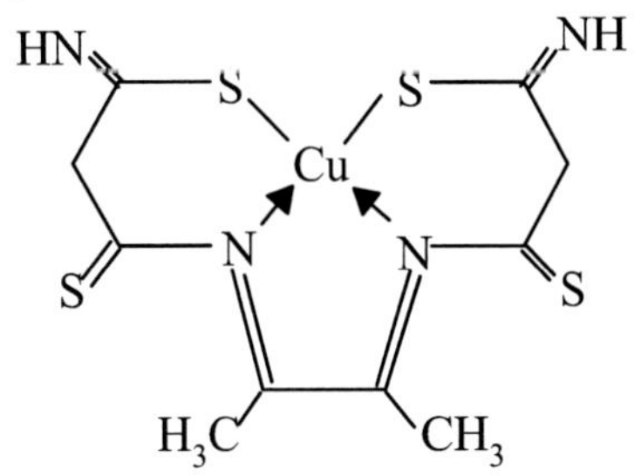
Fe(III)- (L^{I})- (L^{I})	-	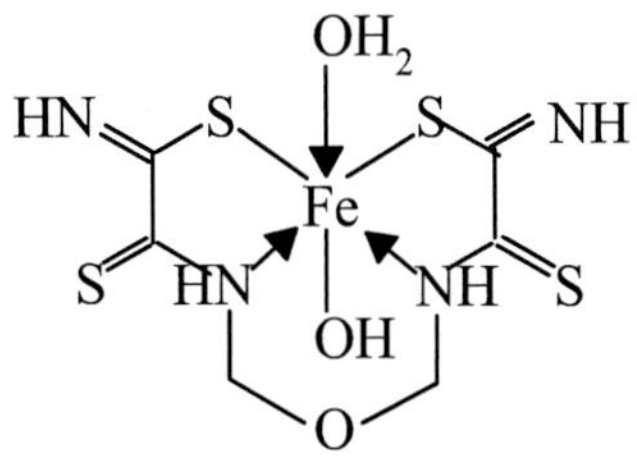
Co(III)- (L^{I})- (L^{I})	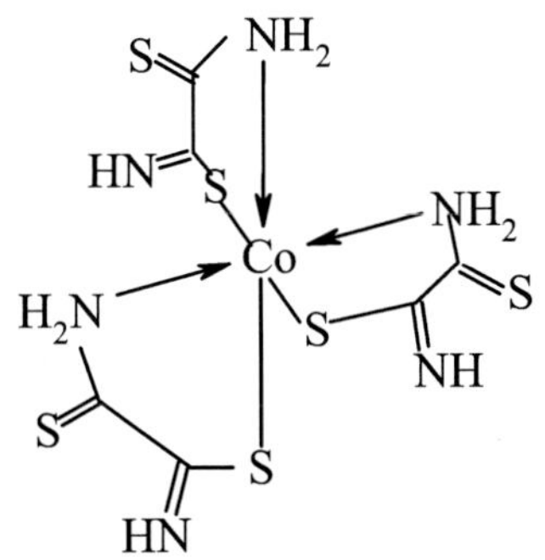	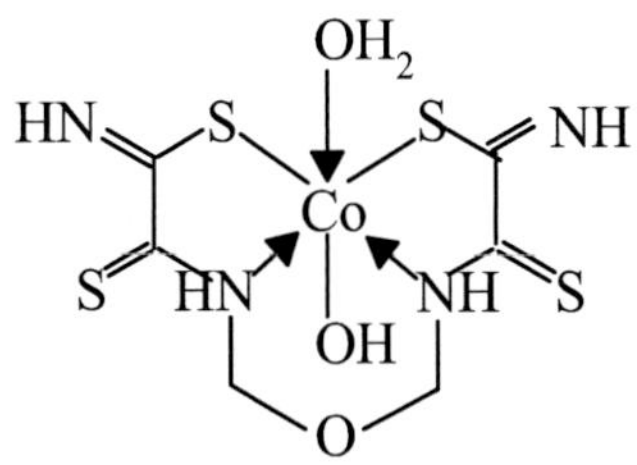
Co(II)- (L^{I})- (L^{III})	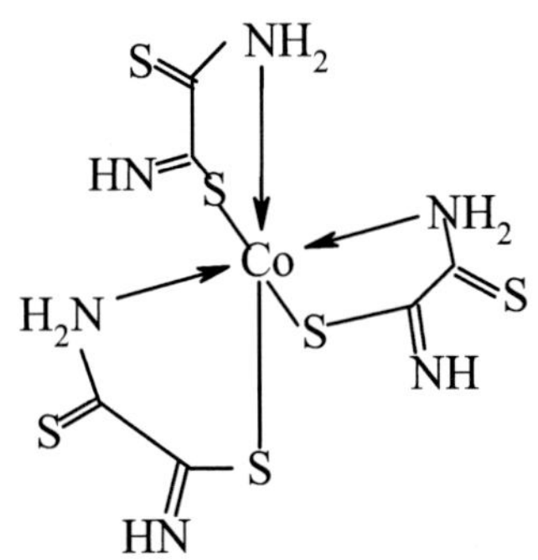	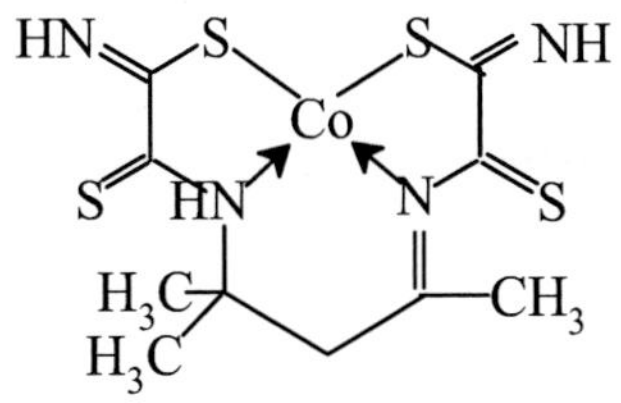

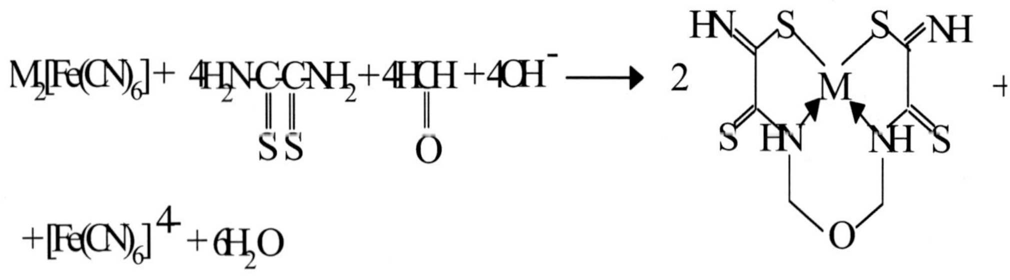

Such coordination is not at variance with a conception of "hard" and "soft" Pearson's acids and bases [13]. The space structures of given template complexes obtained as a result of given calculations, are presented on *Figs. 3-5*. Coplanar coordination of donor centers of ligand related metal ion is more preferable for Co(II) and Cu(II) complexes under examination, with M_S of basic state equal to 4 and 2, respectively; for Ni(II) complex, is more preferable tetrahedral coordination with M_S = 3. The distinctions between energies of structures having M_S value unlike from M_S of basic state [doublet in the case of Co(II), singlet in the case of Ni(II) and quartet in the case of Cu(II)] are 15.6, 30.4 and 117.6 kJ/mole, respectively. The average M–N and M–S bond lengths in complexes with plane N_2S_2 chelate junction [220.93, 227.45 pm (Co), 207.42, 226.74 pm (Cu)], in whole, decrease at transition from Mn to Cu.

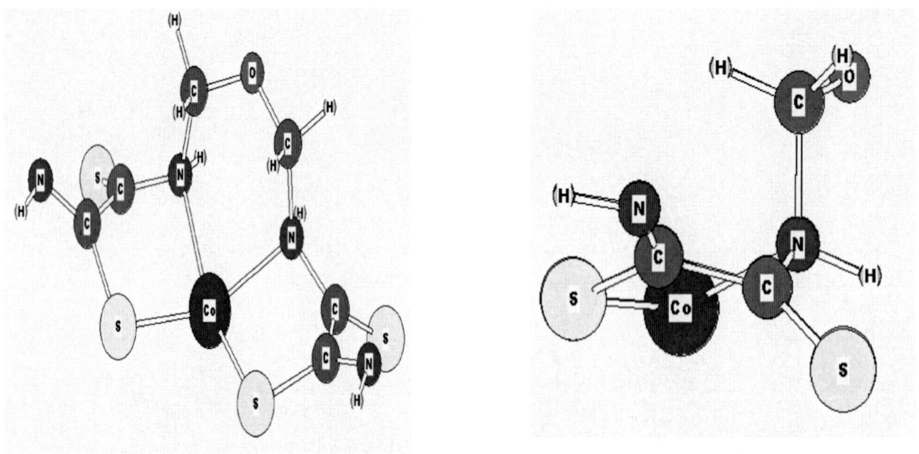

Figure 3. The space structure of Co(II) complex with (N,N,S,S)-coordination of donor centers of template ligand to M(II): "open", front view (*left*) and "shielded", view from the side (*right*).

In the nickel(II) complex with tetrahedral N_2S_2 chelate junction, these bonds are essentially shorter than in Co(II) and Cu(II) complexes. It is significant that they are enough considerably distinguished (for example, the lengths of Cu–N bonds are 207.42 and 207.44 pm whereas the length of each of Cu–S bonds is 226.74 pm). Also, such a circumstance attract its attention that additional six-numbered cycle formed as a result of template "stitching" is not in the same flatness with MN_2S_2 metalchelate junction [the angle of slope between these flatnesses is 67°.4 in the case of Co(II), 47°.2 in the case of Ni(II) and 68°.4 in the case of Cu(II)]. Moreover, this cycle is not plane itself – oxygen atom containing in it, has been deflected from N–CH_2–CH_2–N flatness on angle near to 90°(74°.8 in the case of Co(II), 79°.8 in the case of Ni(II) and 73°.5 in the case of Cu(II)].

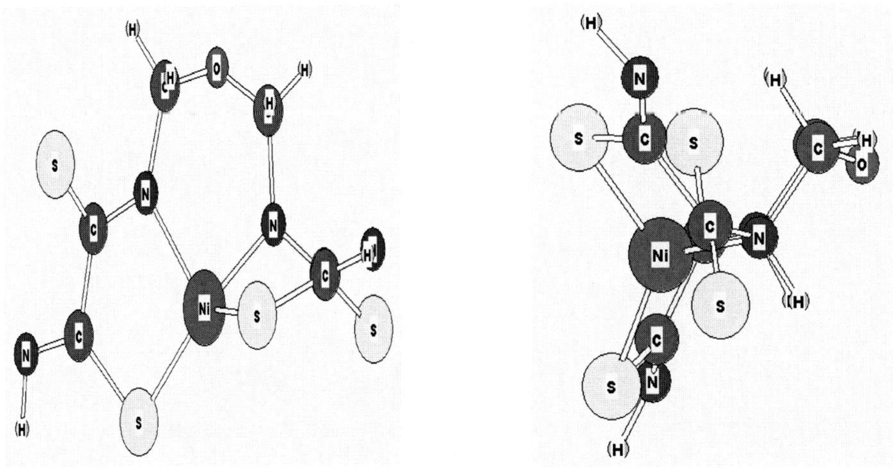

Figure 4. The space structure of Ni(II) complex with (N,N,S,S)-coordination of donor centers of template ligand to M(II): "open", front view (*left*) and "shielded", view from the side (*right*).

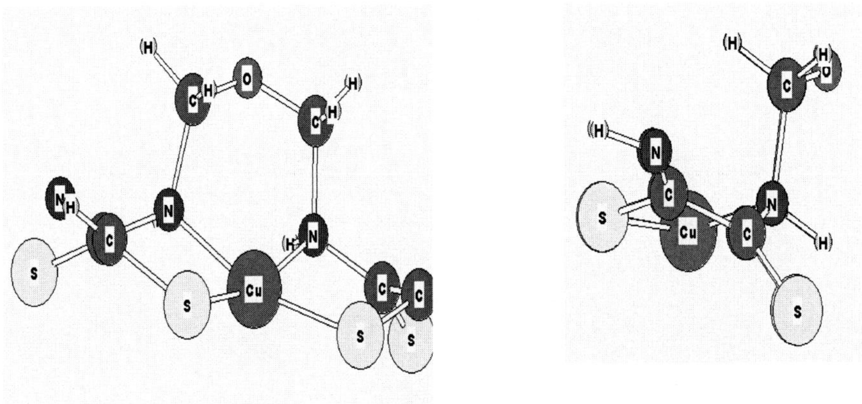

Figure 5. The space structure of Cu(II) complex with (N,N,S,S)-coordination of donor centers of template ligand to M(II): "open", front view (*left*) and "shielded", view from the side (*right*).

It should be noted that a result of calculation of structure and M_S value of nickel(II) complex is not in conformity with the data [14] according to which this complex is diamagnetic and coloured brown that at all is typical for plane Ni(II) complexes but is non-typical for tetrahedral Ni(II) ones [15]. Unfortunately, the data of given calculations does not allow to make of reliable conclusions which of theoretically possible coordinations of «template» ligand – (N,N,S,S) or (N,N,N,N) is more favourable at the complexing in the solution or in these specific conditions that are realized at complexing in the **GIM**. We must note in this connection that analogous calculations for complexes of both these types showed the possibility of their formation (at any rate, at complexing in gas phase). In this connection, it is obvious that further researches are necessary in each of these cases.

CONCLUSION

As can be seen from foregoing, template synthesis in the **MHF-GIM** is specific phenomenon in coordination chemistry of transition metals; undoubtedly, it merits of close attention of chemists working in this specific field of chemical science. That is why, new researches need in given field to understand the nature of this phenomenon, to forecast of compositions of metalcomplexes formed and to learn to guide with it.

ACKNOWLEDGMENT

The Russian Foundation of Basic Researches (RFBR) is acknowledgement for financial support of given work (grant No. 06-03-32003).

REFERENCES

[1] Gerbeleu N.V., Arion V.B. *Template Synthesis of Macrocyclic Compounds*. Kishineu, Stiinta, 1990. 374 P (in rus.)

[2] Lehn J.-M. *Supramolecular Chemistry: Conceptions and Perspectives*. Novosibirsk, Siberian Section of Russian Academy of Sciences, 1998. 334 P.

[3] Gerbeleu N.V., Arion V.B. Burgess J. *Template Synthesis of Macrocyclic Compounds*. Weinheim- New York- Chichester- Brisbane- Singapore- Toronto, Wiley-VCH, 1999. 565 P.

[4] Garnovskii A.D., Vasil'chenko V.S., Garnovskii D.A. *Modern Aspects of Synthesis of Metalcomplexes. The Basic Ligands and Methods*. Rostov, LaPO Publisher, 2000. 355 P (in rus.)

[5] Mikhailov O.V. *Russ Chem Bull*, 1995, *64*, 657-671.

[6] Mikhailov O.V. *Revs Inorg Chem*, 1997, *18*, 287-332.

[7] Mikhailov O.V. *Gelatin-Immobilized Metalcomplexes*. Moscow, Scientific World, 2004. 236 C.

[8] Mikhailov O.V., M.A. Kazymova M.A., Shumilova T.A., Solovieva S.E. *Transition Met Chem*, 2004, *29*, 732-736.

[9] Mikhailov O.V., Kazymova M.A., Chmutova G.A., Shumilova T.A., Solovieva S.E. *Transition Met Chem*, 2005, *30*, 18-21.

[10] Mikhailov O.V., Kazymova M.A., Chmutova G.A., Shumilova T.A., Solovieva S.E. *Transition Met Chem*, 2005, *30*, 299-304.

[11] Chachkov D.V., Mikhailov O.V. *XVI International Conference on Chemical Thermodynamics in Russia (RCCT 2007) (Suzdal, July 1-6, 2007)*. Abstracts, Volume II. P. 4/S-644.

[12] Chachkov D.V., Mikhailov O.V. *XXIII International Chugaev Conference on Coordination Chemistry (Odessa, September 4-7, 2007)*. Abstracts. Kyev, «Kyevskii Universitet» Publisher, 2007. P. 725-726.

[13] Pearson R.J. *J Amer Chem Soc*, 1963, *85*, 3533-3540.

[14] Mikhailov O.V. *Int J Inorg Mater*, 2001, *3*, 1053-1061.

[15] Cotton F.A., Wilkinson G.B. *Advanced Inorganic Chemistry*, 5th Edit., Wiley, New York, 1990.

In: Thermochemistry and Advances in Chemistry Research ISBN: 978-1-60692-376-4
Ed: Anton G. Mirskiy

Short Communications

COMPUTATIONAL THERMOCHEMISTRY OF LARGE HYDROCARBONS AND CARBON CLUSTERS BASED ON TIGHT-BINDING APPROACH

*Alexander A. Voityuk**
Institució Catalana de Recerca i Estudis Avançats (ICREA)
Institute of Computational Chemistry, Universitat de Girona 17071 Girona, Spain

ABSTRACT

We describe an extremely efficient approach based on the non-iterative orthogonal tight-binding scheme to calculate the structure and energetics of large hydrocarbon and carbon systems. The results are reported for an extensive set of molecules and radicals including linear and branched alkanes, polycyclic aromatic compounds, polyene and polyyne systems, and fullerenes. The calculated values of heats of formation are close to the experimental data with the mean absolute error of 2.26 kcal/mol obtained for 113 comparisons. The internal consistency of the calculated data allows for reliable prediction of reaction enthalpies for bond dissociation, isomerization, and cyclization reactions. The method provides accurate structural geometries and good estimates of vibrational frequencies. We conclude that the proposed computational scheme is very promising for atomic calculations and dynamics simulations of hydrocarbons, fullerenes and carbon nanostructures.

INTRODUCTION

The heat (enthalpy) of formation, $\Delta H^0_{f,298}$ is a key thermodynamic parameter of molecules allowing for quantitative description of chemical processes. Nowadays, experimental heats of formation are available for many hydrocarbons that include up to 6 carbon atoms. However, for larger systems especially if they incorporate cyclic structures,

* Email: alexander.voityuk@icrea.es

aromatic rings, double and triple bonds, experimental estimates are often missing. For instance, despite fullerene based compounds are of great interest, only a few values of $\Delta H^0_{f,298}$ have been obtained for these systems.

Experimental measurements of accurate heats of formation are known to be difficult. Because of that, elaboration of computational approaches to reliable predictions of the thermochemical data has attracted considerable attention. The dramatic increase in computer facilities makes quantum chemical (QC) methods a good alternative to experiment. Developments in this area have recently been reviewed [1-3]. Different quantum chemical methods are employed to derive the enthalpies of formation and related data. The Gaussian-3 (G3) method, a composite technique that employs a sequence of *ab initio* calculations, was recently proposed for accurate estimation of thermochemistry for small-sized molecules [4]. The method predicts reaction energies for compounds containing up to 15 heavy atoms to chemical accuracy (±1 kcal/mol).

Density functional theory (DFT) has become the method of choice for computational studies of medium-sized molecules. The B3LYP model is the most popular DFT approach to calculate the energetics of chemical reactions because of its generally good and robust performance at reasonable cost [5,6]. However, recent critical examinations of the B3LYP and other popular functionals have shown certain important limitations of these methods [7,8]. For most of the density-functional methods, the errors increase significantly with increasing number of electrons. In particular, the B3LYP method predicts reaction energies with an error of 9 kcal/mol for molecules having 28 or more valence electrons [9,10]. This functional performs poorly in predicting the C-C bond energies for chain hydrocarbons and hydrogen abstraction reactions [11,12]. Considerable progress has recently been made in elaboration of new exchange-correlation functionals, which allow a chemist to obtain quite reliable thermochemical properties for a wide variety of molecules consisting of up to 100 heavy atoms [13].

The *ab initio* and DFT techniques become inapplicable for extended systems because of huge computational demands. Quantum chemical calculations of molecular systems that contain hundreds of atoms can be performed using much more efficient semiempirical MNDO-like methods [14]. New semiempirical models have been recently developed to obtain accurate data on the ground state structure and energetics [15-17]. In many cases, the results of semiempirical calculations are of the same quality as those of DFT calculations [18].

Another computationally efficient QC approach is based on a tight-binding (TB) model, which bear the close similarity to the extended Hückel method [19]. The use of the parameterized TB approach for exploring the electronic properties of molecule and crystals was suggested in the seminal work of Slater and Koster [20]. In the last three decade, the TB schemes have been widely applied to a variety of chemical systems [21,22]. Two approaches are employed to determine parameters that describe the TB Hamiltonian. One is based on the reference data derived from DFT calculations and usually referred to as DFTB [23,24]. The performance of the DFTB method to predict the structure and energetics of organic molecules has recently been evaluated [25-27]. In general, the accuracy of the DFTB is comparable to that of conventional semiempirical methods. Alternatively, the effective TB Hamiltonian can be parameterized using experimental data [28-34].

Properly parameterized TB models make it possible to calculate accurately molecular geometries and heats of formation of hydrocarbons [35,36]. Two formulations of the TB model, *nonorthogonal* and *orthogonal*, have been considered and compared. The first scheme is based on the conventional TB approach, which explicitly includes the overlap matrix of basis AOs; in the *orthogonal* model, AOs are assumed to be orthogonal and the overlap matrix is replaced by the unit matrix. Usually, one considers the *nonorthogonal* scheme to be more accurate. However, Truhlar et al. comparing calculated data for aluminium clusters, found that the use of the orthogonal TB scheme provides more reliable results than the nonorthogonal model [37]. Comparison of the performance of the *nonorthogonal* versus *orthogonal* TB schemes for hydrocarbons [36] clearly demonstrated the last model shows a significant improvement over the nonorthogonal scheme in line with the conclusion of Truhlar et al. [37].

The purpose of the present work is to describe a modified TB (MTB) scheme and evaluate its performance in predicting the structure and energetics of hydrocarbon and carbon compounds. Based on our previous work [36], we justified the scheme using some new reference data (in particular, $\Delta H^0_{f,298}$ of C_{60}). The paper is organized as follows. In the next section, we briefly outline the tight-binding method and define the effective Hamiltonian and the short-range repulsive potential. Then we consider the performance of the proposed model by comparing the MTB results with the experimental data and results of G3 calculations. Finally, we give conclusions and outline possible extensions of the model.

METHOD

Within the tight-binding model, the total energy of the system can be expressed as a sum of the electronic energy and short-range repulsion energy:

$$E = \sum_i n_i \varepsilon_i + E_{rep} . \tag{1}$$

In Eq. (1), ε_i is an eigenvalues of the effective one-electron Hamiltonian *H,* $\varepsilon_i = \langle \psi_i | H | \psi_i \rangle$, n_i - occupation number of MO ψ_i (it equals 2, 1 and 0, for occupied, half-occupied and vacant orbitals, respectively). The approach employs a minimal valence basis of atomic orbitals. The effective Hamiltonian *H* is defined by using semiempirical parameters. The repulsion energy E_{rep} is usually approximated by a sum of interatomic potentials G_{AB} depending only on the distance between atoms *A* and *B*:

$$E_{rep} = \sum_{A>B} G_{AB} . \tag{2}$$

The two-center potential implicitly includes terms of different physical nature (the core-core repulsion, a correction due to double counting of the two-electron interaction, the exchange and correlation energy) and is determined by fitting to suitable reference data.

In the general case (nonorthogonal model), the orbital energies ε_i are obtained by solving the eigenvalue equations

$$\sum_i (H_{\mu\nu} - \varepsilon_i S_{\mu\nu})\, c_{\nu i} = 0. \tag{3}$$

The computational treatment becomes essentially faster if one employs an orthogonal AO basis (the overlap integrals in Eq. 3 are set to be zero, $S_{\mu\nu} = \delta_{\mu\nu}$). The orthogonal scheme is employed in the MTB model. Let us briefly consider how the effective Hamiltonian and the repulsion energy are defined. The diagonal matrix elements of H are expressed via parameter U_μ, which is the energy of an electron in AO φ_μ ($\mu = s$ and p). The off-diagonal one-center elements are set to be zero. Thus,

$$H_{\mu\nu} = U_\mu(A)\cdot\delta_{\mu\nu} \quad (\mu, \nu \in A) \tag{4}$$

The two-center matrix elements (in the diatomic coordinate system) are defined by using an empirical formula suggested by Kolb and Thiel [38]

$$H_{\mu\nu} = \pm\beta_{\mu\nu}^{AB}\left(R_{AB}/a_0\right)^{1/2}\exp\left(-\lambda_{\mu\nu}^{AB}R_{AB}^2/a_0^2\right), \tag{5}$$

where $\beta_{\mu\nu}^{AB}$ and $\lambda_{\mu\nu}^{AB}$ $(\mu \in A, \nu \in B)$ are adjustable parameters, a suitable phase factor ±1 has to be used to retain the rotational invariance of the scheme. R_{AB} is the distance between atoms A and B in Å, the Bohr radius a_0=0.52917 Å.

The two-center short-range potential G_{AB} consists of two terms:

$$G_{AB} = \gamma_{AB}\exp\left(-\alpha_{AB}R_{AB}\right) + \omega_{AB}\exp\left[-6\left(R_{AB} - r_{AB}\right)^2\right] \tag{6}$$

The first term approximates the repulsion potential between the atoms A and B. The second term is introduced to correct the repulsion at distances close to the Van-der-Waals contact; the r_{AB} is a sum of the Bondi radii of atoms A and B [39], r_{HH}=2.4 Å, r_{CH}= 2.9 Å, r_{CC}= 3.4 Å.

The standard heat of formation at 298 K, $\Delta H^0_{f,298}$, is computed using the same procedure followed in the development of many semiempirical methods. It is expressed via the computed atomization energy and experimental heats of formations of the atoms:

$$,\Delta H^0_{f,298} = E - \sum_A E_{isol}(A) + \sum_A H^0_{f,298}(A) \tag{7}$$

where E is the total energy of the molecule defined by Eq.(1), $E_{isol}(A)$ is the energy of an isolated atom A, which is expressed via one-center parameters U_μ:

$$E_{isol}(A) = n_s U_s(A) + n_p U_p(A)\,. \tag{8}$$

Here, n_s and n_p denote the number of the valence s- and p-electrons in atom A. $\Delta H^0_{f,298}(A)$ is the experimental heat of formation of the atom A, for hydrogen and carbon $\Delta H^0_{f,298}(A)$ is 52.1 kcal/mol and 170.89 kcal/mol, respectively.

The zero-point vibrational energy and the enthalpy to heat the molecule from T=0° K to T=298 K are implicitly incorporated into the total energy E through the parametrization. That is, the semiempirical parameters of the model are fitted using Eq. (7) to reproduce experimental values of $\Delta H^0_{f,298}$ for a set of reference molecules, and therefore, unlike *ab initio* and DFT calculations, the explicit inclusion of zero-point and thermal corrections into $\Delta H^0_{f,298}$ is not required within MTB.

Parameterization.

The parameters of the effective Hamiltonian and the repulsion potential G_{AB} have been obtained using the experimental values of $\Delta H^0_{f,298}$ and geometries for several tens of molecules representing different bonding situations in hydrocarbons. The experimental heats of formation were adopted from the NIST Chemistry WEB book [40] and other compilations.

Because fullerenes play an important role in chemistry, computational modeling of such systems is of great interest. So despite large dispersions among experimental heats of formation of C_{60} (the measured values of $\Delta H^0_{f,298}$ range from 599.1 to 634.9 kcal/mol [40,41]) we include this molecule in the training set. As reference value of $\Delta H^0_{f,298}$ we adopt the value of 604.6 kcal/mol employed by Green et al. to elaborate a group additivity method for thermochemistry of polycyclic aromatic hydrocarbons [41]. Note that the B3LYP/6-31G* calculation predicts $\Delta H^0_{f,298}$ = 618 kcal/mol [42].

A nonlinear least-squares method was used to optimize the semiempirical parameters. U_s for hydrogen was set to -13.605 eV. Several parametrization runs were carried out starting from different parameter values and using different training sets. The resulting parameters were tested in extensive survey calculations in order to choose the set which yields the most balanced results. Table 1 lists the final values of the parameters.

Table 1. Parameters of the MTB model

Atomic parameters			
Parameter	H	C	
U_s^A (eV)	-13.605	-21.516	
U_p^A (eV)		-13.362	
Bond-type parameters			
Parameter	H-H	C-H	C-C
λ_{ss}	0.282	0.275	0.086
λ_{sp}		0.218	0.180
$\lambda_{pp\sigma}$			0.186
$\lambda_{pp\pi}$			0.282
β_{ss} (eV)	-4.438	-8.574	-5.969
β_{sp} (eV)		-6.813	-6.160
$\beta_{pp\sigma}$ (eV)			-8.420
$\beta_{pp\pi}$ (eV)			-7.403
α_{AB} (Å^{-1})	2.823	2.831	3.401
γ_{AB} (eV)	13.75848	98.06733	658.25657
ω_{AB} (eV)	-0.0512	-0.0355	-0.0026

RESULTS AND DISCUSSION

Heats of formation

Table 2 contains the calculated and experimental values of $\Delta H^0_{f,298}$ of 113 molecules and radicals that belong to different classes. In average MTB slightly overestimated the stability of hydrocarbons: a mean error is -0.19 kcal/mol. The mean absolute error (MAE) is 2.26 kcal/mol considerably smaller than those of the conventional semiempirical methods. For the same set of molecules, the AM1 and PM3 MAEs are 15.0 and 9.3 kcal/mol, respectively. When the fullerenes C_{60} and C_{70} are excluded from the statistics (the $\Delta H^0_{f,298}$ of these compounds are substantially overestimated by AM1 and PM3), the MAEs are reduced to 8.4 and 5.7 kcal/mol, however, they remain essentially larger than the error of MTB). The good performance of the MTB method is observed for all classes of compounds. As seen from Table 2, small deviations of the calculated values from experimental data are found both for short and long alkanes. The mean absolute error for 24 normal and branched alkanes listed in Table 2 is 1.20 kcal/mol. MTB provides good estimates for C_nH_{2n+2}

Table 2. Calculated and experimental heats of formation (kcal/mol)

Compound	MTB	Exp.	Error
hydrogen	4.0	0.0	4.0
N-alkanes			
methane	-14.8	-17.8	3.0
ethane	-18.4	-20.0	1.6
propane	-24.2	-25.0	0.8
butane	-29.2	-30.0	0.8
pentane	-34.2	-35.1	0.9
hexane	-39.1	-39.9	0.8
heptane	-44.1	-44.9	0.8
octane	-49.0	-49.9	0.9
nonane	-54.0	-54.5	0.5
decane	-58.9	-59.6	0.7
undecane	-63.8	-64.8	1.0
dodecane	-68.8	-69.2	0.4
tridecane	-73.7	-74.5	0.8
tetradecane	-78.7	-79.4	0.7
pentadecane	-83.6	-84.8	1.2
hexadecane	-88.6	-89.6	1.0
Branched alkanes			
isobutane	-31.5	-32.1	0.6
2-methylbutane	-37.0	-36.7	-0.3
neopentane	-39.8	-40.2	0.4
3-methylpentane	-40.9	-41.1	0.2
2,3-dimethylbutane	-42.5	-42.5	0.0
2,2,3-trimethylbutane	-51.8	-49.0	-2.9
2,2,3,3-tetramethylbutane	-60.1	-56.2	-3.9
tri-t-buthylmethane	-51.5	-56.2	4.7
Cycloalkanes			
Alkenes and Alkynes			
ethylene	15.6	12.5	3.1
acetylene	54.6	54.5	0.1
propene	6.3	4.8	1.5
propyne	45.5	44.2	1.3
allene	46.9	45.5	1.4
1-butene	2.2	0.0	2.2
trans-2-butene	-2.1	-2.7	0.6
cis-2-butene	-3.4	-1.7	-1.7
isobutene	-3.6	-4.0	0.4
1,2-butadiene	38.5	38.8	-0.3
1,3-trans-butadiene	28.7	26.3	2.4
1-butyne	40.4	39.5	0.9
2-butyne	36.5	34.8	1.8
vinylacetylene	68.5	73.0	-4.5
diacetylene	110.3	113.0	-2.7
butatriene	76.2	83.0	-6.8
trimethylethene	-11.7	-9.9	-1.8
tetramethylethene	-20.7	-16.8	-3.9
trans-1,3-pentadiene	20.3	18.2	2.1
cis-1,3-pentadiene	20.4	19.5	0.9
isoprene	19.1	18.0	1.1

Table 2. (Continued).

Compound	MTB	Exp.	Error
1,4-pentadiene	29.0	25.2	3.8
Cycloalkanes and Cycloalkenes			
cyclopropane	17.2	12.7	4.5
cis-dimethylcyclopropane	-2.3	1.3	-3.6
cyclobutane	5.5	6.8	-1.3
cyclopentane	-20.2	-18.3	-1.9
cyclohexane	-30.7	-29.5	-1.2
cyclopropene	63.1	66.2	-3.1
1-methylcyclopropene	52.3	58.2	-5.9
1,2-dimethylcyclopropene	41.6	46.4	-4.8
methylenecyclopropane	46.0	47.9	-1.9
cyclobutene	42.7	37.5	5.2
1,2-dimethylcyclobutene	21.0	19.8	1.2
methylenecyclobutane	27.2	29.0	-1.8
3,4-dimethylenecyclobutene	82.7	80.4	2.3
cyclopentene	6.6	8.1	-1.5
cyclopentadiene	33.4	32.1	1.3
fulvene	51.3	53.5	-2.2
cyclohexene, halfchair	-3.7	-1.2	-2.5
1,3-cyclohexadiene	19.8	25.4	-5.6
1,4-cyclohexadiene	21.4	25.0	-3.6
cycloheptatriene	39.2	43.2	-4.0
cyclooctatetraene	63.4	70.7	-7.3
Bi- and polycyclic compounds			
bicyclobutane	46.6	51.9	-5.3
trans-bicyclopropyl	34.3	30.9	3.4
bicyclo[2.1.0]pentane	37.8	37.8	0.0
spirocyclopentane	44.9	44.3	0.6
norbornane	-8.1	-13.1	5.0
norbornadiene	56.0	58.8	-2.8
cyclo[2,2,2]-Octane	-26.5	-23.6	-2.9
adamantane	-35.0	-32.2	-2.8
cubane	146.5	148.7	-2.2
Aromatic molecules			
benzene	21.3	19.7	1.6
toluene	12.6	12.0	0.6
ethylbenzene	8.0	7.1	0.9
styrene	36.1	35.3	0.8
indene	40.1	39.1	1.0
mesitylene	-4.2	-3.8	-0.4
naphthalene	36.6	35.9	0.7
anthracene	54.9	55.2	-0.3
phenanthrene	52.4	49.0	3.4
azulene	65.4	69.1	-3.7
biphenylene	98.5	100.5	-2.0
pyrene	56.9	53.9	3.0
benzo[c]phenanthrene	71.4	69.6	1.8
benzo[a]anthracene	70.1	68.9	1.2
chrysene	69.2	63.0	6.2
naphthacene	74.0	73.3	0.7

Table 2. (Continued).

Compound	MTB	Exp.	Error
perylene	76.8	76.3	0.5
coronene	78.9	73.5	5.4
corannulene	111.5	110.1	1.4
Fullerenes			
C60	605.1	604.8	0.3
C70	653.9	653.8	0.1
Radicals			
methyl	39.6	34.8	4.8
ethyl	29.1	28.0	1.1
n-propyl	22.8	24.0	-1.2
i-propyl	17.3	22.3	-5.0
n-butyl	17.7	18.0	-0.3
s-butyl	12.7	17.0	-4.3
t-butyl	6.0	11.0	-5.0
vinyl	67.4	63.4	4.0
allyl	40.2	39.0	1.2
HCC	113.3	123.0	-9.7
cyclopropyl	64.6	66.9	-2.3
cyclopropenyl	104.6	105.0	-0.4
cyclopentadienyl	62.0	58.0	4.0
phenyl	73.6	79.0	-5.4
benzyl	47.7	49.0	-1.3

molecules independent of n. For instance, $\Delta H^0_{f,298}$ calculated for propane and hexadecane deviates from experiment by ~1 kcal/mol. By contrast, the B3LYP deviation steadily increases with system size due to an accumulation of error ; the error is found to be of -1.5 and -30.3 kcal/mol for propane and hexadecane, respectively. The G3 enthalpies of formation of the *n*-alkanes deviate with experiment by less than 2 kcal/mol; the errors in the G3 enthalpies range from 0.25 to 1.93 kcal/mol increasing with chain length [12]. Note that the effects of conformational averaging on the enthalpies of the *n*-alkanes are estimated to be small [9]. The MTB shows good performance in computing the energy differences between branched hydrocarbons and their linear isomers. For instance, the conversion energetics of neopentane to pentane and 2,2,3-trimethylbutane to heptane, are predicted to be 5.6 and 7.7 kcal/mol which are in good agreement with experimental values 5.1 and 4.1 respectively. Note that B3LYP shows rather large branching errors (10-15 kcal/mol) [8].

The MTB method reproduces well the heats of formation of branched hydrocarbons including sterically crowded molecules with adjacent methyl groups. Small errors are found for compounds with double- and triple- bonds (for 22 comparisons MAE= 2.05 kcal/mol). The largest deviation of -6.8 kcal/mol is calculated for butatriene. For cyclic and polycyclic molecules, MTB provides somewhat less accurate estimates of $\Delta H^0_{f,298}$ (MAE=3.06 kcal/mol for 30 entries). For "difficult" molecules, like cubane and adamantine, MTB values are found to be in very good agreement with experiment. Reliable heats of formation are predicted by MTB for aromatic compound and fullerenes (for 21 molecules the MTB mean absolute error is only 1.71 kcal/mol). The largest deviation with experiment, 6.2 kcal/mol, we find for chrysene.

Even for open-shell systems the MTB scheme gives quite acceptable estimates of $\Delta H^{0}_{f,298}$. The MAE of 3.33 kcal/mol is obtained for 15 radicals listed in Table 2. The largest deviation is found for the radical HC≡C.

For 16 hydrocarbons of the G3 subset [43], the MAE of MTB, 1.9 kcal/mol, is considerably smaller than the MAE of the B3LYP calculation (9.6 kcal/mol [43]) and the errors of the AM1 and PM3 methods (6.4 and 3.4 kcal/mol). However, the high-level composite methods of the G3 family provide more accurate results (the MAE ranges from 0.7 to 1.0 kcal/mol [44]).Overall, the MTB scheme provides accurate estimates for heats of formation of hydrocarbons and fullerenes.

Polyenes and polyynes.

Thermochemistry of two types of linear carbon chains, acetylenic (C≡C-C≡C)$_n$_ and cumulenic·(:C=C=C=C:) are of special interest for nanoelectronics. Table 3 compares the $\Delta H^{0}_{f,298}$ values of several of polyenes H2C=(C=C=)$_n$CH2 and polyynes HC≡(C-C≡)$_n$CH calculated using MTB with the corresponding reference data derived from G3 calculations [44]. The MTB method reproduces well the G3 results. The mean absolute deviation is 3.6 and 1.6 for polyenes and polyynes, respectively. The performance of B3LYP, with the corresponding errors of 3 and 11 kcal/mol [44], is less satisfactory.

Table 3. Comparison of $\Delta H^{0}_{f,298}$ of polyenes H2C=(C=C=)$_n$CH2 and polyynes HC≡(C-C≡)$_n$CH calculated using MTB and G3

Polyene	MTB	G3	Error	Polyine	MTB	G3	Error
C_2H_4	15.6	12.5	3.1	C_2H_2	54.6	55.1	-0.5
C_4H_4	76.2	77.0	-0.8	C_4H_2	110.3	110.7	-0.4
C_6H_4	133.8	136.5	-2.7	C_6H_2	164.9	164.8	0.1
C_8H_4	189.9	194.4	-4.5	C_8H_2	219.0	219.1	-0.1
$C_{10}H_4$	245.3	251.7	-6.4	$C_{10}H_2$	272.9	269.3	3.6

Fullerenes.

To assess the performance of MTB for fullerenes, we carried out calculations of several systems. Large dispersions among the experimental heats of formation have been already mentioned. For of C_{60} and C_{70} , MTB gives 605.1 and 653.9 kcal/mol which is in good agreement with the most reliable experimental values of 604.8 and 653.8 kcal/mol [41].

We optimized the structure and estimated $\Delta H^{0}_{f,298}$ for all 19 isomers of C_{86}. The isomer numbering system for fullerenes is given in [45]. Figure 1 compares the relative energy of the isomers calculated using MTB with B3LYP data obtained by Morokuma et al. [46]. Both methods predict the isomer 17 to be the lowest energy isomer of C_{86}. According to the MTB calculation, $\Delta H^{0}_{f,298}$ of this isomer is 734.1 kcal/mol. Then follow isomers 16, 11, 18 and 12.

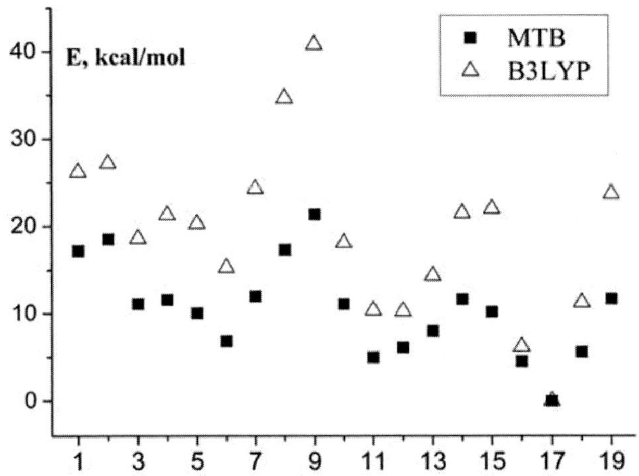

Figure 1. Relative thermochemical stability of 19 isomers of the fullerene C_{86}. Both the MTB and B3LYP [46] methods predict isomer 17 to be most stable.

According to MTB all these compounds lie within 6 kcal/mol. B3LYP gives a larger energy range, *ca.* 10 kcal/mol. As can be seen, MTB and B3LYP suggest very similar order of relative stability of isomers. However, MTB gives remarkably smaller range for the relative energy than B3LYP.

Isomerization reactions.

The variety of hydrocarbons is based on the ability of carbon atoms to form single, double and triple bonds. The changes in the valence state of carbon atoms are associated with remarkable variations of atomic energies. Enthalpies of isomerization reactions can be considered as a good test to assess the performance and inherent consistency of a computational method. In Table 4, we compare relative experimental heats of formation with their computed values for isomers and $C_{10}H_{10}$. B3LYP and MP2 results are taken from [8]. Also the MTB and experimental values of $\Delta H^0_{f,298}$ are listed in Table 4. For C_6H_6 isomers the mean absolute error of MTB, B3LYP and MP2 energies from experimental data are 1.2, 3.0 and 6.8 , respectively. The MTB scheme shows the best performance; also. B3LYP provides good estimates of relative stability of the isomers. The situation changes remarkably, however, for $C_{10}H_{10}$ isomers. In these compounds (see Table 4), the MTB, B3LYP and MP2 errors amount to 4.2, 16.0 and 6.8 kcal/mol, correspondingly. Thus, the MTB scheme provides most accurate data, the estimates obtained within MP2 appear to be acceptable, the relative energies of isomers derived from B3LYP are unreliable. The presented data suggest that MTB provides consistent estimates of $\Delta H^0_{f,298}$ across different classes of hydrocarbons.

Table 4. Comparison of experimental and calculated values of $\Delta H^0_{f,298}$ and relative energies of C_6H_6 and $C_{10}H_{10}$ isomers

	$\Delta H^0_{f,298}$, kcal/mol		Relative energies, kcal/mol			
C_6H_6	Exp	MTB	Exp	MTB	B3LYP	MP2
	19.7	21.3	0.0	0.0	0.0	0.0
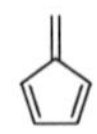	53.5	51.3	33.8	30.0	34.7	36.0
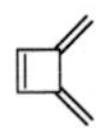	80.4	82.6	60.6	61.4	62.9	67.9
	90.2	91.8	70.4	70.5	68.9	69.5
	94.6	93.1	74.8	71.8	82.1	91.7
$C_{10}H_{10}$	Exp	MTB	Exp	MTB	B3LYP	MP2
	53.5	51.0	0.0	0.0	0.0	0.0
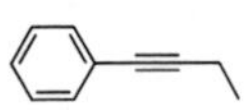	59.4	63.2	5.9	12.2	-8.5	1.2
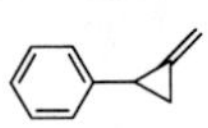	70.0	67.5	16.5	16.5	10.7	12.4

Table 4. (Continued).

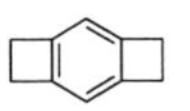	74.0	74.9	32.3	27.5	3.1	10.1
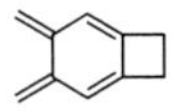	85.8	78.4	32.3	27.5	17.6	31.1
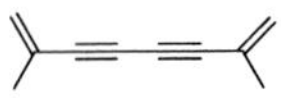	118.1	119.7	64.6	68.8	48.8	63.0

[2+2] cyclization of tetramethylethene and various alkenes.

Table 5 compares our MTB results with G3MP2, MP2 and B3LYP data calculated using 6-31++G** basis set in [12]. As reference data for the reaction energy, we use G3MP2 results. All the reactions are exothermic, although the formation of crowded cyclobutanes has slightly smaller reaction energy. The MTB estimates are in better agreement with G3MP2 than the MP2 calculations, the corresponding MAEs are 4.8 and 8.9 kcal/mol. The MTB scheme tends to overestimate the stability of cyclobutanes. The B3LYP method performs poorly predicting the reaction energies with MAE of 17.2 kcal/mol.

Table 5. Predicted energies (kcal/mol) for the [2+2] cyclization of tetramethylethene and various alkenes

Alkene	G3MP2	MTB	MP2	B3LYP
Ethene	-19.3	-24.0	-27.2	-7.2
Propene	-17.4	-23.0	-26.9	-4.4
Isobutene	-15.7	-20.6	-23.4	2.2
Z-butene	-18.0	-23.0	-25.8	-0.1
E-butene	-16.7	-22.4	-27.7	-2.0
Trimethyethene	-15.5	-20.4	-25.5	4.1
Tetramethylethene	-15.8	-18.3	-23.3	9.4

Bond-dissociation energies.

Since MTB provides quite accurate $\Delta H^{0}_{f,298}$ values for both molecules and radicals, the method can be used to estimate the enthalpy of bond-breaking and bond-forming reactions. In Table 6, we compare energies for 25 bond-breaking reactions. As can be seen, the experimental C-H and C-C bond dissociation enthalpies are well reproduced by the MTB method. The mean absolute error is 2.9 kcal/mol. Thus, the scheme can be employed for computer simulations of processes that involved C-C and /or C-H bond breaking reactions, e.g. oil cracking.

Table 6. Bond dissociation enthalpies, in kcal/mol

Reaction	MTB	Exp.
$CH_4 \rightarrow CH_3$+H	106.4	104.7
$C_2H_6 \rightarrow C_2H_5$+H	99.6	100.1
$C_2H_4 \rightarrow C_2H_3$+H	103.9	103.0
$C_3H_8 \rightarrow$ n-C_3H_7+H	99.1	101.1
$\rightarrow$ i-C_3H_7+H	93.6	99.4
$\rightarrow C_2H_5+CH_3$	92.8	87.8
$C_3H_6 \rightarrow C_3H_5$+H	86.0	86.3
n-$C_4H_{10} \rightarrow$ n-C_4H_9	99.0	100.1
$\rightarrow$ s-C_4H_9	94.0	99.1
$\rightarrow 2C_2H_5$	87.3	86.0
$\rightarrow$ CH3+C_3H_7	92.8	87.8
i-$C_4H_{10} \rightarrow$ t-C_4H_9+H	101.3	102.2
$\rightarrow$ i-$C_3H_7+CH_3$	88.3	89.2
Neo-$C_5H_{12} \rightarrow$ t-$C_4H_9+CH_3$	99.5	106.3
Neo-$C_5H_{12} \rightarrow$ t-$C_4H_9+CH_3$	99.5	106.3
Reaction	MTB	Exp.
2 methylbutane $\rightarrow$ i-$C_3H_7+C_2H_5$	83.2	87.0
$\rightarrow$ t-$C_4H_9+CH_3$	89.2	88.5
c-$C_3H_6 \rightarrow$ c-C_3H_5+H	99.5	106.3
c-$C_3H_4 \rightarrow$ c-C_3H_3+H	93.5	90.9
Ph-H $\rightarrow$ Ph+H	112.8	111.4
Ph-$CH_3 \rightarrow PhCH_2$ + H	87.1	89.0
$\rightarrow$ Ph + CH_3	100.5	101.7
Ph-$C_2H_5 \rightarrow PhCH_2 + CH_3$	79.3	76.9
$\rightarrow$ Ph + C_2H_5	94.6	99.9
Ph-CH=$CH_2 \rightarrow$ Ph + C_2H_3	104.8	107.1

Geometries.

Molecular geometries of some selected hydrocarbons and fullerenes are listed in Table 7. As seen, the method reproduces well different types of bonds and bond angles. Statistical evaluation of structural parameters for 37 systems shows that the mean absolute errors in bond lengths and bond angles are quite small, 0.011 Å (109 comparisons) and 1.3° (37 comparisons), respectively.

Let us briefly consider the structural parameters of fullerenes calculated using MTB. In C_{60} there are two types of C-C bonds with length 1.39 Å and 1.44 Å. The MTB values, 1.399 Å and 1.454 Å, are in excellent agreement. In C_{70} (D_{5h} symmetry), there are five circles of atoms labeled with a, b, c, d, and e from the capping pentagon to the equator. Table 7 compares the calculated and experimental (gas-phase electron diffraction [47]) bond lengths in C_{70}. As seen, the MTB results are in good agreement with experimental data.

Table 7. Geometries of selected molecules (bond lengths in Å, bond angles in deg.)

Molecule	Variable	Calc.	Exp.
Hydrogen	H-H	0.736	0.741
Methane	C-H	1.083	1.087
Ethane	C-C	1.517	1.535
	C-H	1.091	1.094
	HCC	111.4	111.2
Ethylene	C=C	1.341	1.339
	C-H	1.082	1.087
	CCH	122.3	121.3
Acetylene	C≡C	1.229	1.202
	C-H	1.063	1.063
Propene	C=C	1.342	1.336
	C-C	1.493	1.501
	C-H	1.082	1.081
	C-H	1.092	1.098
	CCC	123.8	124.3
	CCH	121.2	121.5
Propyne	C≡C	1.230	1.206
	C-C	1.470	1.459
	C_1-H	1.062	1.056
	C_3-H	1.092	1.105
	CCH	111.3	110.2
Allene	C=C	1.322	1.308
	C-H	1.083	1.087
	HCH	116.2	118.2
Neopentane	C-C	1.534	1.539
	C-H	1.089	1.120
	HCC	112.3	110.0
Cyclopropane	C-C	1.493	1.510
	C-H	1.089	1.074
	HCH	113.0	115.9
Cyclopropene	C=C	1.316	1.296
	C-C	1.481	1.509
	HCC	149.3	149.9
	HCH	111.3	114.6
Cyclopentadiene	C=C	1.357	1.345
	C-C	1.455	1.468
	C-C	1.496	1.506
Benzene	C-C	1.395	1.397
	C-H	1.085	1.083
Naphthalene	C_1-C_2	1.380	1.381
	C_2-C_3	1.409	1.417
	C_1-C_9	1.423	1.422
	C_9-C_{10}	1.423	1.412
Fullerene C_{60}	C-C	1.454	1.440
	C-C	1.399	1.390
Fullerene C_{70}	C_a-C_a	1.454	1.461
	C_a-C_b	1.399	1.388
	C_b-C_c	1.452	1.453
	C_c-C_c	1.392	1.386
	C_c-C_d	1.452	1.468
	C_d-C_d	1.436	1.425
	C_d-C_e	1.418	1.405
	C_e-C_e	1.454	(1.538)

Vibrational frequencies.

In Table 8, we compare the calculated and experimental vibrational frequencies [48]. Although we did not employ these quantities as reference functions by fitting the parameters, vibrational spectra are well reproduced by MTB. The calculated frequencies tend to be somewhat too low. The C-H stretching frequencies are about 10% higher than the observed values. The MTB predictions for C-C, C=C and C≡C bond-stretching modes in C_2H_6, C_2H_4 and C_2H_2 are calculated to be 1470, 1729, and 2075 cm^{-1}, respectively. These values agree within 6% with the experimental frequencies, 1388, 1623, 1974 cm^{-1}. MTB provides a quite accurate estimate for torsion motion in C_2H_6 (289 cm^{-1}).

Because the zero-vibrational correction is implicitly included in the calculated $\Delta H^{0}_{f,298}$ value, the MTB frequencies are used only to estimate the entropy and heat capacity of the system. The accuracy of MTB predicted vibrations appears to be acceptable for such thermochemical calculations.

Table 8. Comparison of MTB vibrational frequencies (in cm^{-1}) with experimental values (in parenthesis)

Molecule	Symmetry of vibration	MTB (Exp.)
Methane	A_1	3247 (3158)
	T_2	3322 (3019); 1268 (1357)
	E	1360 (1534)
Ethane	A_{1g}	3249 (2896); 1470 (1388); 1131 (995)
	A_{2u}	32445(2915); 1385 (1370)
	E_u	3292 (2974); 1330 (1460); 937 (822)
	E_g	3271 (2969); 1363 (1468): 1120 (1190)
	A_{1u}	289 (289)
Ethylene	A_{1g}	3297 (3026); 1729 (1623); 1332 (1342)
	A_{1u}	858 (1023)
	B_{1u}	3299(2990); 1344 (1444)
	B_{2g}	978 (943)
	B_{2u}	3374 (3106); 899 (810)
	B_{3g}	3356 (3103); 1125 (1236)
	B_{3u}	673 (949)
Acetylene	Σ_g	3476 (3374); 2075 (1974)
	Σ_u	3422 (3289)
	Π_u	470 (730)
	Π_g	525 (612)

Computational performance.

The limiting step of semiempirical calculations is the diagonalization of the Hamiltonian matrix. Usually a single-point calculation requires about 20 iterations. It means that for closed-shell systems 20 diagonalizations are needed, while 40 such steps are required for radicals treated using the spin-unrestricted scheme. Moreover, the number of iterations for self-consistent treatment of π-conjugated systems like carbon nanotubes may remarkably increase with the size of the model. In such situations, the semiempirical calculations become very time-consuming.

By contrast, within the MTB method based a non-iterative orthogonal model, only a single matrix diagonalization of the Hamiltonian matrix is required independent of whether a closed- or an open-shell system is calculated. This makes MTB calculations extremely efficient and allows for studying extended systems containing up to a few thousand atoms. Also, the MTB approach appears very promising for dynamic simulations of nanostructures.

Conclusion

We described the semiempirical MTB approach that provides accurate estimates for thermochemistry and structural parameters of carbon and hydrocarbon systems. Because the method is based on the orthogonal non-iterative scheme, it is computationally very efficient, at least an order of magnitude faster than the standard semiempirical methods. The model provides good estimates of heats of formation and reaction enthalpies. The mean absolute error of 2.26 kcal/mol is obtained for the $\Delta H^0_{f,298}$ values across different classes of molecules and radicals (113 comparisons). The suggested scheme is more accurate than the MP2, B3LYP and standard semiempirical methods. The MTB approach reproduce well molecular geometries with the MAEs of 0.011 Å and 1.3° for the bond lengths and bond angles, respectively.

Thus, the proposed scheme, which combines the very good accuracy and high computational efficiency, appears very promising for estimation of thermochemical data of large carbon and hydrocarbon systems and for atomic simulations of new materials.

References

[1] *Computational Thermochemistry*, ed. K. K. Irikura and D. J. Frurip, ACS Symposium Series No. 677, Washington, DC, 1998.

[2] Bond, D. *J. Org. Chem.* 2007, *72*, 5555.

[3] *Quantum-Mechanical Prediction of Thermochemical Data*, Understanding Chemical Reactivity Vol. 22, edited by J. Cioslowski, Kluwer Academic, Dordrecht, 2001.

[4] Curtiss, L.A.; Raghavachari, K. *Theot. Chem. Acc.* 2002, *108*, 61.

[5] W. Koch, W.; M. C.;Holthausen, M. C. *Chemist's Guide to Density Functional Theory*; Wiley-VCH: Weinheim, 2001.

[6] Curtiss, L. A.; Raghavachari, K.; Redfern, P. C.; Pople, J. A. *J. Chem. Phys.* 1997, *106*, 1063.

[7] Curtiss, L. A.; Raghavachari, K.; Redfern, P. C.; Pople, J. A. *J. Chem.Phys.* 2000, *112*, 7374.

[8] Schreiner, P. R.; Fokin, A. A.; Pascal, R. A.; de Meijere, A. *Org. Lett.* 2006, *8*, 3635.

[9] Redfern, P. C.; Zapol, P.; Curtiss, L. A.; Raghavachari, K., *J. Phys. Chem. A* 2000, *104*, 5850-5854

[10] Curtiss, L. A.; Raghavachari, K.; Redfern, P. C.; Pople, J. A. *J. Chem.Phys.* 2005, *123*, 124107.
[11] Wodrich, M. D.; Corminboeuf, C.; Schleyer, P. v. R. *Org. Lett.*2006, *8*, 3631.
[12] Check, C. E.; Gilbert, T. M. *J. Org. Chem.* 2005, *70*, 9828.
[13] Zhao Y, Truhlar D.G. *Acc. Chem. Res.* 2008, *41,* 157.
[14] Thiel, W. *Adv. Chem. Phys.* 1996, *93,* 703.
[15] Repasky, M. P.; Chandrasekhar, J.; Jorgensen, W. L. *J. Comput. Chem.* 2002, *23*, 1601.
[16] Giese, T. J.; Sherer, E. C.; Cramer, C. J.; York, D. M. *J. Chem. Theory Comput.* 2005, *1*, 1275.
[17] Tuttle, T.; Thiel, W. *PCCP,* 2008, *10,* 2159.
[18] Tirado-Rives, J.;Jorgensen, W. L. *J. Chem. Theory Comput.* 2008, *4,* 297.
[19] Hoffman, R. *J. Chem. Phys.* 1963, *39*, 1397.
[20] Slater, J.C.; Koster, G.F. *Phys. Rev. B* 1954, *94*, 1498.
[21] Selvam, P.; Tsuboi, H.; Koyama, M.; Kubo, M.; Miyamoto, A. *Catalysis Today* 2005, *100*, 11.
[22] Elstner, M. *Theor. Chem. Acc.* 2006, *116*: 316.
[23] Porezag, D.; Frauenheim, T.; Kohler, T.; Seifert, G.; Kaschner, R. *Phys. Rev. B* 1995, *51*, 12947.
[24] Elstner, M.; Porezag, D.; Jungnickel, G.; Elsner, J.; Haugk, M.; Frauenheim, T.; Suhai, S. Seifert, G. *Phys. Rev. B* 1998, *58,* 7260.
[25] Krüger, T.; Elstner, M.; Schiffels, P.; Frauenheim, T. *J. Chem. Phys.* 2005, *122*, 114110.
[26] Sattelmeyer, K. W.; Tirado-Rives, J.; Jorgensen, W. L. *J. Phys. Chem. A* 2006, 110, 13551.
[27] Otte, N.; Scholten, M.; Thiel, W. *J. Phys. Chem. A* 2007, *111,* 5751.
[28] Xu, C. H.; Wang, C. Z.; Chan, C. T., Ho, K. M. *J. Phys. Condens. Matter* 1992, *4,* 6047.
[29] Wang, C. Z.; Chan, C.T.; Ho, K. M. *Phys. Rev. Lett.* 1991, *66*, 189.
[30] Wang Y.; Mak, C. H. *Chem. Phys. Lett.* 1995, *235,* 37.
[31] Horsfield, A. P.; Godwin, P. D.; Pettifor, D. G.; Sutton, A. P. *Phys. Rev. B* 1996, *54,* 15773.
[32] Winn, M. D.; Rassinger, M.; Hafner, J. *Phys. Rev. B* 1997, *55*, 5364.
[33] Pan, B. C. *Phys. Rev. B* 2001, *64*, 155408.
[34] Zhao J. J.; Lu, J. P. *Phys. Lett. A* 2003, *319*, 523.
[35] Voityuk, A. A. *J. Chem. Theory Comput.* 2006, *2*, 1038.
[36] Voityuk, A. A. *Chem. Phys. Lett.* 2006, *433,* 216.
[37] A.W. Jasper, N.E. Schultz, D.G. Truhlar, *J. Chem. Theory Comput.* 2007, *3*, 210.
[38] Kolb, M.; Thiel, W. *J Comput. Chem.* 1993, *14*, 775.
[39] Bondi, A. *J. Phys. Chem.* 1964, *68,* 441.
[40] *NIST Chemistry WebBook*, NIST Standard Reference Database, No. 69; Mallard, W. G.; Linstrom, P. J.; Eds.; National Institute of Standards and Technology: Gaithersburg, http://webbook.nist.gov.
[41] Yu, J.; Sumathi, R. Green, W.H. *J. Am. Chem. Soc.* 2004, *126*, 12685.
[42] Cioslowski, J.; Rao, N.; Moncrieff, D. *J. Am. Chem. Soc.* 2000, *122*, 8265.
[43] Curtiss, L.A.; Raghavachari, K.; Redfern, P. C.; Pople, J. A. *J. Chem. Phys.* 2000, *112*, 7374.

[44] Rodriguez, K. R.; Williams, S. M.; Young, M. A.; Teeters-Kennedy, S.; Heer, J. M.; Coe, J. V. *J. Chem. Phys.* 2006, *125*, 194716.

[45] Fowler, P. W.; Manolopoulos, D. E. *An Atlas of Fullerenes*; Clarendon: Oxford, 1995.

[46] Zheng, G. S.; Irle, S.; Morokuma, K. *Chem. Phys. Lett.* 2005, *412,* 210.

[47] Hedberg, K.; Hedberg, L.; Bühl, M.; Bethune, D. S.; Brown, C. A.; Johnson, R. D. *J. Am. Chem. Soc.* 1997, *119*, 5314.

[48] *NIST Computational Chemistry Comparison and Benchmark Database,* NIST Standard Reference Database Number 101 Release 12, Aug 2005, Editor: R. D. Johnson III http://srdata.nist.gov/cccbdb

In: Thermochemistry and Advances in Chemistry Research ISBN: 978-1-60692-376-4
Ed: Anton G. Mirskiy

Short Communications

THEORETICAL THERMOCHEMISTRY OF THE FSO_X (X=1 TO 5) RADICAL SERIES

María P. Badenes, Adela E. Croce and Carlos J. Cobos
Instituto de Investigaciones Fisicoquímicas Teóricas y Aplicadas (INIFTA), Departamento de Química, Facultad de Ciencias Exactas, Universidad Nacional de La Plata, CCT La Plata-CONICET, Casilla de Correo 16, Sucursal 4, (1900) La Plata, Argentina. E-mail address: cobos@inifta.unlp.edu.ar

ABSTRACT

Standard enthalpies of formation at 298 K for the FSO_x (x=1 to 5) radical series are determined at the B3LYP, B3PW91, mPW0, B98, B97-2 and PBE0 levels of density functional theory and at the *ab initio* Gaussian-3 models G3MP2B3, G3B3, BAC-G3MP2B3 and BAC-G3B3. Enthalpies of formation were calculated from total atomization energies and from isodesmic reaction energies. The B98, B97-2 and PBE0 hybrids lead to the most reliable results of the tested functionals. However, the Gaussian-3 and mainly the computations with bond additivity corrections (BAC) combined with the 6-311+G(3df) basis set provide the more accurate results. The best values obtained for FSO, FSO_2, FSO_3, $FS(O_2)OO$ and $FS(O_2)OOO$ radicals are -66.6, -94.0, -120.4, -109.8 and -126.3 kcal mol^{-1}, with a stated error limit of ±2 kcal mol^{-1}. A comparison with previous values is reported.

INTRODUCTION

Thermodynamic stability of radical species plays a relevant role in understanding and predicting their reactivity. This property is most useful in the interpretation of the mechanisms of chemical reactions such as the modeling of systems relevant in the chemistry of the atmosphere, of the combustions and of many industrial processes [1-3]. The standard enthalpy of formation is an important thermodynamic property which can be accurately predicted and has been a central objective of a large number of recent electronic structure calculations. In particular, sulfur-containing compounds play a role in some regions of the atmosphere. In this context, a typical sulfur compound as SO_2 has been detected in large amounts in active volcanic areas, in the marine biosphere, and in industrialized regions. In this direction, interesting couplings between the chemistry of SO_2 with the chemistry of some atmospheric fluorinated radicals as CF_3O and FO_2 have been recently proposed [4]. As a result, ozone depleting catalytic cycles with the participation of the FSO, FSO_2 and FSO_3 radicals might be operating. However, no atmospheric measurements of these radicals have been reported so far, probably due to the low abundances or the lack of characterization of these species. These and recently proposed larger radicals conforms an FSO_x (x=1 to 5) radical series. At our laboratory the chemistry of most of these species from the experimental and theoretical point of view has been explored. Stationary [5-7] and, more recently, laser flash photolysis techniques in conjunction with quantum chemical and chemical kinetic theoretical treatments [8-23] have been employed to study a large number of reactions of these radicals.

For chemical compounds with validated experimental data, estimations may not be necessary, but for reactive substances with insufficient or unavailable experimental data as the present, estimation methods are of prime importance. In this context, we report the theoretical results of our most recent enthalpies of formation of the sulfur oxyfluoride radicals of the FSO_x (x=1 to 5) series. For this, a number of hybrid functionals in conjunction with extended basis sets and the high-level Gaussian-3 model chemistry were employed. In addition, calculations using the recent implementation which accounts for deficiencies in the last treatments by applying empirical bond-additivity atomic, molecular and bondwise corrections to electronic energies were also carried out.

RESULTS

Theoretical methods

Different approaches of the density functional theory (DFT) and the *ab initio* G3MP2B3 and G3B3 model chemistries [24] as implemented in the Gaussian 03 software package were employed in this work [25]. The DFT methods, although not truly *ab initio*, include electron correlation using functionals of electron density. The exact form of the exchange correlation-functional remains elusive. For this reason, the most popular hybrid B3LYP method [26,27] and the more recent formulations B3PW91 [26,28], mPW0 [28,29], B98 [30], B97-2 [31,32] and PBE0 [33,34] were used.

In the G3B3, the optimized molecular structure and harmonic vibrational frequencies (scaled by a factor 0.96) are calculated at the B3LYP/6-31G(d) level [24]. Based on the optimized structure, the energy at 0 K is obtained from a set of single-point energy evaluations at the levels MP4/6-31G(d), MP4/6-31+G(d), MP4/6-31G(2df,p), QCISD(T)/6-31G(d) and MP2(full)/G3Large. Spin-orbit, higher-level and zero-point vibrational energy corrections (ZPE) to calculate the total electronic energy at 0K are performed *a posteriori*. Normally, the final energies approach very well those computed at the quadratic configuration interaction QCISD(T, full)/G3Large level with a strong reduction in the computational resources. On the other hand, a reduced order of perturbation theory and only valence electrons are used in the G3MP2B3 treatment [24].

The total energies computed with the aforementioned methods were used to derive standard enthalpy of formation values for gaseous species at one atmosphere and 298.15 K, $\Delta H_{f,298}$. Firstly, we derive the total atomization energy $\sum D_0$ from the energy of the molecule (including ZPE) and their constituent elements. The enthalpy of formation for the molecular species at 0 K, $\Delta H_{f,0}$ is simply calculated by subtracting $\sum D_0$ from the experimental heats of formation of the atoms. For the present case, values of 58.99±0.02, 18.47±0.07 and 65.66±0.06 kcal mol^{-1} were employed for oxygen, fluorine and sulfur atoms respectively [35]. Then $\Delta H_{f,298}$ is obtained using the expression

$$\Delta H_{f,298} = \Delta H_{f,0} + [H^{\circ}_{298}\text{-}H^{\circ}_{0}]_M - \{x_O [H^{\circ}_{298}\text{-}H^{\circ}_{0}]_O + x_F [H^{\circ}_{298}\text{-}H^{\circ}_{0}]_F + x_S [H^{\circ}_{298}\text{-}H^{\circ}_{0}]_S\} \quad (1)$$

Here, $[H^{\circ}_{298}\text{-}H^{\circ}_{0}]_M$ is the enthalpy correction for the molecule which is calculated with the harmonic vibrational frequencies and the last terms in brackets account for atomic enthalpy corrections. The $[H^{\circ}_{298}\text{-}H^{\circ}_{0}]_{atom}$ values for oxygen, fluorine and sulfur atoms are 1.04, 1.05 and 1.05 kcal mol^{-1} respectively [35].

Due to the fact that the atomization process is not isogyric (equal spin), an accurate determination of the energy of all the species is required to perform a reliable estimation of $\Delta H_{f,298}$. In particular, systematic errors arisen in the G3B3 treatment have been recently accounted for by correction of the molecular electronic energy with a bond additivity corrections (BAC) treatment conformed by simple empirical expressions which efficiently account for atomic ($E_{BAC\text{-}atom}$), molecular ($E_{BAC\text{-}molecule}$) and bondwise ($E_{BAC\text{-}bond}$) corrections [36]. The first term corrects deficiencies in the interaction electron correlation, the second corrects for errors due to the molecular electronic structure and the third takes into account electron pairing corrections. In the last one all bond distances in the molecule are required as input. The total correction evaluated as $E_{BAC\text{-}correction} = E_{BAC\text{-}atom} + E_{BAC\text{-}molecule} + E_{BAC\text{-}bond}$, is afterwards subtracted from the G3B3 enthalpy of formation to obtain the corrected value. The average difference between calculated and an extended test set of neutral molecules is 0.44 kcal mol^{-1} for the BAC-G3B3 method with a maximum error of 1.76 kcal mol^{-1} [36].

The second method used relies on the construction of isodesmic (equal bond) and isogyric reaction schemes [37]. In this procedure, the $\Delta H_{f,298}$ is calculated by combining the computed enthalpy change of the selected reaction with well-established enthalpies of formation for other reaction components. Due to the fact that in these reactions the reactants and products contain the same type and number of bonds, systematic errors due to truncation of the number of configurations used for treating electron correlation energy and the incompleteness of the one-electron basis sets mostly cancel. Owing to the presence of lone pair electrons on O atoms, the correlation effects are expected to be here important. For these reasons, the $\Delta H_{f,298}$ computed *via* isodesmic energies are often more accurate than those obtained by the atomization procedure.

Molecular structures

In the Kohn-Sham DFT, the particular contributions of exchange and correlation functionals that will give the best results in a given instance are difficult to predict [38]. For this reason, as in recent works from this laboratory we have employed several combinations of exchange and correlation functionals for computing structures, harmonic vibrational frequencies and thermochemical properties [39-42]. For all radicals, fully optimized geometries were computed using analytical gradient methods, and harmonic vibrational frequencies *via* analytical second derivative methods. The molecular structures predicted at the B3LYP/6-311+G(3df) level are listed in Table 1 and illustrated in Figure 1.

Table 1. Geometrical parameters for FSO_x (x=1 to 5) radicals at the B3LYP/6-311+G(3df) level of theory (bond lengths in Angstroms, angles in degrees).

Parameter	*FSO*	FSO_2	FSO_3	$FS(O_2)OO$	$FS(O_2)OOO$
r(O-O)	-	-	-	1.312	1.193
r(O-O)	-	-	-	-	2.542
r(S-O)	-	-	-	1.705	1.473
r(S-O)	-	-	-	-	-
r(S=O)	1.456	1.439	1.441	1.409	1.424
r(S=O)	-	1.439	1.445	1.410	1.442
r(S=O)	-	-	1.445	-	-
r(S-F)	1.627	1.606	1.560	1.561	1.566
∠(O-O-O)	-	-	-	-	111.6
∠(S-O-O)	-	-	-	112.8	115.0
∠(O-S=O)	-	-	-	110.1	114.2
∠(O-S=O)	-	-	-	103.9	107.6
∠(O-S-F)	108.3	106.4	105.3	96.6	104.0
DIH (S-O-O-O)	-	-	-	-	166.7
DIH (O-O-S=O)	-	-	-	40.3	31.8

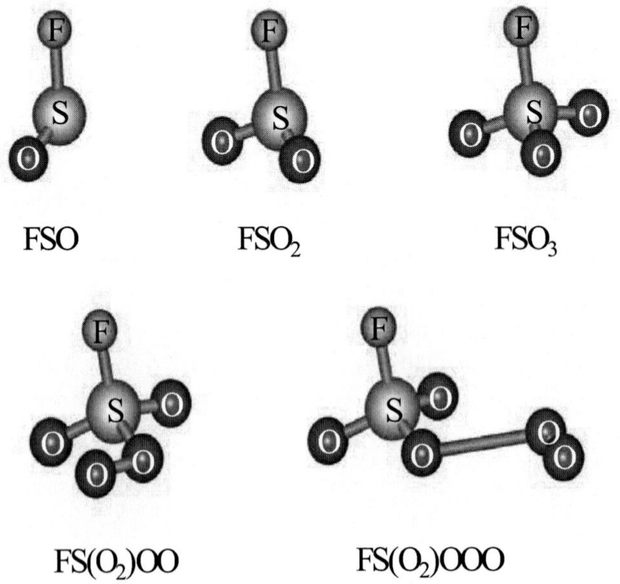

Figure 1. Molecular structures for FSO_x (x=1 to 5) radicals optimized at the B3LYP/6-311+G(3df) level.

No experimental data are available for comparison. However, the computed values agree satisfactorily with measured parameters in other FSO_3 group containing species [43-45]. For instance, values of 1.545, 1.409, 1.606 Å and 94.0° have been reported for the S-F, S=O and S-O bond lengths and FS-O angle in $FS(O_2)OF$ [45]. At the calculated structures we compute harmonic vibrational frequencies. The derived zero-point vibrational energy for FSO, FSO_2, FSO_3, $FS(O_2)OO$, $FS(O_2)OOO$ and FSO_5 are 3.4, 6.4, 8.2, 11.2, 11.5 and 11.7 kcal mol^{-1} respectively. Quite similar molecular parameters were obtained using the other employed functionals and, for sake of simplicity, are not given here.

Enthalpies of formation determined from total atomization energies

Enthalpies of formation for FSO_x radicals estimated from atomization energies were calculated using equation (1). The resulting $\Delta H_{f,298}$ values are listed in Table 2. The B3LYP calculations show that the addition of polarization functions to the 6-311+G(d) basis set reduces strongly the calculated enthalpy values. In this context, the importance of including high d- and f-polarization is well-documented for the SF_x (x=1 to 6) series [46]. The use of diffuse functions in the basis set grants radial flexibility to represent electron density far from the nuclei, while the p, d and f-type polarization functions confer angular flexibility to represent regions of high electron density among the bonded atoms. Due to the magnitude of the correlation energy is typically 10% of the exchange energy, differences in the $\Delta H_{f,298}$ values obtained with the different generation of hybrid functionals can be mostly attributed to specific particularities in the exchange functionals. A first inspection of Table 2 shows that the Gaussian-3 models predict a higher stability for all radicals. As Table 2 also shows, an additional stabilization is achieved after BAC corrections.

Table 2. Enthalpies of formation at 298 K for FSO_x (x=1 to 5) radicals calculated from atomization energies (in kcal mol^{-1}). A=6-311+G(d); B=6-311+G(3df).

Level	*FSO*	*FSO_2*	*FSO_3*	*$FS(O_2)OO$*	*$FS(O_2)OOO$*
B3LYP/A	-42.0	-48.0	-55.9	-43.0	-55.5
B3LYP/B	-64.5	-87.5	-111.0	-96.9	-114.1
B3PW91/B	-66.3	-92.5	-118.3	-105.2	-122.4
mPW0/B	-63.0	-86.9	-113.8	-98.8	-111.8
B98/B	-67.7	-93.4	-119.6	-106.8	-124.6
B97-2/B	-68.4	-95.3	-122.6	-110.0	-129.6
PBE0/B	-65.6	-92.7	-120.5	-107.4	-123.3
G3MP2B3	-64.1	-87.9	-110.8	-99.0	-102.7
BAC-G3MP2B3	-65.6	-91.7	-117.0	-105.9	-110.7
G3MP2//B3LYP/B	-64.2	-89.3	-112.5	-102.0	-117.5
BAC-G3MP2//B3LYP/B	-65.9	-93.7	-119.4	-109.7	-126.3
G3B3	-63.6	-88.2	-113.3	-100.6	-
BAC-G3B3	-65.4	-91.7	-118.5	-105.9	-
G3//B3LYP/B	-63.5	-88.8	-114.0	-102.2	-
BAC-G3//B3LYP/B	-66.0	-94.0	-121.4	-110.2	-

The individual terms in $E_{BAC\text{-}correction}$ are given in Table 3. The term $E_{BAC\text{-}bond}$ depends on the number of bonded atoms and represents the most important contribution to $E_{BAC\text{-}correction}$. In addition, as reported in other studies, Table 2 shows that more realistic $\Delta H_{f,298}$ are obtained improving the molecular geometry by using larger basis sets in the Gaussian-3 models [22,41,42].

The above considerations indicate that the BAC-G3//B3LYP/6-311+G(3df) (the BAC-G3MP2//B3LYP/6-311+G(3df) for $FS(O_2)OOO$) enthalpies of formation represent the more precise results. Thus, in the absence of experimental information, DFT predictions were confronted against these values. This analysis shows that the B98, B97-2 y PBE0 methods lead to the best DFT results with mean deviation errors of 1.8, 1.7 and 1.7 kcal mol^{-1} while the functional B3LYP/6-311+G(3df) yields the poorest result with 8.8 kcal mol^{-1}. Then, the more realistic results obtained by atomization are: -66.0 (FSO), -93.9 (FSO_2), -120.4 (FSO_3), -110.0 ($FS(O_2)OO$) and -126.3 kcal mol^{-1} ($FS(O_2)OOO$).

Table 3. Bond additivity corrections for FSO_x (x=1 to 5) radicals (in kcal mol^{-1}). [a]At the G3//B3LYP/6-311+G(3df) level; [b]At the G3MP2//B3LYP/6-311+G(3df) level.

Correction	*FSO* [a]	*FSO_2* [a]	*FSO_3* [a]	*$FS(O_2)OO$* [a]	*$FS(O_2)OOO$* [b]
$E_{BAC,atom}$	-0.641	-0.641	-0.641	-0.641	-1.702
$E_{BAC,molecule}$	-0.636	-0.954	-1.272	-1.590	-1.716
$E_{BAC,bond}$	3.276	5.562	7.769	8.181	12.223
$E_{BAC,crrection}$	1.999	3.967	5.856	5.950	8.805

Enthalpies of formation determined from isodesmic reaction energies

Errors in absolute quantities from quantum chemical calculations are frequently systematic. To compensate for some of these errors, isodesmic reactions in which the number of broken and formed bonds are conserved, normally lead to more accurate results [37]. Therefore, we compare the $\Delta H_{f,298}$ derived from atomization energies with those resulting from isodesmic reaction schemes. For a given set of target molecules, frequently many different choices for these hypothetical reactions are possible. However, they are normally limited by the accuracy of the enthalpy data of the reference compounds. For the present purposes the working reactions listed in Table 4 were selected. Isodesmic reaction enthalpies for reactions (1) to (10) were calculated by adding the zero-point energies and the thermal corrections to the total energy of each molecule. These reaction enthalpies are very different from zero, ranging from -70 to 7 kcal mol^{-1}. For instance, values as large as -73.7 and -47.3 kcal mol^{-1} were calculated at the G3B3 level for reactions (3) and (4), indicating that bond energies are far from being additive. The $\Delta H_{f,298}$ were then obtained using the computed isodesmic reaction enthalpies in combination with the following well-established enthalpies of formation (in kcal mol^{-1}): -70.55 (SF_2) [46], -180.0 ($HFSO_3$) [47], -175.1 (H_2SO_4), 8.91 (HO), -57.8 (H_2O), 4.1 (CH_3O), -48.0 (CH_3OH), 3.3 (HO_2), 2.15 (CH_3O_2) and -32.5 (HOOH) [2]. The resulting $\Delta H_{f,298}$ at all employed levels of theory are summarized in Table 5. It is observed that despite of the abovementioned differences in the isodesmic reaction enthalpies, the differences in the calculated $\Delta H_{f,298}$ are much smaller. In fact, averaging all DFT results computed with the 6-311++G(3df,3pd) basis set the following values (in kcal mol^{-1}) are obtained: -66.0±2.8 (FSO), -95.1±2.8 (FSO_2), -124.4±2.5 (FSO_3), -109.0±3.2 ($FS(O_2)OO$) and -128.0±4.7 ($FS(O_2)OOO$). Similarly, the Gaussian-3 methods lead to the values: -67.1±0.8 (FSO), -94.0±0.8 (FSO_2), -120.3±1.0 (FSO_3), -109.4±1.0 ($FS(O_2)OO$) and -126.2±0.8 ($FS(O_2)OOO$)

Table 4. Isodesmic reaction schemes for FSO_x (x= 1 to 5) radicals.

FSO

1) $SF_2 + H_2SO_4 + OH \rightarrow FSO + HFSO_3 + H_2O$
2) $SF_2 + H_2SO_4 + CH_3O \rightarrow FSO + HFSO_3 + CH_3OH$

FSO_2

3) $SF_2 + H_2SO_4 + 2\ OH \rightarrow 2\ FSO_2 + 2\ H_2O$
4) $SF_2 + H_2SO_4 + 2\ CH_3O \rightarrow 2\ FSO_2 + 2\ CH_3OH$

FSO_3

5) $HFSO_3 + OH \rightarrow FSO_3 + H_2O$
6) $HFSO_3 + CH_3O \rightarrow FSO_3 + CH_3OH$

$FS(O_2)OO$

7) $HFSO_3 + HO_2 \rightarrow FS(O_2)OO + H_2O$
8) $HFSO_3 + CH_3O_2 \rightarrow FS(O_2)OO + CH_3OH$

$FS(O_2)OOO$

9) $HFSO_3 + CH_3O_2 + H_2O_2 \rightarrow FS(O_2)OOO + CH_3OH + H_2O$
10) $HFSO_3 + HO_2 + H_2O_2 \rightarrow FS(O_2)OOO + 2\ H_2O$

Table 5. Enthalpies of formation at 298 K for FSO_x (x=1 to 5) radicals calculated from isodesmic reactions (in kcal mol^{-1}). A=6-311++G(d,p); B=6-311++G(3df,3pd).

Level	*FSO*		*FSO_2*		*FSO_3*		*$FS(O_2)OO$*		*$FS(O_2)OOO$*	
	1	2	3	4	5	6	7	8	9	10
B3LYP/A	-59.7	-58.3	-90.7	-89.4	-123.8	-122.5	-112.9	-114.6	-128.4	-126.8
B3LYP/B	-66.1	-64.4	-95.8	-94.1	-125.9	-124.2	-109.6	-111.1	-128.2	-126.7
B3PW91/B	-70.0	-68.6	-99.3	-97.9	-128.3	-126.9	-109.0	-114.1	-136.1	-131.0
mPW0/B	-68.3	-61.7	-96.8	-90.2	-122.8	-121.7	-109.7	-109.1	-128.7	-129.3
B98/B	-64.0	-62.2	-93.3	-91.5	-123.2	-121.3	-103.5	-108.2	-124.4	-119.8
B97-2/B	-65.6	-63.5	-94.7	-92.6	-123.5	-121.4	-103.4	-108.4	-126.9	-121.9
PBE0/B	-69.2	-68.0	-97.9	-96.7	-127.4	-126.2	-108.5	-113.6	-134.5	-129.5
G3MP2B3	-66.8	-68.4	-93.4	-95.0	-119.2	-120.7	-108.2	-109.5	-126.6	-125.3
G3MP2//B3LYP/B	-66.6	-67.8	-93.8	-94.9	-118.9	-120.1	-109.4	-110.7	-127.1	-125.7
G3B3	-66.2	-67.5	-93.0	-94.3	-120.2	-121.6	-108.0	-109.2	-	-
G3//B3LYP/B	-66.2	-67.3	-93.3	-94.4	-120.4	-121.4	-109.3	-110.5	-	-

Comparison of enthalpy of formation values with previous data

A comparison between the more accurate results derived from atomization and isodesmic energies shows that, after BAC corrections, the Gaussian-3 model predicts $\Delta H_{f,298}$ values in very good agreement with those estimated *via* balanced isodesmic reactions. Therefore, both sets of data were averaged to obtain our best $\Delta H_{f,298}$ values. In this way, the final results for FSO, FSO_2, FSO_3, $FS(O_2)OO$ and $FS(O_2)OOO$ radicals are -66.6, -94.0, -120.4, -109.7 and -126.3 kcal mol^{-1}. From the above considerations, a conservative error estimate of ±2 kcal mol^{-1} seems to be reasonable.

We now confront our best value for FSO with those reported in previous studies. A large difference between the $\Delta H_{f,298}$ values reported by Herron [48] and by Li [4] may be recognized. In fact, from the experimental value of 87.5±4.0 kcal mol^{-1} for the F-FSO dissociation enthalpy and an estimate of -118±8 kcal mol^{-1} for the $\Delta H_{f,298}$ of F_2SO [47], a value of -50±8 kcal mol^{-1} has been calculated in Ref. 48. This result strongly contrasts with -81.9 kcal mol^{-1} obtained from the O atom extrusion from FSO_2 of 59.5 kcal mol^{-1} and an estimate of -81.8 kcal mol^{-1} for the $\Delta H_{f,298}$ of FSO_2 radical (see below) [4]. More, recently, from a large number of isodesmic reactions analyzed at the CBS-q level, the value -69.9±2.0 kcal mol^{-1} was obtained in Ref. 19. The present value of -66.6 kcal mol^{-1} supports this last result.

FSO_2 is the most extensively studied radical of the FSO_x series. A value of -117±5 kcal mol^{-1} has been estimated by Takacs [49] at 0 K assuming for the S-F bond energy in F_2SO_2 a value equal to the activation energy measured for the F_2SO_2 thermal decomposition in the falloff range, 81±2 kcal mol^{-1}, [50] and the value -179.3±2 kcal mol^{-1} for the enthalpy of formation of F_2SO_2. A more realistic value of 98±3 kcal mol^{-1} for the S-F bond strength recommended by Herron [48], leads to -102±4 kcal mol^{-1} for FSO_2. More recently, Sudlow and Woolf [51] suggested a value of -110 kcal mol^{-1} on the basis of an extrapolated F-(F)SO_2 bond energy of 93 kcal mol^{-1} obtained from the S-F bond energies in the SF_2, SF_4 and SF_6 molecules. An F-SO_2 bond energy of 29.8 kcal mol^{-1} founded on QCISD/6-31G(d) calculations [4] lead to the value -81.8 kcal mol^{-1} for FSO_2. A recent estimation of ΔH^0_{298} is provided by computed CBS-q isodesmic energies [19]. The reported value of -96.2±3.0 kcal mol^{-1} [19] agrees reasonably well with the present estimation of -94.0 kcal mol^{-1}.

FSO_3 radical has been extensively studied from the kinetic point of view [5-18.20-23]. Our enthalpy of formation for FSO_3 of -120.4 kcal mol^{-1} agrees nicely with the determined at our laboratory, -121.3±3.0 kcal mol^{-1}, from an analysis of isodesmic energies calculated at a lower level of theory (B3LYP/6-311++G(3df,3pd)//B3LYP/6-311++G(d,p)) [19]. Both studies suggest that FSO_3 radical is about 6 kcal mol^{-1} less stable than assumed previously. In fact, a value of –127 kcal mol^{-1} has been derived from the measured enthalpy for the equilibrium $FS(O_2)OO(O_2)SF \leftrightarrow 2\ FS(O_2)O$ and an estimation for the enthalpy of formation of $FS(O_2)OO(O_2)SF$ of -276±3.0 kcal mol^{-1} [47]. A similar value, based on bond additivity methods has been also reported [52].

The $FS(O_2)OO$ radical has been suggested long time ago to explain the pronounced O_2 influence in the photodissociation of $FS(O_2)OF$ in the presence of SO_2 and SO_3 [5] and in the reaction $FS(O_2)OF$ with CO [7]. The first proposed value estimated using bond additivity schemes of -120.2 kcal mol^{-1} [52] contrasts with -106.4±3.0 kcal mol^{-1} obtained from

B3LYP/6-311++G(3df,3pd)//B3LYP/6-311++G(d,p) isodesmic energies [19]. The present value, -109.8 kcal mol^{-1}, lies between these estimates, within the stated error limits of Ref. 19.

Finally, the larger member of the series, $FS(O_2)OOO$, has been postulated as an intermediate in the mechanism of the 193-nm pulsed photolysis of $FS(O_2)OF$ in the presence of molecular oxygen [18]. This species has been recently confirmed by low-temperature matrix isolation techniques and quantum chemical calculations [53]. No previous data to confront with our predicted value of -126.3 kcal mol^{-1} are available.

CONCLUSIONS

Clearly, direct and precise experimental determinations for the enthalpy of formation of the studied radicals are desirable. In the absence of such information, the values -66.6, -94.0, -120.4, -109.8 and -126.3 kcal mol^{-1} determined for FSO, FSO_2, FSO_3, $FS(O_2)OO$ and $FS(O_2)OOO$ in the present quantum chemical research provide valuable information of this important thermodynamic property.

ACKNOWLEDGMENTS

This research was supported by the Universidad Nacional de La Plata, the Consejo Nacional de Investigaciones Científicas y Técnicas (CONICET), the Agencia Nacional de Promoción Científica y Tecnológica and the Max Planck Institute for Biophysical Chemistry Göttingen through the "Partner Group for Chlorofluorocarbons in the Atmosphere".

REFERENCES

[1] R. Atkinson, R; Baulch, D. L.; Cox, R. A., Crowley, J. N.; Hampson, R. F.; Hynes, R. G.; Jenkin, M. E.; Rossi, M. J.; Troe, J. *Atmos. Chem. Phys.* 2006, *6,* 3625-4055.

[2] Sander, S. P.; Friedl, R. R.: Golden, D. M.; Kurylo, M. J.; Moortgat, G. K.; Wine, P. H.;Ravishankara, A. R.; Kolb, C. E.; Molina, M. J.; Finlayson-Pitts, B. J.; Huie, R. E.; Orkin, V. L. *Chemical Kinetics and Photochemical Data for Use in Atmospheric Studies; NASA/JPL Data Evaluation, JPL Publication 06-2 Evaluation No. 15; NASA*: Pasadena, CA, November 20, 2006.

[3] Baulch, D. L; Bowman, C. T.; Cobos, C. J.; Cox, R. A.; Just, Th.; Kerr, J. A.; Pilling, M. J.; Stocker, D.; Troe, J.; Tsang, W.; Walker, J.; Warnatz, J. *J. Phys. Chem. Ref. Data,* 2005, *34,* 757-1397.

[4] Li, Z. *Chem. Phys. Lett.* 1997, *269,* 128-137.

[5] Basualdo, W. H.; Schumacher, H. J. *Z. Physik. Chem., NF,* 1965, *47*, 57-64.

[6] Gatti, R.; Sicre, J. E.; Schumacher, H. J. *Z. Physik. Chem. NF*, 1965, *47*, 323-336.

[7] Vasini, E.; Schumacher, H. J. *Z. Physik. Chem, NF,* 1977, *104*, 219-228.

[8] Croce de Cobos, A.E.; Cobos, C. J.; Castellano, E. *J. Phys. Chem.* 1989, *93*, 274-278.

[9] Cobos, C. J.; Croce de Cobos, A. E.; Hippler, H.; Castellano, E.; *J. Phys. Chem.* 1989, *93*, 3089-3094.

[10] Croce, A. E.; Cobos, C. J.; Castellano, E. *Chem. Phys. Lett.* 1989, *158*, 157-160.
[11] Cobos, C. J.; Croce, A. E.; Castellano, E. *Int. J. Chem. Kinet.* 1990, *22*, 289-297.
[12] Croce, A.E.; Cobos C. J.; Castellano, E. *J. Photochem. Photobiol., A: Chem.* 1990, *55*, 135-144.
[13] Croce, A.E.; Cobos C. J.; Castellano, E. In *Laser and Quantum Optics", CIF Series*-Vol 13, L. M. Narducci, E. J. Quel y J.R. Tredicce (Eds.), *World Scientific, Singapore,* 1990, pp. 379-390.
[14] Cobos, C. J.; Croce, A. E.; Castellano, E. *J. Photochem. Photobiol., A: Chem.* 1991, *59*, 143-151.
[15] Croce, A. E.; Cobos, C. J.; Castellano, E.; *J. Photochem. Photobiol., A: Chem.* 1992, *64*, 15-24.
[16] Cobos, C. J.; Croce, A. E.; Castellano, E. *J. Photochem. Photobiol., A: Chem.* 1994, *84*, 101-104.
[17] Cobos, C. J.; Croce, A. E.; Castellano, E. *J. Fluorine. Chem.* 1996, *79*, 157-160.
[18] Croce, A. E.; Cobos C. J.; Castellano, E. *J. Fluorine Chem.* 1997, *82*, 91-97.
[19] Tucceri, M. E.; Badenes, M. P.; Cobos, C. J. *Z. Physik. Chem.* 2000, *214*, 1193-1208.
[20] Tucceri, M. E ; Badenes, M. P.; Croce, A. E.; Cobos, C. J. *Chem. Commun.* 2001, 71-72.
[21] Tucceri, M. E.: Badenes, M. P.; Croce, A. E.; Cobos, C. J. *Phys. Chem. Chem. Phys.* 2001, *3*, 1832-1839.
[22] Tucceri, M. E.; Castellano; E.; Croce, A.E.; Cobos, C. J. *J Argent. Chem. Soc.* 2005, *93*, 29-41.
[23] Tucceri, M. E.; Croce, A. E.; Cobos, C. J. *Chem. Phys. Lett.* 2005, *404*, 232-236.
[24] Baboul, A. G.; Curtiss, L. A.; Redfern, P. C.; Raghavachari, K. *J. Chem. Phys.* 1999, *110*, 7650-7657.
[25] Frisch, M. J.; Trucks, G. W.; Schlegel, H. B.; Scuseria, G. E.; Robb, M. A.; Cheeseman, J. R.; Montgomery, J. A, Jr..; Vreven, T.; Kudin, K.N.; Burant, J. C.; Millam, J. M.; Iyengar, S. S.; Tomasi, J.; Barone, V.; Mennucci, B.; Cossi, M.; Scalmani, G.; Rega, N.; Petersson, G. A.; Nakatsuji, H.; Hada, M.; Ehara, M.; Toyota, K.; Fukuda, R.; Hasegawa, J.; Ishida, M.; Nakajima, T.; Honda, Y.; Kitao, O.; Nakai, H.; Klene, M.; Li, X.; Knox, J. E.; Hratchian, H. P.; Cross, J. B.; Bakken, V.; Adamo, C.; Jaramillo, J.; Gomperts, R.; Stratmann, R. E.; Yazyev, O.; Austin, A. J.; Cammi, R.; Pomelli, C.; Ochterski, J. W.; Ayala, P. Y.; Morokuma, K.; Voth, G. A.; Salvador, P.; Dannenberg, J. J.; Zakrzewski, V. G.; Dapprich, S.; Daniels, A. D.; Strain, M. C.; Farkas, O.; Malick, D. K.; Rabuck, A. D.; Raghavachari, K.; Foresman, J. B.; Ortiz, J. V.; Cui, Q.; Baboul, A. G.; Clifford, S.; Cioslowski, J.; Stefanov, B. B.; Liu, G.; Liashenko, A.; Piskorz, P.; Komaromi, I.; Martin, R. L.; Fox, D. J.; Keith, T.; Al-Laham, M. A.; Peng, C. Y.; Nanayakkara, A.; Challacombe, M.; Gill, P. M. W.; Johnson, B.; Chen, W.; Wong, M. W.; Gonzalez, C.; Pople, J. A. *Gaussian 03*, revision D.01; Gaussian, Inc.: Pittsburgh, PA, 2004.
[26] Becke, A. D. *Phys. Rev. A*, 1988, *38*, 3098-3100.
[27] Lee, C.; Yang, W.; Parr, R. G. *Phys. Rev. B,* 1988, *37*, 785-789.
[28] Perdew, J. P.; Burke, K.; Wang, Y. *Phys. Rev. B* 1996, *54*, 16533-16539.
[29] Adamo, C.; Barone, V.; *J. Chem. Phys.* 1998, *108*, 664-675.
[30] Schmider, H. L.; Becke, A. D. *J. Chem. Phys.* 1998, *108*, 9624-9631.

[31] Hamprecht, F. A.; Cohen, A. J.; Tozer, D. J.; Handy, N. C. *J. Chem. Phys.* 1998, *109*, 6264-6271.
[32] Wilson, P. J.; Bradley, T. J.; Tozer, D. J. *J. Chem. Phys.* 2001, *115*, 9233-9242.
[33] Adamo, C.; Barone, V. *J. Chem. Phys.* 1999, *110*, 6158-6170.
[34] Hoe, W. -H.; Cohen, A. J; Handy, N. C. *Chem. Phys. Lett.* 2001, *341*, 319-328.
[35] Curtiss, L. A.; Redfern, P. C.; Frurip, D. J. in Lipkowitz K. B. and Boyd D. B. (Eds.), *Reviews in Computational Chemistry*, Vol 15, Wiley-VCH, New York, 2000, pp. 147-202.
[36] Anantharaman, B.; Melius, C. F. *J. Phys. Chem. A*, 2005, *109*, 1743-1747.
[37] Hehre, W. J.; Dichfield, R.; Radom, L.; Pople, J. A. *J. Amer. Chem. Soc.* 1970, *92*, 4796-4801.
[38] Kohn, W.; Sham, L. J. *Phys. Rev.* 1965, *140*, A1133-A1138.
[39] Cobos, C. J. *J. Mol. Struct. (Theochem)* 2002, *581*, 17-29.
[40] Tucceri, M. E.; Badenes, M. P.; Cobos, C. J. *J. Fluorine Chem.* 2002, *116*, 135-141.
[41] Sicre, J. E.; Cobos, C. J. *J. Mol. Struct. (Theochem)* 2003, *620*, 215-226.
[42] Badenes, M. P.; Croce, A. E.; Cobos, C. J. *Phys. Chem. Chem. Phys.* 2004, *6*, 747-755.
[43] Oureshi, A. M.; Levchuck, L. E.; Aubke, F. *Can. J. Chem.* 1971, *49*, 2544-2551.
[44] Carter, H. A.; Shreeve, J.M. *Spectrochimica Acta,* 1973, *29A,* 1321-1328.
[45] Aubke, F.; Casper, B.; Müller, H. S. P.; Oberhammer, H.; Willner, H. *J. Mol. Struct.* 1995, *346*, 111-120.
[46] Bauschlicher, C. W. Jr.; Ricca, A. *J. Phys. Chem.* 1998, *102*, 4722-4727.
[47] Benson, S. W. *Chem. Rev.* 1978, *78*, 23-35.
[48] Herron, J. T.; *J. Phys. Chem. Ref. Data,* 1987, *16*, 1-6.
[49] akacs, G. A. *J. Chem. Eng. Data,* 1978, *23*, 174-175.
[50] Wray, K. L.; Feldman, E. V. *J. Chem. Phys.* 1971, *54*, 3445-3449.
[51] Sudlow, K.; Woolf, A. A. *J. Fluor. Chem.* 1993, *64*, 269-277.
[52] Grela, M. A.; Colussi, A. J. *J. Phys. Chem.* 1996, *100*, 10150-10158.
[53] Beckers, H.; Garcia, P.; Willner, H.; Argüello, G. A.; Cobos, C. J.; Francisco, J. S. *Ang. Chem. Int. Ed.* 2007, *46*, 3754-3757.

In: Thermochemistry and Advances in Chemistry Research ISBN: 978-1-60692-376-4
Ed: Anton G. Mirskiy

Commentary

CURRENT TRENDS IN THE THERMOCHEMISTRY OF ATMOSPHERIC NUCLEATION CLUSTERS

Alexey B. Nadykto** and Fangqun Yu
Atmospheric Sciences Research Center, State University of New York at Albany,
251 Fuller Rd., Albany, NY12203, USA.

The critical importance of clear understanding and insight of the thermochemistry of nucleating clusters/particles for quantitatively assessing the climate-related, health and environmental impacts of atmospheric aerosols is commonly accepted. Traditionally, the thermochemistry of nucleation clusters has been studied using the bulk luquid/ capillarity approximation, which is a cornerstone of the classical nucleation theory (CNT) [1-2]. The simplicity and convenience of the CNT is largely offset by moderately poor performance and excessive sensitivity to input parameters, whose experimental determination is quite difficult [3]. At the present time, there are two classes of methods that are imposed as an alternative to the classical theory: molecular based methods [4-9] and computational quantum methods [10-17]. Molecular-based models are more advanced theoretically than the CNT and they compute nucleation kinetically. However, interactions between vapour monomers and *i*-mers in the existing molecular studies are usually approximated using empirical (e.g. TIP4P, TIP5P, Dang–Chang) potentials parameterized to the bulk liquid properties. The aforementioned empirical potentials "were never intended to treat small clusters" [4] and their applicability to the pre-existing and nucleating clusters is questionable. In our view, the existing molecular-based nucleation approaches should be classified as semi-classical because they implicitly treat molecular clusters as liquid-like droplets. The application of inappropriate potentials has lead to the failure of molecular-based methods to predict the sign preference in aqueous systems [10, 13] and dimerization free energy for water [10, 18]. It is important to note that molecular-based methods using empirical potentials are generally not invalid or defective. However, the further improvement of the quality of their model predictions will require more realistic potentials that could be obtained using e.g. quantum methods or experiments.

---

* Email: alexn@asrc.cestm.albany.edu

Quantum methods have been progressing continuously since Schrödinger's original work and their development has reached the stage when "chemical properties can often be calculated with wave function-based methods as well or better than they can be measured" [19]. However, at the present time, quantum computations are quite demanding, and, thus, the size of a system that can be treated using computational quantum methods is limited. A perfectly logical question, "Why the application of quantum methods can systematically improve the nucleation theory?" leads to very simple answers: (a) Nucleation rates are very sensitive to free energies of the formation of initial (or pre-critical) clusters that can not be predicted adequately by both CNT and molecular-based methods; (b) The pre-nucleation clusters are essentially quantum systems, and must be treated using a quantum approach; (c) The number of H_2SO_4 molecules in critical clusters in typical atmospheric nucleation events are not excessively large and are generally smaller than ~ 10 [20]; (d) Recently developed kinetic nucleation models can effectively use the clustering energy changes obtained from quantum calculations [21].

A number of recent studies have shown how the atmospheric science can benefit from using quantum methods e.g. [10-17]. They have been used to solve the classical sign preference problem in the ion-induced nucleation [10, 12], analysis of relative stability of binary and ternary clusters [11, 13-15,17] and effect of trace organic species on the cluster free energies [13]. Given the ability of the quantum theory to treat the thermochemistry of multicomponent clusters explicitly, with accounting for the actual chemical identity of clusters/complexes and without referring to assumable/hypothetic bulk liquid properties, one can conclude that the further perspectives of the quantum theory as a principle tool for studying the thermochemistry of nucleation clusters seem to be very promising. Future research in this promising field includes (but not limited to): (1) Identification of new species involved in the atmospheric nucleation or new nucleation pathways; (2) Extension of the quantum calculations to larger clusters; (3) Assessment and reduction of uncertainties in the clustering thermochemistry calculations; (4) Improvement of existing nucleation theories and development of new nucleation models using the quantum-derived thermochemistry.

REFERENCES

[1] Volmer M., A. Weber (1926) Z. Phys. Chem. 119, 277

[2] Becker, R., W. Doering (1935) Ann. Phys. 24, 719

[3] Oxtoby, D. W. (1992) J. Physics: Condensed Matter 4, 7627

[4] Kathmann, S.M., Schenter, G.K., Garrett, B.C. (2002). J. Chem. Phys. 116, 5046 Brodskaya, E.P., A.P. Lyubartsev, A. Laaksonen (2002) J. Phys. Chem. B 106, 6479

[5] Oh, K. J. , G. T. Gao X. C. Zeng (2001) Phys. Rev. Lett. 86, 5080

[6] Warshavsky, V.B., X.C. Zeng (2002) Phys. Rev. Lett. 89, 246104

[7] Kathmann, S.M., G. K. Schenter, B. C. Garrett (2007) Phys. Rev. Lett. 98, 109603

[8] Schenter, G. K. , Kathmann, S.M., Garrett, B.C. (1999) Phys. Rev. Lett. 82, 3484

[9] Nadykto, A. B. et al. (2007) Phys. Rev. Lett. 98(10), 109604.

[10] Nadykto, A.B., F. Yu (2007) Chem. Phys. Lett., 43514

[11] Nadykto, A. B. et al. (2006), Phys. Rev. Lett. (2006) 125701.

[12] Kurten, T. et al. (2006) J. Geophys. Res. 112, D4, D04210,(2007a) 10.1029/ 2006-JD007391
[13] Kurten, T. et al. (2007) Boreal Env. Res. 12 (3), 431.
[14] Kurten, T. et al. (2007) Atmos. Chem. Phys. 7 (10), 2765
[15] Dunn , M. et al. (2006) J. Phys. Chem. A, 110, 303
[16] Nadykto et al. (2008), Adv.Quant.Chem. 55, 449
[17] J. Liu , L.Yang, D.J. Doren (2006), Chem. Phys. Lett. 417, 63
[18] Garrett, B.C. et al. (2005) Chem. Rev. 105 (1), 355
[19] Yu, F., and R. P. Turco (2008), Atmos. Chem. Phys. Discuss., 8, 5683
[20] Yu, F. (2006) Atmos. Chem. Phys. 6 (12) 5193

INDEX

A

B

C

D

E

H

K

L

M

N

O

P

Q

R

S

T